T-LEVELS
THE NEXT LEVEL QUALIFICATION

SECOND EDITION

BUILDING SERVICES ENGINEERING FOR CONSTRUCTION

CORE

Mike Jones
Stephen Jones
Tom Leahy
Peter Tanner
David Warren

hachette LEARNING

The teaching content of this resource is endorsed by WJEC Eduqas to support the T Level Technical Qualification in Building Services Engineering for Construction, 610/5780/0.

This resource has been reviewed against WJEC Eduqas' endorsement criteria. As this resource belongs to a third party, there may be occasions where a specification may be updated and that update will not be reflected in the third party resource. Users should always refer to WJEC Eduqas' specification and Sample Assessment Materials to ensure that learners are studying the most up to date course.

It is recommended that teachers use a range of resources to fully prepare their learners for the exam and not rely solely on one textbook or digital resource.

WJEC, nor anyone employed by WJEC has been paid for the endorsement of this resource, nor does WJEC receive any royalties from its sale.

The publisher would like to acknowledge Mike Jones for his permission to reproduce text from *Foundation in Construction and Building Services Engineering: Core (Wales).*

To order, please visit www.HachetteLearning.com or contact Customer Service at education@hachette.co.uk / +44 (0)1235 827827.

ISBN: 978 1 0360 2098 9

First published in 2022

This edition published in 2026 by Hachette Learning (a trading division of Hodder & Stoughton Limited),

An Hachette UK Company

Carmelite House

50 Victoria Embankment

London EC4Y 0DZ

www.HachetteLearning.com

The authorised representative in the EEA is Hachette Ireland, 8 Castlecourt Centre, Dublin 15, D15 XTP3, Ireland (email: info@hbgi.ie)

Impression number 10 9 8 7 6 5 4 3 2 1

Year 2030 2029 2028 2027 2026

Typeset in India by Integra Software Services Ltd.

Printed and Bound in Great Britain by Bell & Bain Ltd, Glasgow

A catalogue record for this title is available from the British Library.

Contents

Acknowledgements iv

About the authors v

Introduction vii

Guide to the book viii

1 Health and safety in construction 1
2 Construction science principles 47
3 Construction design principles 86
4 Construction and the built environment industry 109
5 Sustainability principles 146
6 Measurement principles 179
7 Building technology principles 195
8 Information and data principles 229
9 Relationship management in construction 247
10 Digital technology in construction 269
11 Construction commercial/business principles 282
12 Building services engineering (BSE) systems 301
13 Maintenance principles 365
14 Tools, equipment and materials 377

About the exams and employer-set project 394

Answers 396

Glossary 407

Index 413

Acknowledgements

Thank you to Tom Stottor who gave me the opportunity to contribute to this important textbook and my thanks to the copy-editors, proof-readers, illustrators, and editors who have put so much time and effort into making this new textbook a reality.

Many thanks to Matthew Sullivan who's been so supportive and provided lots of suggestions and ideas on how to shape the chapters I've produced into something of value for T-level learners.

Imogen Miles and Sylvia Bukowski have demonstrated their skills in creating a book that is visually appealing as well as of practical value. Thank you.

Finally, thanks to my lovely wife Sue who has patiently supported me through the many hours I've spent at my desk.

Mike Jones

I would like to thank my family and the team at Hodder Education for their continued and relentless support on this project, without which it would not have been possible to successfully complete the book.

I would also like to acknowledge the readers of this book and hope that it provides some support in achieving your personal and professional goals.

Stephen Jones

I would like to give special thanks to my partner Charlotte Geileskey for her grammatical and typing skills. I would also like to thank Debbie Coomes, Assistant Principal at West Suffolk College, for her support, and my family for their encouragement. Lastly, I would like to thank Hodder Education for this opportunity, my first foray as an author within vocational education.

David Warren

About the authors

Mike Jones

Working in the construction industry has been the mainstay of Mike's working life over a long period. He has worked as a skilled tradesman, a supervisor and a site manager on projects ranging from small extensions to multi-storey contracts worth millions of pounds. For a number of years, he employed a small team of skilled workers in his own construction company working on contracts for selected customers.

The skills Mike developed over the years allowed him to design and successfully build his own family home in rural Wales, which he has always viewed as a highlight of his construction career.

After over 30 years working out on site, Mike moved into the education sector, first as an NVQ trainer and assessor for three years and subsequently as a college lecturer at Cardiff and Vale College in South Wales for 10 years. After leaving his post at college, he continued to work in education and training, producing teaching and learning resources for City and Guilds, along with his work as a technical author writing textbooks. He is also involved in the writing, reviewing and editing of vocational examinations.

Mike's aim during his time teaching others has been to impart to learners the great job satisfaction that can be gained from becoming a skilled practitioner. Put maximum effort into developing your skills and knowledge, and you will be able to take full advantage of the vast range of career opportunities in the construction industry.

Stephen Jones

Stephen Jones has been writing books about the construction industry, site carpentry and architectural joinery since 2009. Besides being a published author, he is also a lecturer in the construction department at South Devon College. Stephen has Qualified Teaching and Learning Status (QTLS) and is a highly experienced lecturer with a Post Graduate Certificate of Education (PGCE), Higher National Certificate in Construction (HNC) and Nebosh National Certificate in Construction Health and Safety. Prior to teaching in further education, Stephen had his own business and worked in the construction industry as a carpenter, joiner and shopfitter since leaving school at 16 years of age, where he gained a wealth of subject knowledge and skills.

Stephen lives in Torquay on the south coast of Devon with his wife, children and dog, Maisy. His main interest is his family; he is also a football fanatic and enjoys working out in the gym.

Tom Leahy

Tom's journey started over 20 years ago when he served a four-year apprenticeship with a local authority, working on both commercial and domestic plumbing/heating systems whilst attending college for one day a week. This experience would give him the underpinning knowledge and practical skills to make a successful plumber.

In 2005 Tom started work for another local authority, working on a variety of authority-owned buildings, from libraries to town halls and community centres, as part of a reactive maintenance and installation team.

After successfully contributing to the training and development of several apprentices, Tom moved into further education and achieved teaching, assessing and quality assurance qualifications. He continues to work in further education as a Programme Leader training the next generation of Construction and Building Services professionals.

Peter Tanner

Having started in the electrotechnical industry while still at school, chasing walls for his brother-in-law for pocket money, Peter was keen to progress in the industry. After a spell in the armed forces, Peter gained a place as a sponsored trainee on the Construction Industry Training Board training scheme.

On completion of his apprenticeship Peter worked for a short time as an intruder alarm installer, then for a company involved in shop-fitting and restaurant and pub refurbishments. Peter was later seconded to the Property Services Agency, designing major installations within some of the best-known buildings in the UK.

A career-changing accident took Peter into teaching. Peter has worked with City & Guilds for over 25 years and has represented them on industry committees, such as JPEL/64 which is responsible for the production of BS 7671. He is passionate about using his experience to maintain the high standards the industry expects.

David Warren

David is currently working at West Suffolk College as Head of Department for Building Services Engineering. He also works part time for City and Guilds as a Moderator and TQA for the new T Levels. David has been in education for the last 17 years. Prior to this, he successfully ran two businesses, one specialising in plumbing and heating; the other in ceramic floor and wall tiling and project management.

David is the proud dad of three children. He is a dedicated Spurs fan and has great interest in architecture, motoring, Formula One and interior design.

Introduction

A T Level in Building Services Engineering for Construction will provide you with broad and deep foundations for a career in the Building Services Engineering industries, including in areas such as electric installation or maintenance and plumbing or heating. The Level 3 technical qualification, which was developed by WJEC/Eduqas in collaboration with employers and industry bodies, will enable you to progress to a range of careers and further education opportunities. You can find out more about the qualification on the WJEC/Eduqas website.

This book will help you to develop the knowledge and understanding you need to complete the core component of your T Level. Each of the chapters in the book follows the topics required for the core component's knowledge outcomes, which you can find on the WJEC/Eduqas website. Headings in the book follow those in the qualification handbook, so that you can check your learning. Although the book focuses on the core component only, the content that you cover and the skills that you learn will also be useful in your chosen occupational specialism(s).

The book will introduce you to the Building Services Engineering workplace, and the fundamental principles that underpin it, across a range of engaging and real-world examples. The book covers the key content in detail and includes hundreds of photos and technical drawings, as well as industry tips to support you in your T Level work placement. There are also lots of activities and learning features; you can find out more about these and how to use them on the next page.

Your teacher or lecturer might decide to use the book in your classroom, either as a central learning tool or to supplement their teaching. The book will also prepare you for your final exams and the employer set project. (You can find out more about how you will be assessed on page 394 of this book.) If you have any questions about how you should be using the book, you should consult your teacher or lecturer.

Guide to the book

The following features can be found in this book.

Learning outcomes

Core knowledge outcomes that you must understand and learn

Key term

Important terms that you should understand

Industry tip

Useful tips and advice to help you in the workplace

Research

Research-based activities: either stretch and challenge activities enabling you to go beyond the course, or industry placement based activities encouraging you to discover more about your placement

Case study

Placing knowledge into a fictionalised, real-life context, to introduce dilemmas and problem solving

Test yourself

A knowledge-consolidation feature containing questions and tasks to aid understanding and guide you to think about a topic in detail

Health and safety

Important points to ensure safety in the workplace

Improve your maths

Short activities that encourage you to apply and develop your functional maths skills in context

Improve your English

Short activities that encourage you to apply and develop your functional English skills in context

Assessment practice

Knowledge-based practice questions to help prepare you for the exam (answers found at the back of the book)

Project practice

Short scenarios and focused activities, reflecting one or more of the tasks that you will need to undertake during completion of the employer-set project

Chapter 1 Health and safety in construction

This chapter looks at health and safety legislation and approved standards, and how they protect workers and others who may be affected by work activities.

We will also consider the legal responsibilities of people working in construction and building services engineering (BSE) and the implications of not following health and safety law.

Learning outcomes

By the end of this chapter, you will understand:

1. construction legislation and regulations
2. public liability and employer's liability
3. approved construction codes of practice
4. development of safe systems of work
5. safety conscious procedures
6. safety inspection of a work environment
7. implications to those working within the BSE industry of not following health and safety legislation
8. safe working practices for the safe isolation of systems
9. implications of poor health and safety on building performance and individual stakeholders
10. recording and reporting of safety incidents and near misses
11. emergency procedures for unsafe situations
12. types of PPE
13. first-aid facilities
14. warning signs for the main groups of hazardous substance
15. safe practices and procedures for the use of access equipment and manual handling
16. safe practices and procedures for working in excavations and confined spaces.

1 Construction legislation and regulations

1.1 The role of legislation and regulations in the construction industry

Working in the construction industry can be extremely hazardous. In 1974, the Health and Safety at Work etc. Act (HASAWA) was introduced. It replaced many older laws and enabled new regulations to be passed in Parliament to protect workers and the general public from work activities.

The main objectives of the HASAWA are to:

- secure the health, safety and welfare of people at work
- protect people other than those at work (for example the general public or visitors) from risks to health or safety arising out of or in connection with work activities
- control the possession and use of highly flammable, explosive and dangerous substances.

Everyone has a moral responsibility to protect the health and safety of themselves and others at work. However, under HASAWA some parties – known as **duty holders** – have legal responsibilities. Employers, employees, the self-employed, manufacturers and people in control of premises are all duty holders.

Key term

Duty holders: people with a legal responsibility under health and safety law

Shortly after HASAWA was introduced, the UK government established the Health and Safety Executive (HSE). The HSE is an independent regulator that aims to prevent workplace ill health, injury and death by targeting industries with the greatest risks and worst risk-management records, which include construction and manufacturing.

The HSE provides advice and guidance to employers so that they can identify hazards and manage risks correctly. Its emphasis is on the prevention of accidents and ill health by raising awareness in workplaces, so that workers and other people can stay safe and well.

Research

Visit the HSE website (www.hse.gov.uk) to find out the responsibilities of duty holders under HASAWA.

The HSE has legal powers to hold people or companies to account when risks are not managed and legislation is breached. We will look at how the HSE enforces HASAWA later in this chapter.

Legislation made under HASAWA is divided into a number of different regulations. Table 1.1 outlines the main regulations that control health, safety and welfare in the construction industry.

▼ Table 1.1 Main regulations that control health, safety and welfare in the construction industry

Regulation	Overview
Reporting of Injuries, Diseases and Dangerous Occurrences Regulations (RIDDOR) 2013	These regulations place legal duties on employers, the self-employed and people in control of premises to report to the HSE serious accidents, dangerous occurrences and occupational diseases resulting from workplace activities.
Control of Substances Hazardous to Health (COSHH) Regulations 2002	These regulations state that employers must control substances hazardous to health by preventing or reducing workers' exposure. This can be achieved by: • finding out what the health hazards are • deciding how to prevent harm to health by assessing the risks (**risk assessment**) • providing control measures, such as secure storage for chemicals • making sure that control measures are followed • maintaining control measures • providing training, information and instruction for employees and others • monitoring and providing health surveillance when necessary • planning for emergencies. Note: asbestos is not covered by COSHH Regulations.

Regulation	Overview
Control of Asbestos Regulations 2012	There are several different types of **asbestos** that have been used in the construction industry, some of which are more hazardous than others. In most cases, asbestos can only be removed by a licensed contractor. However, small quantities of lower-risk asbestos can be removed by non-licensed contractors, providing they are competent and have effective controls in place. The law states that before doing any building work or maintenance on premises or on plant and equipment that contain asbestos, a risk assessment has to be carried out to manage and control the risks. It is mandatory for anyone who is likely to be exposed to asbestos fibres at work to receive training.
Provision and Use of Work Equipment Regulations (PUWER) 1998	These regulations place duties on people and organisations that operate or have control over work equipment. They state that: • equipment must be suitable for its intended use • equipment must be safe, maintained and regularly inspected • people who use the equipment must have received training, information and instruction • equipment must be used accompanied by suitable health and safety measures, for example guarding or emergency stop devices.
Manual Handling Operations Regulations (MHOR) 1992	These regulations define manual handling as 'any transporting or supporting of a load ... by hand or bodily force'. They set out a hierarchy of control measures for dealing with the risks posed by manual handling: 1 Avoid hazardous manual handling operations so far as is reasonably practicable. 2 Assess any hazardous manual handling operations that cannot be avoided. 3 Reduce the risk of injury so far as is reasonably practicable.
Personal Protective Equipment (PPE) at Work Regulations 1992 (Note: the current legislation which refers to the supply of PPE is the Personal Protective Equipment Regulations 2016/425, which is enforced by the Personal Protective Equipment (Enforcement) Regulations 2018)	These regulations relate to the use of personal protective equipment (PPE). Employers have a responsibility to provide PPE free of charge to employees to control the hazards identified in risk assessments. They should also provide information, instruction and training on how to use it and take care of it.
Work at Height Regulations 2005	Under these regulations, people who are in control of others working at height have a duty to make sure work is properly planned, supervised and carried out by competent people. All work at height must be risk assessed.
Control of Noise at Work Regulations 2005	Exposure to high levels of noise can cause temporary or permanent hearing loss. Employers have a duty to reduce the risk by: • assessing the risks • eliminating or controlling the noise • providing hearing protection • providing hearing checks for those at risk • providing employees with information and training • maintaining any noise-control equipment and ensuring it is being used.
Control of Vibration at Work Regulations 2005	These regulations place a duty on employers to reduce the risk to employees' health from exposure to vibration caused by work equipment, machinery and tools, for example when using a core drill. Employers must calculate the amount of vibration that employees may be exposed to at work. At a specific level (referred to as the 'exposure action value'), employers must introduce technical and organisational measures to reduce the risk of personal injury to an acceptable level. The law also prevents workers from being exposed to a higher exposure limit value of 5.0 m/s^2 A(8).

▼ Table 1.1 Main regulations that control health, safety and welfare in the construction industry

Regulation	Overview
Confined Spaces Regulations 1997	The aim of these regulations is to prevent people working in confined spaces whenever possible, because of the serious risk to health and safety from collapse, drowning, low oxygen levels and exposure to natural gases. If the activity is unavoidable, a **safe system of work** must be followed to control the risk.
Management of Health and Safety at Work Regulations 1999	These regulations apply to every work activity. They explain what employers need to do to manage health and safety at their place of work under HASAWA. The main requirement is for employers to complete a risk assessment and record significant findings when they have five or more employees.
Electricity at Work Regulations 1989	These regulations outline the responsibilities of those involved in the design, operation, construction and maintenance of electrical equipment and systems. They are relevant to most work activities and premises, except certain offshore installations and particular ships.
Environmental legislation	There are numerous environmental laws in the UK that protect wildlife, the countryside, listed buildings, national parks and monuments, for example the Control of Pollution Act 1974, the Wildlife and Countryside Act 1981 and the Planning (Listed Buildings and Conservation Areas) Act 1990. The environment is protected further by the Planning and Energy Act 2008, Energy Act 2020 and Climate Change Act 2008. The Climate Change Act 2008 is an environmental law designed to meet the UK government's target to reduce **carbon emissions** recorded in 1990 by at least 80 per cent by 2050.
Waste management legislation	There are several requirements for handling waste, including the Waste Electrical and Electronic Equipment (WEEE) Regulations 2013, as well as regulations on waste carriers. For more information, see Chapter 5, section 7.
Lone working	Lone working refers to situations where someone works by themselves without direct supervision, and UK regulations require employers to ensure these workers are safe. Under the Health and Safety at Work Act and the Management of Health and Safety at Work Regulations, employers must assess risks, provide proper training, and ensure lone workers can communicate and get help if needed.

Key terms

Risk assessment: a formal process of identifying significant workplace hazards, whom they affect and control measures that could be used to eliminate or reduce risk to an acceptable level

Asbestos: a naturally occurring mineral that used to be mixed with other construction materials to create insulation, pipe lagging and flooring; it has been banned in the UK since 1999 because it is extremely hazardous to health, however workers may still be exposed to it in older buildings

Safe system of work: a formal set of procedures that must be followed when hazards cannot be eliminated completely

Carbon emissions: carbon dioxide released into the atmosphere; scientists believe this is a cause of climate change

Health and safety legislation is regularly reviewed and updated to reflect changes in the construction and BSE industries. These changes are often made with guidance and support from the following bodies:

- employers
- unions
- trade associations
- professional bodies
- academics.

Test yourself

How can you reduce the risk of personal injury caused by manual handling?

Health and safety

The use of PPE should be a last resort after all other methods of controlling a hazard have been considered. For example, the use of guardrails, safety nets and airbags should be considered for use while working at height before a lanyard. This is because PPE only protects the user, whereas the other methods protect others as well.

Research

Find out the short- and long-term effects that vibration can cause on different parts of the human body. Suggest ways to reduce or control the risks to comply with the Control of Vibration at Work Regulations 2005.

1.2 Regulations relating to the provision of welfare facilities during construction work

Under HASAWA, employers have a duty to provide basic welfare facilities for their employees at their place of work. The Construction (Design and Management) (CDM) Regulations 2015 outline the minimum facilities that should be provided on construction sites:

- Drinking water:
  - An adequate supply of fresh drinking water should be appropriately labelled and accessible in suitable places.
  - An adequate supply of cups or other drinking vessels must be provided, unless the water is supplied from a fountain for workers to drink easily.
- Toilets:
  - These should be separate for men and women, with doors that are lockable from the inside.
  - They should be clean and maintained with adequate ventilation and lighting.
- Washing facilities:
  - These should be located in the vicinity of the toilets and changing rooms, with clean cold and hot or warm running water if possible.
  - Soap or another means of cleaning must be provided, along with towels or another suitable method of drying.
  - Showers may also need to be provided if the nature of the work requires it.
- Rest facilities:
  - These should be equipped with tables and seating with backs for the number of people expected to use them.
  - The rest area must be able to maintain an appropriate temperature and have facilities to prepare and eat meals and boil water.
- Changing rooms with lockers:
  - These must be provided if workers have to wear special clothing for their job, unless they can be reasonably expected to change elsewhere.
  - Changing rooms should be provided with seating and facilities to dry and store clothing.
  - Separate rooms must be provided for men and women.

Health and safety

The CDM Regulations state that, if necessary, employers have to provide facilities for pregnant or nursing mothers to rest lying down.

▲ Figure 1.1 Welfare facilities on a construction site

1.3 Implications of not adhering to legislation

Health and safety law is made up of Acts of Parliament and statutory regulations. Failure to comply with them is a criminal offence. HSE inspectors have a number of powers to enforce the law, including carrying out inspections, issuing simple cautions and notices to duty holders, and sometimes initiating prosecutions.

The law states that HSE inspectors can enter a workplace without notice at any reasonable time to conduct an inspection or investigation. During a visit, they may want to speak to workers, look at possible health risks, and inspect equipment and machinery. They can also take samples, such as sound and dust levels, photographs and measurements, and make copies of records or other documentation needed as part of their investigation.

If the inspector thinks that the employer has broken the law, they will issue a 'notification of contravention'. This document outlines what laws have been broken and how, and what needs to be done to put things right.

An HSE inspector may issue an improvement notice when one or more laws have been breached; this means that faults have to be remedied within a specific period of time, which should be no less than 21 days.

If there has been a serious breach of the law and people are at risk of immediate harm, the HSE will issue a prohibition notice, preventing work from continuing.

Failure to comply with improvement or prohibition notices can result in prosecution, fines and imprisonment.

Research

What are the maximum legal penalties for employers that have not adhered to health and safety legislation?

Case study

Amelia is a self-employed painter and decorator who subcontracts work from medium-sized developers in the local area.

Several months ago, she was visited by an HSE inspector. During the visit, the inspector noticed that a large amount of hazardous paint was not being stored correctly and was a potential source of fuel for a fire; therefore, they issued an improvement notice.

The enforcement action taken by the inspector has been published on the HSE website. What impact do you think this will have on Amelia's business?

When people suffer loss or injury as a result of an accident at work, they may seek compensation. If health and safety legislation has been breached, the duty holder is usually fined because they have broken criminal law. Magistrates and the Crown Court have a discretionary power to award compensation to an injured person. However, claims are usually pursued by the injured person because of a breach of **common law**.

Employers should have liability insurance to cover any claims against them for a breach of common law; however they cannot insure against fines when they have broken criminal law.

Key term

Common law: legislation made in the civil courts rather than statute law that is made in Parliament

Research

Research an employer's common-law duty of care and compare this to their duties under HASAWA.

1.4 Statutory and non-statutory documents in construction

The UK construction industry is subject to a number of laws which cover many different areas, including health, safety and welfare. Legislation comprises Acts of Parliament and regulations (statutory legislation) which have legal status and must be complied with. However, there are also many non-statutory guidance documents, which are not compulsory but offer advice on good practice and compliance with the law, for example the Approved Codes of Practice (ACOPs) produced by the HSE which describe preferred methods and standards. Unless stated, you do not need to follow the ACOP guidance, but in doing so you will meet the requirements of HASAWA. The HSE states that other practical methods can be used, but they must meet or exceed the standards in the ACOP.

▲ Figure 1.2 An Approved Code of Practice

Test yourself

State one example of statutory legislation that is applicable to a construction activity and an ACOP to support it.

2 Public liability and employer's liability

2.1 Current requirements

We have already seen that the HSE has the power to issue fines to employers and other duty holders if they neglect the law. The fines issued by the HSE are intended to punish employers financially; therefore they cannot insure against them.

However, employers can protect themselves with insurance against claims made by employees through the common-law courts in the pursuit of compensation. In this section, we will look at the different types of insurance that employers may have.

Public liability insurance

Public liability insurance protects a business against claims for compensation for any loss or injury to members of the public, customers or employees in connection with work activities.

There is no legal requirement for employers to have public liability insurance. However, if they engage with the public, or the public and their property are likely to be affected by the employer's work activities, then they should have this insurance to protect themselves from any potential claims made against them.

The level of insurance protection is determined by the nature and scale of the work undertaken. Some clients will stipulate the minimum level of insurance cover needed by a contractor, and have it agreed in writing before awarding any contracts or allowing construction work to start.

Employers' liability insurance

Common law is legislation that places a duty of care on employers to protect their employees from any unnecessary risks, and to ensure a safe system of work. If a person is injured or suffers a loss in the workplace, they may seek financial compensation from the employer if they are **liable**.

Key term

Liable: legally responsible

Under the Employers' Liability (Compulsory Insurance) Act 1969, all employers are required by law to insure against liability for injury or disease to their employees arising out of their employment and any financial claims made against them.

The amount of liability insurance an employer needs is usually determined by the number of employees and level of risk involved with their business. However, the minimum amount of cover for most insurance policies starts between 5 and 10 million pounds. If the injured person was successful with their claim against the employer, the compensation would go towards the recovery of their lost earnings, medical treatment, ongoing care and any legal fees.

Research

Find out the other types of insurance that a contractor or subcontractor may take out to safeguard themselves and their assets.

Improve your maths

The HSE reported that there were 142 work-related fatal injuries in the UK in 2020–21; 39 fatalities occurred in the construction industry. What percentage of fatalities occurred to construction workers?

3 Approved Codes of Practice (ACOPs)

3.1 The HSE Legal (L) Series

In 2015, the Construction (Design and Management) (CDM) Regulations came into force. Their aim was to manage health, safety and welfare on all construction projects by defining roles for duty holders. People with responsibilities under the regulations include clients, designers and contractors. If there is more than one designer or contractor working on the same construction project, a **principal designer** or **principal contractor** must be appointed (named). Any other designers or contractors must work under the control of the principal designer or contractor.

The CDM Regulations state that where possible the designer (or principal designer) should plan for health and safety by identifying then eliminating or minimising all foreseeable risks at the planning stage of a project. They also state that all construction work, regardless of size, should have a 'construction phase plan'. This is a key document that identifies all the potential hazards for a project and the measures that will be used to remove or mitigate them.

Contractors and principal contractors have a duty under the CDM Regulations to:

- provide site inductions for all workers and visitors to the site
- provide suitable welfare facilities
- secure the site to prevent unauthorised access and protect non-workers (for example members of the public).

The HSE has published a range of documents known as the HSE Legal (L) Series (also referred to as the CDM Series). Their primary purpose is to support duty holders in meeting their obligations under the CDM Regulations. They are available online to download for free, or a hard copy can be purchased through the HSE's website. The L Series contains both Approved Codes of Practice (ACOPs) and guidance on various health, safety and welfare regulations.

Key terms

Principal designer: a designer appointed by the client to take the lead in planning, managing, monitoring and co-ordinating health and safety during the pre-construction phase of a project involving more than one designer

Principal contractor: a contractor appointed by the client to take the lead in planning, managing, monitoring and co-ordinating health and safety during the construction phase of a project involving more than one contractor

Improve your English

Include the word 'mitigate' in a short sentence about health and safety hazards that you may face at your training centre.

3.2 Legionnaires' disease: the control of *Legionella* bacteria in water systems (L8)

Legionnaires' disease is a type of pneumonia caused by *Legionella* bacteria that naturally occur in fresh-water environments such as streams and lakes. It can be contracted by inhaling infected airborne water droplets and can be fatal.

'Legionnaires' disease: The control of *Legionella* bacteria in water systems (L8)' is an example of an ACOP and part of the HSE's L Series. It outlines the responsibilities that employers and further duty holders have towards others in relation to *Legionella*. The latest (fourth) edition of the document provides revisions to simplify some of the text contained in the regulations (for example COSHH).

The ACOP provides practical advice to prevent the growth and spread of *Legionella* bacteria in manmade water systems. The first step in managing the risk is to complete a suitable and sufficient *Legionella* risk assessment (LRA), to identify the source of any risk. The ACOP recognises the importance of the LRA and states that it must be completed by a trained and competent person who fully understands the water systems being evaluated. Risk assessments must be regularly reviewed and updated when changes are necessary as a result of a review, to keep records up to date and current.

The ACOP states that duty holders must appoint a competent person (known as a responsible person) with suitable authority to manage the day-to-day operational procedures, to make sure 'effective and timely' actions are taken to manage the hazards identified in the risk assessment.

In addition to risk assessments, the L8 ACOP describes other methods of implementing control schemes and measures, including managing, monitoring precautions and inspection.

3.3 (IET) Codes and Guidance

The Institute of Engineering and Technology (IET) Codes and Guidance, including the IET Wiring Regulations (BS 7671), serve as authoritative standards for electrical installation and safety in the UK construction industry. While not legally binding in themselves, they are widely recognised as the benchmark for compliance with statutory requirements under the Electricity at Work Regulations 1989. These documents provide detailed technical guidance on design, installation, inspection, testing, and maintenance of electrical systems, helping professionals ensure safe and efficient practices.

4 Development of safe systems of work

4.1 Development and use of safe systems of work in construction projects

There are different types of management system that can be used to create safe systems of work. However, the HSE favours the following approach, which can be applied to most construction businesses:

- **Plan** – for specific health and safety objectives
- **Do** – implement the plan
- **Check** – that the plan is working and measure performance, for example the number of accidents and near misses
- **Act** – learn from any mistakes and put them right.

The law states that every employer must have a policy for managing health and safety, and that it must be in written form if they have five or more employees. A health and safety policy sets out the general approach an employer has towards managing health and safety in their business and is divided into three sections:

- general statement of intent (the employer's commitment)
- arrangements (what the employer intends to do and how)
- responsibilities (areas of responsibility within their business and defining roles).

The Management of Health and Safety at Work Regulations 1999 contain a schedule known as the 'General principles of prevention'. This provides a hierarchy of control measures for employers to manage risks to health and safety in the workplace (see Figure 1.3).

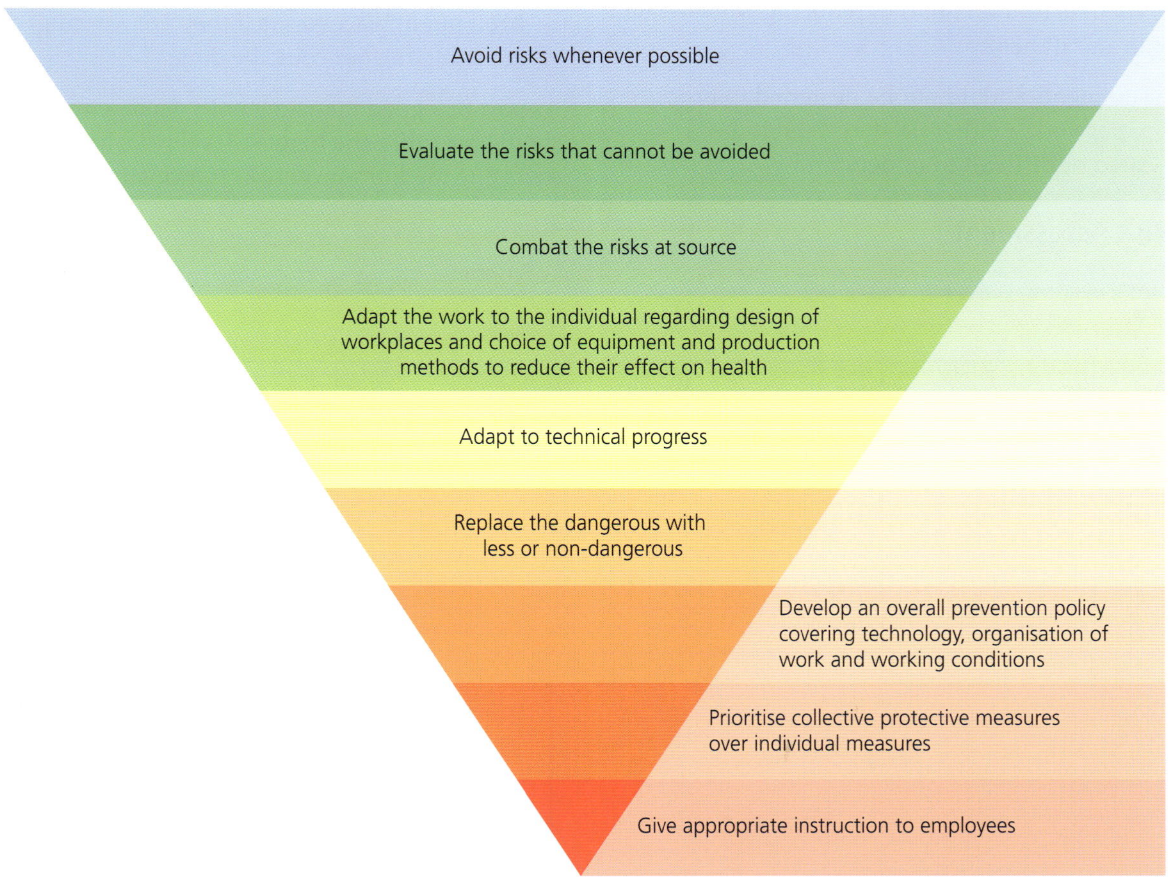

▲ Figure 1.3 Hierarchy of control measures for managing health and safety risks (source: www.legislation.gov.uk)

Test yourself

What is the least effective method of controlling health and safety risks?

Research

Find out about permits to work as part of a safe system of work. Explain:

- when they are used
- who issues them
- how they are controlled

You should also make a list of as many work activities as you can think of that would benefit from a permit-to-work system and justify your reasons for choosing these tasks.

4.2 How to complete risk assessments

The Management of Health and Safety at Work Regulations 1999 place a legal duty on employers to manage the risk of potential harm to people from work activities. A risk assessment that identifies hazards and determines measures to eliminate or control them is fundamental to reducing work-related accidents and ill health.

Risk assessments should not be designed to produce lots of paperwork, neither do they need to be complicated or difficult to understand.

A risk assessment should be a structured examination of workplace activities, appropriate and proportional to the level of risk and the nature of the hazards. It should be site and task specific; therefore generic risk assessments would not be considered suitable or sufficient.

Completing a risk assessment is a five-step process:

1. Identify hazards and potential hazards (what could cause injury or illness).
2. Decide who might be harmed and how.
3. Assess the risks by looking at the likelihood of the hazard causing harm and the severity of that harm.
4. Establish control measures to remove or reduce the risk of harm. (Look at what you are already doing to protect people, what future actions need to be taken and who needs to take those actions).
5. Record any significant hazards found (when there are five or more employees) and review the assessment after the control measures have been put in place to see if they are working, or if anything further could be done.

As part of a good management system, employers should prioritise the highest-level risks before assessing medium-level risks. Once all of these

Risk Assessment

Activity / Workplace assessed: Return to work after accident
Persons consulted / involved in risk assessment
Date:
Reviewed on:

Location:
Risk assessment reference number:
Review date:
Review by:

Significant hazard	People at risk and what is the risk Describe the harm that is likely to result from the hazard (e.g. cut, broken leg, chemical burn etc.) and who could be harmed (e.g. employees, contractors, visitors, etc.).	Existing control measure What is currently in place to control the risk?	Risk rating Use matrix identified in guidance note. Likelihood (L) Severity (S) Multiply (L) * (S) to produce risk rating (RR)				Further action required What is required to bring the risk down to an acceptable level? Use hierarchy of control described in guidance note when considering the controls needed.	Actioned to: Who will complete the action?	Due date: When will the action be completed by?	Completion date: Initial and date once the action has been completed.
			L	S	RR	L/M/H				
Uneven floors	Operatives	Verbal warning and supervision	2	1	2	M	None applicable	Site supervisor	Active now	Ongoing
Steps	Operatives	Verbal warning	2	1	2	M	None applicable	Site supervisor	Active now	Ongoing
Staircases	Operatives	Verbal warning	2	2	4	M	None applicable	Site supervisor	Active now	Ongoing

Severity \ Likelihood	1 Unlikely	2 Possible	3 Very likely
1 Slight/minor injuries/minor damage	1	2	3
2 Medium injuries/significant damage	2	4	6
3 Major injury/extensive damage	3	6	9

1 – Low risk: action should be taken to reduce the risk if reasonably practicable.
2, 3, 4 – Medium risk: is a significant risk and would require an appropriate level of resource.
6 & 9 – High risk: may require considerable resourced to mitigate. Control should focus on elimination of risk, if not possible control should be obtained by following the hierarchy of control.

◀ Figure 1.4 A risk assessment

hazards have been identified and eliminated or controlled in some way to a satisfactory level, lower-risk activities should be risk assessed.

Risk assessments should be completed by a competent person with a satisfactory level of knowledge and experience. For more complex activities, the person completing them may seek advice from other sources, such as the HSE or trade organisations. Employers with five or more employees are required by law to record significant findings from their risk assessments.

Health and safety

Some regulations require risk assessments to be carried out for specific tasks, for example the Work at Height Regulations 2005 and the Control of Substances Hazardous to Health (COSHH) Regulations 2002.

4.3 How to write method statements

Method statements are documents prepared by employers that describe a logical sequence of steps to complete a work activity in a safe manner. They reflect the hazards identified in risk assessments and describe the way in which the job should be undertaken.

These documents are used as part of a safe system of work to communicate vital health and safety information and guidance to people completing the activity; together with risk assessments they are referred to as RAMS (risk assessments and method statements).

A typical method statement describes:

- hazards identified
- safe access and **egress**
- supervision needed
- hazardous substances and how to control them
- permit-to-work systems (if applicable)
- personal protective equipment
- emergency procedures
- environmental controls
- health and safety monitoring
- workforce details.

Although method statements are an effective way of keeping people safe at work, there is no legal requirement for employers to produce them. However, they are often requested by principal contractors.

Key term

Egress: an exit or way out

METHOD STATEMENT			
Revision Date:	**Revision Description:**		**Approved By:**
Work Method Description	Risk Assessment	Risk Levels	Recommended Actions* (Clause No.)
1.			
2.			
3.			
4.			
RISK LEVELS: Class 1 (high)	Class 2 (medium)	Class 3 (low)	Class 4 (very low risk)
Engineering Details/Certificates/Work Cover Approvals:		Codes of Practice, Legislation:	
Plant/Equipment:		Maintenance Checks:	
Sign-off			
Print Name: Signature:	Print Name: Signature:	Print Name: Signature:	Print Name: Signature:
Print Name: Signature:	Print Name: Signature:	Print Name: Signature:	Print Name: Signature:

▲ Figure 1.5 A method statement

4.4 How to complete a COSHH assessment

Hazardous substances can take many forms, including chemicals, dust and biological agents, and are not just products labelled as hazardous.

People at work are often exposed to hazardous substances and may be at risk of long-term (**chronic**) health problems such as asthma, cancer and skin disease.

Employers and the self-employed have a legal duty under the COSHH Regulations to assess the risk to their employees and others from exposure to products and processes that may contain or create substances harmful to health. If they have five or more employees, these assessments must be written down.

The main legal requirement is to use appropriate control measures to prevent harm from exposure, and this can be achieved by applying the hierarchy of control measures outlined in Figure 1.6.

Key term

Chronic: continuing for a long time

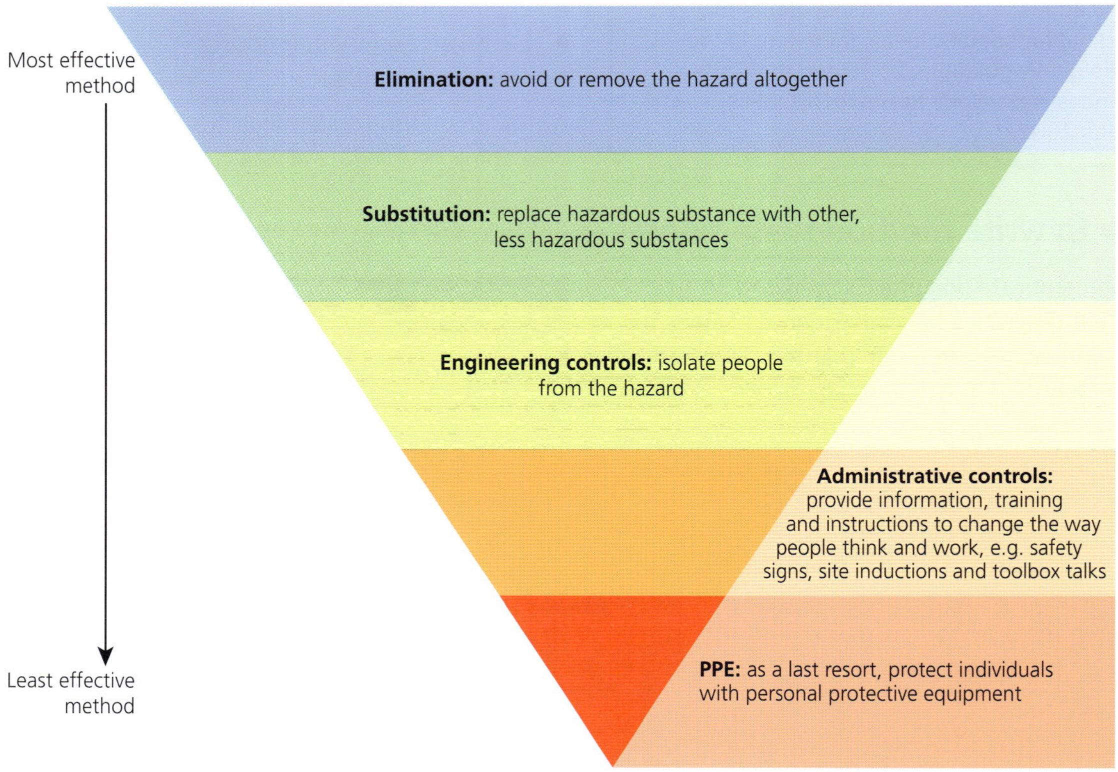

▲ Figure 1.6 Hierarchy of control measures to prevent exposure

By law, suppliers must provide up-to-date **safety data sheets** for any hazardous substances. These contain the supplier's information about a product and should not be confused with COSHH assessments.

Key term

Safety data sheets: written documents produced by manufacturers and suppliers of hazardous substances that contain important information about how products should be transported, used, stored and safely disposed of after use, any special conditions you should be aware of and how to deal with the substance in an emergency

Steps to follow to complete a COSHH assessment:

1. Identify the hazardous substance, who is likely to be harmed and how.
2. Evaluate the risk of the hazard causing harm by considering frequency of exposure to the substance and what effects it could have.
3. Decide what reasonably practicable measures are necessary to prevent or control any exposure to the hazard and how these will be maintained; make a plan for emergencies.
4. Record the assessment.
5. Decide if and when the assessment needs to be reviewed, and by whom.

Company name: Patio path and paving **Department:**		**Date assessment made:** **Date discussed with employees:**				
Step 1 **Substance**	**Step 2**	**Step 3**		**Step 4** **Action**		
What is the hazard?	**What harm, and who?**	**What are you doing already?**	**What improvements do you need?**	**Who**	**When**	**Check**
Breathing in dust from cutting paving	Long-term lung damage, e.g. bronchitis and silicosis. Everyone involved in cutting and anyone working nearby	Water to suppress dust. Protective goggles	Get enough water. Those involved should wear P3 respirators, ear plugs			
Breathing in exhaust fumes from cutter	Carbon monoxide poisoning. Everyone nearby	Always use cutter outdoors	Make sure area is not enclosed and fumes can disperse			
Ready-mix cement – skin contact – splashes	Skin burns when laying cement. Anyone – cement on skin	Avoid skin contact. Use protective gloves	Get access to running water. Wash off splashes immediately. Use skin care products			
Moss killer – skin irritation	Mixing concentrate. Anyone – splash on skin	Use protective gloves. Wash out applicator after use	Get access to running water. Use skin care products			
Also:		**Action taken**	**Action needed**			
Thorough examination and test – COSHH		None	Check water suppression			
Supervision		When available				
Instruction and training		Yes	Respirator training for P3 and fit test			
Emergency plans		Mobile phone				
Health surveillance		None	Ask doctor about lung function tests			
Monitoring		None	None			
Step 5 **Review date:**		1. Review your assessment – make sure you are not sliding back 2. Any significant change in the work? Check the assessment and change it if necessary				
Other hazards needing attention: lifting and handling heavy weights, noise, reversing vehicles						

▲ Figure 1.7 A COSHH assessment

Test yourself

Which substances are covered by the COSHH Regulations?

Categories of safety signs

There are four main categories of safety signs:

- Mandatory
- Prohibition
- Information
- Warning

Please see section 13 on page 32 for more information on safety signs.

5 Safety conscious procedures

5.1 Construction Skills Certification Scheme

The Construction Skills Certification Scheme (CSCS) is one of many industry-recognised card schemes accredited by the Construction Industry Training Board (CITB). CSCS cards prove that the card holder has a satisfactory level of health and safety awareness; they also show relevant qualifications the holder has achieved to confirm they are working in the correct job role on site.

There are various types of CSCS card, designed to suit people working at different levels in the construction industry, from labourers to academically and professionally qualified managers. CSCS cards last a maximum of five years before the holders have to reapply, however once an apprentice or trainee card has expired it cannot be renewed. There is no legal requirement for workers to hold a CSCS card, although most principal contractors and clients require people to have them as a way of managing health and safety on sites under their control.

To qualify for a basic CSCS card, a candidate must first successfully complete the CITB (Site Safety Plus) Health and Safety Awareness course and test. They must then pass the online CITB Health, Safety and Environment test before they are allowed to apply for an appropriate CSCS card.

5.2 Site Management Safety Training Scheme

Principal contractors and clients usually require people with planning, organising, controlling and monitoring responsibilities (for example site managers) to hold the Site Management Safety Training Scheme (SMSTS) qualification. This qualification ensures that the holder is up to date with health, safety, welfare and environmental legislation, so that they can fulfil their legal responsibilities. Every five years, the holder has to complete the Site Management Safety Training Scheme Refresher (SMSTS-R) to retain the qualification and to make sure they are aware of any changes in legislation, new guidance and industry best practice.

5.3 Site Supervision Safety Training Scheme

The Site Supervision Safety Training Scheme (SSSTS) is a similar qualification to the SMSTS. It is designed for people with supervisory responsibilities or those preparing to start in this role and provides an understanding of topics such as **site inductions**, risk assessments and method statements. Every five years, the Site Supervision Safety Training Scheme Refresher (SSSTS-R) course has to be completed, to keep up to date with any changes that affect the supervisory role and to retain the qualification status.

Key term

Site induction: an occasion when all new visitors and workers on a construction site are informed about potential hazards and the control measures used to protect people from harm

5.4 Procedures that promote and support safety consciousness

HASAWA states that employers have a duty to provide information, instruction and training to their employees, so that they know how to work safely without risks to health. Employees and the self-employed should be made aware of:

- the hazards and risks they may face
- control measures that are in place to deal with hazards
- what to do in the event of an emergency, for example assemble at the designated muster point (assembly point).

Employers must pay particular attention to certain groups, such as young people, new recruits and people changing roles, because they are less likely to be aware of potential risks and therefore require additional training and support.

Under the Construction (Design and Management) (CDM) Regulations 2015, principal contractors must engage with workers about their health, safety and welfare, and provide a site-specific induction and any other information and training that they need. Before entering a workplace, people should be made aware of potential hazards and the measures they need to take to protect themselves and others. On construction sites, this is usually achieved by following the site rules displayed on notice boards and safety signs. All workers and visitors to site have to sign in, so that the site manager is aware of who is on site at all times, especially in the event of an emergency. Visitors and new workers will have to attend a site induction, where they will be informed about:

- hazards
- site rules
- personal protective equipment (PPE)
- welfare facilities
- first aid
- fire extinguishers
- emergency procedures
- key personnel (for example site manager, assistant manager, supervisors, first aider)
- pedestrian routes
- parking
- working hours
- waste and the environment
- boundaries
- working at height
- traffic-management plans
- **permits to work**
- security
- risk assessments and method statements (RAMS).

Key term

Permits to work: documents issued by site managers to workers undertaking high-risk activities as a method of control; they authorise certain people to carry out specific work tasks within a given timeframe and set out the precautions needed to complete the work safely

Research

Make a list of activities that should be controlled by a permit-to-work system. Choose one activity from your list and create a detailed permit to work using a template from the HSE's website.

▲ Figure 1.8 A hot works permit-to-work activity

The duration of a site induction will depend on the size and nature of the work and the personnel undertaking the training. Experienced workers or designers (architects and engineers) may only need to attend a brief site induction, whereas someone that is newly qualified will need to attend a full induction. Inductions must be completed for every new site that is visited, regardless of whether the person is working for the same employer or principal contractor.

It is likely that the conditions and hazards on a site will evolve as work progresses, after people have attended a site induction. Workers will need to be updated on any matters of health and safety when necessary. Short presentations known as **toolbox talks** are often used to provide this information to workers. These are usually delivered to small groups of workers, in an area of the workplace where they should not be disturbed. These talks usually cover a single aspect of health and safety, such as good housekeeping (keeping areas clean and clear).

Key term

Toolbox talks: short training sessions arranged at regular intervals at a place of work to discuss health and safety issues; they give safety reminders and inform personnel about new hazards that may have recently arisen

5.5 Benefits of procedures that promote and support safety consciousness

It is important to remember that employees have duties under HASAWA, as well as employers, and that every effort must be made to comply with the law.

The law states that employees must:

- take care of their own health and safety and that of others
- report any work situations that present a serious and imminent risk
- co-operate with their employer to help them comply with health and safety legislation
- follow any instructions or health and safety training provided by their employer
- inform their employer of any shortcomings they identify in their health and safety arrangements.

When an employee fails in these responsibilities or does not follow their employer's health and safety procedures, they are essentially breaking the law and could be putting themselves and others at risk of injury, illness or death.

Accidents and near misses in the workplace often cost the employer in terms of time, effort and money. There may also be penalties imposed on them by the HSE and civil claims made against them.

A workplace accident could also result in indirect costs for the employer, such as:

- project timescales slipping and missed deadlines
- time spent training new staff
- increased insurance premiums.

Employers with a poor health and safety record usually have difficulty retaining staff, because they do not feel safe in their place of work. It also leads to low morale and a negative health and safety culture, which in turn makes it challenging to recruit new employees.

Poor standards of health and safety can also have a detrimental effect on future contracts and potentially result in losing business.

Safety consciousness within construction sites/environments/workshop areas include:

- reporting of potential hazards, following the company's reporting procedure and RIDDOR.
- procedures that must be followed when hazards cannot be eliminated completely.
- construction site inductions for all new visitors and workers including identification of risks and hazards and the control measures used to mitigate them.
- on and off the job health and safety training and related qualifications.
- toolbox talks which involve short training sessions at a place of work to discuss health and safety issues and inform personnel about new hazards that may have recently arisen.
- good housekeeping including systematic ways of working and keeping areas clean and clear.

Test yourself

1. What powers does the HSE have to enforce HASAWA?
2. What impact do you think an accident at work could have on an injured employee?

Improve your maths

In 2020, the HSE published a profile of non-fatal workplace injuries by accident kind:

- falls from a height 8%
- struck by a moving object 11%
- slips, trips or falls on the same level 29%
- acts of violence 9%
- handling, lifting or carrying 19%
- other 24%.

Produce a pie chart on a computer to illustrate this data.

6 Safety inspection of a work environment

6.1 Methods used to inspect a workplace to ensure it is safe

Employers have a responsibility to monitor health and safety arrangements in the workplace, to ensure they are effectively controlling risks.

There are two types of monitoring system that are typically used in the construction industry:

- **Active monitoring** is the monitoring of people, procedures, premises and plant in the workplace in order to identify potential hazards before an accident or incident occurs, so that control measures can be taken to prevent harm.
- **Reactive monitoring** is an inspection completed after an incident has taken place, for example a review of accident forms or statistics which might identify an area of concern, such as skin burns from welding equipment or falls from height.

Although reactive monitoring is a review of performance standards, it is a useful method of identifying problems to prevent further accidents and incidents occurring and should be used alongside active monitoring.

Key terms

Active monitoring: monitoring people, procedures, premises and plant in the workplace in order to identify potential hazards before an accident or incident occurs, so that control measures can be taken to prevent harm

Reactive monitoring: an inspection completed after an incident has taken place to prevent further incidents occurring

6.2 Monitoring health and safety in the workplace

As part of a good health and safety management system, employers should plan a programme of statutory, routine, periodic and pre-use inspections of equipment and their workplace. The frequency of the inspections will be determined by a number of factors, for example pre-use inspections of safety equipment may be completed daily by employees, whereas a periodic inspection of lifting equipment could be every six months, as determined by the Lifting Operations and Lifting Equipment Regulations (LOLER) 1998.

▲ Figure 1.9 A construction crane covered by LOLER

Active monitoring should involve structured, well-planned and organised examinations carried out by competent, experienced inspectors. Employers must decide:

- what type of inspections need to be completed
- who needs to be involved
- what analytical equipment may be needed, even if this means consulting specialists.

There are various types of health and safety inspection that can be implemented in the workplace:

- Health and safety audits (also known as desktop audits) are used to inspect health and safety documentation used in the workplace.
- Safety sampling is used to focus an investigation on a representative sample of a workplace standard, for example a random selection of portable electrical equipment could be selected to check if portable appliance testing (PAT) is up to date. If a big enough sample is selected, then it is likely to reflect any issues that may be identified on a wider scale within the company.
- Safety surveys are often used to focus a detailed health and safety investigation on a particular topic or issue, rather than the workplace as a whole.
- Safety tours are full inspections of entire workplaces to identify any hazards or shortcomings in employers' or employees' health and safety responsibilities.
- Incident inspections are carried out after an accident, a near miss or a case of reported ill health to the HSE.

Routine health and safety inspections are often guided by checklists and templates of set items that must be covered. These provide excellent prompts and are easy to complete with little training. They are also cheap, easy to reproduce and provide a consistent approach for every inspection, regardless of who completes it. The disadvantage is that hazards can often be overlooked if they are not listed, and therefore this form of inspection is not always preferred.

A random visual or sensory inspection of a work area by a team of representatives is sometimes a better way of identifying all potential hazards, because inspectors are not restricted by a checklist or template.

6.3 Types and use of recording documentation

Employers and their safety representatives are responsible for recording all inspections in a clear written report, as soon as possible after they have taken place. In some cases, this is a legal duty.

The report should contain:

- the date the inspection took place
- any potential hazards or defects found
- when the next inspection is due.

Inspection reports are then used to inform an action plan to put right any problems identified. An action plan should detail:

- any corrective actions needed
- people with responsibilities for carrying out the corrective actions
- the timescale in which the corrective actions must be completed.

Research

Research HSE forms F2534 and F2533 – documents that can be used to record the results of inspections and notify an employer of any unsafe or unhealthy working practices and unsatisfactory welfare facilities.

Under the Provision and Use of Work Equipment Regulations (PUWER) 1998, employers have specific duties to ensure that equipment provided for use at work is:

> *'... maintained in a safe condition and inspected to ensure it is correctly installed and does not subsequently deteriorate.'*

Other legislation, such as the Lifting Operations and Lifting Equipment Regulations (LOLER) 1998, the Pressure Systems Safety Regulations (PSSR) 2000, the Work at Height Regulations 2005 and the Personal Protective Equipment Regulations 2002, also give specific duties to employers to examine work equipment and keep an up-to-date register of all inspections.

It is important that employers maintain these inspection records because they may be referred to in the event of an investigation into an accident or near miss in the workplace. Inspection records, maintenance schedules and service plans could also be requested by HSE inspectors or required by an employer's insurance company.

7 Safe working practices for the safe isolation of systems

During the installation and maintenance of Building Services Engineering systems, engineers are required to isolate a range of supplies:

- water supplies (hot and cold)
- gas supplies
- electrical supplies.

It is important that isolation is carried out safely and in accordance with relevant guidance documents and set procedures.

Prior to isolating any building services system, the responsible person should be informed and should also be advised as to how long the system will be out of action (**decommissioned**). Where applicable you should always arrange for alternative services to be provided.

Key term

Decommissioning: the process of isolating a system and taking it out of action either permanently or temporarily

7.1 Water supplies (hot and cold)

When working on both cold and hot water supplies, it is important to know where to safely isolate the system. Isolation valves turn off (isolate) either complete systems, parts of systems or individual appliances. They can be divided into the following types:

- those that isolate high-pressure systems, such as stop valves
- those that isolate low-pressure systems, such as full-way gate valves
- those that isolate appliances and terminal fittings on either high- or low-pressure systems.

Where isolation valves are not in place, a pipe freezing kit can be used to isolate the system.

▲ Figure 1.10 Service valve supplying a cold water storage cistern

Direct cold water systems can be completely isolated at the main stop valve where the supply enters the building. If this is used to isolate the water supply, both the hot and cold water will be isolated. However, the CWSC and cylinder will need draining down before beginning work if it is an open vented system.

Indirect cold water systems can be isolated at the gate valve located on the distribution pipe connected to the cold water storage cistern. If the cold water storage cistern requires replacement, it will need to be isolated at the service valve supplying it.

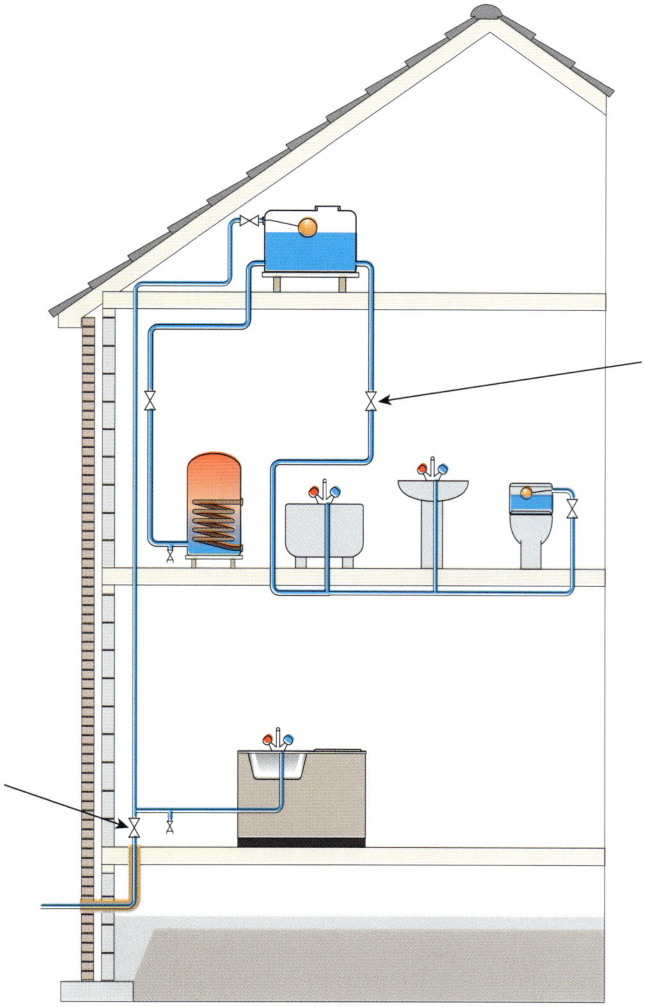

▲ Figure 1.11 Isolating an indirect cold water system

Unvented hot water systems can be isolated at the service valve on the cold water supply to the cylinder.

Where hot water in a property is supplied from a combination boiler or water heater, this should be isolated at the service valve located under the boiler. (Remember to isolate the electrical supply.)

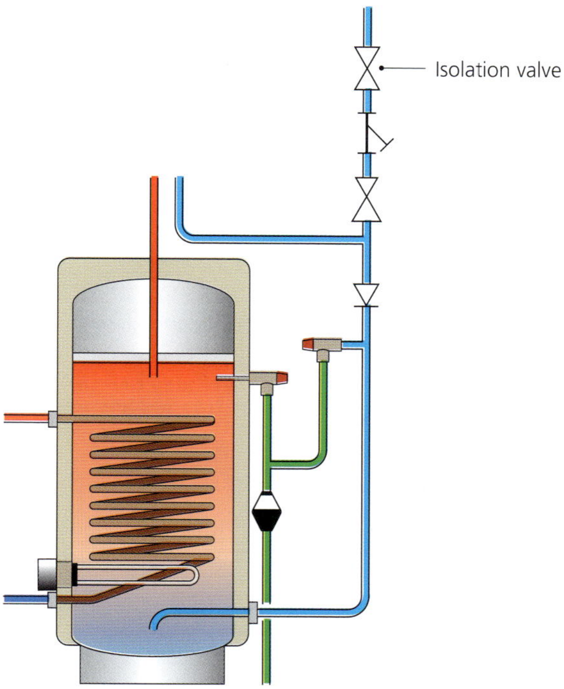

▲ Figure 1.12 Isolating an unvented hot water system

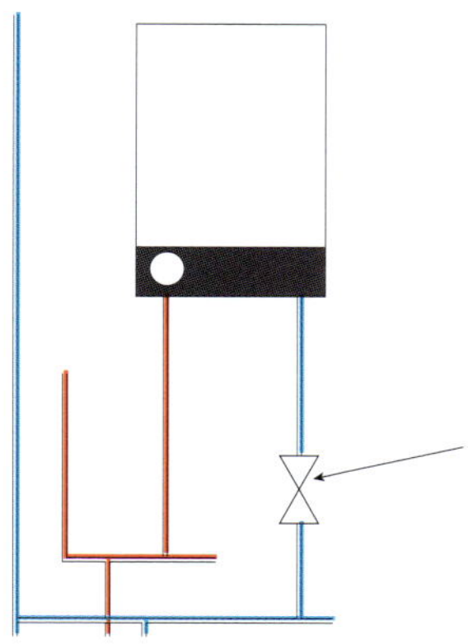

▲ Figure 1.13 Isolating a combination boiler or water heater

Vented hot water systems are isolated at the gate valve supplying the hot water cylinder.

Health and safety

Care should be taken when draining down hot water systems to prevent scalding.

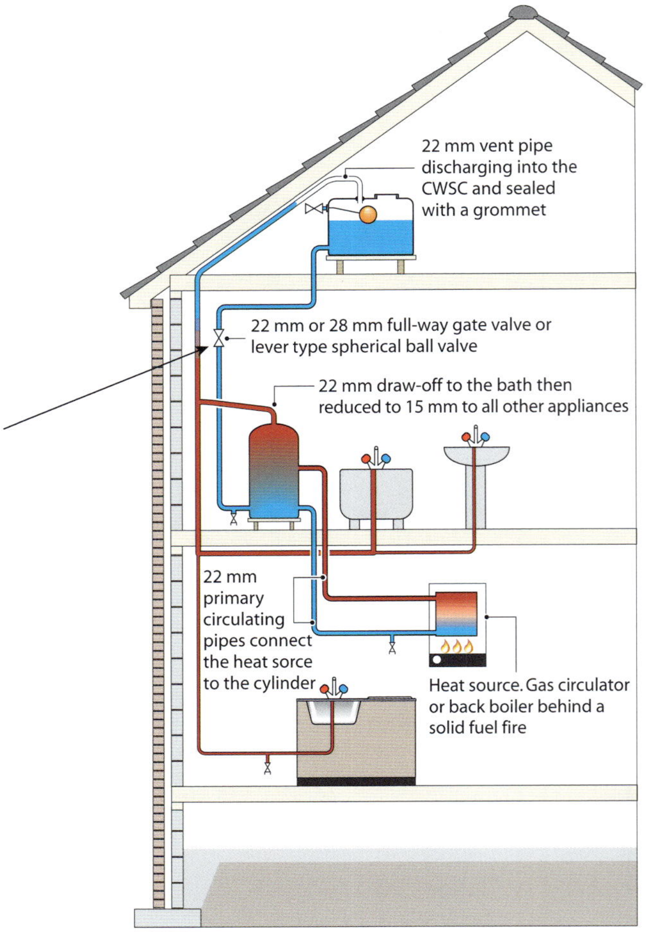

▲ Figure 1.14 Isolating a vented hot water system

At the point of isolation, a warning notice should be displayed informing people that the system is out of order and not to use it. This will prevent the system being turned on while it is being worked on.

▲ Figure 1.15 Warning notice

Where there are open ends on hot and cold water supply pipework, these should be capped off using a suitable fitting.

Test yourself

1. Where is the cold water supply to the entire property isolated?
2. Where hot water is supplied from a combination boiler or water heater where do we isolate it?
3. Where is the hot water supply from a vented hot water system isolated?

7.2 Gas supplies

Anyone can turn an ECV off in the event of an emergency, but capping it off has to be completed by a Gas Safe registered engineer.

Located at the inlet of the gas meter is an emergency control valve (ECV) which is used to isolate the gas supply to a property. The ECV should:

- fall to the 'off' position
- have on/off tape indicating the direction to close the valve
- move freely.

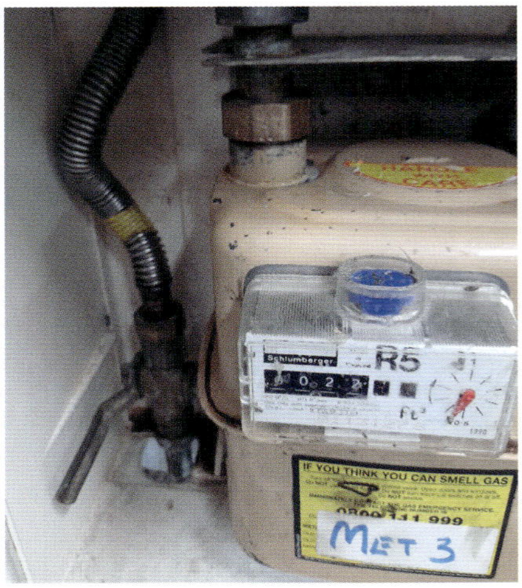

▲ Figure 1.16 Gas meter showing the position of the ECV

In the event of an emergency gas situation, the following procedure should be followed:

- Turn off the ECV
- Open doors and windows to ventilate the property
- Call the National Gas Emergency Service number
- DON'T turn any power or light switches on or off
- DON'T light any sort of flame within the property
- DON'T use any appliances that could cause a spark.

As stated in the Gas Safety (Installation and Use) Regulations 1998, there should be an isolation valve at the inlet of all gas appliances to aid localised isolation of the supply.

▲ Figure 1.17 Appliance gas isolation valve

Industry tip

It is an offence to leave an open end on gas supply pipework. All open ends must be capped using a suitable stop end.

7.3 Electrical supplies

Before any work is carried out on electrical systems, the isolation procedure should be carried out to ensure the installation is safe to be worked on.

When you isolate an electricity supply, there will be disruption. Careful planning should precede isolation of circuits. For example, when isolating a section of a nursing home where elderly residents live, you will need to consult the nursing home staff to consider the possible consequences of isolation and to prepare a procedure.

The following questions are useful:

1. How will the isolation affect the staff and other personnel? For example, think about loss of power to lifts, heating and other essential systems.
2. How could the isolation affect the residents and clients? For example, some residents may rely on oxygen, medical drips and ripple beds to aid circulation. These critical systems usually have battery back-up facilities for short durations.
3. How could the isolation affect the members of the public? For example, fire alarms, nurse call systems, emergency lighting and other systems may stop working.
4. How can an isolation affect systems? For example, IT programs and data systems could be affected; timing devices could be disrupted. In this scenario, you must make the employers, employees, clients, residents and members of the public aware of the planned isolation.

Alternative electrical back-up supplies may be required in the form of generators or uninterruptable power supply systems.

The main incoming electrical supply can be isolated at the consumer unit or distribution board. This component contains a main switch which will isolate all the power within a property. Each individual circuit will have its own miniature circuit breaker (MCB) and means of isolation allowing isolation of individual circuits such as lighting, sockets, cooker points, immersion heaters and appliances.

▲ Figure 1.18 Distribution board

▲ Figures 1.19 and 1.20 Switched fuse unit (left) and locked fuse (right)

The electricity supply to appliances and components is usually provided by a fused spur, either switched or unswitched. These appliances can be isolated at this point by isolating and removal of the cartridge fuse.

Industry tip

The electricity supply to a macerator should be provided via an unswitched fused spur.

Safe isolation of electrical systems

The following equipment is required to undertake the safe isolation of electrical systems:

- a voltage indicator which has been manufactured and maintained in accordance with Health and Safety Executive (HSE) Guidance Note GS38
- a proving unit compatible with the voltage indicator
- a lock and/or multi-lock system
- warning notices which identify the work being carried out
- relevant personal protective equipment (PPE) that adheres to all site PPE rules.

Procedure

1 Obtain permission to start work. In some environments a permit to work may be needed
2 Locate and identify circuit or equipment to be worked on
3 Identify means of isolation
4 Ensure isolation of circuit or equipment by switching off and:
   - withdrawing fuses
   - locking off
   - isolating switches or circuit breakers
   - fitting warning notice at point of isolation

▲ Figure 1.21 Warning notice at point of isolation

5 Select an approved test lamp or voltage indicating device

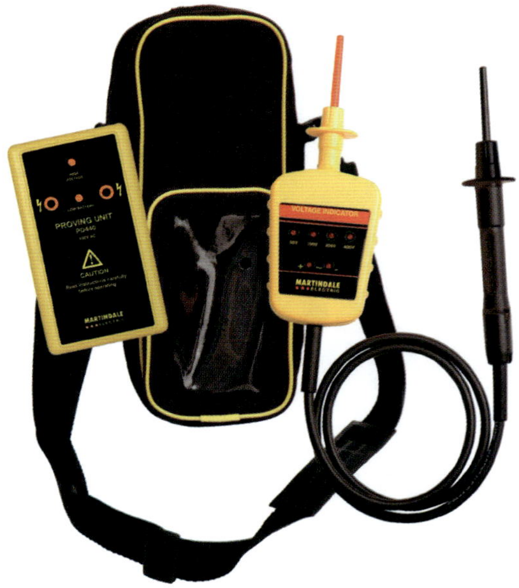

▲ Figure 1.22 Voltage proving meter to indicate voltage

6 Verify that the device is functioning correctly on a known supply or proving unit
7 Verify that the circuit or equipment to be worked on is dead using a voltage indicating device, testing between all line conductors (three phase):
   - Line – Earth
   - Line – Neutral
   - Neutral – Earth
8 Recheck that the voltage indicating device is functioning correctly on a known supply or proving unit
9 Begin work

Test yourself

1 Prior to working on an electrical system, what should be carried out?
2 How can we inform people that a system has been isolated?

8 Implications of poor health and safety on building performance and individual stakeholders

The implications of poor health and safety and its impacts at different levels i.e. employee, employer/business, client/customer/public, include:

- Accidents: incidents that happen unexpectedly and unintentionally, typically resulting in damage or injury.

- Injuries: harm or damage to someone caused by an accident.
- Fatalities: death by accident.
- Slips, trips, falls resulting in accidents, injuries or fatalities.
- Down time: a pause in work due to a health and safety issue.
- Financial: a loss of income or work due to a health and safety issue or financial penalties.
- Reputation: poor business reputation as the result of poor health and safety practices; this often leads to financial implications.
- Environmental: an event that results in environmental or equipment damage.
- Near misses: an event that has the potential to cause but does not actually result in human injury.

9 Recording and reporting of safety incidents and near misses

9.1 Reporting an incident or near miss in the workplace

Workplace accidents and incidents must be reported to employers so that they can be properly dealt with and investigated to help develop potential solutions and reduce the risk of them reoccurring.

Under the Social Security (Claims and Payments) Regulations 1979, employers must record details of any workplace accident in an **accident book**. It is best to complete the entry as soon as possible after the accident has occurred, so that important information is not left out.

Key term

Accident book: a formal document used to record details of accidents that occur in the workplace, whether to an employee or visitor

As this book contains sensitive information about the accident and personal details about the injured person, it must be stored safely and securely to prevent a breach of confidentiality.

By making reports and keeping records of accidents on site, it is possible to see patterns that may be emerging, possibly due to bad habits or incorrect work practices. These records can assist in planning future work activities to reduce the occurrence of accidents and may be used when legal matters arise related to an accident or emergency. Accident records can be requested and looked at by an employer's insurance company, if a claim is made against them, or the HSE as part of an investigation into an accident or dangerous occurrence. The law states that employers must keep accident records for at least three years after the date they were completed.

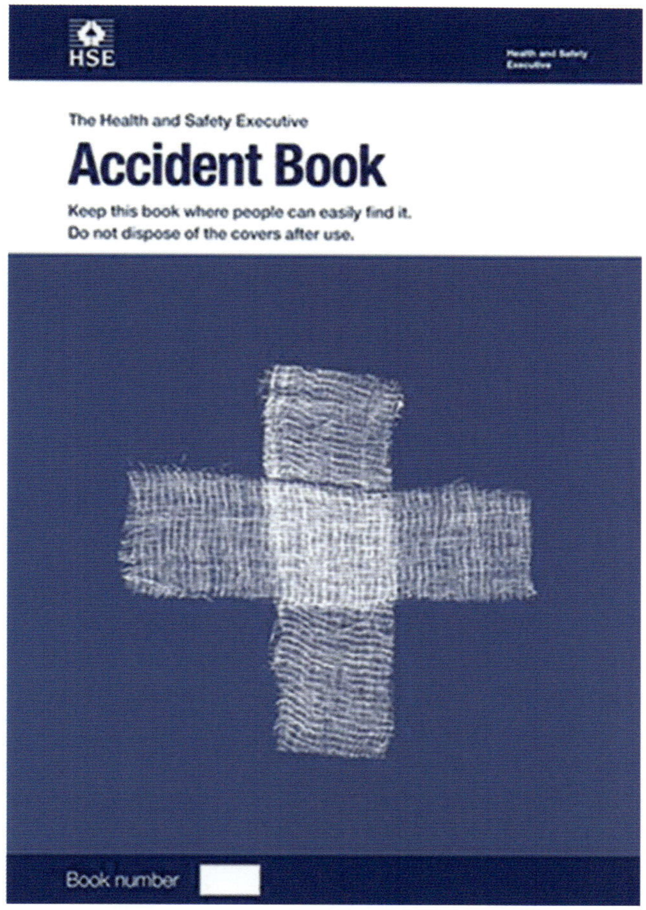

▲ Figure 1.23 An accident book

Employers have a duty to report the following to the HSE under RIDDOR:

- the death of any person caused by a workplace accident
- dangerous occurrences (also known as 'near-miss accidents')
- gas incidents where someone has died, lost consciousness, or been taken to hospital for treatment

- non-fatal accidents to non-workers, for example a member of the general public; however, this only needs to be reported if the person received treatment at hospital
- occupational diseases:
  - asthma
  - cancer
  - carpal tunnel syndrome
  - dermatitis
  - disease attributed to exposure to a biological agent
  - hand–arm vibration syndrome
  - severe cramp of the forearm or hand
  - tendonitis of the forearm or hand
- specified injuries caused by workplace accidents:
  - amputations
  - injuries arising from working in an enclosed space
  - any scalping requiring hospital treatment
  - crush injuries to the torso or head
  - fractures, other than to fingers, toes and thumbs
  - loss of consciousness caused by asphyxia or a head injury
  - permanent loss of sight or reduction in sight
  - serious burns (including scalding)
- over seven consecutive days' absences when a worker is unable to perform their duties as a result of a workplace accident.

If an employee has been absent from work for three days as a result of an accident, it does not need to be reported to the HSE; however the employer must keep a record. A copy of a report in an accident book is usually sufficient.

▲ Figure 1.24 A dangerous occurrence

Test yourself

When should an employee be made aware of their employer's accident and incident reporting policy?

Research

Find out how to report a dangerous occurrence to the HSE under RIDDOR.

10 Emergency procedures for unsafe situations

10.1 Procedures to follow in unsafe situations

An emergency situation in the workplace can often result in delays with construction work, missed deadlines and additional costs for the principal contractor, not to mention the risk to workers' lives. To reduce the potential impact of an unforeseen incident, employers have a duty under the CDM Regulations to plan for emergencies on construction sites as part of their pre-construction work. Under the Regulatory Reform (Fire Safety) Order 2005, employers also have a duty to plan for emergencies on other sites such as offices, factories and warehouses.

▲ Figure 1.25 A fire on a construction site

Examples of unsafe or emergency situations in a workplace include:

- fire, for example welding pipes using a blow torch providing a source of ignition
- gas leaks, for example groundworks with a multi-purpose excavator causing damage to a buried gas pipe
- terrorist threats, for example while undertaking construction work on government buildings
- water leak, for example boring (drilling) into the ground and coming into contact with a mains water pipe

- carbon monoxide, for example from a poorly installed or maintained boiler
- electric shock, for example an operative cutting into a masonry wall with a disc cutter coming into contact with buried services such as electrical wires.

As part of their health and safety preparations, an employer (as a **responsible person**) must nominate a competent person to make an emergency plan, complete a fire risk assessment and inform workers about the findings.

Key term

Responsible person: (in law) usually the employer (assuming control of the workplace), a person with control of the premises or an owner; that person can then nominate other competent people to perform legal responsibilities, for example to manage fire safety

Fire risk assessments are intended to identify fire hazards in a workplace and those at risk of harm. The level of risk should then be evaluated and a plan should be prepared by the responsible person. Where an employer has five or more employees, the fire risk assessment should be recorded.

Workers on construction sites are usually informed about plans for emergencies and fire risk assessments during their site induction. The actions to follow are often reinforced on safety notices and information boards displayed in work areas.

An emergency plan should be prepared at the start of every project and cover the following areas:

- safe evacuation of the site in the event of an emergency
- emergency escape routes
- safety signage (for example directional signs and assembly points)
- safe storage of hazardous and flammable materials, including waste
- site security to minimise the risk of **arson**
- assembly points and registers of workers on site
- fire wardens
- fire detection and alarms
- designated spill kits to deal with hazardous chemical, fuel or oil spillages and leaks
- training for workers on emergency equipment and fire safety (for example use of fire extinguishers, spill kits, and how to raise the alarm)
- procedures for calling the emergency services
- emergency lighting
- fire extinguishers
- maintenance of escape routes and fire doors to keep them clear at all times
- **hot works** permit systems
- arrangements for smoking
- arrangements for vehicles and plant, for example refuelling
- emergency plant shut down and isolation to make processes safe.

▲ Figure 1.26 A fire marshal

Key terms

Arson: the criminal act of deliberately setting fire to property

Hot works: any construction work involving open flames or generating heat or friction; hot works are particularly high-risk activities that have the potential to cause fires if they are not controlled properly with a permit-to-work system

Research

Research permits to work and list as many construction activities as possible that may need one.

Research

Research the role of a fire warden and list their responsibilities in the event of a fire in a workplace.

10.2 Actions to be taken when dealing with fire situations

In the event of a fire in the workplace, workers must follow their employer's procedures. The main steps in a fire procedure are outlined in Figure 1.27.

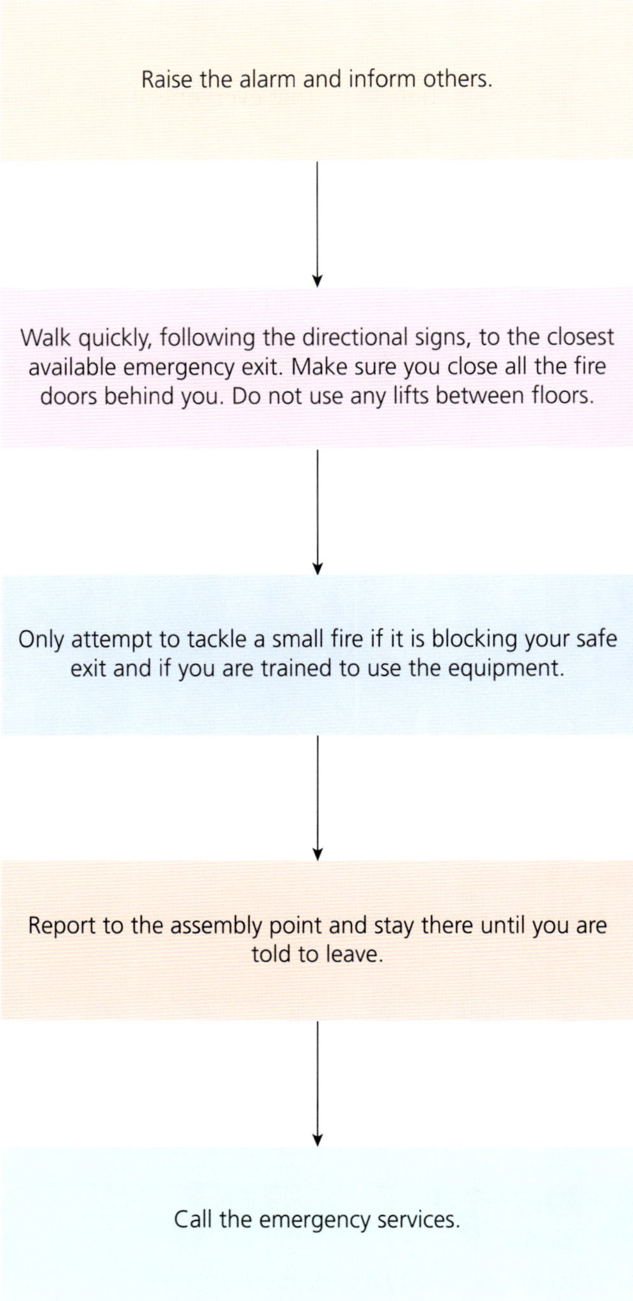

▲ Figure 1.27 Fire procedure

Employers should prepare for an emergency by practising evacuation procedures, without notice, to make sure everyone understands their roles to act promptly and appropriately. The control measures should be monitored, maintained and reviewed by the appointed person to make sure they are effective and do not deteriorate over time, for example a fire drill that is over-rehearsed could be disregarded by workers in a real emergency situation.

Research

Look up the Gas Industry Unsafe Situations Procedure (GIUSP) and Gas Safety (Installation and Use) Regulations (GSIUR) 1998. Write a short report on the role of an engineer when an appliance or installation poses an immediate danger to life or property.

▲ Figure 1.28 A first aider attending to a casualty on a construction site

Research

What are the emergency procedures to deal with electric shock?

10.3 Fire extinguishers and their uses

Fires need three elements to burn:

- heat – such as a spark or naked flame
- fuel – any material or substance that is combustible
- oxygen – a gas that occurs naturally in the air.

If all three are present, a fire is unavoidable. If one element is missing, a fire cannot occur. This is often referred to as the 'fire triangle'.

Fire extinguishers are designed to put out fire by removing one or more of these elements; however, if the wrong extinguisher is used, it could make the fire worse or increase the risk of harm.

The fire risk assessment should identify the appropriate fire extinguishers for a workplace and where they should be situated (known as a fire point). They are usually placed close to exits and where there is an increased risk of a fire, for example hot works.

Fires are classified according to the type of fuel that is burning:

- Class A (flammable solids) – paper, cardboard, fabric
- Class B (flammable liquids) – paints, adhesives, petrol
- Class C (flammable gases) – liquid petroleum gas (LPG), natural gas
- Class D (flammable metals) – magnesium, lithium, aluminium
- Electrical (not classed as E because electrical fires are a source of ignition, not fuel)
- Class F (cooking fats and oils) – chip pan or grease fires.

The CDM Regulations state that everyone at work should be instructed on the safe use of fire extinguishers and understand the colour-coding system used to distinguish different types. All fire extinguishers should have a red body with a colour-coded panel to identify their content and the type of fire that they can safely be used to extinguish.

Table 1.2 describes the range of fire extinguishers.

▼ Table 1.2 Fire extinguishers

Type of extinguisher	Colour of label	Fire classification	Special considerations	How it works
Water	Red 	Class A	Do not use on Class B or electrical fires	Cools the fire to remove the heat
Dry powder	Blue 	Class A Class B Class C Class D Electrical	Extinguishers leave a residue that may be harmful to sensitive electronics	Smothers the fire to remove the oxygen
Foam	Cream 	Class A Class B	Do not aim the extinguisher directly at liquid fires Do not use on electrical or Class F fires	Floats on the surface of the burning liquid to create a seal, starving the fire of fuel

▼ Table 1.2 Fire extinguishers

Type of extinguisher	Colour of label	Fire classification	Special considerations	How it works
Carbon dioxide (CO_2)	Black 	Class B Electrical	Horn can become very cold when it is discharged and may freeze, although frost-free versions are available Not to be used in a confined space	Displaces (shifts) the oxygen in the air
Wet chemical	Yellow 	Class A Class F	Prevents splashing of hot fats and oils	Cools the fire to remove the heat

Research

Research other types of equipment that can be used to extinguish a fire and explain why they would be preferred to fire extinguishers.

11 Types of PPE

11.1 The purpose and correct use of PPE

When the principles of prevention are applied to mitigate the risk of harm, personal protective equipment (PPE) is always considered a last resort because it only protects the user. Only after employers have considered all other methods of protecting their employees, for example eliminating or controlling risks at source, should they provide suitable PPE or **respiratory protective equipment (RPE)**.

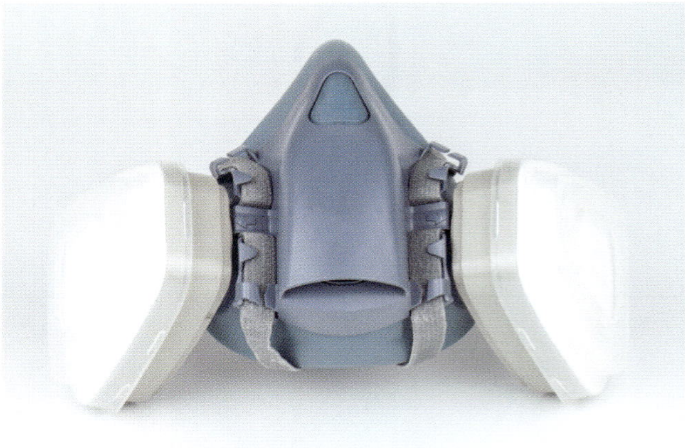

▲ Figure 1.29 A respirator

Key term

Respiratory protective equipment (RPE): personal protective equipment that protects the user's respiratory system

Under the Personal Protective Equipment (PPE) at Work Regulations 1992, employers have legal duties to provide adequate information, instruction and training to their employees on the safe use of PPE. The law also states that employers should provide PPE free of charge, and that their employees should understand how to use and take care of it.

PPE is a last line of defence and will only work properly if the correct type has been selected, it fits correctly and it is in good condition. Therefore, employees must:

- wear PPE when instructed to do so by their employer
- wear and adjust PPE correctly to work safely as trained
- store PPE after use and maintain it
- report any defects in, damage to or loss of PPE.

If more than one item of PPE has to be worn at the same time, you must make sure that they are compatible with each other and that this does not affect the level of protection they provide. PPE that is uncomfortable to wear and does not suit the user can sometimes create further hazards, and discourage workers from using it.

Table 1.3 illustrates types of PPE and the hazards they protect against.

▼ Table 1.3 Types of PPE

Body part protected	Hazards	Types of PPE	Correct use
Ears	Noise – average weekly exposure above 85 dB and other stated levels.	• Ear defenders • Ear muffs • Ear plugs • Canal caps/semi-insert ear plugs	Ear protection should reduce noise levels to an acceptable level, so that you are still able to communicate while wearing it. Ear protectors are manufactured with a single number rating (SNR) system, which allows the acoustic pressure on your ears to be calculated. For example, ear plugs with an SNR of 38 dB (decibels) will reduce a noise of 87 dB by 38 dB to a safe level of 49 dB. To put this into context, a hammer drill produces in excess of 100 dB. Disposable foam ear plugs should be fully inserted in the ear to work properly and disposed of after each use.
Eyes	• Sparks • Dust • Chemicals • Debris	• Goggles • Safety spectacles • Face screens • Face shields • Full-face visors • Sunglasses	Eye protection should be compatible with other PPE worn, adjustable and stored correctly to prevent damage to the lenses.
Feet and legs	• Slips • Falling objects • Objects penetrating the sole (for example nails sticking out of timber)	• Safety trainers, shoes, boots and wellingtons with toecaps and protective mid-soles • Chainsaw and foundry boots • Knee pads • Kneeling pads	Footwear should have a good grip for different surfaces, for example oil and mud, and be replaced when it becomes damaged. The risk assessment will identify which footwear should be worn.
Hands and arms	• Cuts and abrasions • Impacts • Chemicals • Temperature extremes • Biological agents • Hand-arm vibration syndrome (HAVS)	• Anti-vibration gloves • Nitrile foam coated gloves • Gloves with cuffs • Gauntlets • Protective arm sleeves • Elbow pads	Care should be taken to select the correct type of gloves to protect against hazards. They must not create further risks, such as entanglement in machinery, when they are being used.

▼ Table 1.3 Types of PPE

Body part protected	Hazards	Types of PPE	Correct use
Head and neck	• Falling objects • Hair entanglement • Chemicals • Adverse weather	• Hard hats • **Bump caps** • Snoods • Hair nets	PPE should be worn as directed by the manufacturer. In general, hard hats should be square on your head with the peak facing forwards. Avoid wearing caps or beanies underneath hard hats, unless they are designed for this. Avoid marking hats with paint or pens, because the chemicals damage them. Bump caps are no substitute for hard hats and should only be worn when there is a very low risk of bumping your head.
Lungs (respiratory system)	• Dust • Vapours • Mists • Gases • Atmospheres with low or no oxygen	Respiratory protective equipment (RPE): • disposable half-mask respirators • reusable half-mask respirators with a filter • full-face mask respirators • powered respirators with a mask/hood or helmet • breathing apparatus (BA)	Masks should form a good seal around the user's face to protect them properly. Employers often assess the performance and suitability of masks provided for their employees by undertaking face-fit testing. The type of masks and filters used should reflect the hazards. Employees should understand when and how to replace respirator filters. When masks are not in use, they should be stored correctly to prevent them being contaminated with hazardous substances.
Whole body	• Chemicals • Temperature extremes • Adverse weather • Dust • Metal splashes	• Aprons • Overalls • Boiler suits • Chemical suits • High-visibility clothing	Whole-body protective equipment must be worn according to the manufacturer's instructions and should not cause a risk of entanglement with equipment or machinery. Contaminated PPE should be cleaned or disposed of properly and never mixed with personal clothing.

Key term

Bump caps: a type of PPE designed to protect the user's head when there is a low risk of bumping it; it is not designed to take an impact from an object falling from height

Emergency equipment may be required for those working near water, in confined spaces, at height or in oxygen-deficient atmospheres. This includes life jackets, safety harnesses, safety ropes and breathing apparatus. You must be trained in the safe use of emergency equipment and report any damage to it.

▲ Figure 1.30 A construction worker wearing PPE

Health and safety

Always check the expiry dates of safety helmets to ensure they offer the correct level of protection in the event of an accident. Your employer should tell you how to find this information as part of your PPE training.

Health and safety

The Personal Protective Equipment (Enforcement) Regulations 2018 state that PPE must be carefully selected to make sure it is manufactured to European standards. Always check your PPE before using it to make sure it bears **CE marking**. If it does not, inform your employer immediately and replace it. Note that rules on CE marking on PPE are due to be replaced with UK Conformity Assessed (UKCA) marking on 1 January 2023.

Key term

CE marking: a mark on a product that shows it has been designed and manufactured to meet EU safety, health or environmental requirements; CE is an abbreviation of a French term '*conformité européenne*', meaning 'European conformity'

▲ Figure 1.31 PPE with CE marking

12 First-aid facilities

12.1 First-aid facilities that must be available in the work area

The Health and Safety (First Aid) Regulations 1981 place legal duties on all employers to provide adequate and appropriate first-aid equipment, facilities and people to assist their employees if they are injured or taken ill at work. To establish what is considered adequate and appropriate, employers must first carry out a workplace-specific first-aid assessment to determine their needs.

A first-aid assessment should take into consideration:

- the nature of work being carried out and any specific hazards
- the number of staff on site at any one time
- the spread of the workforce
- **lone workers**
- the amount of first aiders and cover needed for holidays/absences
- the number of first-aid boxes and their contents
- first-aid resources needed, for example a first-aid room, stretcher and defibrillator
- staff training, information and instruction.

Key term

Lone workers: those who work by themselves without close or direct supervision by their employer or work colleagues; they may work in a remote location or in a workplace with members of the public

Employers responsible for work activities in low-risk environments have to provide a first-aid kit for their workers. They must also appoint a person to take charge of their first-aid arrangements, for example checking first-aid boxes are adequately stocked and calling the emergency services when necessary.

In workplaces where there is a significant risk to employees' health and safety, employers must also have a trained first aider. In the event of the first aider's absence, due to any unforeseen circumstances, the appointed person should be able to provide emergency cover for them.

The HSE publication 'First aid at work' recommends the contents of a basic first-aid box. However, an employer's first-aid assessment will usually determine their exact needs.

The first-aid equipment and facilities provided by employers should be kept clean and dry and made easily accessible. They should also be clearly signposted, in accordance with the Health and Safety (Safety Signs and Signals) Regulations 1996.

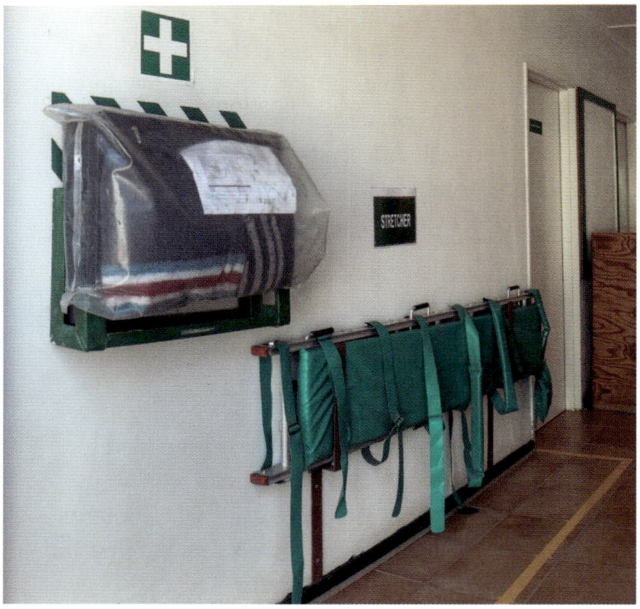

▲ Figure 1.32 First-aid facilities on a construction site

Employers have a legal duty to inform their employees about their first-aid arrangements, including who their first aiders are, how to contact them and the location of the first-aid provision. This information is usually shared with workers when they first arrive at work, as part of their site or workplace induction.

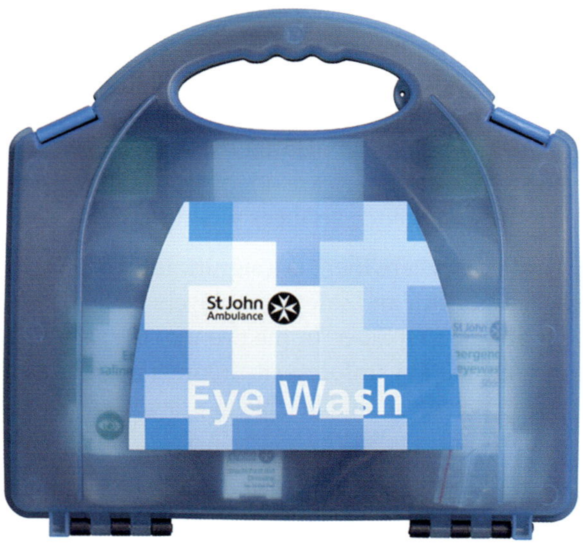

▲ Figure 1.33 First-aid eyewash kit

Research

What provision has your industry placement employer made for lone workers?

Health and safety

The HSE recommends that tablets and medicines are not kept in a first-aid box.

13 Safety signs

13.1 Categories of safety sign

Safety signs and notices are often displayed around work areas and construction sites to convey important health and safety information. The Health and Safety (Safety Signs and Signals) Regulations 1996 categorise safety signs by their geometrical shape, colour and a simple pictogram. They may also be supported with supplementary text in the same colour as the sign.

The main categories of safety sign are listed in Table 1.4.

▼ Table 1.4 Categories of safety sign

Type of safety sign	Description
Mandatory 	A circle with a blue background and white symbol or text Tells you that something *must* be done, for example eye protection must be worn
Safe condition 	A square or rectangle with a green background Shows directions to areas of safety and medical assistance in case of emergency
Prohibition 	A circle with a red outline and a red line from the top left to the bottom right Tells you that something *must not* be done, for example no smoking or no pedestrian access

Type of safety sign	Description
Warning 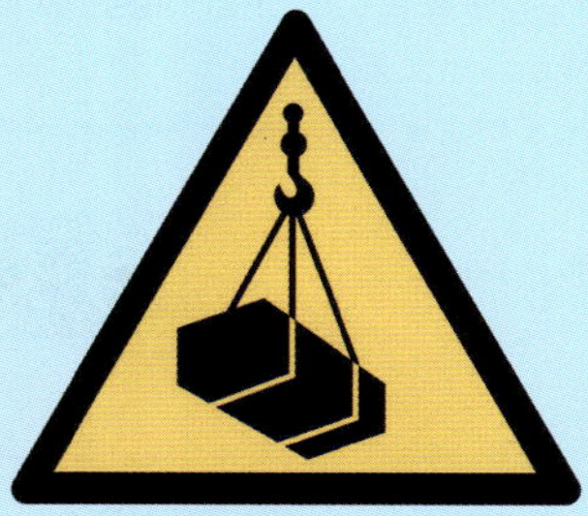	A yellow triangle with a black outline Makes you aware of nearby danger
Fire fighting 	A red rectangle or square with a white pictogram Marks the location of fire-fighting equipment and fire-alarm activation points

The law states that safety signs should only be used when there is a significant risk to health and safety that cannot be removed or controlled in other ways, or when they can reduce a risk further. If too many signs are displayed in close proximity to each other, they can cause confusion and the main points may be overlooked; therefore employers must select and position them carefully.

Health and safety

Safety signs should not be used as a substitute for other control methods, for example engineering controls such as safety guards.

Test yourself

What are the five main classifications of safety sign?

Case study

The principal contractor for a construction project is in the pre-construction stage of planning for work on a new building site. The project is expected to last longer than 30 working days and have more than 20 workers on site at any one time during the construction phase.

Explain what safety signs and notices would legally have to be displayed around the site under the Construction (Design and Management) Regulations 2015.

What do you think are the advantages of using safety signs and notices, rather than other safety control measures?

13.2 Symbols for hazardous waste

Manufacturers, importers, distributors and other users of chemicals in the UK have legal duties under the Classification, Labelling and Packaging (CLP) Regulation 2010 to use appropriate safety signs for the classification, labelling and packaging of hazardous substances and waste. These signs are explained in Table 1.5.

▼ Table 1.5 CLP Regulations safety signs

Safety sign	Meaning	Encountered when using ...
Explosive 	Explosive, self-reactive	Gas

Safety sign	Meaning	Encountered when using ...
Flammable 	• Flammable gases, solids, liquids and aerosols • Self-heating, self-reactive • Contact with water creates flammable gas	• Expanding foam • Nail-gun canisters • Solvent cement • Paint stripper
Oxidising 	• Oxidising gases, liquids and solids • May cause fire or explosion • May intensify fire	Chemicals
Gas under pressure 	• Contains gas under pressure • May explode if heated • Contains refrigerated gas which may cause cryogenic burns	Carbon-dioxide cylinders used in welding
Corrosive 	• Corrosive to metals • Causes severe skin burns and eye damage	• Portland cement • Hydrated lime • Brick cleaner • Batteries

▼ Table 1.5 CLP Regulations safety signs

Safety sign	Meaning	Encountered when using ...
Acute toxicity 	• Toxic from single or multiple exposure • Toxic/fatal if swallowed, in contact with skin or inhaled	• Materials containing formaldehyde • Hazardous air pollutants
Health hazard/hazardous to the ozone layer 	• May cause respiratory, eye or skin irritation • May cause drowsiness or dizziness • Harmful if swallowed, inhaled or in contact with skin • Harms the environment by destroying the ozone layer	• Expanding foam • Grab adhesive • Wood adhesive • Solvent cement • Portland cement • Paint stripper
Hazardous to the environment 	Toxic to the surrounding natural environment, especially aquatic life	• Wood preservative • White spirit • Diesel, petrol and paraffin oils • Epoxy resin • Bitumen paint
Serious health hazard 	• May be fatal if swallowed or enters airways • May cause damage to organs • May damage fertility or cause genetic defects • May cause cancer • May cause allergy, asthma or breathing difficulties if inhaled	• Expanding foam • Grab adhesive • Paint stripper • Wood dust • White spirit • Asphalt • Silica dust

Research

Find one example for each of the CLP Regulation classifications listed in Table 1.5 and describe how they should be safely handled and disposed of as hazardous waste.

The Carriage of Dangerous Goods and Use of Transportable Pressure Equipment Regulations 2009 deal with potentially hazardous substances that are transported by road or rail. These regulations have a slightly different set of diamond-shaped safety signs, designed to suit their classifications of predominant hazards.

Research

Research the classifications of safety sign under the Carriage of Dangerous Goods and Use of Transportable Pressure Equipment Regulations 2009.

Make a list of some of the materials you may have used that would require these signs to be displayed on the vehicles transporting them, and explain how waste materials and packaging were disposed of after use.

14 Safe practices and procedures for the use of access equipment and manual handling

14.1 Types of access equipment

The HSE reported that 111 workers died as a result of falling from height in 2019–20 in the UK, and 40 of those people worked in the construction industry.

If there is a risk of people falling any distance above or below ground that could result in an injury, the employer must take necessary precautions to eliminate the hazard completely or reduce the risk of harm to an acceptable level by assessing the risks.

Where a risk remains, employers should use equipment or other measures to minimise the distance and consequences of a fall, for example safety nets or air bags, or PPE such as a safety harness and lanyard. In the event of an accident, employers should have a well-rehearsed plan and equipment designed to rescue and evacuate people from height.

Equipment specifically designed for working safely at height is referred to as access equipment. It should only be used by trained, competent and authorised people in accordance with the manufacturer's instructions.

Ladders

▲ Figure 1.34 Ladder

A ladder should be:

- set at an angle of 75° or a ratio of 1:4 (the distance between the wall and the base of the ladder should be one quarter of the ladder's height)
- placed on firm, level ground
- placed against a stable surface
- extended 1 m above a working platform
- secured to prevent slipping/moving
- used for light work and short durations.

The user should have three points of contact with the ladder at all times and never overreach.

Mobile scaffold towers

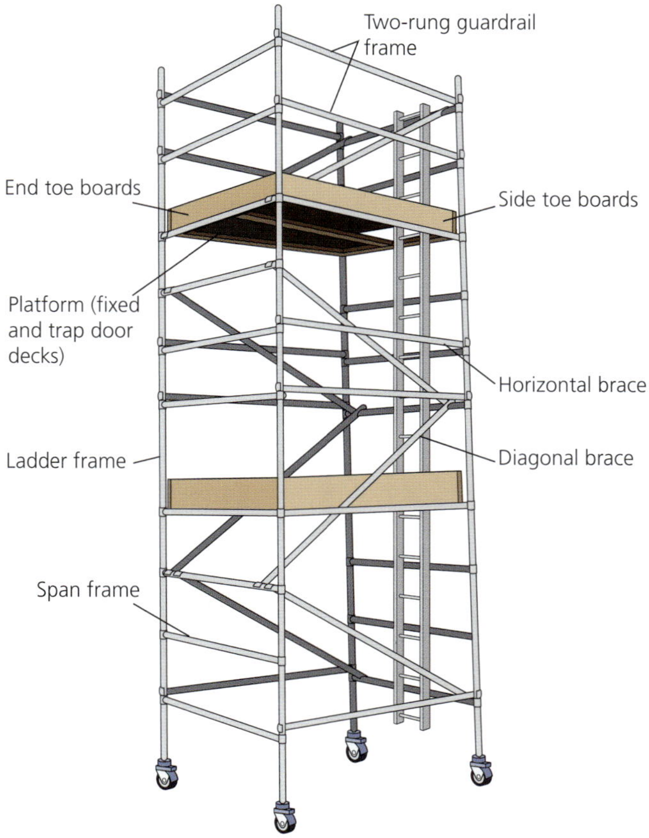

▲ Figure 1.35 Mobile scaffold tower

A mobile scaffold tower should:

- be set up on firm, level ground
- be erected in accordance with the manufacturer's instructions
- not be overloaded
- use guardrails and toe boards
- have brakes applied on castors before use
- have its outriggers correctly positioned and secured when necessary to gain extra height
- not be repositioned with people, materials or equipment still on it.

Scaffolding

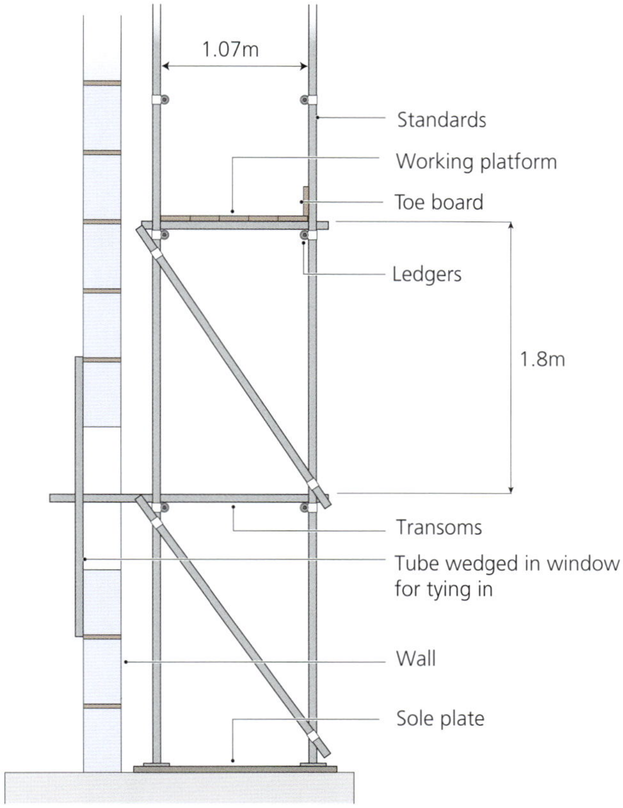

▲ Figure 1.36 Scaffolding

Scaffolding should:

- only be erected, inspected, adjusted and dismantled by trained and competent scaffolders
- be designed and erected in accordance with British Standards and the Work at Height Regulations 2005
- have handrails 950 mm high, with no more than a 470 mm gap between guardrails
- have toe boards 150 mm high
- have platforms kept clean and clear.

Steps

▲ Figure 1.37 Steps

Steps should be:

- fully opened
- placed on firm, level ground
- positioned facing the work, not sideways
- used for light work and short durations.

Podiums

▲ Figure 1.38 Podiums

Podiums should:

- be set up on firm, level ground
- be erected in accordance with the manufacturer's instructions
- have the gate locked while working
- have brakes applied on castors before use.

Staging boards

▲ Figure 1.39 Staging boards

Staging boards are designed to create a safe working platform between two supports, with guard rails on one or both sides.

You should never exceed the safe working load of staging boards with materials and people.

Trestles

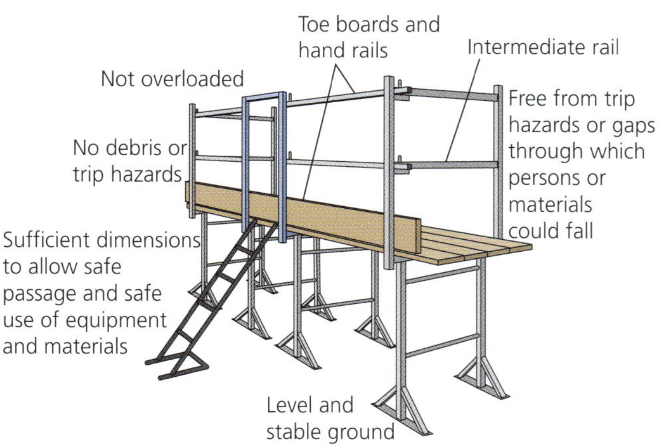

▲ Figure 1.40 Trestles

Trestles should:

- ▶ be set up on firm, level ground
- ▶ be erected in accordance with the manufacturer's instructions
- ▶ not be overloaded
- ▶ have staging boards kept clean and clear
- ▶ allow safe access and egress.

Mobile elevating work platforms (MEWPs)

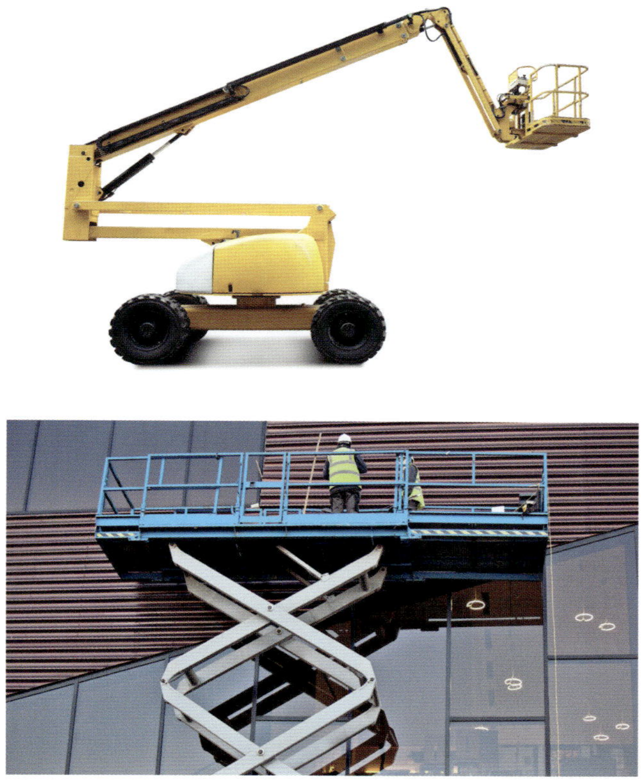

▲ Figure 1.41 Boom lift (top) and scissor lift (bottom)

There are a number of different types of mobile elevating work platform, each one designed for access, safety and work at height. Only suitably trained, experienced and competent workers should operate a MEWP.

14.2 Safety checks on access equipment

Under the Provision and Use of Work Equipment Regulations (PUWER) 1998, employers must ensure that access equipment is suitable for its intended use, safe and well maintained.

They must also regularly inspect the components of access equipment for signs of wear and damage, to ensure they have not deteriorated. Records should be kept of weekly, monthly and annual inspections by the people that supply the access equipment, for example employers and hire companies. A tagging system may be adopted, with tags attached to different types of access equipment as a way of recording inspections and identifying when it is safe to use.

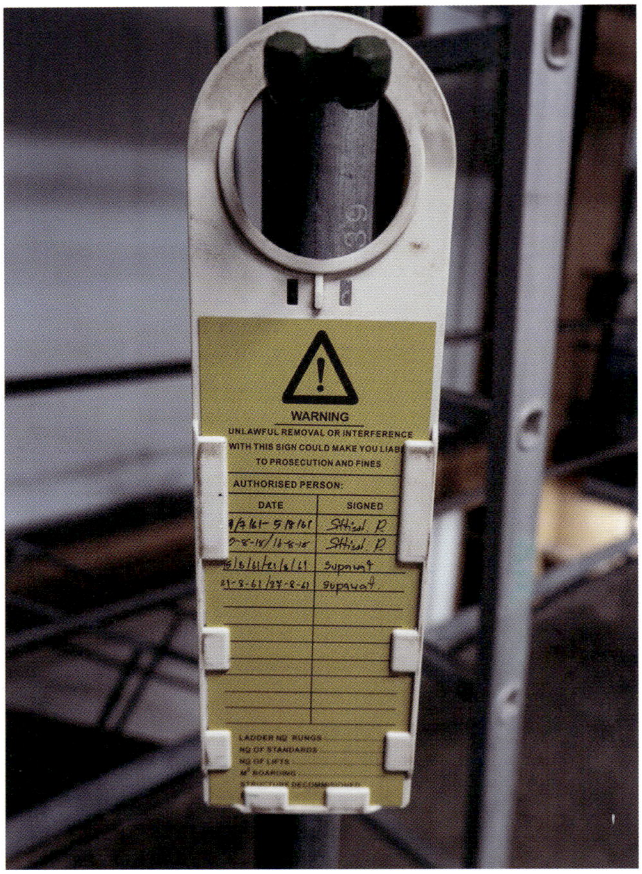

▲ Figure 1.42 A tag on access equipment

Some work equipment may be subject to other regulations which have specific requirements regarding inspections, such as LOLER. Scaffolding, for example, should be inspected before use and every seven days after that. However, further inspections may be necessary if there has been adverse weather or an accident/incident.

Daily visual pre-use inspections of access equipment should be completed by competent employees, to ensure there are no defects and the equipment is safe to use. These inspections should include checking that the non-slip feet on ladders and step ladders are not missing or damaged, and the stiles, steps and rungs are not bent or missing. Even the smallest amount of damage to access equipment can compromise its strength; therefore it should always be reported as soon as possible and damaged equipment should never be used.

14.3 Selecting access equipment

Access equipment must be appropriate for the work intended. This will be determined by the risk assessment and method statement, not its availability and cost.

Before working at any height, employers must consider the safest way to complete the task, its duration and the frequency with which access equipment will be used.

Other points to consider when selecting access equipment include:

- whether it is for internal or external use
- the ground conditions, for example uneven, sloped or soft
- the weather conditions, for example high winds or snow/ice
- safe access and egress
- the height at which it will be used
- the space needed for people and materials
- how materials and equipment will be loaded onto it
- the weight it will carry
- nearby hazards, for example overhead power lines and fragile roofs
- the type of work being undertaken
- training, information, instruction and supervision.

The risk of a fall from ladders and steps is relatively high; therefore they should only be used for a short duration and no longer than 30 minutes at a time. If this is not possible, the risk must be reduced by using a safer type of equipment, such as a podium.

Certain types of access equipment have action points for heights; for example a tower scaffold needs outriggers to support the equipment when it exceeds heights specified in the manufacturer's instructions. Further advice and guidance on the safe use of access equipment can be found on the HSE website.

14.4 Manual handling operations

Manual handling is defined as any lifting, carrying, supporting or moving of a load using bodily force. Employees working in the construction industry will inevitably be involved in carrying out manual handling operations and therefore have an increased risk of an injury or chronic muscular damage. These types of injury are referred to as musculoskeletal disorders (MSDs) and affect different parts of the body, including the back, shoulders, arms, legs, feet and toes. The HSE reported in the year 2019–20 that half a million people were suffering with a new or existing work-related musculoskeletal disorder, and this figure does not include workers who have not reported injuries under RIDDOR.

Where employees are expected to undertake manual handling operations in the workplace, their employers must take reasonably practicable measures to protect them from harm. The Manual Handling Operations Regulations 1992 state that this should be done by:

- avoiding carrying out manual handling if possible – consider completing the task another way, for example with a forklift, telehandler or crane
- assessing the hazards when it is not possible to avoid manual handling activities by completing a suitable and sufficient manual handling risk assessment
- reducing the risk as much as is reasonably practicable.

If this is not possible, workers should use good **kinetic lifting** techniques or be provided with mechanical lifting aids, such as a pallet truck or sack truck. Where this is not possible, consider changes to the task, load and working environment.

Key term

Kinetic lifting: the act of manual handling

▲ Figure 1.43 Safe manual handling sequence for a single person

Manual handling injuries often occur because activities are badly performed or planned. Employers have a duty to make sure their workers have the necessary information, instruction and training. They must consider the following points when assessing manual handling activities:

- **Task:**
  - What does the task involve?
  - Does the task involve more than one person?
  - How far will the load have to be carried?
  - How often will the task take place?
- **Individual:**
  - Is the worker physically and mentally able to carry out the task?
  - Does the worker have the necessary knowledge to complete the task safely?
  - Has the worker received the necessary training?
  - Does the worker have the right PPE, for example gloves or footwear with a good grip?
- **Load:**
  - Consider the size, shape, weight, centre of gravity and temperature of the load.
  - Is the load sharp?
  - Is the load likely to move?
- **Environment:**
  - Consider the weather and ground conditions.
  - Will the route be clear or are there likely to be obstructions or obstacles, such as stairs or scaffolding?

▲ Figure 1.44 Manual handling as a team

Employees have a legal duty to follow their employer's safe system of work, to protect themselves and others from the risk of injury from manual handling operations. They must also use lifting equipment and machinery provided by their employer, if they are trained to use it, and report any unsafe conditions they identify to their supervisor as soon as possible.

▲ Figure 1.45 Examples of mechanical lifting aids: a pallet truck (top) and a sack truck (bottom)

Improve your English

Think of a word that can be used to describe a condition that develops suddenly or in a short space of time. Use this word in a sentence to explain a particular risk to health in your industry.

Health and safety

There are *no* safe working loads when you are manual handling. Even light loads carry a risk of injury.

15 Safe practices and procedures for working in excavations and confined spaces

15.1 Dangers associated with excavations

Excavations are often created on construction sites to form trenches and holes for building foundations, or to gain access to underground services and drainage. Pedestrians, materials and vehicles are all at risk of falling into exposed excavations unless satisfactory preventative measures are taken to keep them at a safe distance.

Working in an excavation can be extremely dangerous because of the risk of flooding or collapse causing people to be crushed. There could be reduced levels of oxygen and poisonous or explosive atmospheres.

Excavations are most likely to fail if they are too deep and unsupported, or if the soil conditions are weak. The walls of an excavation may also become unstable if there is vibration in the ground caused by vehicle movement around the site or nearby public roads. Additionally, they can weaken and collapse if the soil of an excavation dries out in hot weather, expands after a heavy frost or becomes saturated with rain water.

▲ Figure 1.46 An excavation on a construction site

Plant operators must be competent and take precautions while excavating with machinery to ensure they do not come into contact with buried services, such as gas pipes or live electricity cables. Special care should also be taken when digging next to adjacent structures and buildings, to make sure they are not disturbed or undermined, causing them to become unstable or collapse.

If the construction site has previously been used for manufacturing, industrial or agricultural purposes, the soil may have been contaminated with hazardous waste or chemicals that have been buried or have naturally occurred in the ground.

Under the CDM Regulations, the client has a duty to provide information to the contractor about the location of buried services, any underground structures or any other relevant information before work starts on site. This will allow the contractor to prepare a pre-construction health and safety plan, reducing the risk of an accident or dangerous occurrence.

Key term

Plant operators: people in control of heavy construction machinery and equipment

▲ Figure 1.47 A plant operator

15.2 Safety when working in excavations

All excavations should be individually planned and carried out in a safe way by competent workers and supervisors. Wherever possible, working in an excavation should be avoided and only considered as a last resort.

Employers have a duty to reduce the likelihood of an accident or injury in an excavation by assessing the risks and introducing control measures such as the following:

- Isolate the hazard by erecting suitable edge protection barriers, for example Heras-style fencing (temporary metal fence panels), to prevent unauthorised access and to protect the general public.
- Install 'stop blocks' a safe distance away from the excavation to prevent vehicles getting too close and reduce the risk of hazardous exhaust fumes entering the excavation.
- Display safety signs and notices to raise awareness of the hazards and the precautions to be taken.
- Provide adequate lighting.
- Provide safe access (entry) to and egress (exit) from the excavation, for example using ladders or temporary stairs.
- Use a temporary support system to prevent the excavation from collapsing, for example a trench box (metal sidewalls held apart with struts).
- Provide safe designated crossing points.
- Test the soil for contamination.
- Adopt a permit-to-work system to dig and work in the excavation.
- Provide adequate ventilation to workers in the excavation.
- Identify and protect exposed services.
- Carry out daily inspections of the excavation.
- Ensure a competent person carries out atmospheric testing at various points within the excavation before anyone enters, at the start of each shift and after breaks, and continuously thereafter if indicated by the risk assessment and permit to work. Personal monitors may also be worn by individuals working in the excavation.
- Use appropriate portable gas-detection equipment.
- Ensure good communication methods between workers in the excavation and the supervisor.
- Put in place a rescue plan to recover an injured person and provide appropriate rescue equipment.

▲ Figure 1.48 A trench box supporting an excavation

15.3 Dangers associated with confined spaces

The definition of a confined space is a workplace which may be substantially but not always entirely enclosed, where there is a foreseeable serious risk of injury because of the conditions or from hazardous substances. Excavations, loft spaces, sewers or wells could be described as confined spaces, because they are enclosed with restricted access and egress.

Confined spaces may already contain hazards, such as a lack of oxygen or a risk of an explosion. Hazards can also be created by workers with electrical equipment, machinery, materials and substances such as petrol and solvents. Some confined spaces may also contain toxic atmospheres created by gases such as naturally occurring methane, nitrogen or carbon dioxide. Carbon monoxide and carbon dioxide are particularly dangerous because they are colourless, odourless and tasteless, making them more difficult to detect without specialist testing equipment.

The air we breathe contains a mixture of different gases, with oxygen making up about 21 per cent. If the oxygen in the atmosphere is reduced just below this level, it affects our ability to function normally; if depleted further to 6 per cent or below, it will result in almost certain death.

Additional hazards that may be found in confined spaces include:

- fire
- extremes of heat and cold
- dust, fumes and vapours
- flooding resulting in drowning
- free-flowing solids causing suffocation, for example sand
- entrapment.

▲ Figure 1.49 A confined space

Research

Which solvents that you use in your trade would be particularly hazardous in a confined space? Consider what alternative substances you could use.

Test yourself

Which regulations control the use of hazardous substances?

15.4 Safety when working in confined spaces

The Confined Spaces Regulations 1997 state that working in a confined space should be avoided wherever possible, and work should be completed in another way without entering the space. If this cannot be done and there is still a significant risk of injury, then the work must be properly planned and organised with appropriate control measures in place before it starts.

The Management of Health and Safety at Work Regulations 1999 state that employers and the self-employed must complete a suitable and sufficient risk assessment for work in a confined space. This is used to identify hazards and determine what precautions need to be taken to reduce the risk of injury. It will also help employers to prepare a safe system of work.

Everyone involved in working in a confined space must be competent and specifically trained to undertake their tasks. This includes managers, supervisors and emergency personnel.

If a permit-to-work system is adopted, it would limit the number of people and time spent in a confined space and safely control work activities. Before a permit is issued, workers are told about:

- the risks to health and safety and how to control them
- methods used to communicate from the inside to their supervisor or **sentry** on the outside
- how to raise the alarm in an emergency
- testing and monitoring of the atmosphere
- PPE or RPE to be worn
- arrangements to recover them from the confined space if necessary (a rescue plan).

Key term

Sentry: a person who supervises workers from the access/egress points of a confined space

Assessment practice

Short answer

1 Which organisation enforces the Health and Safety at Work etc. Act 1974?
2 What are the five main categories of safety sign?
3 Which regulations require specific accidents, injuries and near misses to be reported to the HSE?
4 What should not be kept in a first-aid box?
5 When should you be informed about the site rules at work?

Long answer

6 Explain the duty every employer has under the Control of Vibration at Work Regulations 2005 to protect people from the harmful effects of exposure to vibration.
7 Explain the term 'safe system of work'.
8 Explain what a toolbox talk is and when it might be used.
9 Give a definition of 'confined space'.
10 Explain the best way to avoid the hazards of working in a confined space.

Project practice

Your industry placement employer has reported a significant rise in the number of employees absent from work in the last twelve months due to musculoskeletal disorders (MSDs). At the last health and safety meeting, the management team decided to conduct a health and safety audit of the reported accidents and injuries over the past year and a safety survey, with a view to reducing the risk of MSDs.

Your task is as follows:

- Work collaboratively with your workplace mentor or employer to gather as much information as you can about the number of people at work that may have had an MSD. This could be achieved by studying health and safety records, interviewing workers or conducting a survey.
- Review the current manual handling risk assessment and method statements and identify any areas that could be improved.
- Produce a digital toolbox talk for a manual handling activity and present this to a small group of workers.

Chapter 2 Construction science principles

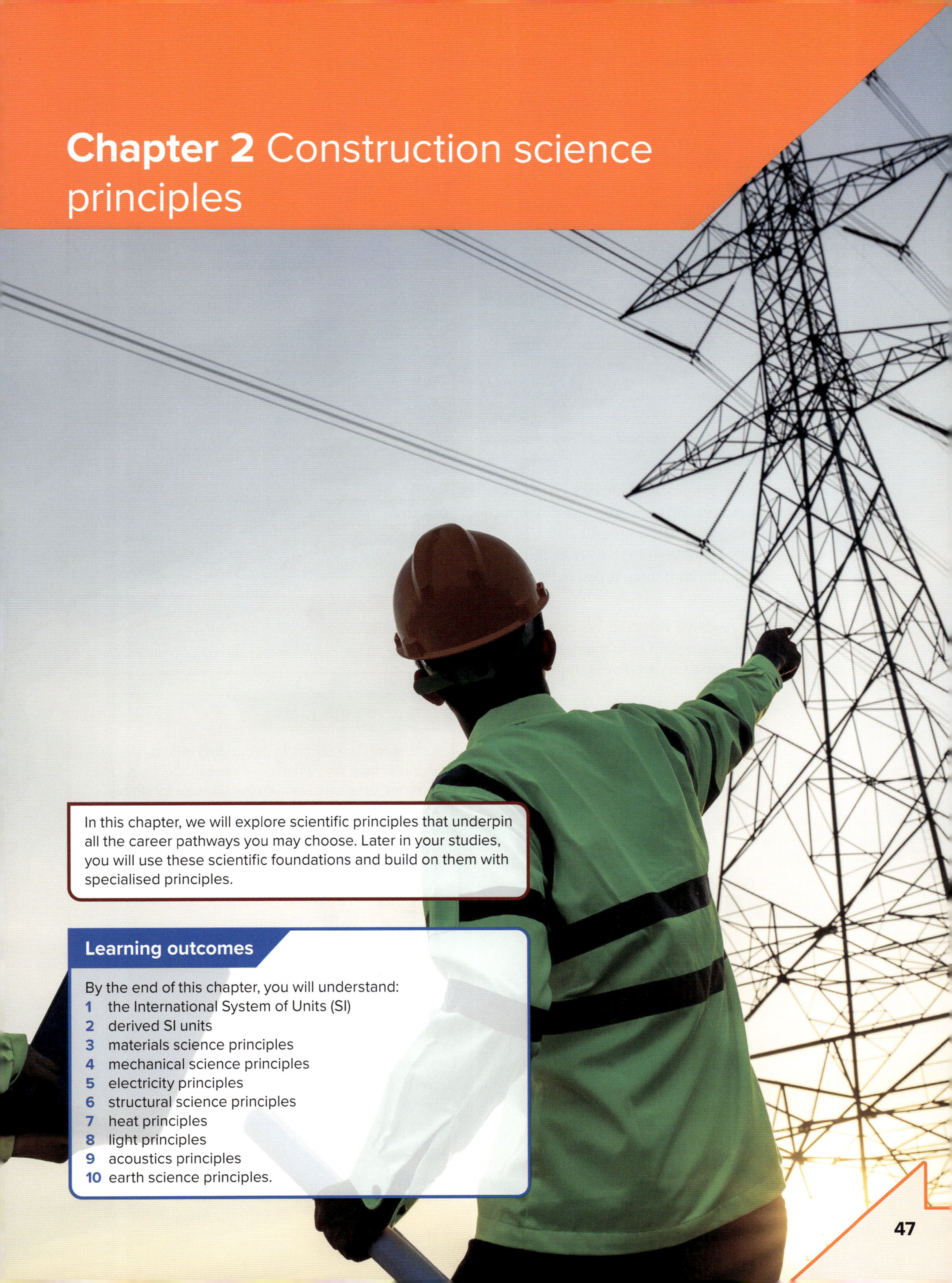

In this chapter, we will explore scientific principles that underpin all the career pathways you may choose. Later in your studies, you will use these scientific foundations and build on them with specialised principles.

Learning outcomes

By the end of this chapter, you will understand:

1. the International System of Units (SI)
2. derived SI units
3. materials science principles
4. mechanical science principles
5. electricity principles
6. structural science principles
7. heat principles
8. light principles
9. acoustics principles
10. earth science principles.

1 The International System of Units (SI)

Imagine a job building a house where the carpenter measures everything in inches, the plumber measures in centimetres, the bricklayer uses feet, and the electrician uses metres. What do you think the outcome would be?

Equally, in science, if someone performed a calculation using one unit of measurement, and another person used a different unit, they would produce different results which could lead to confusion.

As a result, the International System of Units, known as SI units, was introduced. The system sets out what units of measurement, symbols and indices are used to calculate quantities.

Before we look at the SI units, it would be a good idea to understand what indices are. Indices are used to replace repetitive multiplications. For example, 10 × 10 × 10 = 1000. This calculation can be written easily as 10^3, which means ten multiplied by itself three times, or three lots of ten multiplied together.

Where indices are negative, the value becomes a fraction, because the value moves to the right of the decimal point, becoming smaller. For example:

$$5^{-1} = \frac{1}{5} \text{ or } 0.2$$

$$5^{-2} = \frac{1}{25} \text{ or } 0.04$$

$$5^{-3} = \frac{1}{125} \text{ or } 0.008$$

Most calculators have an x^2 button to square a number, and scientific calculators also have an x^y button, which allows a number to be raised to any power or index. For example, to calculate 5^5, use buttons 5 x^y 5 = 3125. This is much easier than keying 5 × 5 × 5 × 5 × 5 into a calculator.

Generally, in construction science and principles, large values are used, such as thousands of **watts** or millions of joules. Other aspects of building services engineering or construction work deal with tiny amounts, such as millionths of a metre or thousandths of an ohm. This can become a problem in calculations, as errors may occur if the correct number of zeros is not entered into the calculator.

Instead of inserting the actual number with lots of zeros, we use 'to the power of ten'.

The 'power of' numbers are given names that are explained in Table 2.1. There is less chance of making an error using this method.

▼ Table 2.1 Numbers expressed as indices (to the power of 10)

Actual number	Number shown to the power of 10	Prefix used
1 000 000 000 000	10^{12}	tera (T)
1 000 000 000	10^9	giga (G)
1 000 000	10^6	mega (M)
1000	10^3	kilo (k)
100	10^2	hecto (h)
10	10^1	deca (da)
0.1	10^{-1}	deci (d)
0.01	10^{-2}	centi (c)
0.001	10^{-3}	milli (m)
0.000 001	10^{-6}	micro (μ)
0.000 000 001	10^{-9}	nano (n)
0.000 000 000 001	10^{-12}	pico (p)

Test yourself

What is 30 kilometres multiplied by 500 millimetres?

Improve your maths

Calculators are different depending on the manufacturer and when they were made. Using the internet, your calculator instructions, or by simply experimenting with your calculator, find out what the following functions do. Note that what appears on the button may differ depending on the calculator.

- x^{-1} or $1/x$
- ENG and shift ENG
- $x10^x$ or EXP
- x^2 and x^3

SI units are broken down into two different categories:

- base SI units
- derived SI units.

Base SI units are the main category used, whereas derived units require the base functions to determine them. Base SI units are shown in Table 2.2.

Key term

Watt: SI unit of power

▼ Table 2.2 Base SI units

Quantity	Unit of measurement	Identification symbol
Mass	kilogram (kg)	m
Length	metre (m)	l
Time	second (s)	t
Temperature	kelvin (K)	T
Electric current	ampere (A)	l
Luminous intensity	candela (cd)	l

Note: the amount of a substance in moles is also a base unit but not relevant to the subjects in this book.

As well as knowing the base SI units of measurement, you may be required to carry out simple calculations using them.

Calculations using base SI units obtain other values called derived SI units. For example, length multiplied by width of a rectangle gives the rectangle's area (or $m \times m = m^2$).

2 Derived SI units

There are many derived SI units of measurement. Table 2.3 shows those relevant to construction-based activities, including the formulae used to calculate them.

▼ Table 2.3 Derived SI units of measurement and associated calculations

Quantity	Unit of measurement	Identification symbol	Base formula (where relevant)
Area (or cross-sectional area)	Square metre (m^2)	A	For squares or rectangles: length × width For circles: $\pi \times \text{radius}^2$ or πr^2
Volume	Cubic metre (m^3)	V	For cuboids: length × width × depth or area × depth For cylinders: $\pi r^2 \times \text{depth}$ or area × depth
Flow	Kilograms per second (kg/s) or commonly litres per second (l/s)	mdot ($\dot{m}$)	$\frac{\text{volume}}{\text{time}}$ or $\frac{\text{litres}}{\text{second}}$
Velocity (speed)	Metres per second (m/s)	v	$\frac{\text{distance}}{\text{time}}$ or $\frac{\text{metres}}{\text{second}}$
Acceleration	Metres per second squared (m/s^2)	a	$\frac{\text{velocity}}{\text{time}}$

Quantity	Unit of measurement	Identification symbol	Base formula (where relevant)
Electromotive force	Volt (V)	ε	$\varepsilon = BLv$ Based on the length of conductor (L) in a field of magnetic flux density (B) and the velocity of movement of the conductor through the field (v)
Electrical resistance	Ohm (Ω)	Ω	$\frac{\rho L}{A}$ Based on material resistivity (ρ), length of the conductor (L) and cross-sectional area of the conductor (A)
Illuminance (light on a surface)	Lux (lx)	E	Lumens per m^2, where lumens are the measure of visible light emitted by a source
Internal pressure	Pascal (N/m^2)	Pa	$\frac{N}{m^2}$ or $\frac{J}{m^3}$
Atmospheric pressure	Bar	bar	Equal to 100 000 Pa
Energy (work)	Joule (J)	E	force × distance or f × d or power × time or watts × seconds
Force	Newtons (N)	F	mass × acceleration
Density	kg/m^3	ρ	$\frac{\text{mass}}{\text{volume}}$
Power	Watts (W)	P	$\frac{\text{energy}}{\text{time}}$ or $\frac{\text{joules}}{\text{seconds}}$ In electrical circuits: volts × amperes or V × I
Specific heat capacity	Joules per kilogram per degree Celsius (J/kg/°C)	C_p	$J \times kg \times (t_2 - t_1)$ Total energy (J) required to raise the temperature of a mass (kg) from temperature 1 (T_1) to temperature 2 (T_2)

Industry tip

You will need to apply indices when calculating room sizes from plan drawings, as plans use mm ($\times 10^{-3}$) and area and volume SI units are m^2 and m^3 respectively.

Improve your maths

A room measures 5 m wide by 8 m long and has a ceiling height of 2.5 m. The air mass in the room is required to calculate heating needs. If air has a density of 0.946 kg/m^3, what is the mass of air in the room?

3 Materials science principles

3.1 Materials and their properties

Construction work and BSE installations use many different materials based on their properties. Some of these materials are described in Tables 2.4 and 2.5.

▼ Table 2.4 Properties and common uses of metals

Material	Properties	Common uses	Reasons for use	Other information (where relevant)
Iron	• Produced by melting mined iron ore • Very heavy **ferrous metal** with a high melting point	• Versatile material used for decorative fences and gates, as well as stoves • Widely used for structural supports before steel	• Common element which is mined worldwide • Resistant to high temperatures but can be formed and shaped when heated, providing decorative structures	• Prone to oxidising (rusting) when exposed to the weather if not suitably treated, such as by painting
Steel	• Technically an alloy, made from iron and carbon • Ferrous metal that is very rigid	• Structural supports • **Rebar** • **Catenary wires** • Support cables • Sheet materials, such as stainless-steel sinks or splashbacks • Cable reinforcement	• Very strong dense material which provides long-lasting structural support and expands at the same rate as concrete, so used widely for reinforcement	• Prone to rusting if not treated during or after production
Copper	• Fairly soft but rigid non-ferrous metal which is mined as ore • Often not pure and mixed to make an alloy • Corrosion resistant • Good electrical conductor	• Pipes • Cable conductors • Architectural features	• Durable, corrosion-resistant metal with low **thermal** expansion • Anti-microbial, meaning it remains clean from bacteria	• Common material which is easily recycled (which is why it may be an alloy, due to impurities during the recycling process) • Can form a green coating known as patina
Aluminium	• Very lightweight and durable non-ferrous metal • Corrosion resistant to weather	• Sometimes used as a cable conductor where lightweight heavy-duty cables are needed • Used widely in construction for facias/cladding and window frames • Used extensively for flexible ducting systems	• Low thermal expansion • Easily formed or fabricated • Lightweight material	• Can react with other materials • Not suitable for applications such as piping, due to reaction with salt and other chemicals

Material	Properties	Common uses	Reasons for use	Other information (where relevant)
Alloys	• Customised metals made by adding different metals together • Common examples include nickel, tungsten and manganese • Can be ferrous if iron or steel is used in the process	Made specifically for an application by mixing metals with the desired properties, for example adding chromium to steel produces stainless steel	Ability to produce a material suitable for an intended application, such as low thermal expansion materials or corrosion-resistant metals	

Key terms

Ferrous metal: a metal that contains iron and is magnetic

Rebar: reinforced steel bar commonly used in concrete to act as a frame to stop it moving and cracking

Catenary wires: strong wires which are tied at each end and used to support other objects, such as cables which may stretch or break under their own weight when hung between two buildings

Thermal: related to heat or temperature

Table 2.5 identifies other common materials used in construction, together with their properties and common uses. None of these contain metal, so they are not ferrous.

▼ Table 2.5 Properties and common uses of non-metals

Material	Properties	Common uses	Reasons for use	Other information (where relevant)
Thermoplastics and thermosetting plastics	Soft plastics which are more flexible and suitable for higher temperatures than general PVC	Cable insulation and sheathing	• Flexible • Can operate at high temperatures while remaining good insulators	
Unplasticised polyvinyl chloride (uPVC)	More rigid material than PVC but has a lower thermal expansion rate	• Window and door frames • Guttering, facias and cladding	• Does not expand at the same rate as PVC • Durable, non-corrosive product which can be moulded into any shape during production	Shatters easily when hit, as far less flexible than plastic
Rubber	Very flexible material which can maintain electrical insulation at higher temperatures, such as 125°C	• Cable insulation • Flexible joints • Water seals	Although a natural resource, can be replicated using a manufacturing process to provide a highly durable yet very flexible material that is corrosion and water resistant	

Material	Properties	Common uses	Reasons for use	Other information (where relevant)
Ceramics	• Include a broad range of materials manufactured from **minerals**, sometimes with added metals to provide high degrees of strength • Hard, durable and resistant to heat and fire	• Heat-resistant bricks (firebricks), tiles, sinks and worktops • Retain and slowly release heat in storage heaters • Electrical insulators in high-temperature applications	• High durability when exposed to heat • Corrosion resistant, tough materials	
Plaster	Made from gypsum, lime or cement mixed with water and sand	• Provides a smooth finish to walls and ceilings • Can be moulded to make decorative features • Can be pre-manufactured into sheets for boarding and partitioning	• Cheap to manufacture • When in a wet state, easy to apply to uneven and out-of-line surfaces to provide a smooth, level finish • Can provide a reasonable fire barrier, slowing the spread of fire in a building	• Only suitable for interior uses • Requires cladding or coating if used in generally wet areas, such as shower rooms
Concrete	• Contains **aggregate**, sand and a binder such as cement • Sometimes contains additives to give flexibility or quick-drying properties	Building foundations or supporting structures when used with reinforcement bars such as steel	• Cheap to produce and easy to lay in a wet state, as it flows into place • When **tamped down**, leaves no air gaps or unfilled spaces	• Can be strengthened with steel reinforcement • Can be pre-stressed during manufacturing, where the concrete is tightly compressed to form blocks or decks used as floors in high-rise buildings

Key terms

Mineral: a solid, naturally occurring, inorganic substance

Aggregate: material in the form of grains or particles, such as sand, gravel or crushed stone

Tamped down: pressed down by a succession of blows

Voltage: the amount of potential energy between two points in an electrical circuit, expressed as volts or V

Other environmental effects on materials to consider in more detail include:

- electrolytic and galvanic corrosion (metals)
- dissimilar metals
- thermal expansion
- UV radiation.

Electrolytic and galvanic corrosion (metals)

This is a chemical reaction where two metals are separated by an electrolyte which causes one metal to corrode at an accelerated rate. Without suitable protection, these types of corrosion can cause structural parts of a building to fail.

Electrolytic corrosion occurs where electrical current is induced into nearby metals from other influences, such as buried high-voltage cables. This causes a **voltage** difference between the metals, which in turn creates a circuit. As current is a flow of electrons, the electrons from one metal are drawn to the other, which means the metal is losing electrons and corroding. The electrolyte is whatever is between the two metals, such as soil or clay, and allows electric current to flow through it.

Galvanic corrosion is where two dissimilar metals are in close proximity and separated by an electrolyte or the two metals are touching. If one metal is more chemically active than the other, such as stainless steel to zinc, the proximity is similar to creating a battery and this causes current to flow between them, which again results in loss of electrons and corrosion.

Research

Aluminium can be used as a cable conductor as it has good conductive properties but is a lightweight material, meaning that the cables are significantly lighter than equivalent copper cables. However, aluminium also has some disadvantages; for example it is prone to electrolytic corrosion if preventative measures are not taken.

Research where this problem of electrolytic corrosion occurs and why, and suggest measures to reduce the risk of it occurring.

Dissimilar metals

Although two dissimilar metals in close proximity can cause electrolytic or galvanic corrosion, by carefully selecting the two metals we can use their different properties – such as thermal expansion – to an advantage.

By fixing two dissimilar metals together, such as iron and brass, a **bimetallic strip** is created. When the metals are heated, one expands more than the other and this causes them both to bend.

Figure 2.1 shows how this process can be used in a heat-controlled switch, such as a thermostat controlling room temperature for a cooling system. When the room gets too hot, the metal bends, closing the switch which activates the cooling system.

Key terms

Bimetallic strip: a temperature-sensitive component comprising two different metals bound together; when heated, each metal expands at a different rate, causing the strip to bend and activate a switch

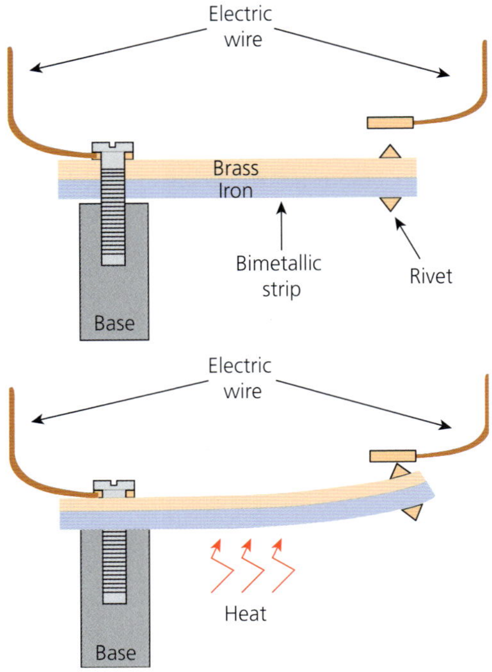

▲ Figure 2.1 Bimetallic strip used as a heat-operated switch, such as in a thermostat (top: in normal state; bottom: when heat is applied and the strip bends)

Thermal expansion

Different materials expand at different rates. This can cause failure of building structures and services if correct measures are not taken.

One example is PVC conduit used to protect electric cables. It is usually fixed to a wall using saddles. The amount of expansion, in metres, for PVC conduit is, on average, 52×10^{-6} for every metre length of conduit, per °C rise in temperature (52×10^{-6} m/m/°C). This is known as its linear temperature expansion coefficient.

If a 6 m run of conduit was installed at 20°C but the temperature of the space was increased to 45°C (as might occur in a loft space), by how much would it expand?

$$52 \times 10^{-6} \times 6 \text{ m} \times (45° - 20°) = 0.0078 \text{ m or } 7.8 \text{ mm}$$

If the conduit was installed without taking the expansion into consideration, the additional 7.8 mm would cause it to bend between supports, making it look wavy instead of straight.

As a result, PVC conduit systems should have expansion couplers evenly spaced in the installation to provide a space for the conduit to expand into and stop it from buckling or waving.

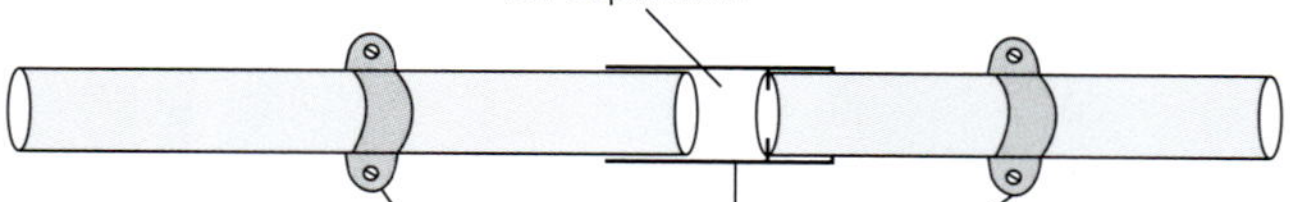

▲ **Figure 2.2 PVC conduit with expansion coupler to stop it buckling**

Concrete and steel expand at similar rates. If they did not, many structures made of concrete and reinforced with steel would crumble, with one material expanding more quickly than the other.

Other linear temperature expansion coefficients for common materials are shown in Table 2.6.

▼ **Table 2.6 Linear temperature expansion coefficients for common materials**

Material	Linear temperature expansion coefficient (m/m/°C)
Aluminium	21–24 × 10^{-6}
Brick (masonry)	5 × 10^{-6}
Cast iron	10 × 10^{-6}
Concrete	12–14 × 10^{-6}
Copper	16–17 × 10^{-6}
Plate glass	9 × 10^{-6}
Plaster	17 × 10^{-6}
PVC	54–110 × 10^{-6}
Steel	10–12 × 10^{-6}
Wood across its grain	30 × 10^{-6}
Wood along its grain	3 × 10^{-6}

Ultra-violet radiation

Ultra-violet (UV) radiation comes from direct sunlight and can cause some plastic-based materials such as white PVC to degrade.

The radiation breaks down polymers in the plastic which normally allow it to be flexible. When the polymers break down, the plastic becomes brittle and can break easily. This is why white PVC pipes or cables should never be used outdoors in direct sunlight.

Some plastics are specially treated to withstand UV radiation, such as black plastics. This is why most outdoor plastic pipes are black or grey. Unplasticised PVC (uPVC) is also UV resistant and is commonly used as a plastic material where white colouring and strength are required, such as for guttering or windows and window sills.

4 Mechanical science principles

It is important to understand the ways in which mechanical loads are measured. You need to know the difference between mass and weight, as well as energy and power. Energy and power can be mechanical as well as electrical.

The fundamental relationship between mass and weight is defined by Newton's second law. According to this law:

$$F = ma$$

where:

- F = force (N)
- m = mass (kg)
- a = acceleration (m/s^2).

Mass is a measure of the quantity of matter in an object. It is not dependent on gravity and is therefore different (but proportional) to weight.

Weight is the downwards gravitational force acting on a mass. Newton's second law can be transformed to express weight as a force by replacing acceleration (a) with acceleration due to gravity (g):

$$W = mg$$

where:

- W = weight (N)
- m = mass (kg)
- g = acceleration due to gravity (on Earth, this is 9.81 m/s^2).

4.1 Acceleration

Acceleration is the rate of change of velocity with time:

$$a = \frac{v}{t}$$

where:

- a = acceleration (m/s^2)
- Δv = change in velocity (m/s)
- t = time (s).

As acceleration is a measurement of the rate of change of speed in metres per second (m/s) each second (s), the unit is metres per second per second, or metres per second squared (m/s^2).

4.2 How levers work

Levers are tools that exert a large force when applying a lesser force. They come in all shapes and sizes, such as crowbars, claws on claw hammers, wheelbarrows or even seesaws in playgrounds.

While levers give an advantage in terms of the force applied, the distance travelled is increased, meaning the energy used is essentially the same. This is because the turning moments (torque) applied to the ends of a lever must be equal but opposite. Therefore:

$$\text{force} \times \text{distance} = \text{force} \times \text{distance}$$

Torque is measured in newton-metres (Nm). This is the product of the force (N) applied to a lever and the distance (m) of the force from the **fulcrum**.

Key term

Fulcrum: the pivot point of a lever

Levers are categorised into three classes.

Class 1 levers

In a class 1 lever, the force and the load are on different sides of the fulcrum. The effectiveness of the force depends on its distance from the fulcrum; the greater the distance, the greater the effect of the force.

In Figure 2.3, the force applied downwards can be four times less than the load, as the lever on the force (effort) side of the fulcrum is four times longer than the lever on the load side.

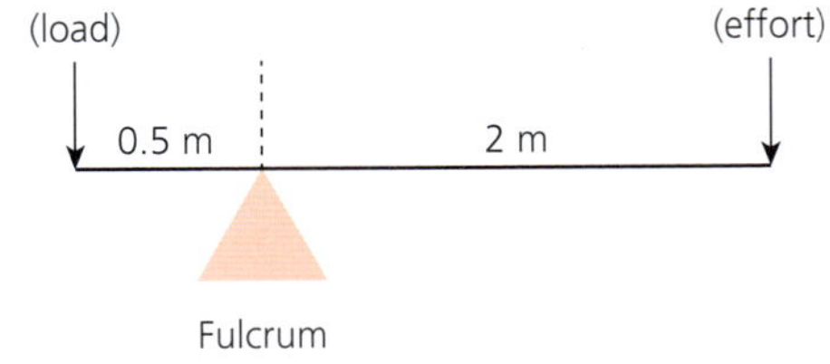

▲ Figure 2.3 Class 1 lever

Examples of class 1 levers are seesaws, crowbars and claw hammers when used to lift nails. The advantage gained by a lever is referred to as mechanical advantage (MA), which is calculated as:

$$MA = \frac{\text{load}}{\text{effort}}$$

So, for the lever in Figure 2.3, the mechanical advantage is:

$$\frac{400}{100} = 4$$

Class 2 levers

In a class 2 lever, the load is between the force and the fulcrum.

The calculations remain the same, but the load is more limited because it is between the two points, and the force and the desired movement are both in the same direction. An example of a class 2 lever is a wheelbarrow.

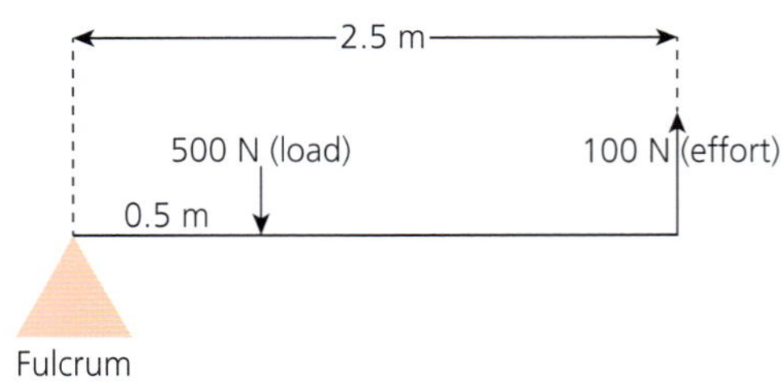

▲ Figure 2.4 Class 2 lever

In the example in Figure 2.4, the force of effort of 100 N is applied 2.5 m from the fulcrum, and the load is applied 0.5 m from the fulcrum.

The load that can be lifted is calculated as follows:

$$\text{force of load} \times \text{distance from fulcrum} = \text{force of effort} \times \text{distance from fulcrum}$$

Therefore, the force of the load in Figure 2.4 can be calculated by:

$$\text{force of load} \times 0.5\text{ m} = 100\text{ N} \times 2.5\text{ m}$$

This can then be represented as:

$$\text{force of load} = \frac{100\text{ N} \times 2.5\text{ m}}{0.5\text{ m}} = 500\text{ N}$$

Class 3 levers

In a class 3 lever, the force is between the load and the fulcrum. Examples of class 3 levers include fishing rods, tweezers, tongs and the human arm.

The advantage with class 3 levers is not so much the effort needed, but the distance travelled.

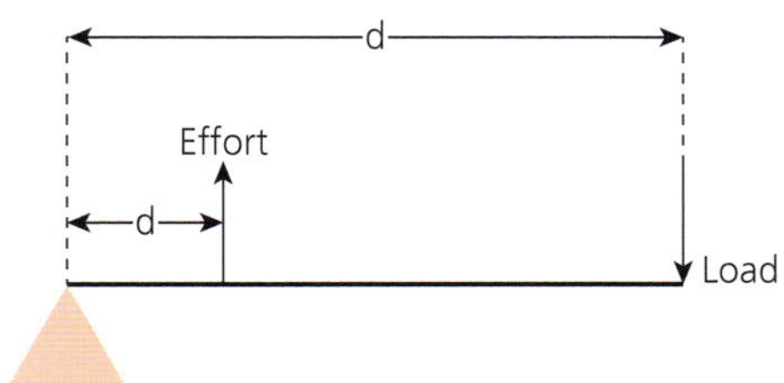

▲ Figure 2.5 Class 3 lever

4.3 Pulleys

A pulley is an effective way of gaining a mechanical advantage when lifting an object. By running a rope through a four-pulley system, a mechanical advantage of four is gained.

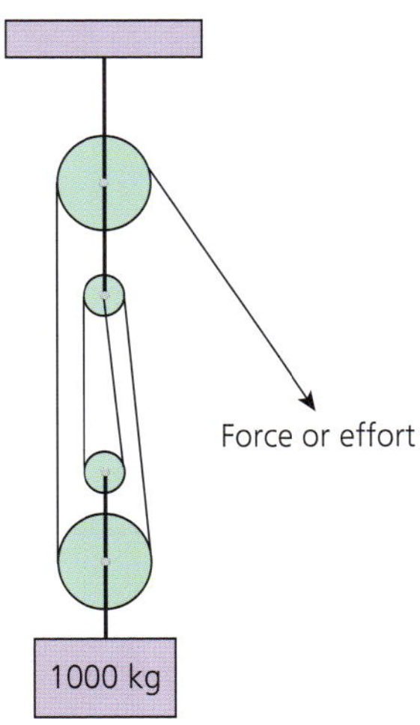

▲ Figure 2.6 A pulley system lifting a mass of 1000 kg against gravity

Looking at Figure 2.6, we can determine the force required to raise the mass of 1000 kg. As the load has been described as a mass, first determine the downward force of the load:

$$\text{force} = \text{mass} \times \text{gravity}$$

$$f = 1000 \times 9.81 = 9810\text{ N}$$

Remember: gravity is acceleration, which averages 9.81 m/s^2.

As the system has four pulleys compared to one pulling rope, the mechanical advantage is 4:1.

Remember:

$$\text{MA} = \frac{\text{load}}{\text{effort}}$$

Rearrange the formula for mechanical advantage to find the effort required:

$$\text{effort} = \frac{\text{load}}{\text{MA}}$$

$$\frac{9810}{4} = 2452\text{ N}$$

Therefore, a downward force (or effort) of 2452.5 N is required to raise the load.

Although the pulley gives a mechanical advantage, the pulling rope will need to be pulled four times further than the load is raised. This means to raise the load by 1 m, the pulling rope needs to be pulled 4 m.

Test yourself

A mass of 2575 kg needs to be raised 2 m by a pulley with a mechanical advantage of six. How much effort is required?

4.4 Work

If the force applied to a body results in movement, then work has been done. This applies to forces that lift, push or twist objects.

When an object moves in the same direction as the force exerted, the work done is equal to the force exerted multiplied by the distance moved:

$$\text{work} = \text{force} \times \text{distance}$$

Or, to include the values used to determine force:

$$\text{work} = \text{mass} \times \text{gravity} \times \text{distance}$$

Mechanical work is measured in **joules** (J). (Newton-metres (Nm) can be used for mechanical work, but are also used as a measurement for torque.) Other units of work or energy, which are not SI units but commonly used for specific applications, include:

- kilowatt hour (kWh), used by electricity supply companies to measure electrical energy
- calorie, often used as a measure of food energy
- BTU (British Thermal Unit), often used for heat source applications such as burning gas.

Key term

Joules: the unit of measurement for energy. Where energy is expressed as mechanical energy, it is known as work

Let us work through an example. If a mass of 100 kg is lifted 10 m, calculate the work done:

$$\text{Force} = \text{weight} = m \times a = 100 \times 9.81 = 981 \text{ N}$$

$$\text{Work done} = \text{force} \times \text{distance} = 981 \times 10 = 9810 \text{ J}$$

If the mass is doubled, the work done is:

$$\text{Force} = \text{weight} = m \times a = 200 \times 9.81 = 1962 \text{ N or } 1.962 \text{ kN}$$

$$\text{Work done} = \text{force} \times \text{distance} = 1962 \times 10 = 19620 \text{ J}$$

Energy can exist in many forms but is categorised into two main groups:

- kinetic energy (the energy an object has due to its motion, such as a rotating machine)
- potential energy (stored energy held by an object because of its position or state, for example a coiled spring).

The potential energy of gravity keeps a mass on the ground. If the mass is raised, then a machine uses kinetic energy. If the input of kinetic energy ceases, the potential energy tries to bring the mass back down to the ground.

4.5 Power

Power is defined as the rate of doing work. It is calculated by dividing work done by the time taken to carry out that work:

$$\text{average power} = \frac{\text{work done}}{\text{time taken}}$$

The unit of power is joules per second (J/s), and 1 J/s is equivalent to 1 watt (W).

The output (mechanical) power required for a motor to raise a mass of 1000 kg to a height of 5 m above the ground in one minute is calculated as:

$$P = \frac{m \times a \times d}{t}$$

So:

$$P = \frac{1000 \times 9.81 \times 5}{60} = 817.5 \text{ W}$$

If the same motor raised the same load in 10 seconds, the output power required by the motor would be:

$$P = \frac{1000 \times 9.81 \times 5}{10} = 4905 \text{ W}$$

The amount of energy used is the same, no matter how quickly the task is carried out, but more power is required to do the work in a shorter time.

Test yourself

How much power is needed to raise a load of 90 kg a distance of 150 m in 20 seconds?

4.6 Efficiency

The law of conservation of energy states that energy cannot be created or destroyed, only transformed from one form to another. However, during the transformation process, some of the energy may be turned into unwanted forms, such as noise or heat, known as energy losses. Such losses are common in mechanical processes.

The **efficiency** of a mechanical system is the ratio of output power compared to input power expressed as a percentage:

$$\%\ \text{efficiency} = \frac{\text{output power}}{\text{input power}} \times 100$$

Key term

Efficiency: the ratio of output power compared to input power expressed as a percentage

It is more common for efficiency to be expressed in terms of power rather than energy, although energy could replace power in this formula.

If a machine with a 200 kW output has an input power of 220 kW, the machine efficiency is:

$$\% \text{ efficiency} = \frac{200}{220} \times 100 = 90.9\%$$

Test yourself

An electric motor drives a hoist which is to raise a weight of 400 N to a height of 25 m in 120 seconds.

If the motor and mechanical losses are 20 per cent, calculate the electrical power input to the motor.

As well as occurring in mechanical systems such as machines, losses in efficiency can also occur in specific BSE systems:

- Electrical systems: the passage of an electric current represents a flow of power or energy. When current flows in a circuit, power loss occurs in the conductors due to resistance. This causes heat dissipation and voltage drop, which are also losses.
- Heating systems: pipework can lose heat. Long systems can dissipate a lot of heat if measures such as lagging are not used. Pipe systems with small diameters or many bends can also lose pressure, meaning more power is required to pump fluids through the system.
- Forced air systems: these can lose pressure if ducting contains many bends or reduced diameters, which in turn will require more power to force the air through.

4.7 Centre of gravity

The centre of gravity of an object is an imaginary point where the weight of the object is concentrated.

Figure 2.7 shows a uniform object, where the centre of gravity is in the centre of the object. If a pivot is placed under the centre of gravity, the object remains balanced, but if the pivot is placed under any other part of the object, the object is unstable and falls.

When lifting or manually handling, it is easier to carry an object if it is supported under the centre of gravity. This state, where the object is perfectly balanced, is known as equilibrium.

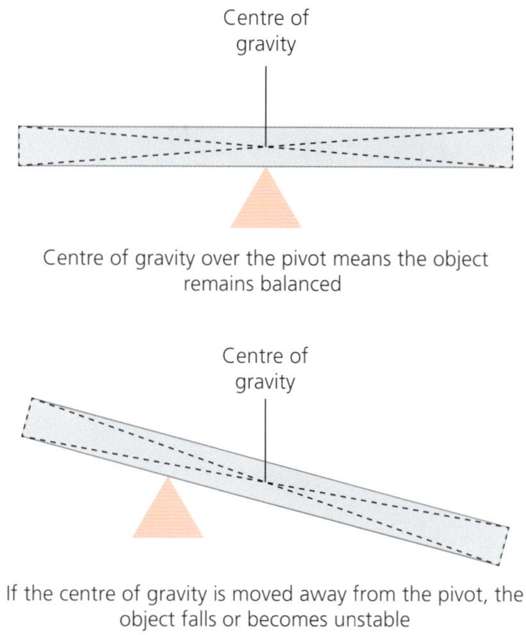

▲ Figure 2.7 Effects of centre of gravity

4.8 Moments or torque

Torque is a term often used in building services and mechanics. It refers to how tight something needs to be, such as a nut, bolt or screw. Too tight and it may break the thread or snap, not tight enough and it may come loose.

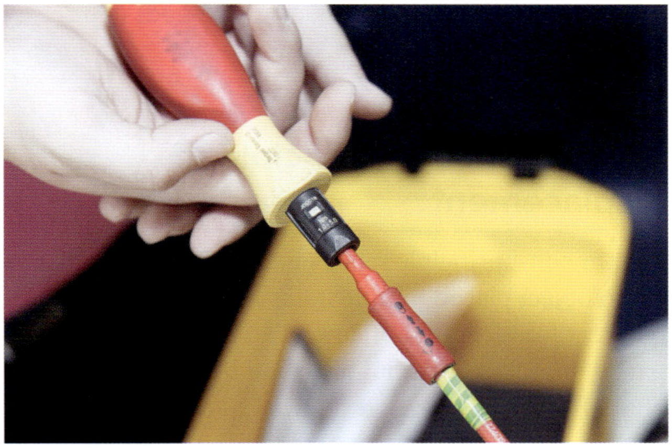

▲ Figure 2.8 A torque screwdriver with adjustable settings

Torque, or moment, is measured in Newton metres (Nm) and determined by:

$$M = \text{force } (F) \times \text{distance } (d)$$

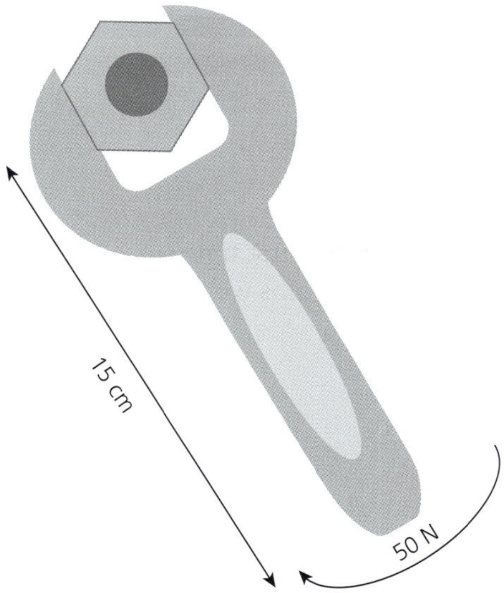

▲ Figure 2.9 Force applied to a spanner in the form of torque

The torque applied by the spanner in Figure 2.9 can be calculated, first ensuring the 15 cm is converted to 0.15 m:

$$\text{Moment or torque} = 50\ \text{N} \times 0.15\ \text{m} = 7.5\ \text{Nm}$$

Test yourself

A nut requires 12 Nm torque by a wrench that has a length of 320 mm. How much force needs to be applied?

4.9 Archimedes' principle of displacement

Archimedes' principle relates to how fluid is displaced by an object placed in it. It states that an upwards buoyant force exerted on an object immersed or partly immersed in a fluid is equal to the weight of the fluid it displaces. If the weight of the fluid displaced is equal to or greater than the weight of the object, the object will float; if the object is heavier, it will sink.

Key terms

Renewable resources: resources that can be replaced over time by natural processes, for example wind energy or solar energy

Fossil fuels: fuels such as coal, oil and gas that are mined from the earth and burned to produce energy

Figure 2.10 shows a body of fluid before a weight is placed into it. The 500 kg weight displaces 400 kg of water, so the object sinks. But when a 300 kg object displaces 400 kg of water, the object floats due to the buoyant forces acting on it.

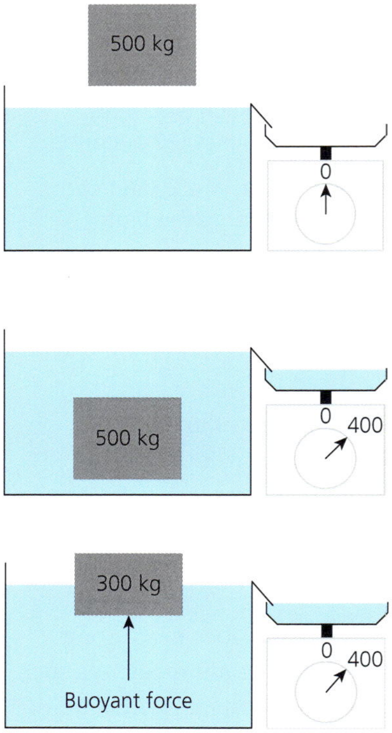

▲ Figure 2.10 When an object is heavier than the displaced fluid, the object sinks; when it is lighter, it floats

5 Electricity principles

Whichever career path you take in construction, you will work on or near electricity, particularly in building services engineering. Knowing basic electrical principles will help you to work safely.

There are three key areas for delivering an electrical supply to an installation:

- generation
- transmission
- distribution.

5.1 Generation

Electricity can be generated in many ways due to a large-scale move towards **renewable resources**. The use of **fossil fuels** in the UK is now in decline, due to the need to protect the environment, although they are better suited to times of peak demand.

Peak demand refers to times when consumer demand for electricity is at its highest, for example on cold, dark days when more heat and light are needed. Another example would be at half-time during a big televised football match, when large numbers of people boil a kettle to make tea.

Fossil-fuel sources

Fossil fuels all work in the same way. Fuel is burned, which heats water to form high-pressure steam. This steam turns a **turbine**, which rotates a generator at high speeds to produce electricity.

Fossil fuels are not renewable. They are very polluting and release carbon dioxide and other greenhouse gases when burned, leading to **climate change**.

> **Key term**
>
> ***Turbine:*** a machine that uses a moving stream of air, water, steam or hot gas to turn a wheel and generate power
>
> ***Climate change:*** a large-scale, long-term change in the Earth's weather patterns and average temperatures

Gas

Gas is the most widely used fossil fuel. It can produce instant heat, so gas generators can heat water to steam much more quickly than coal generators.

While its use is in decline, due to the need to reduce greenhouse gases produced when it is burned, it remained the largest energy source in the UK in 2021. Location of the power plant is not critical, as gas can be piped to most mainland areas of the UK.

Oil

Oil is used for many regional generators, which power local areas at times of peak demand. The generators can deliver electricity immediately.

Oil is also used for private standby generators, for buildings such as hospitals that require a continuous power supply during mains power cuts.

Traditionally, oil-fired power stations are located near coastal oil refineries, to minimise transportation of the oil.

Coal

Coal used to be the most widely used fuel to produce high-pressure steam for turbines. However, it is highly polluting and its use has therefore been scaled down in the UK, with only a few coal-fired power stations remaining. Most of these are only used during periods of peak demand. Coal-fired power stations were traditionally located near to UK sources of coal, such as South Wales, Yorkshire and the North East of England.

In 2017, the UK went 24 hours without using a single coal-fired power station, for the first time since 1882. In 2020, the entire summer months went without coal being burned. This is a huge step forward in reducing pollution and climate change.

Renewable fuel sources

Wind

Wind power turns a propeller, which directly drives a generator. Wind is a natural resource, so no pollution is created or resources are required after production of the turbine.

The use of wind energy is ideal for the UK, as an island with a lot of wind from the sea. Electricity generation from wind power is increasing each year, with many offshore wind farms being constructed several miles out to sea.

Wave

The sea produces a huge amount of energy, both in the form of waves and as the tide moves in and out each day. This movement of water can be used to turn electrical generators through a rotating waterwheel effect or using a back-and-forth motion.

Hydro

There are three types of hydro generation:

- Run of river: this uses the natural downward flow of rivers and harnesses the water's energy by using existing **weirs**.
- Storage: watercourses are held back by a dam and released through pipes that divert the water through turbines.
- Pumped storage: a large volume of water is held in a reservoir high up on hills or mountains until needed. When there is high demand for electricity, the water is released from the reservoir through turbines. When the station is offline, the water is pumped back up to the reservoir, ready for the next time it is needed.

> **Key term**
>
> ***Weirs:*** low dams across a river, which increase the force of the water as it flows over the top; sections of a weir can be raised or lowered to regulate the force of the water

Photovoltaic

Photovoltaic (PV) cells convert solar energy into electricity without any moving parts. You can see these on the roofs of some houses and in fields called solar farms.

> **Research**
>
> Biofuel is increasingly being used to generate electricity and provide a fuel source for boilers in offices and houses. What is biofuel and is it carbon neutral?

Electricity is generated at power stations using a three-phase system. A three-phase system is one where the rotation of the generator is used to maximum effect by having three windings, so one turn of the generator gives three times the current distributed over three wires. A single-phase generator would only distribute current in one wire with a neutral return path. This means much of the magnetic field is unused during one revolution of a large generator. To maximise output, most generators are three-phase, with only small-scale portable generators being single-phase.

5.2 Transmission systems

In the UK, the transmission system is called the **National Grid**. It is a network of mainly overground cables used to send electricity all around the UK from the generator stations. As transmission systems are hundreds of miles long and send vast amounts of electrical power, they use a range of very high voltages:

- 400 kV (known as the super grid)
- 275 kV
- 132 kV.

There are two main reasons why high voltages are used:

- High voltages mean reduced current, so smaller conductor sizes can be used.
- As the cables travel vast distances, the voltage lost due to cable resistance has less of an impact on high voltages than it would on lower voltages.

As an example, a locality consumes 80 MW of electricity. Calculate the current demand at 400 kV and 400 V.

To calculate the current demand based on power and voltage values, use:

$$I = \frac{P}{V}$$

So, at 400 kV and remembering that 80 MW is 80×10^6 watts:

$$\frac{80 \times 10^6}{400000} = 200 \text{ A}$$

So, the cables need to be big enough to carry 200 A to the area of demand.

To calculate the current demand at 400 V, the same process is used, so:

$$\frac{80 \times 10^6}{400} = 200000 \text{ A or } 200 \text{ kA}$$

At this level of current, the cables would need to be huge. The benefit of increasing the voltage is to reduce the current, making cable sizes much more realistic.

Where cables have resistance (see section 5.5), this resistance leads to power loss in the form of voltage drop. The amount of voltage drop is directly proportional to the amount of current and the value of resistance as:

$$\text{voltage drop} = \text{current} \times \text{resistance}$$

So, if a section of transmission cable had a resistance of 20 Ω and carried a current of 200 A, the voltage drop would be:

$$200 \times 20 = 4000 \text{ V}$$

> **Test yourself**
>
> If a section of a transmission system has a resistance of 30 Ω and supplies a locality with a demand of 25 MW, what is the voltage loss at 400 kV and 132 kV?

> **Key terms**
>
> ***National Grid:*** the network of power lines supplying electrical energy around the country

There are issues with transmitting at higher voltages:

- High voltages can break down insulation. When cables are overground, air is used as an insulator between conductors. In underground cables, materials such as PVC are used as insulators between conductors.
- Conductors need to be suspended high above the ground and several metres away from each other, as high voltages can jump across or break down air, especially when the air has a high water content (i.e. relative humidity).

The towers used to carry electricity cables are known as pylons. The higher the voltage used in the transmission system, the bigger the pylons must be. Pylons used for the 400 kV super grid are as high as a tower block and have six cables suspended from them (three on each side). There is a single cable running between the tops of pylons, which acts as a common earth.

Transformers are used to change voltages within the transmission system, by either stepping up the voltage (for example from 132 kV to 400 kV) or stepping it down. If voltage is increased, current will decrease; if voltage is decreased, current will increase. This relates to the number of primary (input) windings in the transformer compared with the number of secondary (output) windings, shown by:

$$\frac{N_p}{N_s} = \frac{V_p}{V_s} = \frac{I_s}{I_p}$$

Key term

Transformers: devices that convert voltages and current, proportionately, to different values

All transmission voltages are three-phase, so the transformers used to step up or step down the voltage in the transmission system need to be three-phase as well. This means they have three incoming wires into three windings, which step up or step down the voltage into three outgoing wires.

Where transmission systems enter urban areas or pass through the countryside but need to be hidden, they may need to run underground. In this case, the voltage will be stepped down (for example to 132 kV) so it does not damage the insulation between the conductors in the underground cable.

Test yourself

Why are high voltages used over long distances?

Why are high voltages, such as 400 kV, not used all the way up to the local substation?

5.3 Distribution systems

At points around the National Grid, electricity is tapped off to be distributed to the user. These systems are known as distribution systems. They are looked after by distribution network operators (DNOs).

As distribution is much more localised, the voltages can be stepped down to lower values. This keeps the pylon sizes smaller. Lower voltages mean cables in urban areas can be run underground, keeping the supplies invisible. As underground systems are more expensive, rural areas normally use cheaper overhead supplies.

Depending on how far the electricity needs to be distributed, distribution voltages may be:

- 33 kV (three-phase)
- 11 kV (three-phase)
- 400 V (three-phase)
- 230 V (single-phase).

In most cases, the underground cables in towns and cities are 11 kV. These supply the many **substation** transformers.

Key term

Substation: equipment that transforms voltage to a suitable level for consumers

On the outgoing side of the substation (to consumers' installations), the supplies are 400 V three-phase or 230 V single-phase, depending on overall current demand. As a general rule:

- buildings with a demand over 100 A will have a three-phase supply
- buildings with a demand below 100 A will have a single-phase supply.

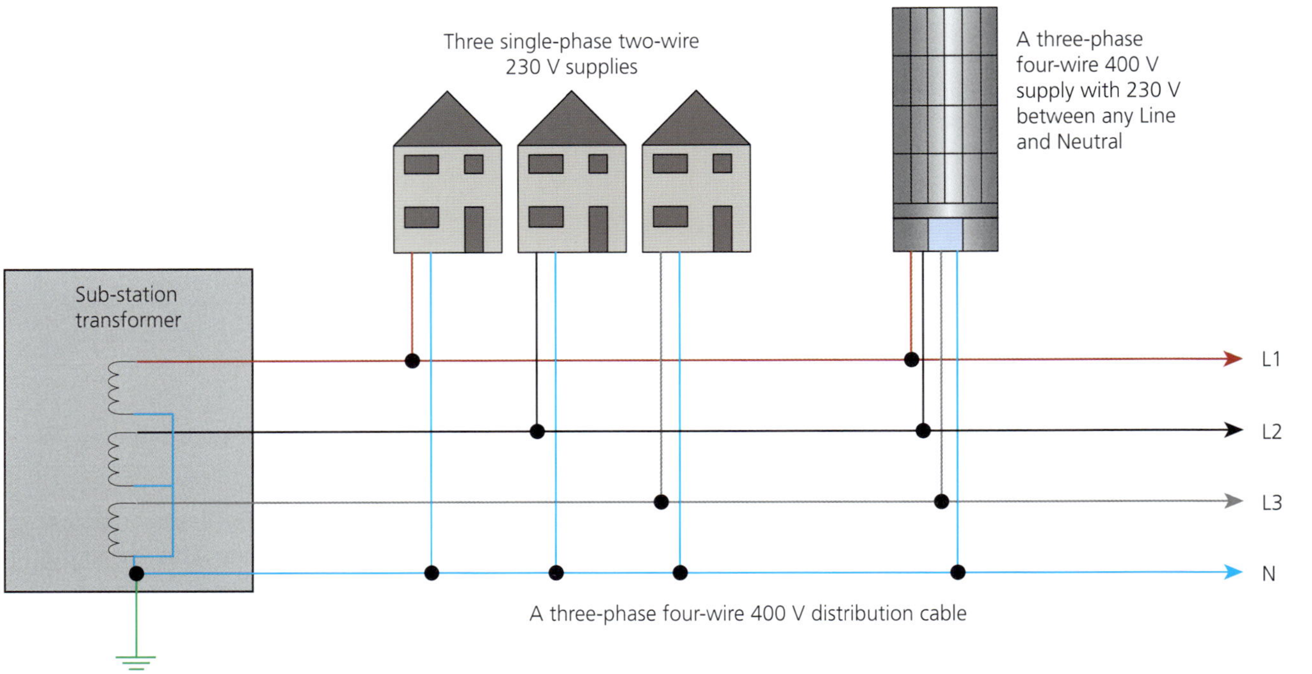

▲ Figure 2.11 Three-phase and single-phase distribution to consumers

Large buildings and facilities may have their own substation transformer. These places may be supplied at 11 kV or 33 kV.

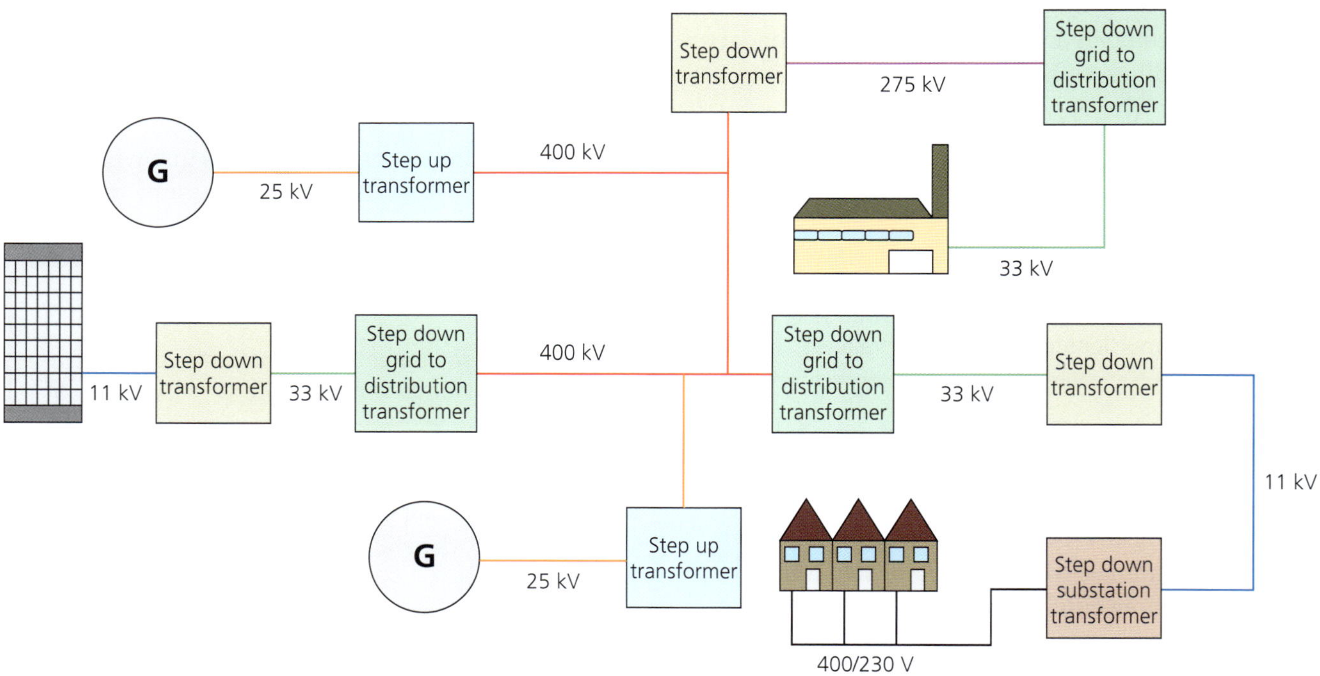

▲ Figure 2.12 Generators (G) supplying the consumer via transmission/distribution systems

5.4 Basic electrical circuit principles

Electricity is the flow of electrons from one atom to another. Materials that are good **conductors** have electrons which move out of orbit from atom to atom. When the material is connected to an electromotive force (emf) such as a battery, the flow can be controlled in one direction. This is because the electrons are attracted to the positive plate of the battery.

This flow of electrons is called **charge** and happens in materials that are good conductors, such as copper, iron and steel.

Materials whose atoms keep their electrons in orbit make good **insulators**, such as rubber or PVC.

What makes materials or elements different is the structure of their atoms. The number of electrons orbiting the nucleus is different for different types of material. For example, copper has 29 electrons but iron has 26. The number of electrons in an atom equals the number of protons.

When 6.24×10^{18} electrons flow in one direction, this is one coulomb of charge.

Key terms

Conductors: materials that have atoms less densely packed together and allow electron flow

Charge: the measure of electron flow (Q) in a material, measured in coulombs (C)

Insulators: materials that have atoms which are densely packed together so that electrons cannot readily move

5.5 Properties of an electrical circuit

Table 2.7 shows key electrical values, their SI units of measurement and formulae used to calculate them.

▼ Table 2.7 Properties of an electrical circuit

What needs calculating	SI unit of measurement	Symbol	Associated formulae
Charge	Coulomb (C)	Q	$1\ C = 6.24 \times 10^{18}$ electrons $Q = It$
Current	Ampere (A)	I	$I = \frac{Q}{t}$ or $I = \frac{V}{R}$
Electromotive force or circuit voltage	Volt (V)	V	$V = I \times R$
Resistivity	Ohm-metre (Ω-m)	ρ	See resistance below
Resistance	Ohm (Ω)	R	$R = \frac{V}{I}$ or $R = \frac{\rho L}{A}$
Power	Watt (W)	P	$P = V \times I$ or $P = I^2 \times R$

Conductors and resistivity

Before we study **Ohm's law** in detail, we need to look at conductors and insulators, and how much resistance a circuit has based on the material used as the conductor. This is based on the resistivity (ρ) of the material.

> **Key term**
>
> ***Ohm's law:*** a law that states the relationship between current, voltage and resistance in an electrical circuit

As different materials have different numbers of electrons, they conduct electricity differently. Each material has a resistivity value based on the measurement of the **resistance** of a 1 m^3 block of the material at 20°C. Table 2.8 shows some common resistivity values.

> **Research**
>
> The elements in the periodic table are all given an atomic number. What do these numbers represent?

> **Key term**
>
> ***Resistance:*** the measure of how well a material conducts electricity in ohms (Ω); the lower the value of resistance, the better it conducts

▼ Table 2.8 Common materials and their resistivity

	Material	Resistivity value (ρ)
Common materials used as conductors	Copper	0.0172×10^{-6} Ω-m
	Aluminium	0.028×10^{-6} Ω-m
	Gold	0.024×10^{-6} Ω-m
	Steel (used in cables)	0.46×10^{-6} Ω-m
Common materials used as insulators	Hard rubber	1×10^{13} Ω-m
	Glass (average value)	1×10^{12} Ω-m
	Dry wood	1×10^{14} Ω-m
	PVC (average)	1×10^{15} Ω-m

To work out the value of resistance, we use the material's length (L) and cross-sectional area (CSA) (A) and apply the resistivity calculation:

$$R = \frac{\rho L}{A}$$

Remember, when you see a formula with no sign between two values, multiply them. So ρL is actually $\rho \times L$.

When calculating resistivity, if you always use a resistivity value in μΩ-m (micro-ohm-metres or $\times 10^{-6}$) and the CSA in mm^2 ($\times 10^{-6}$), then the two values of $\times 10^{-6}$ cancel out. You can simply input the values without using the $\times 10^{-6}$.

What is the resistance, at 20°C, for 30 m of copper cable having a CSA of 2.5 mm^2?

$$R = \frac{\rho L}{A}$$

So:

$$R = \frac{0.0172 \cancel{\times 10^{-6}} \times 30}{2.5 \cancel{\times 10^{-6}}} = 0.21\ \Omega$$

> **Test yourself**
>
> Calculate the resistance of an aluminium cable at 20°C with a length of 100 m and a CSA of 25 mm^2.

Ohm's law

Ohm's law explains the relationship between current, voltage and resistance in any electric circuit. It is applied to work out the quantities of a DC circuit and can be expressed as:

$$V = IR$$

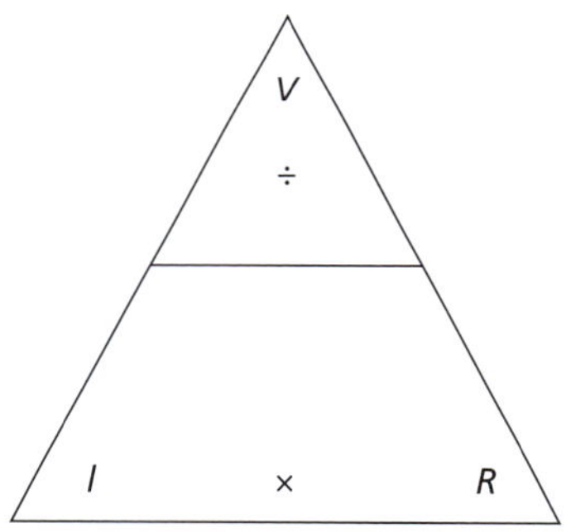

▲ Figure 2.13 Ohm's law

To remember Ohm's law, use the triangle in Figure 2.13 and cover the value you wish to find. It will leave the two other values and tell you whether they need to be divided or multiplied.

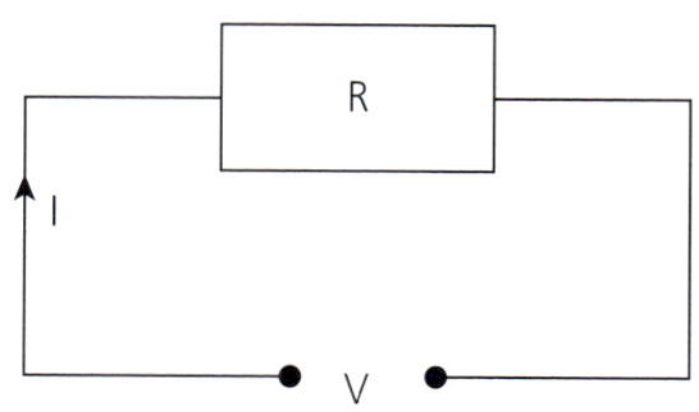

▲ Figure 2.14 Simple DC circuit

Figure 2.14 shows a simple circuit with a resistor (R), a current (I) and an electromotive force (emf) (V). We can apply Ohm's law to calculate values based on those we know. If the circuit in Figure 2.14 has a voltage of 12 V and a resistance of 8 Ω, what is the current?

$$V = IR$$

So, transposed to find the current:

$$I = \frac{V}{R}$$

So:

$$\frac{12}{8} = 1.5\text{ A}$$

Test yourself

If the circuit in Figure 2.14 has a current of 6 A and a voltage of 120 V, what is the circuit resistance?

Series circuits

When resistors in a circuit are connected one after the other, they are connected in series. The total resistance is found by adding all the resistances together:

$$R_{total} = R_1 + R_2 + R_3 \ldots\text{.. and so on}$$

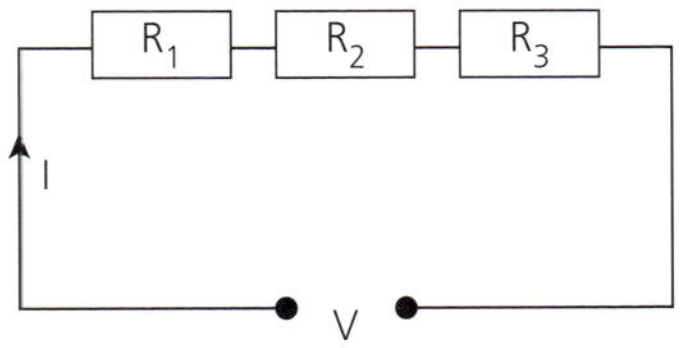

▲ Figure 2.15 Circuit with three resistors in series

The circuit shown in Figure 2.16 has a supply voltage of 200 V and three resistors of 10 Ω, 15 Ω and 25 Ω in series. Let us calculate the circuit current.

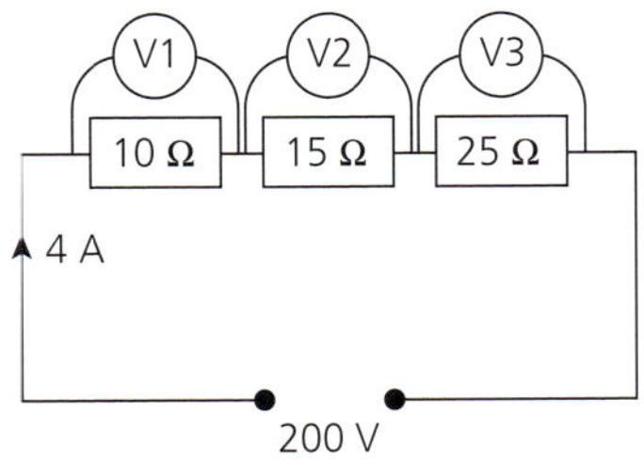

▲ Figure 2.16 Measuring voltage in a series circuit

As the current is based on the total circuit resistance, this needs to be determined first:

$$R_t = 10 + 25 + 15 = 50\ \Omega$$

Now Ohm's law can be applied to determine the circuit current:

$$I = \frac{200}{50} = 4\text{ A}$$

Test yourself

What is the circuit current if a circuit has a supply voltage of 400 V and 12 Ω, 8 Ω, 16 Ω and 22 Ω resistors in series?

In a series circuit, the current remains the same through each resistor, but the voltage across each resistor is different. The voltage across each resistor can be calculated by applying Ohm's law to that resistor. The value of all the voltages must equal the total circuit voltage.

To calculate the voltage across each resistor in Figure 2.16:

$$V1 = 4 \times 10 = 40\text{ V}$$

$$V2 = 4 \times 15 = 60\text{ V}$$

$$V3 = 4 \times 25 = 100\text{ V}$$

And to check the values equal the circuit voltage:

$$40 + 60 + 100 = 200\text{ V}$$

Test yourself

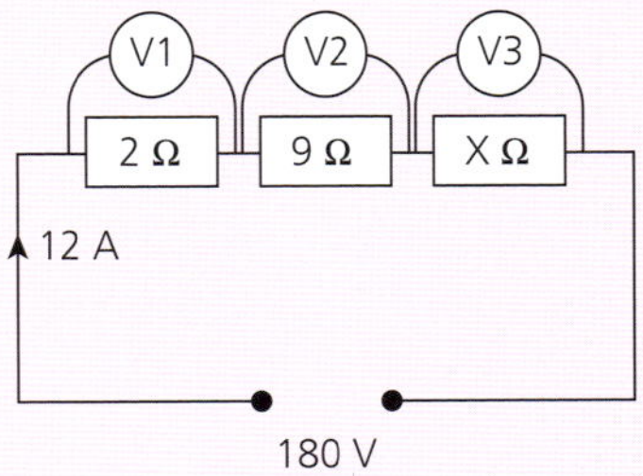

▲ Figure 2.17

For the circuit in Figure 2.17, determine the:

- value of resistor X
- voltages V1, V2 and V3.

Parallel circuits

When resistors are arranged in parallel (see Figure 2.18), the supply voltage remains constant across all resistors, but the current in each branch changes.

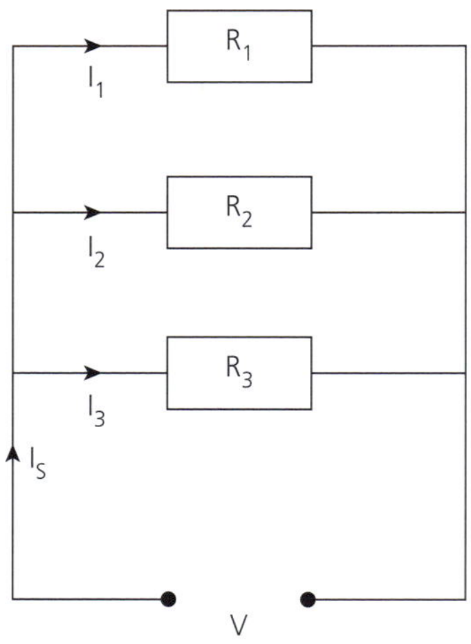

▲ Figure 2.18 Parallel resistors in a circuit

The value of current in each branch is based on the value of resistance and voltage in each branch. The total of each branch will equal the supply current (I_s).

The total resistance for the circuit is calculated by:

$$\frac{1}{R_{total}} = \frac{1}{R_1} + \frac{1}{R_2} + \frac{1}{R_3} \ldots \text{and so on}$$

A golden rule is that the total resistance must be less than the lowest resistor in the circuit.

When calculating parallel resistances, do not forget that the total is divided into 1 at the end.

Let us calculate the total resistance and each value of current, if the circuit in Figure 2.18 has the following values:

- $R_1 = 20\ \Omega$
- $R_2 = 30\ \Omega$
- $R_3 = 40\ \Omega$
- $V = 400$ V.

$$\frac{1}{R_{total}} = \frac{1}{20} + \frac{1}{30} + \frac{1}{40} = \frac{1}{0.108\ldots} = 9.23\ \Omega$$

$$I_1 = \frac{V}{R_1} = \frac{400}{20} = 20\text{ A}$$

$$I_2 = \frac{V}{R_2} = \frac{400}{30} = 13.33\ldots\text{ A}$$

$$I_3 = \frac{V}{R_3} = \frac{400}{40} = 10\text{ A}$$

$$I_s = \frac{V}{R_{total}} = \frac{400}{9.23} = 43.33\ldots\text{ A}$$

And to check:

$$20\text{ A} + 13.33\ldots\text{ A} + 10\text{ A} = 43.33\ldots\text{ A}$$

To work out the total resistance on a calculator, use the x^{-1} button. So, for the values in the worked example, push:

[20] [ ] [+] [30] [x^{-1}] [+] [40] [x^{-1}] [=] [x^{-1}] [=]

Test yourself

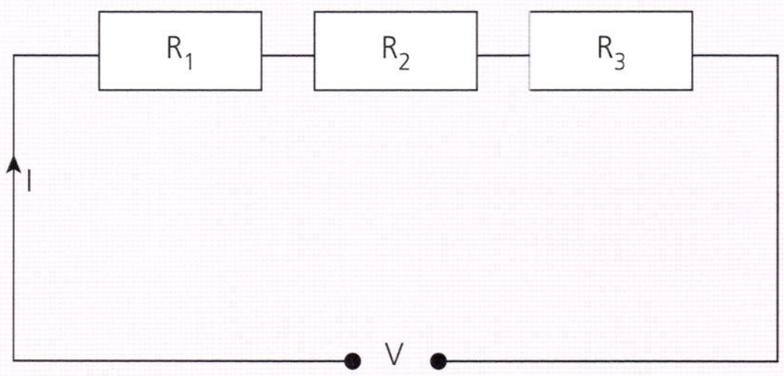

▲ Figure 2.19

The circuit in Figure 2.19 has the following values:

- $R_1 = 5\ \Omega$
- $R_2 = 100\ \Omega$
- $R_3 = 200\ \Omega$
- $V = 100$ V

Calculate the total circuit resistance and each current value.

Improve your maths

A 10 m run of steel conduit contains a 1.5 mm^2 copper circuit protective conductor (CPC), which is connected to the earth of a socket outlet at the end of the conduit run. The conduit has a 20 mm outside diameter and an 18 mm internal diameter (the bit the cables go in).

As the conduit acts as an earth in parallel to the CPC, what is the overall resistance of the conductor?

Power in parallel and series circuits

Let us look at a circuit with series and parallel resistors, then calculate the power.

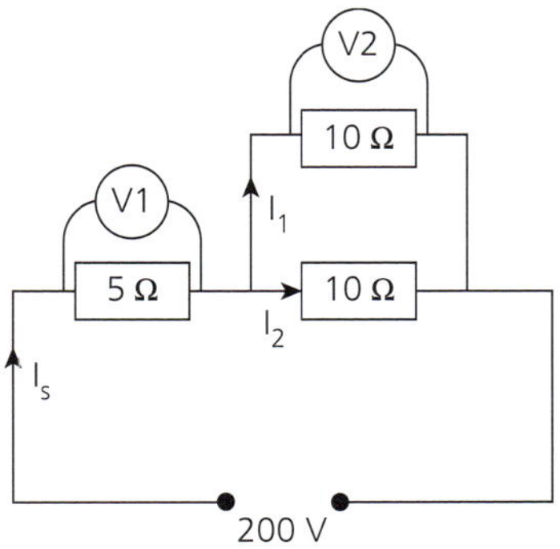

▲ **Figure 2.20 Circuit with series and parallel resistors**

Looking at the circuit in Figure 2.20:

- The current in the 5 Ω resistor is the same as the supply current (I_s).
- The current through each 10 Ω resistor is calculated in the same way as for a parallel circuit and must total the supply current.
- The voltage V2 is the same across each 10 Ω resistor and V1 + V2 must equal the supply voltage.

So, we can calculate:

- total resistance of the parallel section (R_P)
- total circuit resistance
- circuit current I_s
- voltages V1 and V2
- currents I_1 and I_2.

$$\frac{1}{R_p} = \frac{1}{10} + \frac{1}{10} = \frac{1}{0.2} = 5\ \Omega$$

As the parallel branch totalling 5 Ω is in series with the 5 Ω resistance, the total resistance of the circuit is:

$$R_{total} = 5 + 5 = 10\ \Omega$$

So I_s can be found using Ohm's law:

$$I_s = \frac{V}{R_{total}} = \frac{200}{10} = 20\ A$$

Voltage V1 is then found by:

$$V_1 = I_s \times R = 20 \times 5 = 100\ V$$

Voltage V2 will also be 100 V, as the total resistance of the parallel section is also 5 Ω.

Current I_1 is found by applying Ohm's law to that section:

$$I_1 = \frac{V_2}{10} = \frac{100}{10} = 10\ A$$

Current I_2 has the same properties:

$$I_2 = \frac{V_2}{10} = \frac{100}{10} = 10\ A$$

As the total circuit current is 20 A, the current is split evenly over the two equal resistances in parallel.

Power **dissipated** in a circuit can be calculated in two ways:

$$P = V \times I \text{ or } P = I^2R$$

Key term

Dissipated: energy consumed by converting to heat energy

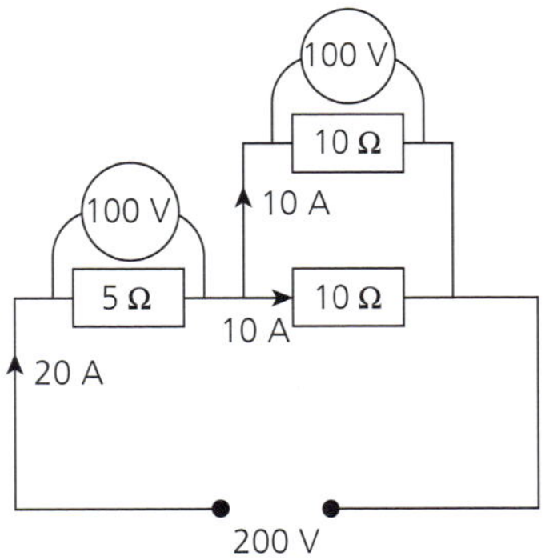

▲ **Figure 2.21 Calculating power in a circuit**

Figure 2.21 shows the circuit from the previous example with calculated values. Remember that the total circuit resistance was 10 Ω.

To work out the power of the whole circuit, we can use the total circuit or supply values, so:

$$P = V \times I = 200 \times 20 = 4000 \text{ W or } 4 \text{ kW}$$

Equally:

$$P = I^2R = 20^2 \times 10 = 4000 \text{ W or } 4 \text{ kW}$$

So, either formula can be used depending on the values known.

The power dissipated by the 5 Ω resistor is:

$$20 \text{ A} \times 100 \text{ V} = 2000 \text{ W or } 2 \text{ kW}$$

The power dissipated by one of the 10 Ω resistors is, using the other formula:

$$10 \text{ A}^2 \times 10\ \Omega = 1000 \text{ W}$$

As the other 10 Ω resistor is the same, a total of 2000 W is dissipated in the parallel part of the circuit.

Test yourself

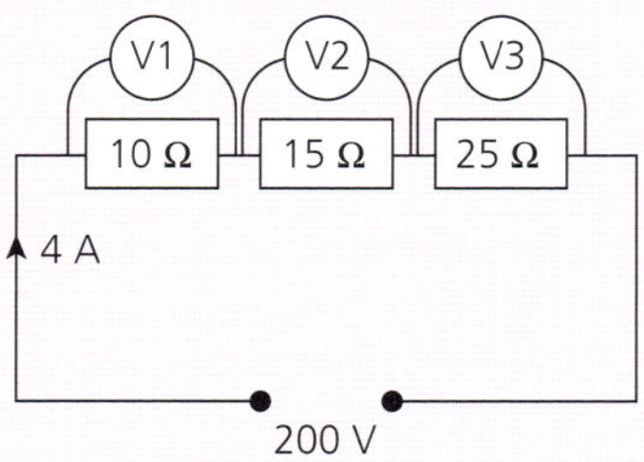

▲ Figure 2.22

Using the circuit shown in Figure 2.22, calculate the:

- total resistance of the circuit
- current I_S
- power dissipated by the 35 Ω resistor
- power dissipated by the parallel section
- total power dissipated by the circuit.

5.6 Measurement of electrical circuits

Instruments used to measure electrical quantities are listed in Table 2.9 and are connected as shown in Figure 2.23.

▼ Table 2.9 Measurement of electrical circuits

Quantity to be measured	Instrument used	Connection	Notes
Voltage	Voltmeter	In parallel to the item being measured	A voltmeter measures voltage difference from one side of a load to the other. This is called **potential difference**.
Current	Ammeter	In series with the load	Ammeters can measure small currents this way, but for much larger currents a current transformer is needed (see section 5.4).
Resistance	Ohmmeter	In parallel to the item being measured	An ohmmeter only works if the circuit or item being measured is disconnected from any power source.
Power	Wattmeter	In both parallel and series	A wattmeter measures the voltage and current and then calculates the resulting power.

Key term

Potential difference: the difference in voltage from one terminal to another

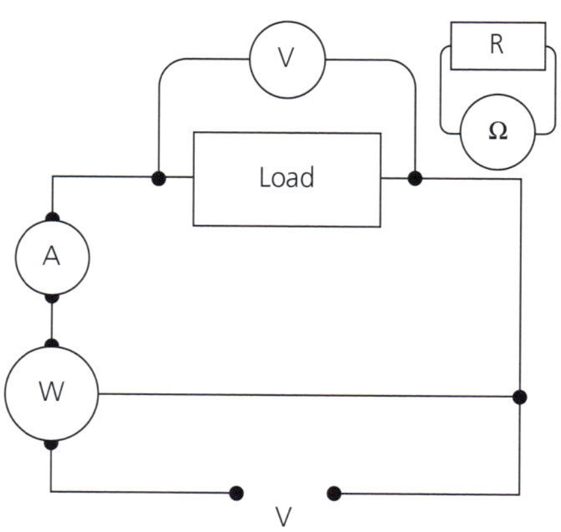

▲ Figure 2.23 How instruments are connected to measure circuit quantities

5.7 Circuit protective devices

There are four main groups of protective devices.

Fuses

Fuses have a wire element which heats up with current.

If the current steadily reaches high values due to overloads, the wire melts over a period of time and the circuit disconnects.

If the current suddenly reaches high values due to a fault, the wire melts very quickly and the circuit disconnects.

Circuit breakers

Circuit breakers have a magnetic coil. When a fault current reaches a pre-set value, the magnetic field causes mechanical movement and rapidly trips a switch, disconnecting the circuit.

They also have a thermal trip (normally a bimetallic strip), which causes gradual mechanical movement when heated by lower overload currents. This also causes the switch to trip, disconnecting the circuit.

Residual current devices (RCDs)

RCDs have either electronic devices or a small **toroidal** transformer. They monitor the current entering a circuit through the live wire and the current returning through the neutral wire.

If the circuit is healthy, the two currents remain identical. If a small fault happens, current flows to earth; the live wire has more current than the neutral wire.

If the imbalance in current exceeds the device's residual current setting, it trips instantly. The most common residual current setting for an RCD is 30 mA or 0.03 A, so these devices are highly sensitive.

Residual circuit breaker with overload (RCBO)

These are a miniature circuit breaker and RCD in the same body. They have the characteristics of both types of device.

> **Key term**
>
> ***Toroidal:*** circular or doughnut shaped

Devices also have two different ratings:

- Nominal rating (I_n) is the current value that the device can tolerate in continued service. For example, a 32 A circuit breaker can carry a 32 A load current for the lifetime specified by the manufacturer, without breaking or deteriorating.
- Activation current (I_a) is the amount of current needed to disconnect the device in the time required by the type of circuit. For example, a socket-outlet circuit must disconnect within 0.4 seconds.

6 Structural science principles

Buildings need to stand up to a lot of stresses and forces. Approved Document A of the Building Regulations 2010 requires a building to be constructed in such a way as to withstand the combined dead, imposed and wind loads without deformation or movement that will affect stability. Therefore, the effects of forces need to be considered.

6.1 Forces acting on a building

Five types of force can act on a building or part of a building. These force are described in Table 2.10 and shown in Figure 2.24.

▼ Table 2.10 Forces

Force	Description	Example	How to resist it
Tension	Force that tries to stretch a building or its components in opposite directions	The tension on vertical cables used in a suspension bridge, which hang the deck from the main cables	Use steel, which has good tensile strength, especially when wound into cables
Compression	Force that tries to crush a component by pushing on both ends	A column in a building with the weight of the building resting on it and the force acting on the ground below	Use materials such as concrete, stone or masonry, which act well under compression
Bending	Force that acts on the centre of a beam that is supported at each end	Beams that provide horizontal support, such as wooden floor joists, or support steels like rolled-steel joists (RSJs)	Use reinforced concrete for long spans, or steel or wooden joists for shorter lengths
Torsion	Force that tries to twist the component in opposite directions	An entire building when subjected to strong wind	Use a closed hollow section, such as a steel box section, or circular structural elements such as poles

Force	Description	Example	How to resist it
Shear	Force that tries to split or divide a component Shearing forces work in opposite directions but do not have the same line of action, although they can be close	Wind in different directions at the top and bottom of a building	Use brick or concrete walls Shearing can occur in cement joints, so steel bracing helps

Health and safety

When you see a beam with 'SWL 500 kg' stamped on it, never suspend any more than 500 kg from it. SWL stands for 'safe working load', and any load beyond this may bend or break the beam.

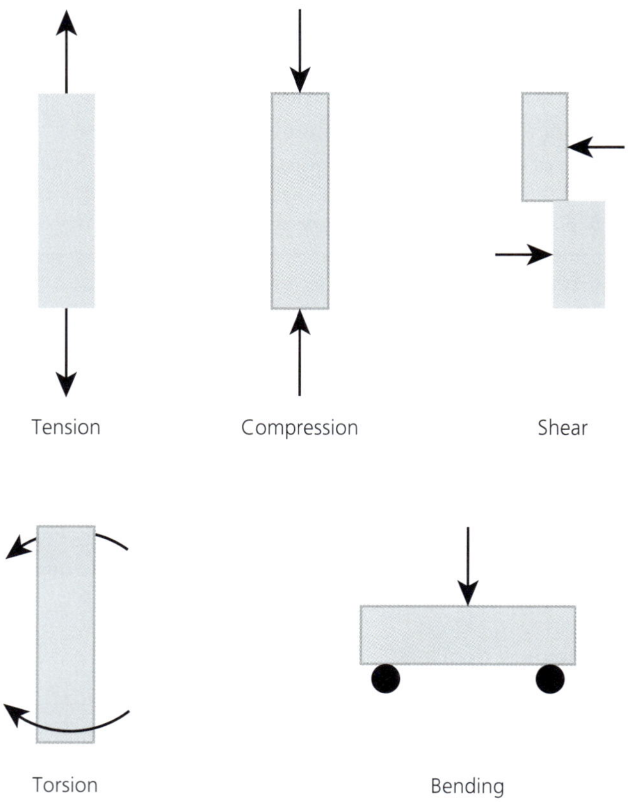

▲ Figure 2.24 Forces acting on buildings and components

There are other terms used to describe forces which act on a building:

- Vertical forces act upwards or downwards on a building and result in tension or compression.
- Horizontal forces act on the sides of a building and result in shear or torsion. They may be:
  - lateral – acting across the width of the building
  - longitudinal – acting along the length of the building.

6.2 Structural members

The supporting parts of a building are known as structural members.

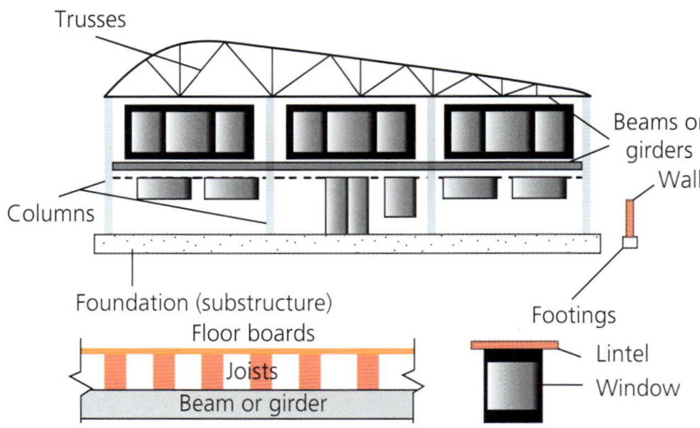

▲ Figure 2.25 Structural members used in a building

In Figure 2.25, you can see the following parts:

- **Foundation:** this is the lowest part of the substructure, supporting the building from sinking into the ground. Metal or concrete piles may be required where a foundation is unable to meet a solid base; these are drilled or driven deep into the ground to make sure the building is on a firm substructure.
- **Substructure:** this is the complete section of a building extending below ground floor level.
- **Footings:** these are the sections of masonry from the foundation to ground-floor level. They are normally linked to a particular member, such as a wall, not to the entire structure.
- **Columns:** these are upright supports and can be steel or concrete. When columns are made from brick, they are often referred to as pillars.
- **Beams or girders:** these are horizontal elements that form the support for floors. In a house, beams are used as main structural supports for first-floor walls that do not have a wall on the ground floor to support them.
- **Joists:** these are similar to beams but are often wooden and many in number. They are used to support a flooring system above them and a ceiling below them. They are supported either by walls or by beams.

- **Trusses:** these are normally associated with roofing systems and form triangles to provide support along large spans. Where trusses are used in walls, they are called braces.
- **Lintels:** these are used to provide structural support above windows and doors.

For a structure to remain sturdy and withstand all expected forces, it is important that the thickness and strength of all members are carefully calculated by structural engineers.

6.3 Drilling, notching and chasing

Where services such as electrical wires, pipes and ducts are installed in buildings, they need to be run through some structural parts.

Building regulations stipulate where notches or holes can be cut into a joist or beam so as not to weaken it, as shown in Figure 2.26.

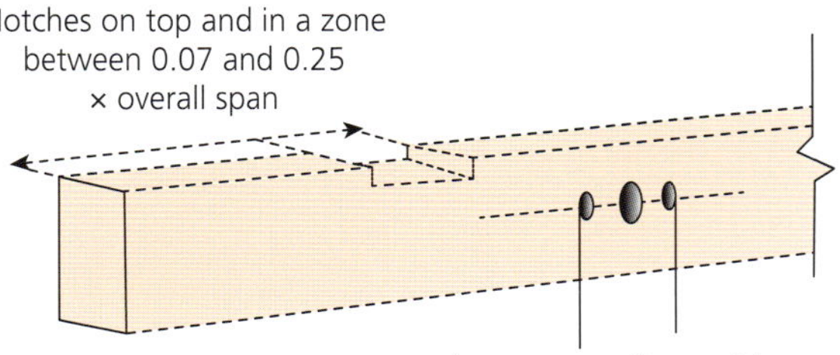

▲ Figure 2.26 Notching and cutting holes in a joist or beam

Other things to note when cutting joists include:

- The maximum diameter of holes should be 0.25 × joist depth.
- Holes in the same joist must be at least three hole diameters apart.
- The maximum depth of notch should be 0.125 × joist depth.

Test yourself

A 30 mm notch needs to be cut into a wooden beam that is 8 m long. At what maximum length from the end of the beam can the notch start?

To avoid weakening the structure:

- vertical **chases** in walls should be no deeper than one third of the wall thickness
- horizontal chases should be no deeper than one sixth of the wall thickness.

If a wall is 400 mm thick, chases should be no deeper than:

$$\text{Vertical} = \frac{400}{3} = 133 \text{ mm}$$

$$\text{Horizontal} = \frac{400}{6} = 66.6 \text{ mm}$$

Improve your maths

What is the maximum depth of a vertical chase in a wall which is 0.2 m deep?

Key term

Chases: cuts in a masonry or plaster wall to contain pipes or cables; once the pipes or cables have been fitted, they are covered with plaster or similar

6.4 Other effects on structures

There are many other factors that can affect a structure and increase the risk of weakening its stability.

Other buildings

Neighbouring buildings can have several effects on each other, such as creating increased wind speeds. Currents of air moving around structures travel at different speeds, and these forces can have damaging effects on buildings.

Foundations from other buildings also have an impact, as the force of the building's weight is transferred into the ground at 45-degree angles. This could cause soil movement beneath or close to adjacent buildings.

Drains and sewers

These underground systems can collapse if buildings are built on top of them, causing buildings to lean as one section of the foundation drops.

Proximity to drains and sewers also influences the depth of a foundation. Building weight is transferred to the ground from the foundation at a 45-degree angle from the footprint. As a result, there is potential for damage to drains and sewers if they are located within that 45-degree area. This means that the foundation is typically excavated to a depth that is at least the same as the deepest part of the drain or sewer.

Trees

The roots of trees, especially fast-growing trees, can damage building foundations. As trees absorb moisture from the ground, this can cause uneven drying of soils under foundations or footings. Shrinkage of soils can cause foundations to collapse or crack.

Different soil structures

Soils shrink and expand depending on rainfall and levels of the water table. Soils which are permeable (allow water to drain through) are more stable than non-permeable soils, such as clay.

Soils can also shift or move. If this happens under part of a building, it can cause **subsidence**, meaning one side of the building drops or moves away from other parts. The initial signs of subsidence are cracks forming in walls.

> **Key term**
>
> ***Subsidence:*** the sinking of a structure into the ground

7 Heat principles

There is much more to heating a building than just considering the source of the heat. Buildings need to be kept warm in winter and cool in summer. Equally, temperature control and air changes are important to prevent condensation, which can lead to damp and mould growth.

> **Health and safety**
>
> You may need to consider the effects of temperature on particular groups of people, such as elderly people.
> - 21°C is the recommended comfortable living temperature.
> - Respiratory problems can start when subjected to long periods at 16°C.
> - If exposed to a temperature of 12°C for more than two hours, elderly people are at risk from raised blood pressure and heart attack.

7.1 Heat transfer

There are three methods of transferring **heat**:
- convection
- conduction
- radiation.

> **Key term**
>
> ***Heat:*** does not, in BSE terms, mean hot; it simply means heat energy, which can be hot or cold

Convection

Convection occurs when air or water is heated and then moves away from the source of the heat, carrying thermal energy.

For example, a wall-mounted convection heater or radiator heats the air around it, causing the hot air to rise. This is moved into the room by cooler air entering the bottom of the heater or radiator, creating convection currents which rotate around a room.

The term 'radiator' is a misnomer, as most of the heat output is via convection. They are not called convectors, however, because a true convector has a heat source that is not exposed.

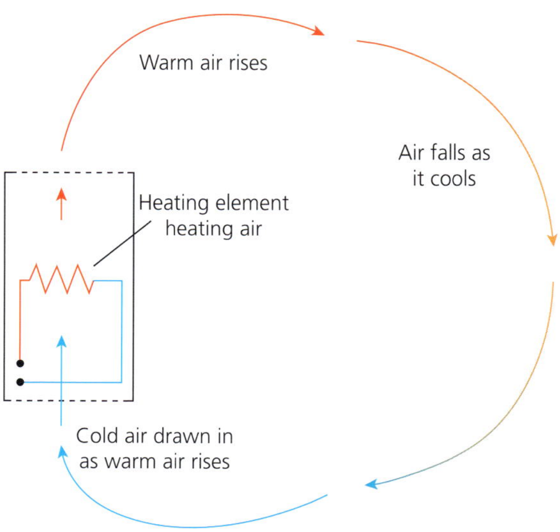

▲ Figure 2.27 The convection cycle

Convection also occurs when an immersion heater conducts heat to the water close to the element. This causes hot water to rise in the vessel, allowing the element to heat cooler water that replaces the hot, rising water.

Convection heaters:
- are a good way to heat most sizes of room
- can be wall mounted
- can use elements or pipes buried in the floor to heat a large space
- are less effective in small spaces with limited air circulation.

Conduction

Conduction is when heat is transferred directly into a material, for example brick in the case of a **storage heater**. When a substance is heated, the molecules gain more energy and vibrate more. They then bump into other nearby molecules and transfer some of their energy to them. Conduction is therefore not appropriate for heating air directly, as the molecules are too far apart.

> **Key term**
>
> ***Storage heater:*** an electric heater that stores thermal energy; it heats up internal ceramic bricks when electricity is cheaper at night and then releases heat gradually during the day, acting in the same way as a convection heater

The following are examples of heating systems using conduction:

- Floor-heating systems conduct heat to the floor, which causes air convection above it.
- Hot-water immersion systems directly heat water around the element.
- Instantaneous water heaters have an element that is wound around a metal water pipe. Heat is conducted through the pipe into the water. As the amount of water in the pipe is small, it heats up quickly.

Radiation

Radiant heaters:

- use infrared radiation to heat bodies, but not the air around them
- are directional – there must be a line of sight in order to heat a person
- are particularly useful in large areas where heating the air is not required or efficient, for example outdoor patio heaters
- require a large amount of power and are therefore less efficient than convection heaters.

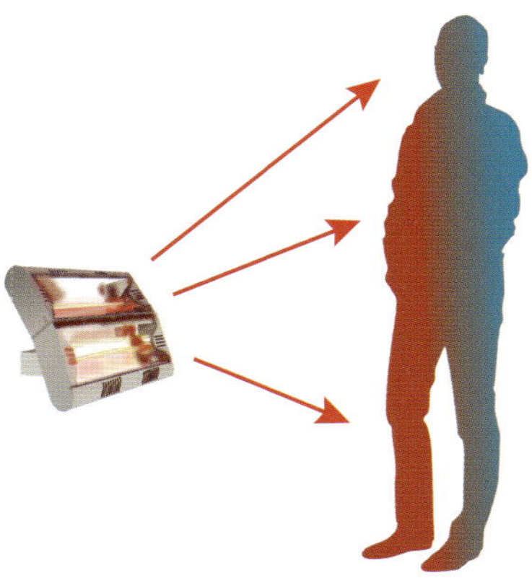

▲ Figure 2.28 A radiant heater

7.2 Characteristics of air

Modern buildings strive to be more energy efficient by keeping heat in, but this can also cause problems if a building does not allow air changes to occur. Air will become stale and contaminated, causing breathing problems.

In addition, if a building is allowed to heat up but moist warm air comes into contact with colder surfaces, such as windows, condensation forms, which in turn can cause mildew or rot. Air contains water vapour, and the amount of water vapour in the air is known as the level of humidity. Warmer air can hold more water vapour, but if that air touches cooler surfaces, or the air is rapidly cooled by mechanical means such as an air-conditioning system, the humid air vapour will condense into liquid water.

▲ Figure 2.29 Mould in the corner of a window

Getting the balance right is difficult, so if a building does not have a forced air change by fans, it needs to occur naturally. This can happen by pressure differences, which can be caused by wind or the convection effect created by differences in temperature or humidity. In either case, the amount of ventilation depends on the size and positioning of openings in the building.

Buildings with a staircase or an open chimney will create air changes by the stack effect, where warm, less-dense air rises, drawing cooler dense air into the base of a building through any opening. As air warms, it expands and becomes less dense, as there is less air occupying a space, but as it cools it contracts, meaning more air occupies a space, so it becomes denser.

Ideally, a normal room in a house should have a minimum of four air changes per hour. A minimum of six changes per hour should occur in bathrooms and toilets, but these are likely to have forced air systems such as extraction fans creating more changes.

In public places, this increases to a minimum of ten changes per hour in normal occupied rooms, and 12 in places such as restaurants.

7.3 Heat loss in buildings

Buildings lose heat through the different construction materials used. Even if walls are well insulated, doors and windows can cause heat loss.

All materials used in construction have ratings of thermal conductivity, known as K-values. These are values given by manufacturers of the materials.

Table 2.11 gives examples of commonly used materials and their average thermal conductivity (K) value in watts per metre Kelvin (W/m K).

▼ Table 2.11 Thermal conductivity of common materials

Material	Average K-value (W/m K)
Light blockwork	0.38
Dense blockwork	1.63
Exposed brick	0.84
Chipboard	0.15
Concrete	1.1
Glass	1.05
Gypsum plaster	0.46
Plasterboard	0.25
Steel	20–40
Granite	2.5
Timber (softwood)	0.14
Timber (hardwood)	0.16
Mineral-wool insulation	0.035
Rigid foam board	0.026
Air in a cavity	0.18

R-values

Once a K-value is known for a material, the amount of heat resistance the material offers can be calculated. This is known as the R-value.

The R-value can be calculated based on the thickness of the material:

$$\text{R value} = \frac{\text{thickness of material (m)}}{\text{K value (W / mK)}}$$

Using Table 2.11, the R-value for 12.5 mm (0.0125 m) thick plasterboard can be determined as:

$$R = \frac{0.0125}{0.25} = 0.05\ \text{W / m}^2\text{.K}$$

Note the units for R-value: watts per metre squared Kelvin or W/m^2.K.

Test yourself

Using Table 2.11, determine the R-values for:

- 25 mm thick chipboard
- 500 mm thick concrete.

U-values

A **U-value** is the rate of heat transfer for a wall, ceiling or floor.

Key term

U-value: a unit of measurement used to record thermal transmittance through a building material

To calculate the total heat loss for a wall, determine the U-value based on the materials used.

For example, consider a wall constructed using the following materials:

- 12.5 mm thick plasterboard (R_1) mounted on 100 mm light blockwork (R_2)
- a 50 mm air cavity (R_a)
- 100 mm exposed brick (R_3).

In addition to building materials in the wall, external air resistance and internal air resistance need to be taken into account. These have predetermined R-values of:

- internal air resistance (R_{si}) 0.13 W/m K
- external air resistance (R_{se}) 0.04 W/m K.

So, to calculate the U-value of the wall:

$$R_1 = \frac{0.0125}{0.25} = 0.05\ \text{W / m}^2\text{.K}$$

$$R_2 = \frac{0.1}{0.38} = 0.26\ \text{W / m}^2\text{.K}$$

$$R_a = \frac{0.05}{0.18} = 0.28\ \text{W / m}^2\text{.K}$$

$$R_3 = \frac{0.1}{0.84} = 0.12\ \text{W / m}^2\text{.K}$$

$$\text{Total U value} = \frac{1}{R_{si} + R_1 + R_2 + R_a + R_3 + R_{se}}\ \text{W / m}^2\text{.K}$$

So, for our wall:

$$U = \frac{1}{0.13 + 0.05 + 0.26 + 0.28 + 0.12 + 0.04} = 1.136\ \text{W / m}^2\text{.K}$$

Some manufacturers provide U-values for their products, such as doors and windows. As an example, these could be:

- 3 W/m²․K for a double-glazed window that has 12 mm spacing between the panes of glass
- 2.5 W/m²․K for an external door which has a standard fire rating.

Using U-values, we can calculate the heat loss in watts through various parts of a building:

$$\text{area of surface} \times \text{temperature difference} \times \text{U-value for surface}$$

The temperature difference relates to the desired room temperature and the average lowest outdoor temperature. So, if a room needs to be heated to 20°C, and the outdoor temperature has an average low of −3°C, the difference is 23°C. Once each surface heat loss has been determined, including the heat loss through air changes, these are added to calculate the overall heat loss.

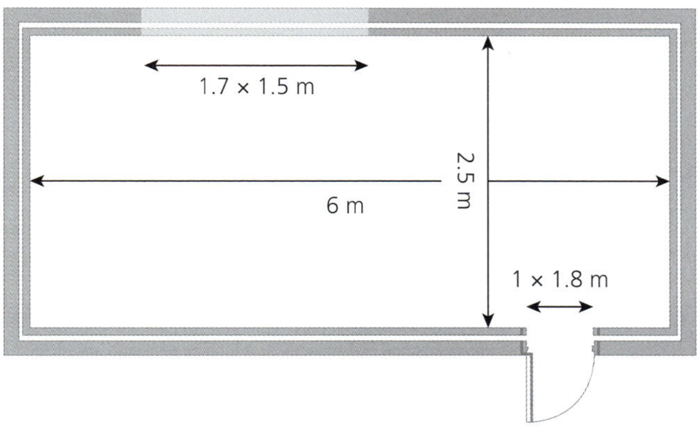

▲ **Figure 2.30** A room with heat loss

Figure 2.30 shows a room with dimensions. Assuming the height of the room is 2.6 m, we can calculate the overall heat loss if the desired temperature is 20°C and the outside average low is 2°C. Let us assume all the R-values have been calculated so that the following U-values are given, and we will assume the room undergoes two air changes per hour:

- Walls: 0.35
- Floor: 0.25
- Ceiling: 0.25
- Windows: 2.9
- Door: 2.9
- Air change: 0.33.

Room volume:

$$6 \times 2.5 \times 2.6 = 39 \text{ m}^2$$

Window heat loss:

$$(1.7 \times 1.5) \times (20 - 2) \times 2.9 = 133.11 \text{ W}$$

Door heat loss

$$(1 \times 1.8) \times (20 - 2) \times 2.9 = 93.96 \text{ W}$$

Wall heat loss:

Wall area:

$$2.5 \times 2.6 = 6.5 \text{ m}^2 \times 2 = 13 \text{ m}^2$$

$$6 \times 2.6 = 15.6 \text{ m}^2 \times 2 = 31.2 \text{ m}^2$$

$$31.2 + 13 = 44.2 \text{ m}^2$$

Less window and door:

$$44.2 - (1 \times 1.8) - (1.7 \times 1.5) = 39.85 \text{ m}^2$$

So, loss is:

$$39.85 \times (20 - 2) \times 0.35 = 251.1 \text{ W}$$

Floor and ceiling heat loss (as both have the same dimension and U-value):

$$6 \times 2.5 \times (20 - 2) \times 0.25 = 67.5 \text{ W} \times 2 = 135 \text{ W}$$

Air-change loss needs to consider the volume of the room, the number of air changes and the U-value for the air loss, which is 0.33:

$$39 \times (20 - 2) \times 2 \times 0.33 = 463.3 \text{ W}$$

So, the total heat loss is:

$$133 + 93.6 + 251.1 + 135 + 463.3 = 1076 \text{ W or } 1.076 \text{ kW}$$

Therefore, the room would require some form of heating system to counter this loss and maintain the required temperature.

7.4 Methods of heating a building

Several types of building services system can be used to provide heat, for example:

- electric wall-mounted convection heaters (including storage heaters)
- electric floor-heating elements
- wet floor-heating systems
- wet heat-emitting radiators
- ground or air source heat pumps.

Electric convection heaters, known as panel heaters, contain electric heating elements. Cool air is drawn into the bottom of the unit, passes over the heating element and rises out of the unit as warmed air.

Electric storage heaters contain bricks that sandwich heating elements. The elements heat the bricks, which then store the heat and gently release it into the room using the convection cycle. The advantage of storage heaters is they can consume electricity at times when costs are lower, such as night time, then release the heat at other times.

Wet heating systems have boilers that heat water and pump it around the building to radiators or floor-heating pipework. The fuel used for the boiler can be:

- gas
- oil
- biomass.

Gas can be delivered by mains pipes into the building, and this is metered at supply. Alternatively, tanks can be used to store gas delivered to the property.

Oil is also delivered to a property and stored in tanks for future use.

Health and safety

Gas components can only be installed and certified by engineers registered with **Gas Safe**.

Key term

Gas Safe: a professional organisation that controls the health and safety of work completed on gas systems in the UK

Biomass heating systems use specially made wood pellets, which are burned to produce the heat source for the pumped water system. They are said to be carbon neutral, or low carbon, as the carbon dioxide (CO_2) released by burning the wood is equal to the CO_2 released by the wood if it were left to decay naturally. We will explore this, ground/air source heat pumps and other technologies used to heat or cool a building, in Chapter 5.

8 Light principles

Lighting design has become a very specialist area because of the vast range of lighting and **luminaires** available. Light-emitting diode (LED) technology is used as the main form of lighting, due to its energy-efficient performance and light quality.

Key term

Luminaires: complete electric lighting units, including the casing, lamp and any internal controlling devices or electronic equipment (known as control gear or drivers)

8.1 Main lighting terms

Table 2.12 considers the main lighting terms.

▼ Table 2.12 Main lighting terms

Term	Definition
Luminous intensity	This is the amount of light emitted per solid angle or in a given direction. It is denoted by the symbol I and measured in candela (cd).
Luminous flux	This is the total amount of light emitted from a source. It is denoted by the symbol F and measured in lumens (lm).
Illuminance	This is the amount of light falling on a surface. It is denoted by the symbol E and measured in lumens per metre^2 (lux).
Efficacy	This is the efficiency of a lamp or luminaire and compares the amount of light emitted to the electrical power consumed. It is denoted by the symbol K and measured in lumens per watt (lm/W).
Maintenance factor (mf) Light loss factor (llf)	These are factors used to derate the light output of a lamp, allowing for dust. The factor used depends on the environment. An average office environment would have a factor of 0.8, whereas a factory where lots of dust accumulates may have a factor of 0.4.
Coefficient of utilisation or utilisation factor (uf)	This is a measure of the efficiency of a lamp in transferring light energy to a surface, such as a wall or ceiling. Emitted light bounces off reflective walls, making more effective use of the light. An average factor for a room is 0.6. The lighter the colour of the room, the higher the factor.
Glare	Glare is a very bright light that is difficult to look at. Poor positioning of luminaires can cause glare, resulting in people squinting, getting headaches or experiencing general discomfort.

Term	Definition
Diffuser/louvre	Diffusers are normally translucent plastic or glass covers that refract harsh light, dispersing it to create an even light from a single source. A louvre is polished metal which reflects light in different directions, dispersing a harsh glare into softer dispersed light.
Colour rendering	This is the appearance of light in terms of colour. Lamps produce either orange light or blue light, depending on the gases used in them. This in turn can make objects under the lights look different in colour. For example, sodium lamps were once commonly used for street lighting, but this gives the appearance that everything lit by them is orange. This was never considered much of a problem, as hazards could still be seen and avoided. In more modern times, whiter colours are preferred as they are better for CCTV images and also make people feel safer.

Test yourself

A lamp has a lumen output of 2500 and consumes 50 W. What is its efficacy?

Research

How does the colour rendering of street and utility lighting affect crime?

8.2 Choosing the right type of lamp

Artificial lighting consumes a lot of energy. Even during daylight hours, many offices, shops and public buildings have lights switched on. It is therefore important to choose lamps that:

- provide a good-quality light with the correct colour rendering
- consume power efficiently.

Table 2.13 shows the different types of lamp available, their basic operating principles and their characteristics including efficacy (an indication of power to light performance).

▼ Table 2.13 Types of lamp

Lamp type	Basic operating principles	Characteristics and applications
General lighting service (GLS)	A filament is suspended in a vacuum. When current flows through the filament, it glows white hot, producing light.	GLS lamps rated above 100 W were banned due to their poor efficacy (approximately 14 lumens/watt). Caps are usually bayonet (BC) or Edison screw (ES).
Tungsten halogen	A filament is suspended in halogen gas, which prevents evaporation. This allows the lamp to run at much higher temperatures than standard GLS lamps, meaning a brighter light.	These were widely available in ratings up to 500 W but have been superseded by LED lighting. Smaller G4 or G9 lamps are still common in table or display lamps. Halogen lamps have an average efficacy of 17 lumens/watt.
High-pressure sodium (SON)	Current strikes across low-temperature sodium gas in a tube, causing it to heat up and ionise, producing light. The lamps are known as discharge lamps.	These were widely used for car park or street lighting, as they are suited to the illumination of large areas. The colour output is slightly orange. Average efficacy is 120 lumens/watt.
Low-pressure sodium (SOX)	These work on a similar principle to high-pressure sodium lamps but they have larger gas-filled tubes, with lower sodium gas pressures, and are usually U shaped.	These were widely used for street lighting due to their very high efficacy but colour rendering is poor, with a dull orange light output. Average efficacy is 180 lumens/watt. Although still around today, they are rarely used in new lighting schemes.
High-pressure mercury (MBFU)	These work on a similar principle to high-pressure sodium lamps, but have tubes filled with mercury gas.	These are commonly used for utility/communal lighting because of their better colour rendering, which is white with a slight blue tinge. Efficacy values are on average 80 lumens/watt.

▼ Table 2.13 Types of lamp

Lamp type	Basic operating principles	Characteristics and applications
Metal halide (HID)	These discharge lamps are similar in operation to high-pressure sodium lamps but they use a mixture of mercury and sodium gases.	These produce a bright white light but efficacy is lower as a result and averages 60 lumens/watt.
Low-pressure mercury (fluorescent)	Commonly known as fluorescent tubes, these mercury-filled tubes have an inner coating of phosphor powder, making a blue ultraviolet light inside the tube and various types of white light on the outside.	These are still widely available but are slowly being replaced by LED tubes. Efficacy is on average 100 lumens/watt.
Compact fluorescent (CFL)	These are small versions of fluorescent tubes, where the tube is often bent to occupy smaller areas.	Designed to replace the GLS lamps banned by EU directives, these often have ES or BC caps. Efficacy is on average 65 lumens/watt.
LED	Made up of several light emitting diodes (LEDs), these lamps require control gear, known as drivers, to reduce and rectify voltage and govern current.	LED lamps and luminaires are now widely used for all applications, from general room lighting to street lighting and stadium illumination. Efficacy is on average 120 lumens/watt but this is improving all the time.

LED lighting is becoming the most common form of lighting due to:

- good efficacy
- the ability to change colour rendering to suit a location, for example warm white, daylight or coloured
- longer-lasting lamps.

9 Acoustics principles

9.1 Acoustics in buildings

Acoustics is a term used to describe the study of mechanical waves that travel through gases, solids and liquids causing vibration and sound. This could be from somebody speaking, a loudspeaker playing music or someone banging on something like a drum.

Key term

Acoustics: the transmission of sound

Different materials and the shape of those materials affect the way sound waves move. In most construction applications, the desire is to provide acoustic comfort rather than to enhance acoustics, such as in a concert hall. Poor acoustic protection can lead to severe health conditions, such as hearing loss, high blood pressure, headaches and sleep deprivation.

9.2 Factors that affect acoustics

Before we consider methods of providing acoustic comfort, we need to understand some terms used in the study of acoustics.

▼ Table 2.14 Terms used in the study of acoustics

Term	Description
Reverberation	Reverberation is the persistence of a sound following its creation. It is caused when sound reflects off surfaces and decays. Even when the initial source of the sound stops, the reflections can continue, with their amplitude decreasing until zero is reached.
Reverberation time	Reverberation time is a measurement, in seconds, of the time it takes for a sound to decay and stop entirely. For small rooms, it should typically be under a second, and no more than two seconds for larger rooms.
Frequency	Frequency is the number of waves that occur in a period of time. For sound waves, this determines the pitch of a sound. Low-frequency sounds (fewer waves per second) are lower in pitch than high-frequency sounds.

Term	Description
Resonance	Materials have different natural resonant frequencies. This is simply the frequency of sound and pressure which causes the material to move easily and vibrate. For example, if a loudspeaker is placed in front of a glass and the frequency of the sound is increased (low pitch to high pitch), there will be a point where the glass shatters because it vibrates to the point of destruction.
Sound absorption	Sound absorption is where sound waves are suppressed or absorbed by an item or structure, rather than being reflected.
Sound transmission class (STC) rating	An STC rating is a numerical value showing how well a structure reduces sound transmission. It is commonly used to rate doors, walls, windows, ceilings and floors. The higher the value, the better the structure is at reducing sound travel. A well soundproofed wall typically has an STC rating above 50.
Sound barrier	A sound barrier is a material that can be placed on a structure (such as a wall, ceiling or floor) to increase its STC rating.
Attenuation	Attenuation is the gradual loss of intensity. It refers to a structure's ability to limit sound transmission.
Baffle	Acoustic baffles are devices which reduce the strength of airborne sound. They absorb sound, reducing echo and lowering reverberation time. When baffles are designed to be suspended from a ceiling, they are referred to as clouds.
Decibel	The decibel is the unit of measurement for sound.
Flanking sound	Flanking sound is sound that travels between two areas, such as through air vents and ducts to reach another room.
Footfall	Footfall is the sound made by walking on a surface.

9.3 Principles of sound

If you have ever been in a large open space with bare, hard walls, you will have experienced a very echoey environment. Soft materials, such as fabrics and foams, create a sound-absorbent surface where sound does not reverberate.

The following can help to absorb sound waves and prevent reverberation:

- cavity wall insulation, such as rock wool or fibre wool
- isolation membrane (a thin barrier in walls, ceiling voids and doors)
- flooring materials, such as carpets or wood
- soundproof foam panels (usually shaped into small pyramids).

When considering the installation of building systems, their location will have a bearing on the noise they produce. If a system component, such as a gas boiler or circulation pump, vibrates in operation, this will create sound. Any materials in contact with the component can also vibrate, which in turn can amplify this sound. In some cases, acoustic hangers or fixings may be required to reduce the transfer of vibrations.

If equipment is located outdoors, such as air source heat pumps, the sound from the fans can cause sound pollution.

▲ Figure 2.31 Soundproof foam panels in use in a recording studio

Health and safety

Noise exposure of 80 decibels (dBA) for two hours can cause hearing damage.

9.4 Laws and regulations on noise restriction

Laws restrict noise levels between 11 p.m. and 7 a.m., and councils must investigate any complaints of noise above permitted levels. As a guide, constant underlying noise levels of 24 decibels (dBA) are permitted, with increases to 34 dBA intermittently.

Approved Document E of the Building Regulations covers resistance to sound and sets out standards for new homes and conversions. While the statutory requirements are barely one page long, the guidance for complying with the requirements is very detailed, including information on:

- soundproofing, including the transmission of sound between walls, ceilings, windows and floors
- prevention of unwanted sound travel within different areas of a building, including flats and connecting buildings
- the structure of materials and formation of elements of the building, including building services, especially where these services penetrate walls, floors and ceilings
- the need to lag pipework or soundproof socket outlets recessed into a partition wall.

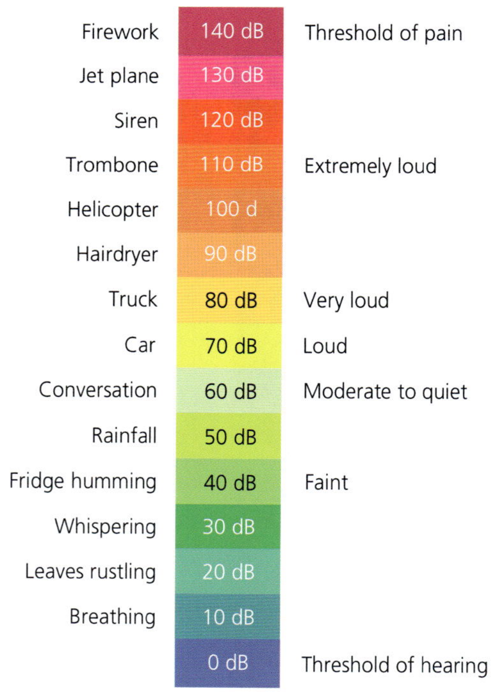

▲ Figure 2.32 Comparison of different noise levels

10 Earth science principles

Earth science is a wide-ranging subject, covering ground structures, watercourses, water cycles and earth forces such as earthquakes. For buildings to remain safe and stable for their lifespan, they rely on the ground being stable below them.

In this section, we will look at basic ground structures, watercourses and the water cycle, and how earth forces can be detrimental to buildings and structures in the UK.

10.1 Ground structure

We usually consider changing soil conditionsin the context of the ground being muddy, hard or good for growing things. However, this only relates to the first layer of topsoil we encounter in our daily lives. When constructing buildings, we need to consider the soil structure deeper below the topsoil; and in the case of high-rise buildings, we must go deeper still. Figure 2.33 shows the different layers of soil. Although the illustrations show the layers in a relatively small depth, in reality the **bedrock** layer can be very deep.

Key term

Bedrock: solid rock usually found beneath the weaker surface materials of the ground

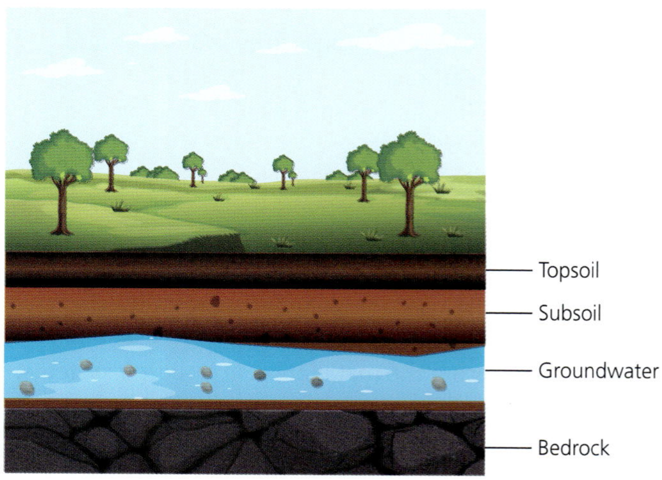

▲ Figure 2.33 Layers of soil

Low-rise buildings with shallow foundations typically have footings or foundations at least 1 m deep in clay-type subsoils, as these soils are prone to shrinkage or freezing due to their water content up to 0.75 m in depth. Stony subsoils, where water is less likely to be held, may allow for shallower foundations, but this depends on the **water table**.

> **Key term**
>
> ***Water table:*** the level below which the ground is saturated with water; this level can rise with rainfall and fall with periods of dry weather

Where buildings require more stability, such as high-rise buildings, foundations need to be in contact with the bedrock layer. This is normally achieved using piles. Figure 2.34 shows piles being drilled into the ground, where reinforced concrete is then poured to give a solid connection to the bedrock.

▲ Figure 2.34 Piles being drilled, with reinforcement being dropped into the pile hole ready for concrete to be poured

In addition, soil cleanliness is important. On all proposed development sites, soil samples must be taken and analysed before any building works begin, to ensure the ground has not been contaminated by previous activities. Many years ago, some locations had uses that were never documented or recorded, such as landfill-type waste tips. In these situations, the decomposing waste could have caused pockets of methane gas to build up in the ground, creating a long-term explosion risk. Other sources of contamination include chemical seepage from old factory sites.

10.2 Watercourses

A watercourse is a channel through which water flows. Examples include a wide river, a small stream, a ditch which is dry for most parts of the year, or an underground river.

Watercourses are important as they drain the land. Without them, water would cause soils to become unstable, leading to subsidence or even landslips. It is therefore important that watercourses of any type are kept free of obstruction and not overloaded.

When new developments are built, **surface water** needs to be drained into local watercourses, which can create flood risks if not managed correctly. Surface-water flooding is now the biggest cause of property flooding. Water that falls as rain on fields takes time to filter through the soils into the natural drain. When water falls on hard surfaces, such as roads, pavements and roofs, this water drains into the watercourses much more quickly, causing **flash floods**. With more developments comes more risk of flash floods.

> **Key terms**
>
> ***Surface water:*** water that collects on the ground or above surface structures and buildings, normally in the form of rain
>
> ***Flash floods:*** floods that appear suddenly following heavy, above-average rainfall

Measures used to manage surface water, known as attenuation, include:

- underground flood tanks that store water and gradually release it into local watercourses
- the creation of ponds or reedbeds to store or absorb surface water or gradually release it into local watercourses such as streams, brooks or rivers
- soakaways, which disperse surface water into the ground where it slowly drains into watercourses.

When considering flood risks, it is worth looking at the water cycle and the way land drains after rainfall. Figure 2.35 shows how rainfall seeps through permeable soils and forms the water table. Water seeps through soils to either aquifers, which are like underground rivers or lakes, or into above-ground rivers or lakes. These eventually drain to the sea, where evaporation starts the water cycle by forming rain clouds.

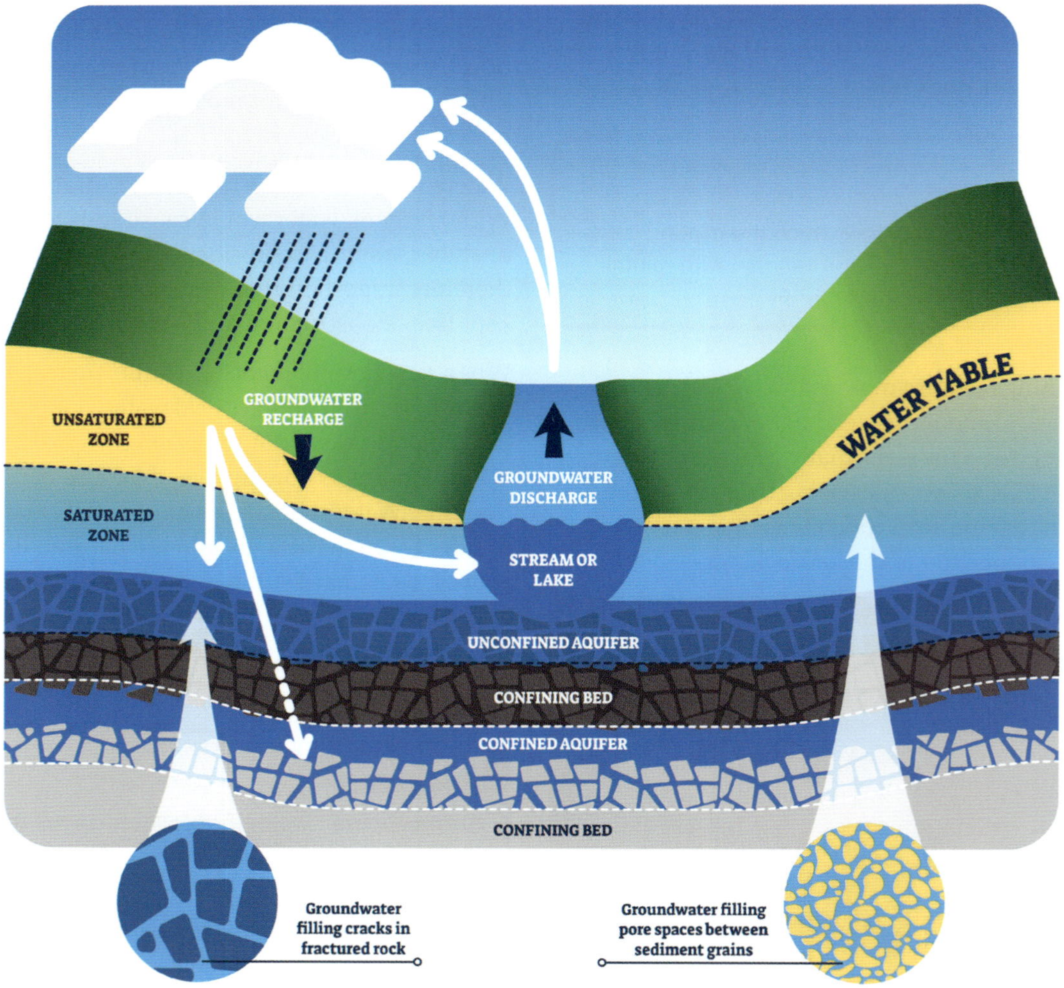

▲ Figure 2.35 How rainfall naturally drains

10.3 Earth forces

The Earth's crust is made up of large, moving pieces known as tectonic plates. At a boundary, plates can move apart, move together or slide past each other. The joins at plate boundaries are known as fault lines, and these are danger zones for earthquakes. On occasion, the moving plates become jammed so tension and forces build. Eventually the forces become too great and the sudden release results in an earthquake, shaking the ground violently. The UK is not close to any fault lines, so the risk of earthquakes is low. That does not mean, however, that they do not happen, as some minor tremors have affected the UK.

Other earth forces that are closer to home are those caused by weather and tides.

While wind can cause serious damage to buildings and structures during storms, persistent wind and rain can also cause slow but damaging erosion to some materials. One example is limestone, which dissolves when it comes into contact with carbonic acid, common in rainwater. Mortar used for pointing between bricks is also prone to weathering, especially in areas such as chimneys. Where building services equipment is located in exposed outdoor areas, it can also become damaged through weathering.

Flooding through storm water is becoming a more frequent occurrence due to global warming.

Storm water is not just a huge amount of rainwater released into drains, as discussed above with watercourses. High tidal conditions, caused by **spring tides** and **storm surges**, also restrict river drainage, because high levels of seawater act as a barrier to the draining river water.

Climate change is increasing the likelihood of heavy rain in the UK, particularly during the winter months, leading to higher risks of flooding and less stable soils.

Key terms

Spring tides: exceptionally high tides that occur twice monthly at the time of the new moon and the full moon, when the moon's orbit aligns with the sun to create a greater gravitational pull; they are known as spring tides because they act like a spring

Storm surges: large-scale rises in sea level caused by storms; high winds push the seawater towards the coast and low pressure at the centre of the storm pulls up the water level

Assessment practice

Short answer

1 **a** State the SI unit of measurement for density.
  b Calculate the area of a circle with a diameter of 840 mm.

2 List three pure metals that are non-ferrous.

3 A lever is to raise a load with a mass of 180 kg. The load is placed 0.5 m from the fulcrum and a force acts 3 m from the fulcrum on the other side. Calculate the force required to raise the load.

4 A weight of 5820 N is to be raised 28 m in 1.5 minutes. Determine the power required.

5 A circuit has a total resistance of 58 Ω and is connected to a 20 V DC supply. Calculate the power dissipated by the circuit.

Long answer

6 Explain how pumped storage hydro plants provide additional power during peak times.

7 Explain how an RCD provides earth fault protection.

8 Explain why glare from artificial lighting should be avoided and describe measures that can be taken to minimise it.

9 Describe how sound attenuation is achieved in a large open room.

10 A room measures 3 m × 2 m and has a height of 3 m. There is one door leading into the room, with a dimension of 2 m × 0.9 m.
The walls have a U-value of 0.4. The room temperature is 28°C, and the outside temperature is −5°C.
Calculate the heat loss from the room through the walls.

Chapter 3 Construction design principles

This chapter compares modern and traditional building methods, and investigates how good design can reduce the impact of construction work on the wider environment. It considers how modern building methods can reduce project durations, lower costs and improve the health and safety of workers by manufacturing off site.

It then explores the role of different disciplines in the design process from conception to completion, together with factors that influence the design of a building project.

Learning outcomes

By the end of this chapter, you will understand:

1 benefits of good design
2 design principles
3 the role of different disciplines involved in design
4 the design process from conception to completion
5 the concept of the 'whole building', including life cycle assessment.

1 Benefits of good design

1.1 Factors of good design

Design function

Before choosing a method of construction, it is important to understand how the building will achieve its design function. All buildings are initially designed to meet the client's **specification** because without them to finance the work, there is no project. However, sometimes the needs of the client have to be compromised to satisfy local restrictions, planning laws, building regulations and environmental requirements. In addition, designers have to consider:

- the aesthetics of the building and how it will fit into its proposed environment
- access and egress
- security
- internal layout and arrangement of rooms/spaces
- energy efficiency.

Once these aspects have been determined, designers and clients must decide what form of construction to use for the **superstructure** of the building. There are several important factors to consider when selecting a method of construction, for example, build speed, familiarity with the building system and cost.

> **Key terms**
>
> ***Specification:*** a detailed description of the materials and working methods that must be used for a project
>
> ***Superstructure:*** the part of a building above ground level, built on the basement or foundation

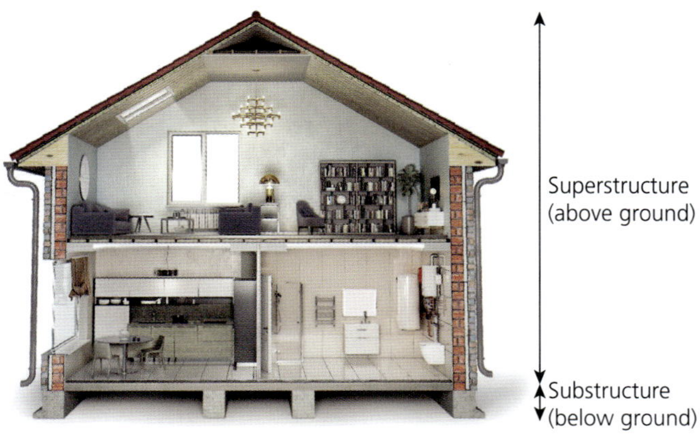

▲ Figure 3.1 Superstructure and substructure of a building

By contrast, the substructure of a building is the part below ground level. For more information on superstructure and substructure, see Chapter 7.

Environmental impact

The way we construct and use buildings for all types of purpose will have an impact on the environment. However, with considerate planning and the use of modern and innovative building materials, this impact can be minimised.

Without tight monitoring systems and control measures, construction work on **greenfield sites** and other types of land such as conservation areas may result in damage or destruction of natural wildlife habitats. (There is further information on greenfield and **brownfield sites** in section 1.3 of this chapter.)

Inconsiderate construction and inappropriate disposal of waste materials can lead to pollution of the land, the air and natural watercourses, such as streams, rivers and lakes. This can damage or destroy whole ecosystems, outside the boundary of the site and beyond the lifetime of the development.

> **Key terms**
>
> ***Greenfield sites:*** areas of land that have not been previously developed or built on, above or below ground
>
> ***Brownfield sites:*** areas of land that have been previously developed or built on, even if there is no physical evidence of earlier use

▲ Figure 3.2 A greenfield site

▲ Figure 3.3 A brownfield site

The environmental impact of a building's design, construction and use is controlled and monitored by:

- an environmental impact assessment (EIA)
- an energy performance certificate (EPC)
- the Code for Sustainable Homes
- a site waste management plan (SWMP)
- building regulations (Part L).

Environmental impact assessment (EIA)

The Town and Country Planning (Environmental Impact Assessment) Regulations 2017 require developers to identify significant effects on the environment of their proposed construction project at the planning application stage. If the type of development is listed in Schedule 2 of the regulations and exceeds the thresholds set out there, the local authority must screen the proposals to determine the likely environmental impact and decide whether an EIA is required. This also allows the public an early opportunity to be involved with decision-making procedures.

The regulations affect only a small proportion of private and public projects that require planning permission; however, they could also apply to some **permitted development**.

Key term

Permitted development: building work that does not require planning permission from the local authority but must still be constructed according to building regulations

Research

Search online for 'Schedule 3 of The Town and Country Planning (Environmental Impact Assessment) Regulations 2017'. You can access it here: **https://legislation.gov.uk/uksi/2017/571/schedule/3/made**

Use the screening criteria to determine the environmental impact of construction work you have recently been involved in.

Energy performance certificate (EPC)

EPCs are a legal requirement for domestic and commercial properties in the UK that are available to buy or rent. However, some buildings may be exempt, for example, residential buildings that are intended to be used for less than four months a year and listed buildings.

To obtain an EPC, an accredited assessor must evaluate the energy efficiency of a building. They will check that loft and wall insulation has been correctly installed, and they will review records of energy use (also referred to as heat demand) for the past two years. Once satisfied, they will award an energy efficiency rating from A (most efficient) to G (least efficient) and a certificate containing information about the property's energy use and average energy costs. The certificate may also make recommendations about how to reduce energy use and save money on energy bills in the future.

EPC assessments are valid for ten years for domestic properties and are available to view on the Landmark Register in England and Wales, and the Scottish EPC Register in Scotland.

▲ Figure 3.4 Loft insulation

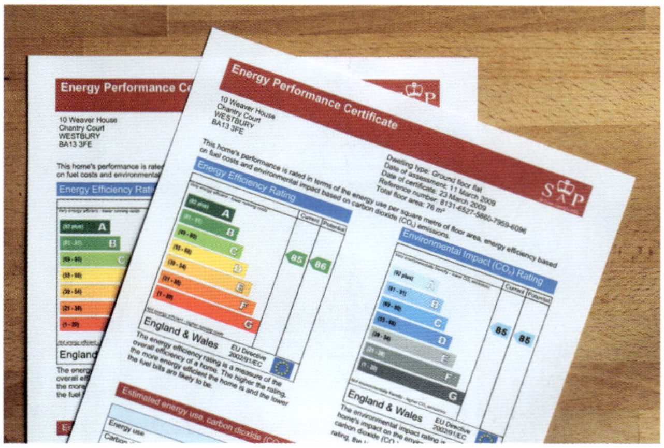

▲ Figure 3.5 Energy performance certificates

Case study

You decide to rent out your home. You do not have an EPC, so you appoint an EPC assessor to inspect your property.

- What energy efficiency rating do you think the assessor would award your property? Explain your answer.
- What recommendations do you think the assessor would make to improve the energy performance of your property and reduce your energy costs?

Code for Sustainable Homes

Launched in 2006, the Code for Sustainable Homes was a government-led environmental assessment method for rating and certifying the performance of new homes in the UK. It set national standards for the design and construction of new domestic properties, with the aim of improving sustainable building methods.

The government withdrew the code in 2015 and consolidated some of the benchmarks into existing building regulations. Although adherence to the code is no longer a condition for new planning proposals, it still endures where outline planning permission was granted before March 2015 or where there are existing contractual arrangements.

Site waste management plan (SWMP)

A site waste management plan (SWMP) sets out how waste will be managed and disposed of during a construction project.

The plan should be prepared by the client, along with the design team, contractor and subcontractor, and it should identify:

- the different types of waste that will be created throughout the project
- where the waste is expected to be found
- the minimum estimated quantities of each type of waste
- the actions that the client and principal contractor intend to take to minimise the amount of waste by reusing and recycling.

For more information on waste management and SWMPs, see Chapter 5, section 8.

▲ Figure 3.6 Construction site waste being recycled in skips

Building regulations (Part L)

Part L of the building regulations is concerned with the conservation of fuel and power in new buildings (Approved Document L1A) and existing buildings (Approved Document L1B). (There is further information on Approved Documents in section 2.1 of this chapter.)

To help the government meet its targets to reduce carbon emissions, the regulations identify measurable rates to determine the energy efficiency of a building.

- The calculated rate of CO_2 emissions from a dwelling (the Dwelling CO_2 Emission Rate, DER) must not exceed the Target CO_2 Emission Rate (TER).
- The Dwelling Fabric Energy Efficiency (DFEE) rate must not exceed the Target Fabric Energy Efficiency (TFEE) rate.

The regulations also provide technical guidance on the quality of materials and standards of work to achieve the TER and TFEE rates in areas such as secondary heating and lighting.

Test yourself

What part of a building is the superstructure?

1.2 The benefits of good design and the potential implications of poor design

A building that has been properly thought out and planned will benefit the end user in many ways. Prospective buyers often look for the following qualities in new homes.

- Aesthetics – the building should have 'kerb appeal' (it should be attractive when viewed from the road) and have modern amenities. It should be in keeping with its surroundings and add value to the community.
- Efficiency – the building should have a good energy efficiency rating, allowing it to be cost-effective to run.
- Sustainability – the use of sustainable materials and alternative energy sources could make the property more desirable.
- Wellbeing and improved quality of life – these can be achieved by maximising natural light and ensuring efficient ventilation within the building, as well as enabling access to outdoor living spaces such as gardens, terraces or balconies.
- Affordable quality – 'affordable' is considered a median household income or lower, determined by the local or national government. Careful selection of building materials, methods and processes without unforeseen delays will bring in the construction project on budget.
- Improved local/community facilities – this could include accessible parks, woodland, play areas, community outdoor gyms, sports clubs, leisure facilities and shops.
- Improved **infrastructure** – this could include footpaths, roads, parking, bus routes and pedestrian bridges.

▲ Figure 3.7 Newly constructed domestic houses for sale

Impact on the local community

A building must meet the needs of both the client financing the project (the 'project sponsor') and the local community, now and for future generations. A development that serves the purpose of the client but is not well designed or managed throughout the construction phase can cause disruption to the local community with noise, light and air pollution, not to mention increased construction traffic and on-street parking.

A poorly planned development could have a negative impact on both the local economy and the way the area functions, for example, consider the effect it might have on local infrastructure. On average, most of the new occupants will drive some form of vehicle that will increase the amount of traffic in the area, particularly at peak times. This may result in delays, wear and tear on the roads, increased pollution and limited parking. The higher population will increase demand on local schools, healthcare facilities, shops and the job market.

Key term

Infrastructure: the basic systems and services that a country or organisation needs in order to function properly, including transportation systems (roads, footpaths, railways, bridges, tunnels and airports), water supplies, sewers and telecommunications systems

▲ Figure 3.8 Infrastructure of a community

Building new houses creates short-term employment opportunities for the development team. However, when the building phase is completed, those houses will be sold and occupied by people with social and economic needs.

To ease the shortage of **affordable housing**, it is often a condition of planning approval that a percentage of all newly built homes are allocated to meet this need. The National Planning Policy Framework states that at least ten per cent of most new developments should be affordable housing.

Key term

Affordable housing: housing for sale or rent, for those whose needs are not met by the market (including housing that provides a subsidised route to home ownership and/or is for essential local workers); this means that low-income households can buy or rent properties at an affordable price, based on their earnings

Improve your maths

Research the average wages for five different trades in your area, then determine the annual median wage.

Impact on parties in the construction chain

A new housing development needs to enable a local community to function and grow, and to bring employment opportunities. A failure to meet these needs will be reflected in slow sales and reduced profitability.

Large housing developments are usually divided into smaller phases by the building management team. Each phase is completed and sold in turn, to finance further phases of the overall development. If the development and buildings have not been well designed and constructed in the right area at the right time, then slow sales at an early stage in the development can have a major impact on a number of different parties in the construction chain:

- The client may be unable to finance further phases or to make repayments to the project sponsor (for example, a bank, a shareholder or an investor). Broken repayment agreements could result in financial penalties for the client. A failed venture could damage the client's reputation and impact on the financing of future developments.
- The project sponsor may not finance any further building work or future developments.
- Construction work might be paused, leading to a reduction in size of the project team until sales pick up and further phases of construction work are started. This could cause difficulties when building recommences in terms of re-employing contractors and subcontractors at the right time, leading to further delays.
- Suppliers may not be paid on time by contractors and subcontractors, because they are no longer employed. This can lead to a loss of confidence and goodwill, making it difficult to find suppliers willing to provide further building materials or credit terms. Material costs could increase the longer the development is delayed.

Impact of poor design and poor quality of work

Poor design and poor quality of work can result in a building failing to perform as expected after transfer of ownership to the end user. This can result in further negative impacts on the customer and client, such as high energy bills and the fabric of the building degrading.

Materials that have been inappropriately specified, selected, stored and used during construction work can often lead to failures at a later stage. Building materials can fail when exposed to certain conditions, as shown in Table 3.1.

▼ Table 3.1 Common failures in building materials

Building material	Failure	Cause
Ferrous metal 	Corrosion (an electrochemical process)	Exposure to moisture and other chemicals
Timber 	Decay/collapse (when the timber contains no chlorophyll)	Fungal attack (dry/wet rot) due to moisture content above 20 per cent Woodworm
	Warps, splits and shakes	Poor storage (unsupported/exposed to the elements/poor seasoning)
	Burning	Unprotected timber positioned too close to a heat source
Brickwork and concrete 	Deterioration and cracking of concrete or mortar joints Splitting of bricks and blocks	Frost attack as a result of water expanding as it freezes in porous materials **Sulphate attack** Incorrect ratios of materials used in the mixing of mortar and concrete
	Efflorescence	Water containing salt in the materials migrating to the surface and drying
Paint 	Flaking/peeling	Ultraviolet light (sunlight) attacking the paint Inadequate paint preparation and poor choice of primers

Building material	Failure	Cause
Plastic 	Becoming brittle	Ultraviolet light (sunlight) attacking the plastic
	Expansion and contraction	Extreme temperatures (heat/cold)
	Discolouration	Low-grade plastic
Glass 	Misting	Failure of seals between double-/ triple-glazed units, allowing the thermal gases to escape and moisture to enter between the layers of glass

Key terms

Ferrous metals: metals that contain iron and are vulnerable to rust when exposed to moisture

Sulphate attack: a chemical process where sulphates cause the cement in concrete to deteriorate, often resulting in cracking

Test yourself

What are the benefits of good design and the implications of poor design?

1.3 Factors that can impact on the profitability of projects

At the planning stage of any development, a **feasibility study** is usually carried out by the client and the management team, to check that the project is both viable and potentially profitable. The study usually considers areas such as:

- finances (profits/losses)
- budgets
- resources (labour, plant and equipment)
- environmental impact
- scheduling
- appraisal of the proposed development site regarding existing services, waterways, boundaries and **topography**
- investigations into the history of the land above and below ground, identifying any unforeseen features that have the potential to make building work more complicated and costly, such as old wells, tunnels and mines
- historical architectural interest or any existing features, structures or buildings that could delay construction work or even prevent it from going ahead.

Key terms

Feasibility study: an investigation to establish the likely success of a proposed project

Topography: the physical features and shape of land surfaces

▲ Figure 3.9 The topography of a piece of land

Greenfield and brownfield sites

Reusing a brownfield site for a new development makes environmental sense. However, from a financial point of view it is more costly, because existing buildings need to be removed and the ground needs to be prepared (for example, by clearing contaminated soil and waste).

For ease of building purposes and to maximise profits, many developers would prefer to use a greenfield site. Building work can usually start quickly on these sites with less preparation, and contractors do not have to contend with any existing hazardous waste or unforeseen development issues. However, building on a greenfield site is controversial, as many people believe we have a moral responsibility to protect the environment. There may also be local planning restrictions in place to preserve greenfield sites and only allow new developments when certain criteria are met.

Over-specification and difficulty of assembly

Developments that are over-specified or over-engineered can result in lower profits, even before construction work has begun. Unique design ideas and complicated assembly of components often result in difficulties during manufacturing and construction, due to the workforce's lack of familiarity, knowledge and skills. These complications can lead to increased timescales, missed deadlines and stretched budgets.

Vernacular construction

The design of the houses in a development may be sympathetic or particular to a region, relying on locally sourced materials and traditional skills that have developed over generations. This method of region-specific house building is referred to as vernacular construction. Traditionally, construction methods did not follow contemporary mainstream architectural designs, until the Victorian period when building materials and styles became increasingly standardised.

Given the distinct regional characteristics of such buildings, it is problematic to design and specify this form of construction anywhere other than where it is usually found. It would be difficult to achieve the same standards of work without employing labour and sourcing materials from further away.

▲ Figure 3.10 This stone house in the Cotswolds, built with local materials, is an example of vernacular construction

Corporate social responsibility (CSR)

Although construction businesses need to make a profit, this should not be at the expense of the world around them. **Corporate social responsibility** (CSR) refers to strategies adopted by businesses to ensure they monitor and manage their social, economic and environmental impact on all aspects of society.

The CSR business model is self-regulated and can take many different forms, depending on the type of industry. It supports companies in being socially accountable to themselves, their stakeholders and the general public, in order to have a positive influence on the world. Businesses using this model not only benefit society but also themselves, by boosting morale between employers and employees and improving the reputation of their corporate brand.

Key term

Corporate social responsibility: the commitment of an organisation to carry out its business activities in a socially and environmentally responsible way

Typical examples of CSR in construction include:

- providing services for the community
- supporting charitable activities
- using ethically and sustainably sourced building materials
- reducing the environmental impact of construction work.

Research

Search online for the International Organization for Standardization's guidance document 'Discovering ISO 26000'.

Research what social and ethical responsibilities could be adopted by your work-placement employer to manage their impact on society and the environment.

Test yourself

What factors could have an impact on the profitability of a building project?

2 Design principles

2.1 Factors to be considered during the design of building services

A client must consider many factors as part of their **concept** for a building, and some of these are more obvious than others, for example:

- building capacity
- number of bedrooms and bathrooms
- internal arrangement of rooms and other spaces
- parking and garages
- aesthetics (design features, choice of materials and use of colour).

Key term

Concept: a principle or idea

One of the most important factors to consider when planning a building project is the availability of mains services, such as electricity, water and drainage. If all these services are not easily accessible at the site (for example, in rural locations), then it could be expensive to connect to them. This usually involves paying service providers to dig long trenches in the ground to run pipes and cables to the boundary of the development, which is not always feasible and may cause damage to the natural environment. Special permission may also need to be obtained if this involves working on someone else's land.

Even if the proposed new development is on a brownfield site, there is no guarantee that the services provided will be in good order, for example, older water pipes may have low pressure or the water supply may be contaminated if the pipes are made of lead or corroded steel.

In remote locations where mains water cannot be accessed, a borehole may have to be drilled into the ground to find a natural water source. This can be very expensive, because the depth of the hole and the ground conditions are relatively unknown until the drilling starts, and any water that is found will have to be filtered and purified before use.

A client may choose to have a self-contained site using alternative energy sources, such as photovoltaic panels for electricity and lighting, ground and air source heat pumps to supply heating and hot water, and storage tanks for sewage. The main benefit of this approach is that it does not rely on the use of finite fossil fuels to generate power, and although the initial installation costs can be quite high, these will be offset with reduced energy bills over a period of time. For more information on energy sources, see Chapter 5.

▲ Figure 3.11 Services being connected

Approved Documents

In many ways, the design of a building and the choice of construction method are influenced by local authority planning restrictions and building regulations. The Ministry of Housing, Communities and Local Government publishes guidance on how to comply with building regulations in the form of Approved Documents. These provide general advice on the performance expected of materials and building work, and practical solutions to some of the more common building situations.

Approved Documents are divided into the key areas listed below:

- Structure: Approved Document A
- Fire safety: Approved Document B
- Site preparation and resistance to contaminants and moisture: Approved Document C
- Toxic substances: Approved Document D
- Resistance to sound: Approved Document E
- Ventilation: Approved Document F
- Sanitation, hot water safety and water efficiency: Approved Document G
- Drainage and waste disposal: Approved Document H
- Combustion appliances and fuel storage systems: Approved Document J
- Protection from falling, collision and impact: Approved Document K
- Conservation of fuel and power: Approved Document L
- Access to and use of buildings: Approved Document M
- Electrical safety: Approved Document P
- Security in dwellings: Approved Document Q
- High-speed electronic communications networks: Approved Document R
- Infrastructure for charging electric vehicles: Approved document S
- Toilet accommodation: Approved document T
- Materials and workmanship: Approved Document 7.

Listed and heritage building regulations

Existing buildings may be protected by listed and heritage building regulations. If a building has special architectural or historical interest, not only will the structure of the building be protected, but also any other attached structures or features, including interiors.

Listed buildings are graded according to their national importance:

- Grade I (one) – buildings of exceptional interest
- Grade II* (two star) – particularly important buildings of more than special interest
- Grade II (two) – buildings of special interest warranting every effort to preserve them.

If a client wants to alter, extend or change the use of a listed building, they must apply to the local authority for 'listed building consent' and be granted permission before any work begins. The grade of listing usually determines what changes can be made and how they should be implemented, although any work will have to be carried out sensitively in order to protect the character and history of the building or structure.

▲ Figure 3.12 A listed building

Traditional versus modern building methods

It is important during the planning stage of any building project to consider carefully which construction method will be most appropriate.

Traditional methods of constructing walls, floors and roofs for **low-rise buildings** have evolved in recent years, in order to meet regulations and building standards on energy efficiency and sustainability. However, this type of construction can be labour-intensive and usually results in longer build times, because the entire superstructure is completed on site with low levels of **mechanisation**. During the initial stages of the construction work, before the building is watertight, the progress made is often weather dependent, for example, bricklayers cannot lay bricks at temperatures below 2°C or during periods of heavy rain. Traditional construction methods also tend to produce a lot of waste.

Key terms

Low-rise buildings: buildings with four storeys or fewer

Mechanisation: the use of machines or automatic devices

Table 3.2 compares some of the most common traditional and modern construction methods.

▼ Table 3.2 Comparison of traditional and modern construction methods

Method of construction	Traditional/ modern	Advantages	Disadvantages
Brick and block 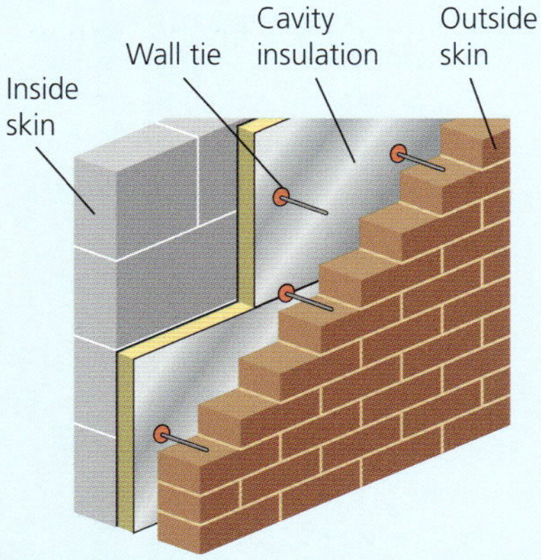	Traditional	• Familiar • Good thermal performance • Good sound insulation • Can be used with concrete upper floors • Good structural performance • Durable • Good fire resistance • Good weather exclusion	• Thick walls due to insulation between inside and outside **skin** • Takes time to dry out • Cannot be carried out in heavy rain or freezing conditions • Production of cement creates high levels of carbon, which is harmful for the environment
Open panel – timber frame 	Traditional	• Sustainable building material if the timber is from a managed forest • Factory-built framework reduces onsite build times • Good thermal performance and sound insulation (although this is reliant on the infill panels between the open frames)	• Can be more expensive than other building methods • Water can stain the exposed timber in the early stages of construction
Straw-bale construction (non-structural/infill system) 	Traditional	• Sustainable • Low cost • Quick to build • Good thermal performance • Good sound insulation • No drying out required	• At risk of damage from vermin • Increased risk of fire • Uneven wall-surface finishes • Straw rots easily if exposed to moisture
Panelised – timber frame 	Modern	• Manufactured off site in a factory, thereby reducing onsite build times, labour requirements/costs and health and safety risks (for example, working at height) • Quick to erect the shell of the building to make it **watertight** • Sustainable building material • Good thermal performance • Reduced waste (factory manufactured) • Accurate, easy to adjust and fix to • Thinner walls compared with other methods	• Unable to support concrete upper floors • Long **lead time** for panels to be made off site • Liable to rot if exposed to moisture

▼ Table 3.2 Comparison of traditional and modern construction methods

Method of construction	Traditional/ modern	Advantages	Disadvantages
Insulated concrete formwork (ICF) 	Modern	• Excellent thermal performance • Good sound insulation • Good weather exclusion • Durable • Excellent structural performance • Good fire resistance	• Inexperienced workers may need training to use the system • Poor preparation of the formwork (concrete mould) can lead to difficult and expensive repairs once the concrete has been poured and cured
Steel-frame construction 	Traditional	• Quick to erect • Excellent weather resistance • Durable • Excellent structural performance • Could support concrete upper floors • Good fire resistance	• Expensive • Heavy • Requires a crane to lift the steel frame into position • Can distort if exposed to extreme heat (for example, in a fire)
Thin-joint blockwork/masonry 	Modern	• No drying out of mortar joints needed • Quick to build • Excellent weather exclusion • Durable • Excellent structural performance • Fire resistant • Excellent thermal performance • Good sound insulation	• Slightly more expensive compared to traditional bricks and blocks • Accurate foundations needed to build on, because it is difficult to correct blockwork courses at a later stage as the walls are built
Structural insulated panels (SIPs) 	Modern	• Good thermal performance • Manufactured off site in a factory, thereby reducing onsite build times, labour requirements/costs and health and safety risks (for example, working at height)	• Precise foundations required to align the SIPs; any deviation could result in costly delays • Requires a crane to lift the panels into position
Volumetric (pod/modular) 	Modern	• Manufactured off site in a factory, therefore reducing onsite build times, labour requirements/costs and health and safety risks • Better quality control and accuracy maintained in a factory because of the level of automation used	• Initially more expensive • Requires a crane to move the pods into position, which can be expensive to acquire if not already on site

Key terms

Skin: a single thick masonry wall

Watertight: when a building is at the stage of construction where the weather cannot penetrate it (for example, the roof is on and the windows and doors have been fitted)

Lead time: the period of time between ordering and receiving goods or materials

Test yourself

What are the advantages of modern construction methods compared with traditional methods?

Buildability

Some modern construction methods involve parts or sections of a building being manufactured away from the site location. Large sections of walls and floors (referred to as cassettes), roof trusses and even whole rooms (known as pods) can be prefabricated more quickly in factories, with better quality control. Production is not delayed due to adverse weather conditions, and off-site factory manufacturing also reduces onsite build times, labour requirements/costs and some health and safety risks.

This method of combining factory-produced, pre-engineered units (or modules) to form major elements of a structure is referred to as modular construction. It results in improved '**buildability**', because potential assembly issues are usually designed out at an early stage for ease of construction onsite. However, the **feasibility** of using heavy lifting equipment and cranes on site to lift the large building sections into place, alongside the increased risks to the safety of workers and others, have to be factored into the planning of a project.

▲ Figure 3.13 A cassette being lifted into position on site

Key term

Buildability: the simplicity of manufacturing or constructing a building through its design

Choosing a method of construction

It can be difficult deciding whether to construct using traditional or modern building methods and materials.

Traditional construction generally uses less-sustainable technologies and materials that produce high levels of CO_2 during manufacture. However, the costs of maintaining modern methods are normally higher over the lifetime of the building. For more information on sustainable methods of construction, see Chapter 5.

Whichever method is chosen, it has to meet the minimum standards determined by building regulations. However, some buildings will naturally perform better than others with regards to heat retention and acoustics. The ability of a building to maintain an ambient temperature for its occupants is influenced not only by the U-value of the walls, floors and roof, but also by the construction of doors, windows and roof lights.

Key term

Feasibility: how easy or difficult it is to do something

Large windows, roof lanterns and bi-folding or sliding doors are popular features of modern homes. These allow natural daylight to illuminate large areas of the building, providing a feeling of more space and a connection between indoor and outdoor living.

The direction that a building faces on a plot and the carefully planned positioning of glazing will allow sunlight to warm a building through solar gain (the sun). This heat can also be stored in the fabric of the building during the day and released during the evening, which will reduce reliance on other sources of energy, decrease bills and maintain a comfortable living environment.

Windows and doors are important design features for access and egress (including for those with disabilities). They are also a means of escape in the event of a fire. In order to maintain the health and wellbeing of the occupants of a building, openings in windows and doors allow for ventilation, whereby stale air is removed from the building and replaced with natural fresh air.

▲ Figure 3.14 Natural light entering a building through a roof lantern

Improve your English

'Luminaire' is a word used to describe a source of artificial light. Write a paragraph to explain the use of different types of luminaire in a building, and explain why natural light is always a better source of energy.

Test yourself

The installation and use of windows and doors are controlled by building regulations. Which Approved Documents provide information and guidance on this?

Research

Search online for the 'Building Research Establishment Environmental Assessment Method' (BREEAM) and explain the role it plays in the design of the built environment.

Test yourself

What factors need to be considered during the design of building services?

2.2 Stages and outcomes of the Royal Institute of British Architects (RIBA) Plan of Work

The RIBA Plan of Work is a design and process management tool used by the UK building industry to bring greater clarity for the client at different stages of a project.

It organises the process of briefing, designing, constructing and operating building projects into eight stages. Each stage has intended outcomes, core tasks and information that should be exchanged with different parties.

In recent years, the Plan of Work has been updated to reflect changes in the industry and to support the government's target to be net zero carbon by 2050.

▼ Table 3.3 RIBA Plan of Work Stages and Stage Outcomes

	Stage	Outcome
Pre-design	0 Strategic definition	Determine the best way of achieving the client's requirements and the most appropriate solution.
	1 Preparation and briefing	Develop the client's concept and make sure it can be accommodated on site. Make sure everything needed for the next stage is in place.
Design	2 Concept design	Make sure the look and feel of the building is meeting the client's expectations and budget.
	3 Spatial co-ordination	Design the spaces within the structure of the building, before preparing detailed information about manufacturing and construction.
	4 Technical design	Develop information received from the design team and specialist subcontractors for the manufacture and construction of the building.
Construction	5 Manufacturing and construction	Manufacture and construct the building.
Handover	6 Handover	Complete the building works and address any defects that have been identified, to conclude the building contract between the client and the contractor.
In use	7 Use	The building should be used, operated and maintained efficiently until the end of its life. At this stage, the client may consider appointing professionals for aftercare activities such as servicing and maintenance.

Source: RIBA Plan of Work 2020 Stages and Stage Outcomes, reproduced courtesy of the Royal Institute of British Architects

Improve your English

The term 'spatial' is used in the RIBA Plan of Work to describe the space, position, area or size of things within a building. Write a sentence that includes the word 'spatial' to describe the arrangement of rooms on a level in a dwelling.

3 Design process from conception to completion

3.1 Key stages of the design process, and factors that may impact or influence design changes

House building can be a relatively straightforward process when good design principles are applied, the project has been well planned and there are no unforeseen problems encountered during the construction phase. However, when you are dealing with the natural environment no two sites are ever the same. It is therefore essential that the land is properly investigated to identify any potential issues early and reduce the risk of additional costs later in the build.

Individuals or organisations responsible for preparing or modifying designs for construction projects have legal duties under the Construction (Design and Management) Regulations 2015.

Designers must:

- ensure the client is aware of their duties under the CDM Regulations and help them to comply with those duties
- take into account any preconstruction information provided by the client
- eliminate foreseeable health and safety risks where possible
- seek to reduce or control any health and safety risk that cannot be eliminated
- provide design information for inclusion in the health and safety file for the project
- co-ordinate with other designers working on the project on matters of health and safety during the construction phase and beyond
- co-ordinate, communicate and co-operate with all contractors working on the project, taking into account their knowledge and experience of building design.

Further information on the CDM Regulations can be found in Chapter 1.

In this section, we will look at the importance of researching and analysing building plots, as well as the process of applying for planning permission, through to the final sign-off of the building work.

Test yourself

What legal duties do designers have under the Construction (Design and Management) Regulations 2015?

Research

Before construction work begins, a **desktop survey** is usually undertaken to identify and record details about previous and current uses of the site. The main objective of this survey is to support the design and construction processes, by allowing correct decisions to be made when planning the site layout.

Key term

Desktop survey: an investigation into a piece of land completed without visiting the site or taking physical samples of soil etc.

A desktop survey will look to establish the following information:

- site history
- waste records
- geology and **hydrology**
- contamination reports
- site boundaries
- position of existing services
- existing structures or buildings
- local roads
- access to the site
- topography of the land
- hedges, trees and fences
- wildlife and habitats.

Much of this information can be gathered from existing plans and records held by the Environment Agency and the Land Registry.

If a specialist or consultant is appointed, they will also produce an **environmental risk assessment**. This document is used to identify potential risks to the natural environment from the proposed building work.

A site of special scientific interest (SSSI) may be identified in the initial survey. This is an area that must be protected from construction activities due to its rare **flora** or **fauna**, or its physiographical or geological features.

▲ Figure 3.15 Some sites must be protected due to their features of special interest, such as their rare flora

▲ Figure 3.16 Some sites can become SSSIs due to their fauna

Key terms

Hydrology: the study of water in the earth and its relationship with the environment

Environmental risk assessment: a document used to identify and control potential risks to the natural environment from construction work

Flora: plants and trees in a particular region

Fauna: animals or insects in a particular region

Case study

A feasibility study for the construction of 75 new homes has identified the existence of fauna on the proposed building site.

Explain in detail what animals are protected and the potential outcomes now that these indigenous species have been discovered on the land.

Site analysis

Once the desktop survey has been completed, a **walkover survey** is usually undertaken. This is a physical inspection of the building site to identify any geological, ecological or topographical issues that may impact the project.

It involves drilling boreholes into the ground at various positions on the site to investigate the:

- composition of the soil
- load-bearing capacity of the ground
- position of the water table.

The water table is the point below the ground where the soil becomes saturated; this depth will vary depending on the location of the site.

The information gathered from the walkover survey is interpreted and used to determine the most appropriate form of foundation to suit the ground conditions.

Key term

Walkover survey: a physical inspection of a building site

▲ Figure 3.17 The water table

Research

Search online for the various methods of soil investigation and explain why they are used.

Planning

Under permitted development rights, you can extend or make certain alterations to an existing domestic property without the need for planning permission from the local authority. However, if the proposed changes are beyond the specific limitations of the rules, or new dwellings are to be constructed, then planning permission must be sought and granted.

Industry tip

Regardless of whether or not planning permission is needed, all building work must be completed in accordance with current building regulations.

The process of making an application for planning approval through to the final decision made by the planning committee can be lengthy and expensive, especially when you consider the investment in time, effort and money in designing the project.

To improve the efficiency of the application process and reduce any potential financial losses, a pre-planning application should be made to the local planning department. This type of application is usually processed much more quickly than a full application, where detailed drawings and specifications are often needed. Local planning departments encourage pre-applications so that they can offer support and guidance to resolve any issues in the proposal, improving the chance of the final application being successful.

If planning permission is required for a project, an application must be made to the local authority by the client or planning team, accompanied by a fee and supporting documentation. The design plans for the site will identify the boundary, **frontage line** and **building line**. These may be stipulated by the local planning department as a condition of planning approval.

At this stage, details of the planning application are advertised by the planning department in local newspapers and at the site of the proposal. Consultation letters are sent to neighbours, informing them about the application and advising how they can view the plans online or raise any valid objections to the project going ahead within a period of 21 days from the date of publishing.

Research

Find out what documents and types of drawing are needed to make a full planning application to a local planning department. Suggest the possible outcome if some of this detail is not provided.

Key terms

Frontage line: the front part of a building that faces a road

Building line: a boundary line set by the local authority beyond which building work must not project

▲ Figure 3.18 A planning application notice

During the consultation period, a planning officer from the local planning department will visit the site, where they will take into account information provided in the application and any responses from the public or their representatives. They may also gather further site-specific information, such as measurements and photographs, in order to make their 'officer's report'. The planning officer is not responsible for making a planning decision at this stage, but will make recommendations for the authorised person or planning committee to base their final decision on.

The final outcome for the majority of domestic planning applications is decided by senior officers, under delegated powers from the planning committee.

There are three possible outcomes for a planning application:

- approved
- approved with conditions that must be complied with (for example, 'No further windows can be added to the proposal on the east elevation of the property')
- refused.

Applications that have been refused can be appealed, however this can be a lengthy and expensive process that is not always successful.

Construction work can be carried out on domestic buildings by submitting a 'building notice' to the local planning department, without the need to submit a full planning application. Although work can start immediately once a building notice has been submitted, lack of detail in the notice could result in the work not complying with building regulations and therefore having to be dismantled or corrected after it has been completed. It is therefore recommended that a building notice is used only for minor building works, or if the contractor has good knowledge of current building regulations.

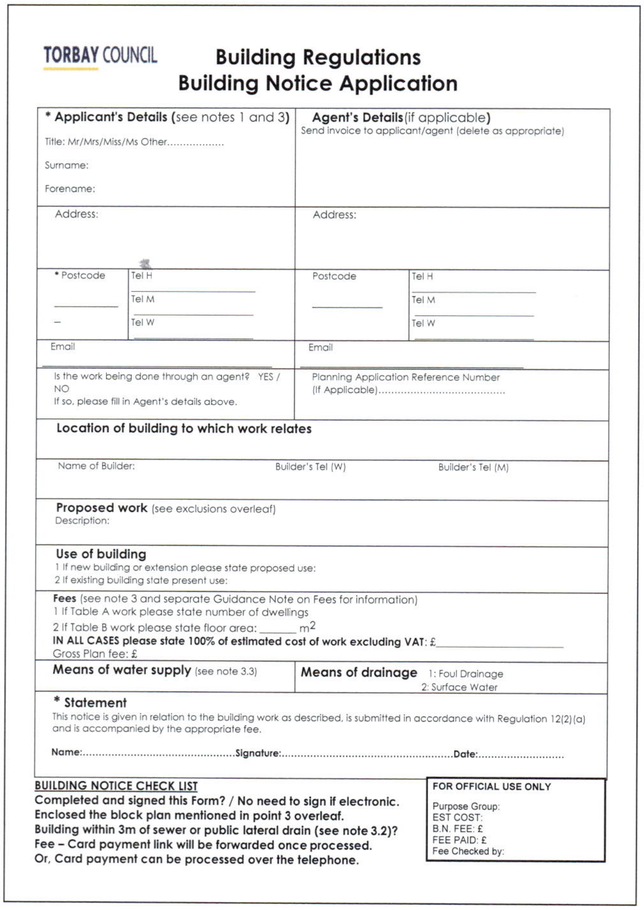
TORBAY COUNCIL

Building Regulations
Building Notice Application

* Applicant's Details (see notes 1 and 3)	Agent's Details (if applicable)
Title: Mr/Mrs/Miss/Ms Other................. Surname: Forename:	Send invoice to applicant/agent (delete as appropriate)
Address:	Address:
* Postcode — Tel H Tel M Tel W	Postcode Tel H Tel M Tel W
Email	Email
Is the work being done through an agent? YES / NO If so, please fill in Agent's details above.	Planning Application Reference Number (If Applicable)..

Location of building to which work relates

Name of Builder: Builder's Tel (W) Builder's Tel (M)

Proposed work (see exclusions overleaf)
Description:

Use of building
1 If new building or extension please state proposed use:
2 If existing building state present use:

Fees (see note 3 and separate Guidance Note on Fees for information)
1 If Table A work please state number of dwellings
2 If Table B work please state floor area: _______ m^2
IN ALL CASES please state 100% of estimated cost of work excluding VAT: £_______________
Gross Plan fee: £

Means of water supply (see note 3.3)	**Means of drainage** 1: Foul Drainage 2: Surface Water

*** Statement**
This notice is given in relation to the building work as described, is submitted in accordance with Regulation 12(2)(a) and is accompanied by the appropriate fee.

Name:...**Signature:**...**Date:**..........................

BUILDING NOTICE CHECK LIST
Completed and signed this Form? / No need to sign if electronic.
Enclosed the block plan mentioned in point 3 overleaf.
Building within 3m of sewer or public lateral drain (see note 3.2)?
Fee – Card payment link will be forwarded once processed.
Or, Card payment can be processed over the telephone.

FOR OFFICIAL USE ONLY
Purpose Group:
EST COST:
B.N. FEE: £
FEE PAID: £
Fee Checked by:

▲ Figure 3.19 A building notice application

Research

Search online to find out how to submit a building notice to your local planning department. Download and complete (without submitting it!) a building notice application for a fictitious extension to your home.

Approval/review

Once building work has commenced on site, it must be inspected at regular intervals by either a building control officer (BCO) from Local Authority Building Control (LABC) or a private building inspector.

The quality and standards of work are checked against building regulations at the following stages:

- excavation of the foundation
- laying of foundation concrete
- installation of damp-proof course (DPC) and damp-proof membranes (DPM)
- laying of drains
- completion of the roof structure
- completion of first-fix installations (before plastering or dry lining)
- testing of drains
- completion of the project.

Where the standard of work falls below that expected, recommendations are made by the inspector to meet legislation.

Building regulations state that any person intending to carry out building work must notify LABC to determine stages of work and when they will be inspected. This is known as an 'inspection service plan'.

Project sign-off

As soon as possible after a building project has finished, the contractor or client must notify LABC, so that it can arrange a final visit to the site. During the inspection, the BCO will check that any outstanding actions from previous visits have been completed and verify that the dwelling meets with building regulations. Once satisfied with the building work, they will issue a completion certificate. Until this certificate is issued, a building is not '**signed-off**' and therefore should not be occupied, and it may be difficult to insure it or secure a mortgage.

Key term

Signed-off: approved by a building control officer

3.2 Project planning

Part of successful planning for a construction project involves scheduling resources, materials and labour for various times throughout the building phase. If this is not given careful consideration at the planning stage, it could result in delays on site, missed completion deadlines and financial penalties for the contractor.

It is inevitable that certain events during construction work will impact on progress, for example, poor weather, equipment failure or accidents. However, these can be factored into a **programme of work** as and when they happen, so that adjustments can be made to reduce the impact further down the line.

▲ Figure 3.20 A site manager monitoring progress against the programme of work

Key term

Programme of work: a document used by construction managers to plan and organise resources for a building project

Construction scheduling software is often used to prepare programmes of work as part of BIM, because it produces documents that are clear, simple to amend and easy to share with all project stakeholders.

The most commonly used programmes of work are:

- Gantt charts
- critical path analysis (CPA).

Gantt charts

Gantt charts are a type of bar chart. They are used to record the project start and completion times, and the sequence in which construction activities are scheduled to take place in between. Different coloured references are used to:

- highlight planned activity durations
- plot the current status of the project
- flag any amendments that need to be made to complete the project on time and within budget.

	Task	Start date	End date	Planned duration (days)
1	Layout and preparation	02/03	06/03	5
2	Excavation	09/03	17/03	9
3	Reinforcement and formwork	16/03	27/03	12
4	Foundation construction	23/03	27/03	5
5	Structural steel	30/03	18/04	20
6	Masonry	14/04	01/05	18
7	Plumbing	21/04	25/04	5
8	Electrical	21/04	23/04	3
9	HVAC	21/04	25/04	5
10	Roofing	28/04	08/05	11
11	Plastering	08/05	12/05	5
12	Carpentry	11/05	22/05	12
13	Installation of windows and doors	18/05	22/05	5
14	Terrazzo	18/05	29/05	12
15	Glazing	01/06	12/06	12
16	Hardware	15/06	19/06	5
17	Painting	15/06	23/06	9
18	Exterior concrete	15/06	25/06	11

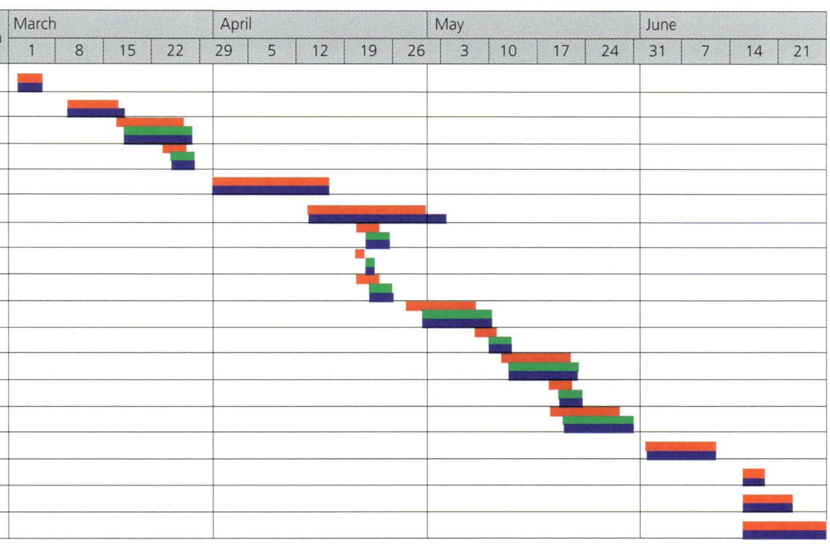

Key:
Red = planned activity durations
Blue = current status of the project
Green = amendments that need to be made to complete the project on time and within budget

▲ Figure 3.21 A Gantt chart

Critical path analysis (CPA)

Critical path analysis (CPA) is a decision-making tool used to plan complex building projects. The order and expected duration of activities are plotted using a networking diagram, connected by a series of **node points** containing critical information.

CPA is used by the project management team to identify when each activity can start and opportunities to relocate resources to improve efficiency; these are referred to as 'float times'.

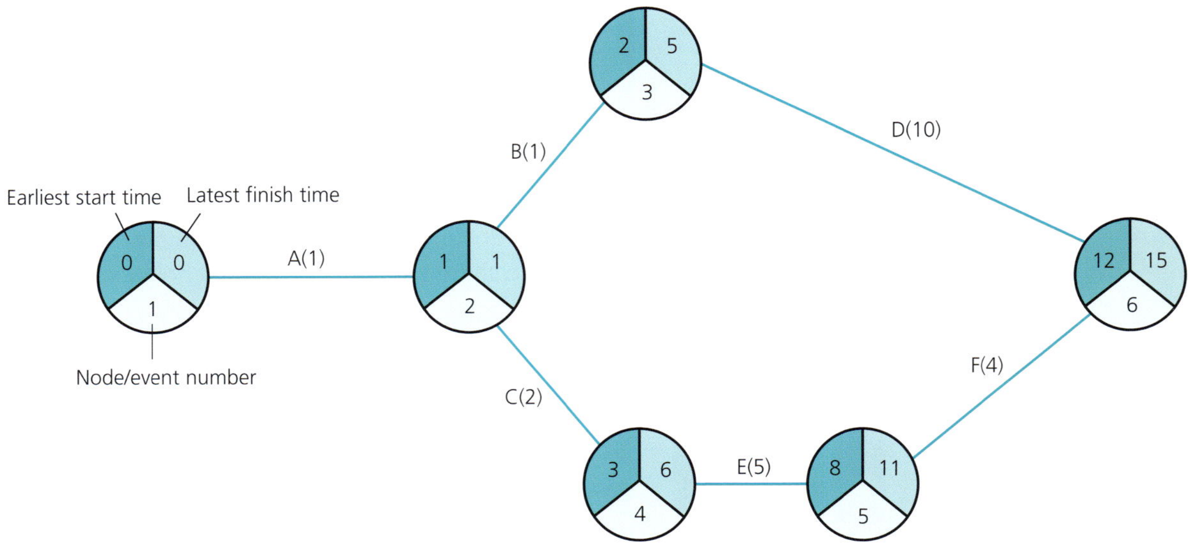

▲ Figure 3.22 Critical path analysis (CPA)

> **Key term**
>
> ***Node point:*** an intersection of lines or pathways in a diagram

> **Improve your maths**
>
> Search on YouTube for a tutorial on how to create a critical path analysis (CPA). Work with one of your peers to design a programme of work on a computer for a simple construction activity.

4 The concept of the 'whole building', including life cycle assessment

4.1 Life cycle assessment

Throughout this chapter, we have looked at how design and construction are influenced by different factors, such as legislation, and at the impact building works might have on the natural environment if measures are not taken to protect it. In order to design and construct '**green buildings**', we need to consider the '**whole building**' and understand how different construction systems work together and how this can determine project planning.

The term 'sustainable home' is sometimes broadly used when designing or planning; however, not all construction materials perform equally with regards to their carbon footprint. There is increasing demand among stakeholders for more sustainable and energy-efficient buildings, in order to reduce the effects on climate change. It is therefore important to understand the life cycle of construction materials, from their creation to their final use/disposal.

The impact that construction materials have on the environment can be calculated using a science-based tool known as 'life cycle assessment' (LCA). LCA accurately evaluates the effect that materials have at each stage of their life cycle, using data from Building Information Modelling (BIM) and other sources of information, and produces a report.

> **Key terms**
>
> ***Green buildings:*** buildings that have a low impact on the environment, during both their construction and use
>
> ***Whole building:*** the impact of a building in terms of resources and effect on the natural environment, from the initial sourcing of raw building materials to manufacturing and construction etc.

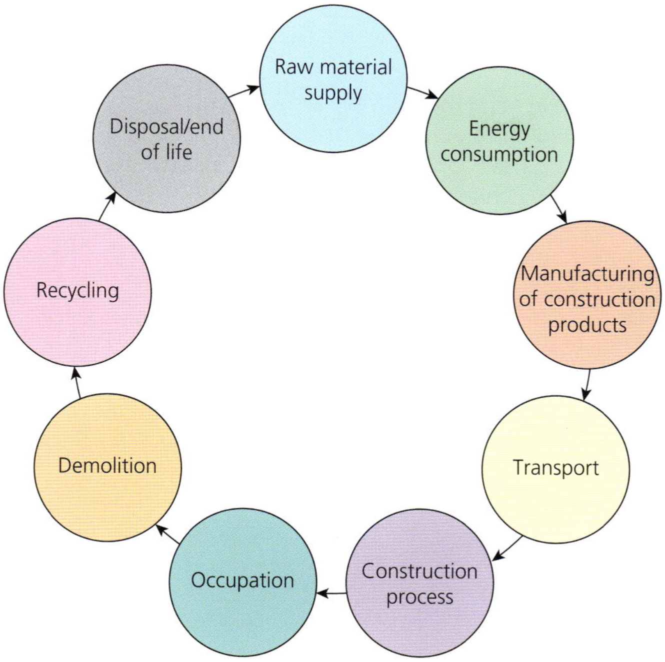

▲ Figure 3.23 Life cycle of building materials

LCA is a reliable source of data, controlled by international standards. Its findings identify areas in the building design that have the biggest impact on the environment. These areas can then be targeted by designers to alter the design or the materials used, to reduce the environmental impact.

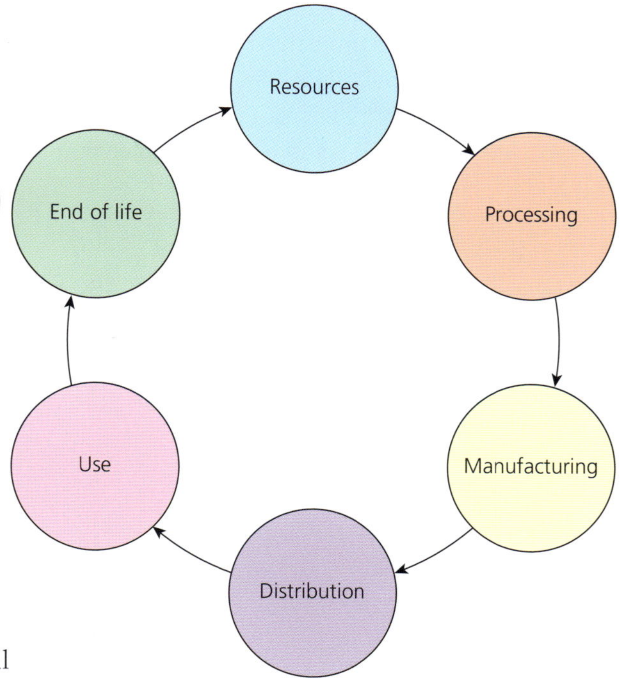

▲ Figure 3.24 Life cycle assessment (LCA)

Assessment practice

Short answer

1. Which Approved Document influences the design of a building's energy sources?
2. Under what rights are contractors and individuals allowed to carry out certain construction projects without planning permission?
3. Identify three stages of building work that are usually inspected by a building control officer.
4. What is the name of the design and process management tool used to bring greater clarity for the client at different stages of a building project?
5. Name one modern building method where components are constructed off site.

Long answer

6. Explain the difference between brownfield and greenfield sites.
7. Explain what is meant by vernacular construction and why it can impact on the profitability of projects.
8. Explain the role of a BIM designer.
9. What factors could influence changes to a building design?
10. List the steps for obtaining planning permission from the local planning department.

Project practice

A principal contractor has been appointed by a client for a 'design and build' project. A rural greenfield site has already been acquired by the client, but it does not have planning permission.

An application has been made to the local planning department, with plans for six new 3- and 4-bedroom low-rise dwellings. However, the plans have been initially refused by Local Authority Building Control.

Discuss in a group the potential grounds for LABC to oppose the planning application.

Bearing in mind the possible reasons you have identified for rejecting the planning application, prepare a new application to address each of the issues. To achieve this, you may need to:

- research construction materials to ascertain their properties and suitability
- consider sustainable construction solutions
- research corporate social responsibility towards the community.

Chapter 4 Construction and the built environment industry

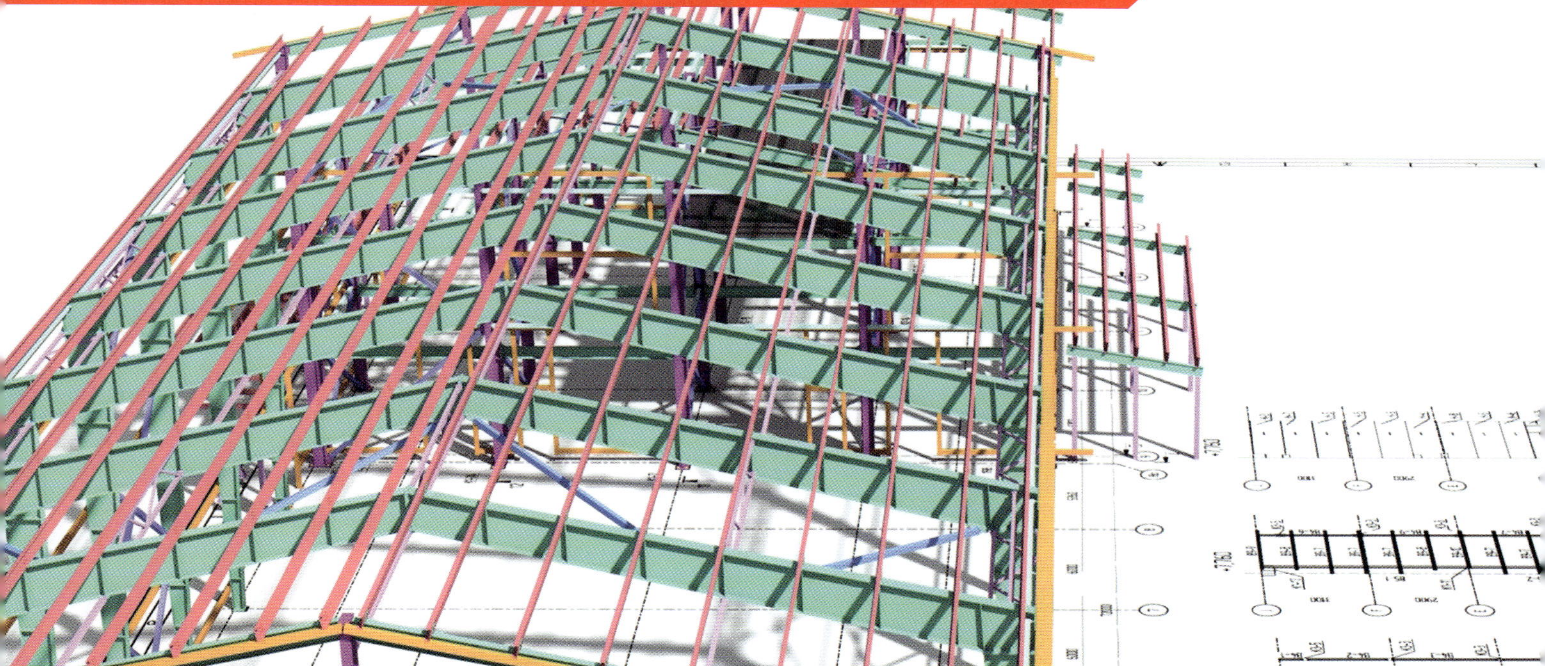

In this chapter, we will look at how the construction industry is structured and its contribution to the UK economy. We will identify the wide range of professionals and operatives who collaborate on various types of residential, commercial and industrial projects and examine the role of continuing professional development (CPD) in keeping their skills current.

We will then analyse the benefits of Building Information Modelling (BIM) for creating and managing information on a construction project, and look at the different types of documentation used.

We will also examine internal and external factors that may influence current and future building projects.

The chapter concludes by looking at handover procedures.

Learning outcomes

By the end of this chapter, you will understand:

1. the structure of the construction industry
2. how the construction industry serves the economy as a whole
3. integration of the supply chain through partnering and collaborative practices
4. procurement of projects within the construction sector
5. managing change requests from various parties
6. roles and responsibilities of construction professionals and operatives
7. the role of continuing professional development (CPD) in developing the knowledge and skills of those working in the sector
8. Building Information Modelling (BIM)
9. PESTLE factors
10. documentation used in construction projects
11. procedures for handing over projects to clients.

1 Structure of the construction industry

1.1 The construction industry

Construction is one of the biggest employment sectors in the UK, employing around 3.1 million people, with some of the largest contractors generating annual turnovers of several billion pounds. In 2018, the construction industry contributed £117 billion to the UK economy.

Construction work is broadly divided into three main categories:

- residential building (housing)
- non-residential building (for example universities)
- infrastructure (**civil engineering**).

▲ Figure 4.1 Residential building

▲ Figure 4.2 Non-residential building

▲ Figure 4.3 Infrastructure

Work in the construction industry involves anything from building and maintaining homes, schools, offices, hospitals and factories to civil engineering projects, such as the creation of roads, bridges and airports. Demand for the construction sector is increasing, with the UK government expected to spend around £600 billion between 2021 and 2031, including £44 billion on residential housing.

Key term

Civil engineering: a profession involving the design, construction and maintenance of infrastructure that supports human activities, for example roads, bridges, airports and railways

Research

Define the following types of client and outline some examples of construction work they might be responsible for:

- government
- public limited company
- commercial
- private.

Improve your maths

Research the number of people working in each trade in the construction industry and produce a graph to illustrate this data. Analyse the information in your graph, and suggest reasons why some trade areas may be underrepresented.

Test yourself

What are the three main categories that construction work is divided into?

1.2 Business types

The size and scale of a project often determine which type of organisation is involved. For example, established and experienced contractors with proven track records of similar projects are usually on the preferred suppliers list for tenders. Private clients, who have a smaller amount of money to invest, may appoint subcontractors to complete their building work based on recommendations from previous customers/clients.

Outlined below are examples of different business types that may be involved in the construction sector. Further information on business types can be found in Chapter 11, section 1.1 (page 283).

Sole traders

A sole trader runs their own business as an individual and is **self-employed**. They are responsible for keeping financial business records of their sales and expenses. They also have to file a self-assessment tax return with HMRC each year, declare any profits and pay any tax due. Sole traders are personally **liable** for any debts or losses incurred by the business.

If annual turnover (sales) exceeds £85,000, the company must become **VAT** (value added tax) registered with HMRC. It must then charge VAT on all taxable sales it makes to its customers. As of August 2021, the VAT rate is 20 per cent, but this is subject to change.

Key terms

Self-employed: the state of working for oneself rather than an employer; a self-employed person is responsible for paying their own tax and National Insurance contributions on any earnings

Liable: legally responsible

VAT: value added tax – a sales duty that businesses have to charge their customers and pay to HMRC if their annual turnover exceeds £85,000

Improve your maths

If the VAT rate is 20 per cent, calculate the amount of VAT that a contractor would have to pay HMRC if they had a turnover of £137,000.

Contractors

Contractors work for a limited company registered with **Companies House**. Although they are responsible for the affairs and day-to-day running of the company, they have no personal liability for any of its financial losses if things go wrong. Limited companies are given more credibility than other types of business, and their owners also have greater control of their business with tax benefits.

There are several disadvantages of managing a limited company, including the number of **shareholders** cannot exceed 50. Setting up a limited company also involves a slightly more complicated process than setting up as a sole trader. To become a limited company, it must first have a suitable business name before paying an administration fee and registering with Companies House.

Key terms

Companies House: a government body that registers and stores information on all the limited companies in the UK and makes it available to the public

Shareholders: investors in a company who have purchased at least one share of a company's stock

Research

Visit **www.gov.uk/topic/company-registration-filing/starting-company** and find out the process for registering a business as a limited company.

Subcontractors

Subcontractors are self-employed tradespeople who are hired by principal contractors to undertake specific work on building projects.

Self-employed workers are responsible for every part of running a business, for example estimating, invoicing, ordering and accounts, although they may appoint an accountant to manage some of their financial affairs. They do not receive some of the benefits that employed people enjoy, for example:

- holiday pay (including bank holidays)
- sick pay
- maternity or paternity leave or pay.

In order to cover the potential losses listed above, subcontractors have to charge their clients more. They also have to consider the days or even weeks that they may be absent through sickness. During these unforeseen periods, the business will cease trading and earning money while the self-employed person recovers.

Small, medium-sized and large organisations

An organisation can be privately owned, a partnership or a **corporation**. Its size can be measured in terms of the number of employees, annual **turnover**, profits or assets.

According to Companies House guidance:

- a small organisation employs on average no more than 50 people and has an annual turnover of £10.2 million or less
- a medium-sized organisation employs on average no more than 250 people and has an annual turnover of £36 million or less
- a large organisation employs at least 250 people and has an annual turnover in excess of £50 million.

Small construction businesses tend to have fewer resources and **overheads** than larger ones, therefore they often recruit local labour to work within a particular area or region of the country. The bigger an organisation becomes, the more resources it has at its disposal and the further it branches out nationally or internationally to compete for lucrative government or commercial construction contracts.

The **hierarchical structure** of personnel within medium-sized and large organisations is much more complex than in smaller businesses. This often causes delays, with information being processed at each level before being filtered down through the organisation. Smaller organisations have a simpler structure, resulting in decisions being made more easily. It is important to establish a hierarchy in an organisation, so that everyone understands lines of communication and areas of responsibility.

Key terms

Corporation: a business owned by its shareholders, who often appoint a board of directors to manage the day-to-day running of its activities; the business is a legal entity and the shareholders have no personal liability for its actions and finances

Turnover: the amount of money that a business or individual has been paid by their clients for services they have provided

Overheads: regular repeated costs associated with the day-to-day running of a business, for example rent and insurances

Hierarchical structure: the arrangement of people and the positions they hold in an organisation according to their level of authority; there is a chain of command from those in the most senior positions to those in subordinate positions

Test yourself

Explain the term 'turnover' in regard to business finance.

1.3 The range of work undertaken in the construction industry

The impact of the construction industry is everywhere you look – from homes to public buildings to infrastructure. While structures may be diverse in their design, method of construction and intended purpose, they all illustrate just how much we depend on the construction industry to live, work and travel.

Construction work usually falls into one of a number of different categories, depending on the client and purpose. Table 4.1 outlines the wide range of work undertaken in the construction industry.

▼ Table 4.1 Range of work undertaken in the construction industry

Category of construction work	Examples of buildings/structures
Commercial	Offices, private hospitals, hotels and restaurants
Residential	Private and affordable housing, flats, apartment blocks, retirement villages and student accommodation
Industrial	Large buildings designed to manufacture goods, for example factories, agricultural buildings, warehouses and power plants
Health	Health centres, hospitals, clinics, treatment centres and GP surgeries
Retail	High-street shops, supermarkets, superstores, shopping centres, retail parks and distribution centres
Recreational	Stadiums, swimming pools, gymnasiums, golf and tennis facilities
Leisure	Public libraries, amusement parks, fitness centres, spas, cinemas and community centres
Utilities	Building services systems, for example water, electricity, gas, telecommunications and sewerage
Transport	Roads, railways, train stations, airports, bridges, footpaths and tunnels
New build	New housing developments
Retrofit	Modernisation or adaption of existing buildings to make them more energy efficient and sustainable

▲ Figure 4.4 Industrial construction

2 How the construction industry serves the economy as a whole

Earlier in this chapter, we highlighted the importance of the construction industry and how the wealth generated from construction developments contributes to the UK economy.

The industry is showing no signs of slowing down in terms of growth, which creates well-paid job opportunities for skilled workers. Where there are skills gaps or shortages of labour, local and national government often invests heavily in further and higher education, training and skills to meet the demand; however, this can be influenced by the government in power at the time and its priorities.

When people are employed and have realistic opportunities to progress their careers, they become financially stable and often spend or borrow money to invest in assets such as housing and cars. These investments create more demand for different types of buildings such as housing, transport and better infrastructure, leisure facilities, education establishments and hospitals, which in turn leads to greater employment and continues the cycle. Improvements of this nature in neglected urban areas or areas previously used for industry often lead to the regeneration of whole areas, which increases the value of property, creates employment and reduces crime in the local community.

The UK government is determined to continue investment in new technologies in the construction sector, to meet the national housing shortage and the need for more affordable homes. It has also made a commitment to tackle climate change and the negative impact that construction has on the environment.

▲ Figure 4.5 Gas hob using a finite fossil fuel

▲ Figure 4.6 Installing solar panels can improve energy efficiency

▲ Figure 4.7 Electric vehicle charging point

The UK government passed the Climate Change Act in November 2008, setting a target to reduce **carbon** emissions to net zero (carbon neutral) by 2050. To meet this target, the government proposed spending caps and total greenhouse gas emission targets for five-year periods, which should not be exceeded. So far, the country has met each target; however, if it were to miss the carbon budget for an agreed period, it could be challenged in a court of law.

Changes are being introduced in the construction industry to improve efficiencies and address the need for a cleaner environment, including:

- increased off-site manufacturing to reduce waste and speed up production
- installing smart technologies in homes
- using sustainable building materials
- creating energy-efficient buildings
- installing electric vehicle charging points
- installing alternative heating systems, for example ground and air source heat pumps
- **retrofitting** existing buildings to improve energy efficiency
- preparing infrastructure for the impact of climate change, for example flood defences
- adapting existing infrastructure to changes in technologies to protect the environment.

Key terms

Carbon: a chemical element that can be released into the atmosphere when fossil fuels are burned

Retrofitting: the process of adding new features and technologies to existing buildings

Test yourself

What are the objectives of the Climate Change Act 2008?

3 Integration of the supply chain through partnering and collaborative practices

Table 4.2 shows the integration and collaboration of partners in the supply chain for a building project.

▼ Table 4.2 Integration and collaboration of partners in the supply chain (continued)

Position	Role		Responsibilities
Customer	Client (individual or organisation)		• Develops the initial concept for a project • Appoints a design, management and building team • Finances the project • Complies with legal duties, for example the Construction (Design and Management) (CDM) Regulations 2015
Client's representative	Clerk of works (CoW)		Checks drawings and specifications, to ensure work is completed to the standards agreed in the building contract
Designers – planning team	Architect		• Interprets the client's design brief to produce detailed drawings and specifications • Complies with CDM duties • Leads the project team at the design stage • Oversees the construction phase • Makes the client aware of their legal responsibilities
	Surveyor	Quantity	• Studies building information and drawings/BIM to prepare tender packages • Controls budgets and costs • Provides professional advice to the client
		Land	• Plots, measures and gathers data on land for construction and civil-engineering projects • Advises the client on planning and construction
		Building	• Undertakes onsite property surveys • Identifies defects and makes recommendations for remedial work • Completes reports and advises the client on legal, planning or environmental issues
	Engineer	Structural	Collaborates with the rest of the design team to prepare structural calculations and designs for construction and civil-engineering projects
		Building services	Designs, plans and supervises the installation of building services, such as heating, water and electricity
		Civil	Designs, plans and supervises the construction and maintenance of public infrastructure, such as roads, railways, bridges and tunnels
	CAD operative/ draughtsperson		• Creates and modifies technical drawings using computer-assisted design (CAD) • Visits sites to co-ordinate with architects, engineers and building services teams
	BIM manager/ technician		• Creates management and planning documentation • Creates 2D and 3D BIM models
Building regulators	Local Authority Building Control (LABC)		• Consults and offers advice on planning applications • Grants or refuses planning applications
	Building control officer (BCO)		• Inspects building work against planning permission granted and building regulations • Monitors health and safety on construction sites • Signs off building work against building regulations
	Health and Safety Executive (HSE)		• Enforces the Health and Safety at Work etc. Act 1974 (HASAWA) in high-risk work environments • Carries out inspections, investigations and enforcement action where necessary
	Fire and Rescue Authority		• Has the same powers as the HSE to enforce HASAWA • Responsible for fire-safety measures in work environments
	Local Authority Environmental Health Department		Enforces HASAWA in low-risk work environments

Position	Role	Responsibilities
Contractor – construction management team	Managing director	• As the most senior person in the company, oversees the direction of the company • Meets with clients • Manages budgets, targets, business growth, personnel and performance • Makes sure the organisation's legal and contractual obligations are met
	Contracts/project manager	• Negotiates/completes tenders • Prepares contracts • Plans and co-ordinates projects • Prepares reports • Liaises with the client • Manages costs and timescales
	Site manager	• Manages progress and budgets on construction projects • Supervises workers and subcontractors on site • Manages health, safety and welfare • Collaborates with the project team and client to ensure work is completed to standard and on schedule • Conducts site inductions
	Assistant site manager	• Helps the site manager to fulfil their responsibilities on site • Monitors progress and supervises workers • Co-ordinates operations and the procurement of materials and other resources • Enforces the contractor's health and safety site rules
	Health and safety advisor	• Advises the contractor on health and safety matters • Prepares risk assessments and method statements • Monitors health and safety performance • Audits and investigates accidents
	Estimator	• Researches costs and sources materials, labour and equipment for contracts • Assists the contracts manager with tender bids • Supports buying activities
	Buyer	• Works closely with the company estimator • Procures resources for construction projects • Maintains accurate records of spending to maximise efficiencies
	Trade foreperson	• Supervises a small team of tradespeople • Liaises with the site manager on topics such as progress, materials and safety
	Fire marshal	• Carries out fire risk assessments • Makes sure everyone leaves the building safely in an emergency via the escape routes • Checks alarms, fire-fighting equipment and safety signage • Performs safety checks • Makes sure workers and visitors are aware of what to do in an emergency on site
	First aider	• Administers first-aid treatment • Looks after first-aid provisions and equipment, for example the first-aid box and eye-wash stations • Reports accidents
Operatives/ subcontractors	Tradesperson (for example carpenter, bricklayer, gas fitter)	• Carries out their specific trade role on site • Produces risk assessments and method statements • Follows health and safety legislation and site rules • Attends toolbox talks • Reports unsafe conditions to the site manager
	General operative	• Assists tradespeople with moving and handling building materials • Carries out general housekeeping on site

▼ Table 4.2 Integration and collaboration of partners in the supply chain (continued)

Position	Role	Responsibilities
	Scaffolder	• Installs, maintains and inspects scaffolding systems to industry standards • Makes sure they do not put themselves and others at risk of harm due to their work activities
	Plant operator	• Inspects and operates heavy plant/equipment on site, for example telehandlers and dumpers • Follows safe working practices
Manufacturers	Manufacturer of building materials, resources and products for suppliers	Carries out manufacturing routinely through the supply chain or makes items to order, for example windows and doors
Suppliers	Supplier of standard building materials, equipment, machinery and tools	• Provides goods to contractors, subcontractors and the general public from a shop, warehouse or yard • Enters into credit agreements with contractors, with special rates and payment terms

▲ Figure 4.8 Land surveyor

▲ Figure 4.9 Building surveyor

If a project team does not collaborate effectively with stakeholders at the planning stage of a project, it often results in disruption during the construction phase, increased building costs and damage to the reputation of the contractor.

Test yourself

Explain the role of a clerk of works.

▲ Figure 4.10 Clerk of works

4 Procurement of projects within the construction sector

4.1 Need/demand

Construction work often requires huge financial investment by clients at each stage of a project, with an element of risk involved depending on the type of work undertaken. A client that has obtained a building plot without planning permission on it and drawn up proposals with an architect may risk the concept being refused by Local Authority Building Control. However, if planning permission has been granted and there is a demand for the project, it has every chance of being successful.

Before investing in construction work, it is important to establish the need for the project, in order to mitigate risk and avoid potential financial loss.

If the proposed project is speculative, with a view to selling or renting the building/s, the client must try to forecast these sales and potential income. It is not uncommon for building work on housing developments to stop if homes are not selling as well as expected, until sales pick up.

If the work is a government contract for a new school or hospital, then it already has a purpose and does not have to be sold upon completion of the building work.

4.2 Tendering and bidding processes

Once the design for a project has been approved by Local Authority Building Control, the process of finding a suitable contractor begins. There are several methods the client can use to establish the right contractor for a project, however the following factors need to be considered when deciding on the most suitable procurement method:

- financing and budgets
- time
- quality
- risk
- project limitations.

The client may already know which contractor they would like to use for their project, based on recommendations from others or having previously appointed them on another project. However, while the client may have total confidence in the contractor's ability to complete the job to the standards expected, they will not have any comparable quotes from other contractors, and therefore may not be getting the best value for money.

Open tendering

The client may invite offers for the contract from anyone, based on the information provided. This is referred to as open **tendering**.

The advantage of open tendering is that the process is fair and without bias. However, the client may have to choose a contractor they are not familiar with or one that may not have experience of completing similar projects. The client may have no alternative but to appoint a contractor based on their quotation and references from previous clients.

> **Key term**
>
> ***Tendering:*** the process of inviting bids from contractors to carry out specific projects

Two-stage tender

If the client would like to secure the services of a contractor early, without having all the relevant design information in place, they can appoint them using a two-stage tender. This involves an agreement with the contractor for work that they are able to price for based on the information provided, so that they are able to start work. At the second stage, when all the design information has been supplied, the contractor is able to negotiate a fixed price for completion of the contract.

Preferred supplier (selective)

Clients with previous experience working on similar projects may have developed successful relationships with a range of contractors. Some of these contractors may be included on the client's list of preferred suppliers. From this list, the client may select a number of contractors and invite them to tender for a contract, although the contractor is not obliged to submit a bid.

Negotiated tendering

Where a client has invited just one contractor to price for a project, this is referred to as negotiated tendering. Before building contracts are signed, the client usually negotiates the best price for the work.

> **Test yourself**
>
> List the different methods of tendering, and explain why they may be used by a client.

> **Industry tip**
>
> Selecting the contractor that has provided the cheapest tender is not always the best option. Clients should also consider:
>
> - the size of the contractor's company
> - the contractor's experience of completing similar projects
> - references from the contractor's previous clients
> - the contractor's health and safety records, including possible HSE enforcement actions.

4.3 Tender documentation

The client or client's representative is responsible for gathering all the documentation needed for the tender package. The same tender package is usually sent to three or four contractors, depending on the tendering process used, so that they can calculate their costs fairly based on exactly the same information, before returning it to the client with their final estimate.

The information contained in a tender package will be specific to each project. However, it may contain some of the following documents:

- Letter inviting the contractor to submit a bid for the work
- Outline of the proposal
- Form of tender and timeline to return the completed bid
- Form of contract and conditions (including the process for payments and interim valuations)
- Programme of work
- Design drawings
- Specifications
- Site-specific information or issues
- **Preliminaries**
- Special planning-permission requirements
- **Bill of quantities** (cost framework)
- Tender return document.

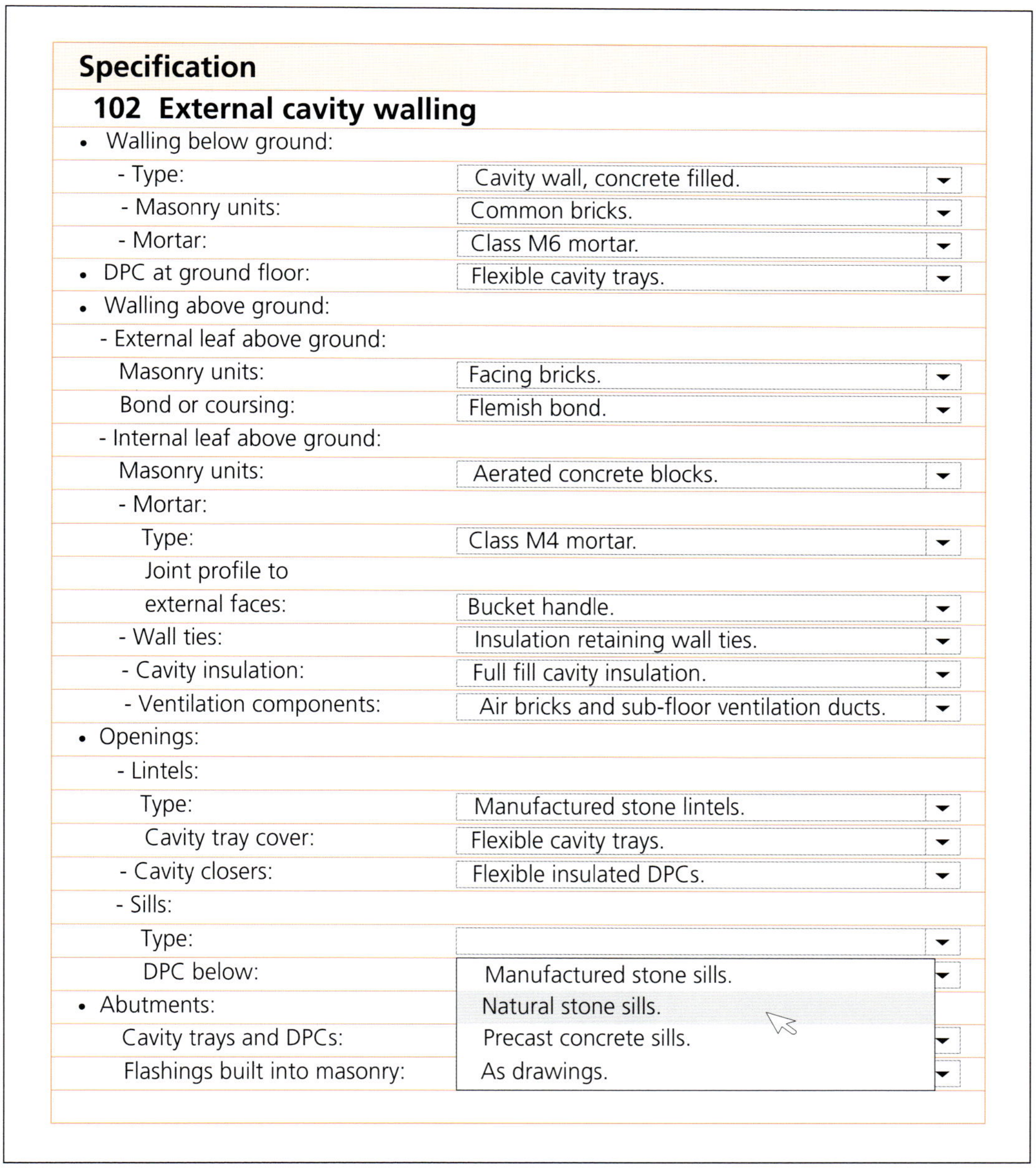

Specification

102 External cavity walling

• Walling below ground:	
- Type:	Cavity wall, concrete filled.
- Masonry units:	Common bricks.
- Mortar:	Class M6 mortar.
• DPC at ground floor:	Flexible cavity trays.
• Walling above ground:	
- External leaf above ground:	
Masonry units:	Facing bricks.
Bond or coursing:	Flemish bond.
- Internal leaf above ground:	
Masonry units:	Aerated concrete blocks.
- Mortar:	
Type:	Class M4 mortar.
Joint profile to external faces:	Bucket handle.
- Wall ties:	Insulation retaining wall ties.
- Cavity insulation:	Full fill cavity insulation.
- Ventilation components:	Air bricks and sub-floor ventilation ducts.
• Openings:	
- Lintels:	
Type:	Manufactured stone lintels.
Cavity tray cover:	Flexible cavity trays.
- Cavity closers:	Flexible insulated DPCs.
- Sills:	
Type:	
DPC below:	Manufactured stone sills.
• Abutments:	Natural stone sills.
Cavity trays and DPCs:	Precast concrete sills.
Flashings built into masonry:	As drawings.

▲ Figure 4.11 Specification

Item No.	Description	Unit	Rate	Amount
1	Setting out works	Lump sum		
2	Temporary works e.g. Welfare facilities, site office and canteen	Lump sum		
3	Scaffolding	Lump sum		
4	Power, heating and lighting	Lump sum		
5	Clear construction site on completion	Lump sum		

▲ Figure 4.12 Preliminaries

Key terms

Preliminaries: pre-construction information that outlines items that are necessary for a contractor to complete the works but are not actually part of the works, for example general plant, welfare facilities and site security

Bill of quantities: a document that breaks down a construction project into an itemised list of work to be carried out, including quality, quantities and costs

Improve your English

Draft a formal letter inviting a contractor to submit a tender bid for a construction project.

The contractor is often asked to provide information about their business and previous work they have completed. This will include proof of employer's insurance and public liability insurance, information about the key personnel that will represent their business on site and references from previous clients.

Research

The most common procurement routes for construction projects are listed below. Research each type and explain why it may be selected by the client:

- Contractor led
- Design and build
- Fast track
- Lump sum
- Single stage
- Two stage.

5 Managing change requests from various parties

It is inevitable that principal design ideas will change as a project develops, until they meet the needs of everyone involved in or affected by the proposal.

When the architect or client makes changes to a design at an early stage in a development, the implications are usually minor because there are few people involved and the development is still at the drawing-board stage. With each amendment to the design, the architect will add references on the plans and other documentation, so that the client is aware of the most up-to-date versions.

The more professionals joining the design team and collaborating on a project, the greater the need for the principal designer to keep everyone aware of developments, especially if collaborative technology is not being used to communicate efficiently or effectively.

Later we will look at the benefits of investing in collaborative working between all stakeholders using technology such as Building Information Modelling (BIM), and how changes in the design can be quickly shared.

As soon as a project has been given approval and a contractor has been appointed, a building contract will be signed between both parties, stating the terms and conditions of the agreement. Once the legal document has been signed and building work has started, changes are often much more difficult for everyone involved. For example, if the client wanted to include an additional window in a design, it could have the following implications:

- A meeting would need to happen with the client and project team to discuss the proposal.
- An architect would need to be instructed to amend the design.
- Building regulations approval may be needed.
- Subcontractors may be unable to complete the walls, floors and roof.
- The contractor may have to reschedule the subcontractors for a later date, and risk them not being available.

- If approval is granted, then the additional window will have to be ordered and made.
- An additional lintel will be required to support the wall above the window.
- Building materials will be wasted.

All of the implications listed above will cause delays, additional building costs and missed completion dates, which could trigger a financial penalty clause in the building contract.

If minor amendments can be agreed between the client, architect and contractor without impacting on the schedule or agreed costs, then the architect will issue a written document known as the 'architect's instructions' to formalise the changes.

If the client or architect makes changes that compromise the terms and conditions of the building contract, then the costs and schedule may have to be renegotiated and agreed in writing before the work is completed. In this situation, an accurate record of the changes and agreement will be made in a document known as a change order, also referred to as a variation order.

Test yourself

Explain the process and documentation used to make changes to a building contract once the construction phase has started on site.

6 Roles and responsibilities of construction professionals and operatives

6.1 Construction professionals

A construction professional is someone who is trained and qualified in their chosen field. Training may involve many years of study in order to gain a recognised qualification. They may also be a member of a professional organisation or body.

For example, an architect will have completed a degree in architecture and be a member of the Architects Registration Board (ARB) or Royal Institute of British Architects (RIBA). The importance of professional membership is explained later in this chapter.

Some of the key professional job roles in the construction industry are outlined below.

Architects

When a client has a concept for a construction project, they usually appoint a professional architect, also known as a designer, to discuss their ideas, needs and budget. A client's project could be a new housing development, an extension or alterations to an existing building. It could also involve the **restoration** or conservation of older buildings of historical interest.

Architects initially translate information provided by clients and surveys into 2D drawings and 3D digital models, using **computer-aided design (CAD)**. While planning a project, they have a responsibility to protect the environment through their designs, choice of materials and construction methods.

Initial designs often evolve and include further building documentation, until the client is satisfied that the proposal is ready to be submitted for planning approval if necessary. Design information provided by an architect also informs the BIM process shared with the project team. For example, information from the architect might allow structural engineers to calculate loads and stresses on different parts of a building to determine the size of structural components such as steel beams and foundations.

An architect could be an individual or organisation; however, where there is more than one architect working on a single project, the client must appoint a principal designer under the Construction (Design and Management) Regulations 2015. The role of the principal designer is to lead the project by planning, managing, monitoring and co-ordinating health and safety during the pre-construction phase.

While there is no requirement to use an architect, their expert knowledge of the construction industry, building regulations and materials makes the design and build process much easier for the client. The client could even extend the services of an architect beyond the design stages to advise them during the building phase and oversee the whole construction project.

Key terms

Restoration: the process of returning a building to its original condition

Computer-aided design (CAD): using computer software to develop designs for buildings and structures

▲ Figure 4.13 Architect

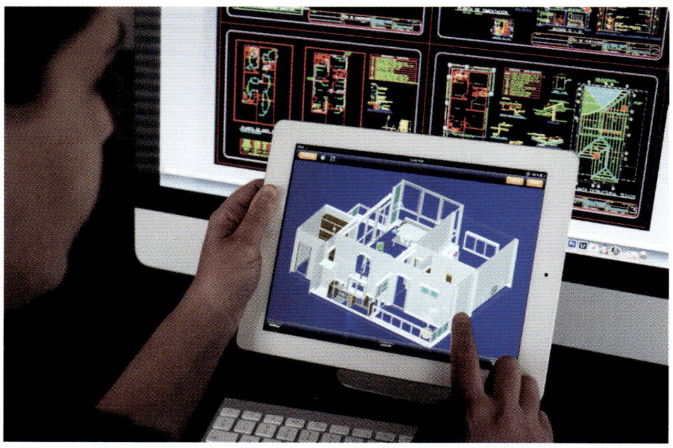

▲ Figure 4.14 Computer-aided design (CAD)

Test yourself

Explain the difference between computer-aided design (CAD) and Building Information Modelling (BIM).

Contractors

Contractors may be given responsibility by the client to design, plan, organise and control a construction project; this is commonly referred to as 'design and build'. Under the Construction (Design and Management) Regulations 2015, a principal contractor must be appointed when there is more than one contractor working on a project. The principal contractor then either appoints designers, or uses one of their own in-house designers, to manage this process as part of their project team.

The main benefit of this approach is that the client has fewer points of contact to communicate with at each phase of the development.

Many contractors start out as tradespeople to gain experience of the construction industry, before they progress into supervisory or principal contractor roles to manage building projects for clients.

Operatives

Operatives support different tradespeople at every stage of a building project by completing manual labour tasks, such as stacking and storing materials, mixing mortar and tidying the site. The availability of skilled labourers for a construction project and their experience of working with a particular building method may influence the design.

Skilled general operatives may progress into a trade, and with further experience into supervisory roles such as trade foreperson (there is further information on this role in Chapter 4).

Civil engineers

Civil engineers are also classified as designers, although they manage the design, construction and maintenance of infrastructure rather than buildings. Typically, they are involved in projects such as:

- roads
- bridges
- tunnels
- airports
- harbours
- railways
- water and sewage systems
- power plants.

▲ Figure 4.15 Civil engineering

There are many different subdisciplines in civil engineering. People usually specialise in one of the following fields:

- transportation engineering
- structural engineering
- environmental engineering
- **geotechnical engineering**
- hydraulic engineering
- construction engineering.

As civil engineers are responsible for the design and implementation of infrastructure in construction projects, their collaboration with other professionals in the design and construction team is vital to the success of any project.

Key term

Geotechnical engineering: a field of civil engineering that deals with the behaviour of earth materials such as soil and rock

▲ Figure 4.16 Geotechnical civil engineering

Planners

The role of planners is to ensure that land in villages, towns, cities, the countryside and commercial sites is used effectively to meet economic, social and environmental needs. Their objective is to achieve a balance between encouraging innovation and growth in the development of housing, industry, agriculture, recreation and transport while trying to preserve the historical environment.

Planners can also work as consultants for clients and provide advice on local planning policy to architects and developers. They can consult with stakeholders to determine planning applications, contest appeals and enforce planning legislation.

Experienced planners may leave the local planning department to become independent, advising contractors and clients on planning and design to meet the local authority's requirements with regards to planning applications.

Building control officers/private building inspectors

Building control officers (BCOs) work for **Local Authority Building Control** (LABC). Private building inspectors work for government-approved building-inspection companies. Private building inspectors do not have the same powers of enforcement as BCOs, which means that their role may have to be transferred to LABC if there is a problem that cannot be resolved without its intervention.

Key term

Local Authority Building Control: local authority department responsible for inspecting building work against building regulations and signing-off completed projects

The role of the BCO is to inspect plans submitted for full planning approval, making sure they meet current building regulations. At intervals agreed between the BCO and the contractor, they will visit the construction site and check that it meets with regulatory standards of design and safety.

The progression route for a BCO might involve becoming a member of the local authority planning committee.

▲ Figure 4.17 A building control officer (BCO) on a construction site

Industry tip

Never deviate from approved working drawings during the construction phase without written consent from either Local Authority Building Control or a government-approved private building inspector. Changes to the design without permission can be expensive to put right if the work fails to meet building regulations approval.

Manufacturers

Manufacturers are responsible for the design, production and sale of building materials and goods. Before their products are sold and distributed to the end user, they are tested and certified to ensure they adhere to industry standards and regulations (for example, the Ecodesign for Energy-Related Products Regulations 2010).

Designers may choose certain materials based on their sustainability, eco-design and energy rating, which means that manufacturers can influence the design and use of buildings through their products.

Mechanical building services (design) engineers

Mechanical building services (design) engineers consult and advise clients on the design, installation, operation and maintenance of building services such as:

- heating systems
- ventilation systems
- air-conditioning systems
- renewable-energy systems
- sustainable technologies.

Their technical knowledge of building services may also be utilised by the client for the production of detailed drawings, specifications and calculations to meet with design standards and regulatory requirements. They may be involved in bids and tenders for work, and mentor the client from the start to the completion of a project.

Mechanical building services engineer

A mechanical building services engineer designs, installs, and maintains the mechanical systems that make buildings safe, comfortable and efficient. Their work is essential to the functionality of residential, commercial, and industrial buildings.

Electrical building services (design) engineers

Electrical building services (design) engineers consult and advise clients on the design, installation, operation and maintenance of electrical systems for domestic, commercial and industrial projects, such as:

- lighting systems
- fire-safety systems
- security systems
- renewable-energy systems, for example, solar photovoltaic (PV) panels and electric-car charging points.

Mechanical design engineers (building services)

Mechanical design engineers use their specialist knowledge of complex mechanical systems to design, manage and supervise projects for the client, from concept to completion.

They often work closely with electrical engineers to ensure the design and installation work is carried out safely and in accordance with industry codes of practice.

Mechanical engineer design co-ordinators

Mechanical engineer design co-ordinators are usually appointed to assist the mechanical design engineers on larger projects, using their technical expertise to create innovative solutions to building-services problems. They may also organise tendering, project management, reporting and scheduling for the client when required.

Mechanical engineer CAD technicians

CAD technicians produce technical information and 2D/3D mechanical diagrams using computer-aided design (CAD). They create manufacturing drawings and work instructions for the building services team. This information is then used to improve the buildability of the project and reduce construction costs for the client.

BIM designers

BIM designers work with clients to implement **Building Information Modelling** (BIM). BIM is an intelligent, streamlined process of sharing **digital** information (for example, 3D models, drawings, specifications and schedules) between all parties and stakeholders. This cost-effective process allows a collaborative approach at every stage of planning, designing, building and managing a project, to improve efficiency and productivity throughout the life cycle of the building.

Key terms

Building Information Modelling (BIM): the use of digital technology to share construction documentation and provide a platform for collaboration

Digital: in electronic form

Retrofit: the process of adding new components to older structures

Retrofit assessors

Retrofit assessors assess of existing buildings in order to compile energy efficiency improvement plans for clients. They may also visit sites to:

- resolve issues (for example, maximising the use of PV panels on a building with a complex roof shape)
- review projects
- carry out audits to ensure health and safety and performance standards are being met.

Retrofit co-ordinators

Retrofit co-ordinators liaise with retrofit assessors, retrofit installers and clients about the installation of renewable-energy technologies in existing buildings, for example, solar thermal, biomass and air/ground source heat pumps.

An important part of their role is to gather data from the retrofit assessors, then check its accuracy and produce an improvement plan for the client. The client may also make arrangements for a co-ordinator to appoint an approved installer to carry out the work identified in the plan.

Retrofit installers

Retrofit installers (also known as retrofit installer technicians) are responsible for the installation, commissioning and final handover to the client of renewable energy systems and measures. Installers must be PAS 2030:2019 certified and meet these standards on all retrofit projects.

Competent and experienced retrofit installers may progress in their careers to become retrofit assessors.

Research

Search online for 'PAS 2030:2019 Specification for the installation of energy efficiency measures in existing dwellings and insulation in residential park homes'.

Identify the standards of PAS 2030:2019 for retrofit installers and explain how they benefit the construction industry.

▲ Figure 4.18 Retrofit installer fitting an energy-efficient boiler

Building services engineer technicians

Building services engineer technicians are responsible for:

- assisting the project team with design solutions, specifications and planning for building services engineering systems (for example water and drainage, lighting and power)
- supervising specialist contractors during installations
- taking the lead on health and safety
- **commissioning** systems while minimising the impact construction work has on the environment
- monitoring quality control, recording progress and reporting to the project team during installations.

Key term

Commissioning: the process of ensuring that a building system is performing or working as it has been designed

Building services engineering site managers

Building services engineering site managers oversee the installation of complex environmental systems in construction projects, such as heating, lighting and electrical power. During the construction phase, they work closely with the project team to make sure the design information is understood, in order to maintain quality and productivity during installation on site.

▲ Figure 4.19 Building services engineer

Facilities managers

Facilities managers take responsibility for the operation, servicing and maintenance of building services once the building work has been completed, signed-off and handed over to the client. When contractors are needed for work on the systems, the facilities manager will prepare tender documents, schedule repairs and monitor work under their control to minimise disruption and protect people from work activities.

Client representatives

It is not always possible or practical for the client to be on site all of the time during the construction phase of a project, and their technical knowledge about the building industry may be limited. Consequently, either the principal designer or the client will appoint a clerk of works (CoW).

The CoW acts as the client's representative on site to oversee the quality and safety of work on the project. In the interests of the client, they work closely with the construction staff, surveyors and engineers to make sure plans and specifications are followed properly. In doing so, the CoW will refer to working drawings, building regulations and health and safety legislation. Where standards have not been achieved in terms of the quality of work, materials or safety, the CoW will report to the site manager and make suggestions for improvement.

As progress is made throughout the construction phase, the CoW will keep the client informed with accurate reports/records from the site, at intervals agreed at the start of the project.

Contract managers

Contract managers are employed by contractors to assist in preparing tenders for clients and securing future business. When tender bids have been accepted by a client, a legal contract is drawn up by the contract manager. The terms and conditions of the contract are negotiated with the client and other stakeholders with regards to budget, the service that will be provided and project timescales. The contract manager is usually the main point of contact for the client and the site/project managers for the duration of the project.

Once construction work has started on site, the contract manager will monitor progress against the agreed schedule and technical standards by attending regular site meetings with the management team. Where unexpected costs arise or the terms of the contract change during the project, the contract manager will have to resolve these issues as quickly and effectively as possible.

Site managers

Site managers, also known as site supervisors, are responsible for organising work on construction sites to ensure it is completed safely, on time and within the client's budget. They are usually based in a temporary site office and will remain on site throughout the construction phase of the project to organise labour, equipment and materials for each stage of the build.

Site managers:

- supervise workers, subcontractors and visitors, for example delivery drivers
- monitor the quality of work and the progress made against the programme of work
- take responsibility for the health, safety and welfare of workers and others that may be affected by work activities; this involves making arrangements to protect people from harm and ensuring workers are suitably trained and informed about risks and control measures, usually through a site induction.

Health and safety standards are closely monitored by the site manager against risk assessments, method statements and health and safety legislation, and where necessary they will take action to maintain or improve safety on site.

Throughout the construction phase, the site manager will liaise with other professionals, such as the quantity surveyor to discuss budgets and the architect or surveyors regarding design issues or amendments. They may also have to deal with accidents, near misses or other emergency situations.

While key information on progress and budgets can quickly be communicated through the project team using collaborative methods, regular site/progress meetings are often chaired by the site manager.

▲ Figure 4.20 Site manager

▲ Figure 4.21 Site induction

6.2a Construction operatives

Most construction operatives are experienced and skilled workers that specialise in a particular area of the building industry. The level of qualifications and registrations needed to become a tradesperson are often determined by the occupation, however the most widely recognised are National Vocational Qualifications (NVQs, SNVQs in Scotland) and Apprenticeship Standards.

In addition to these qualifications, construction operatives may also need a current CSCS card to demonstrate to their employer that they have the minimum level of health and safety knowledge to work safely on site; the CSCS card scheme is covered in more detail in Chapter 1.

Carpenters and joiners

Carpenters and joiners both work with wood:

- Joiners are usually based in a workshop, where they manufacture purpose-made items to order, for example doors, windows, frames and staircases, using a range of machinery, power tools and hand tools.
- Carpenters usually work on construction sites. **First-fix** carpentry includes the installation of floors, roofs, walls and stairs, and **second-fix** carpentry involves the fitting of skirting, architraves, doors and kitchens.

Carpenters and joiners both work from drawings and specifications provided by the design team; however, they may need to gather further information from the management team during a site visit before they begin any manufacturing off site.

▲ Figure 4.22 Carpenter

▲ Figure 4.23 Joiner

▲ Figure 4.24 Gypsum plaster

Key terms

First fix: a phase of construction work completed before plastering

Second fix: a phase of construction work completed after plastering, for example installing kitchens, doors and tiling

Plasterers

Plasterers are tradespeople used by contractors to apply smooth and textured finishes to internal walls and ceilings, using a range of materials such as **gypsum**, cement or lime. Highly skilled plasterers also have the ability to carry out repairs or cast ornamental mouldings in plaster to match existing period features in a property. Plasterers can also apply smooth or textured finishes to external walls, using lime, cement or flexible modern materials such as polymer or acrylic renders.

Plasterers can usually start to apply finishes to the internal walls and ceilings as soon as tradespeople have completed the installation of their respective first-fix items. Where work is needed externally, this is often weather dependent, unless the work is under cover or protected, therefore plasterers will often work closely with the site manager to schedule this type of work when the weather permits.

Key term

Gypsum: a natural mineral, often used in building products such as plaster and plasterboard

Tilers

Tilers may be subcontracted by the principal contractor to lay decorative and protective wall and floor tiles. Tiles can be ceramic, clay, marble, slate or glass and are available in a range of different sizes, shapes, designs and textures.

A tiler will often calculate, cut and lay tiles in kitchens and bathrooms, as areas most exposed to high levels of moisture. A tiler could also be appointed to work outdoors to lay tiles on patios, terraces or swimming pools. These tasks will be completed in the final stages of a project, to protect the areas from damage from other building work.

Bricklayers

Bricklayers set out, build and repair walls, piers and archways for domestic and commercial projects. They work from plans and design specifications to calculate costs for clients and quantities of materials needed.

Bricklayers can build single as well as double **cavity walls** using a range of different bricks, blocks or stone, for internal or external projects. Their job is very physical and often involves working at height using a range of access equipment; therefore, it is essential that they have a strong understanding of health and safety practices.

The stages at which bricklayers are scheduled into a project are determined by the build method chosen. However, the majority of their work is often structural, which would result in them being on site once the foundations have been poured.

Key term

Cavity walls: external, load-bearing, structural walls consisting of two individual leaves (skins) of masonry with a gap (cavity) between them

▲ Figure 4.25 Bricklayer

6.2b BSE operatives

Plumbers

Plumbers install, maintain and repair water, heating and drainage systems in new and existing buildings. This includes cutting, shaping and fixing pipes and fixtures for the installation, testing and commissioning of the following:

- hot and cold running water
- heating systems
- baths
- sinks
- toilets
- showers
- dishwashers and washing machines.

During the first-fix stage of work, plumbers prepare for the installation of services on site by fitting pipework. In domestic projects, pipework is run through floors, walls and ceiling voids where possible, so that it can be hidden from view. On other projects, such as industrial or commercial, the specification may state that pipework has to be located in surface-mounted ducting, so that it can easily be maintained or adapted without causing damage.

At the second-fix stage, plumbers must work closely with other trades to co-ordinate the installation of fixtures and fittings, to protect them from damage and to avoid complications for work that follows.

Electricians

Electricians are responsible for designing, installing, servicing and repairing electrical systems in new and existing buildings. At the first-fix stage, electricians interpret wiring diagrams and the client's specification to run the wiring system throughout the building to power lighting, sockets and alarm systems. Later in the project they will be scheduled by the project manager to return to site to install/connect the second-fix electrical fixtures and fittings, for example the fuse board, light fittings and switches.

Once the system has been installed, the electrician will test and commission it to make sure it is safe to put into service. If a new electrical system has been fitted, the electrician will issue the client with an Electrical Installation Certificate (EIC) to show that it has been installed to a satisfactory standard. If the electrician has carried out work on an existing system, they will issue the client with a Minor Electrical Installation Works Certificate (MEIWC) to demonstrate that the work they have completed is safe.

Heating and ventilation fitters

Heating and ventilation fitters plan, install and maintain large heating and ventilation systems. They follow the designer's plan and specification to ensure work is completed safely to industry standards and to meet the client's requirements.

Once the installation work has been completed, the heating and ventilation fitter will test and commission the system. At the end of the project, they may be appointed by the client to carry out routine servicing of the system, to ensure it is working safely and efficiently. Any work completed by the fitter must be recorded in the client's building log book.

Air conditioning engineer

An air conditioning engineer is responsible for the installation, maintenance and repair of climate control systems in residential, commercial, and industrial settings. Their role ensures efficient operation, energy compliance and occupant comfort by managing ventilation, cooling and air quality systems to meet regulatory and performance standards.

Gas engineers

A gas engineer specialises in the installation, servicing and maintenance of gas systems and appliances. Once all the structural work has been completed on a project, the first-fix installation of the new system can begin.

The gas engineer follows the prepared design drawings and specifications to route their gas pipework safely through the building, where it is least likely to be damaged by other construction work.

At the second-fix stage, the gas engineer connects the client's appliances and pressure tests the system to check for leaks, before connecting to the main gas supply entering the construction site.

Once the gas engineer has completed all the safety checks and is satisfied that the work has been completed safely to building regulations, they will commission the system. It is then the responsibility of the gas engineer to inform their local authority of the installation within 30 days, so that a Building Regulations Compliance Certificate can be issued to the client.

BUILDING REGULATIONS COMPLIANCE

This is a Building Regulations Compliance Certificate:

It confirms the work detailed below has been carried out by a Gas Safe registered business. It is also confirmation from the business that the work:

- was carried out in accordance with the Gas Safety (Installation and Use) Regulations, as well as all other relevant industry standards
- has been self certified as being compliant with Section 4 and 7 of the Building Regulations in England, Wales and Isle of Man.

It is a legal requirement for anyone carrying out gas work on your behalf to be Gas Safe registered, please remember to check your engineer's licence card before you have any gas work carried out.

Registered businesses can self certify that their work complies with building regulations requirements in England, Wales and Isle of Man. This assists you in complying with your obligation to notify your local authority building control as the property owner, and this certificate confirms this notification has been done for you.

Thank you for using a Gas Safe registered business. Gas Safe Register recommends that you have your gas appliances checked for safety annually.

Property Address:
1 Sample Street
SAMPLE TOWN
Sample Shire
WX1 2YZ

Certificate No. 999999

Gas Safe Register has been notified that the work detailed below has been undertaken.

Work completed by: Mr. GAS Fitter
On behalf of: ABC Domestic Appliances
Registration Number: 999999
Date of work: 16-04-2009

Installed a Gas Boiler
Boiler Manufacturer

Installed a Gas Boiler
Boiler Model

Gas Safe Register
PO Box 6804
Basingstoke RG24 4NB

www.GasSafeRegister.co.uk

Gas Safe Register has notified your local authority building control of the work detailed on this certificate.
Gas Safe Register inspects the work of registered businesses to ensure safety and standards are maintained.
Should your property be selected for inspection your co-operation in gaining access to inspect the work will be appreciated.

Please note that the contract for the work carried out is between yourself and the named business. Gas Safe Register can take no responsibility for the standard of work carried out. However, if you have any concerns regarding gas safety please contact your engineer.

GAS safe REGISTER

▲ Figure 4.26 Building Regulations Compliance Certificate

Only heat-producing gas appliances, such as boilers and fires, need a Building Regulations Compliance Certificate and this is only a requirement in England and Wales. The requirements are different in Scotland and Northern Ireland.

Health and safety

Plumbers and gas fitters (or anyone else) are not permitted to work with domestic gas heating systems unless they are on the Gas Safe Register. Gas Safe engineers can only work on the areas that they are qualified for. For example, an engineer may be permitted to work on domestic natural gas boilers, but not on gas fires.

GAS safe® REGISTER

▲ Figure 4.27 Gas Safe logo

Test yourself

Explain the legal responsibility of a gas fitter once they have installed and commissioned a new gas system.

Decorators

Decorators are usually one of the last trades to be scheduled during a construction project. Part of their role is to prepare surfaces such as walls, ceilings, metal and woodwork for decorative and protective finishes. Some of these finishes can be applied internally or externally with brushes, rags, sponges, rollers or spray systems. Skilled decorators can also apply other finishes, such as hanging wallpaper, stencilling, **graining** and creating marbling effects.

Key term

Graining: a method used by decorators to create woodgrain effects on different surfaces

Trade supervisors

Trade supervisors are often experienced construction workers who have progressed from their chosen profession to a supervisory role, where they are responsible for a small team of tradespeople working on a particular project. In most situations, they remain 'on the tools' throughout the building phase, while managing the standard of work, resources and health and safety of operatives under their control.

▲ Figure 4.28 Non-skilled operative

Trade supervisors may be required to attend scheduled site meetings with the construction management team to represent their trade area, report their progress and raise any concerns that they are unable to resolve themselves.

Non-skilled operatives

Non-skilled operatives usually undertake manual-labouring tasks to support tradespeople and other skilled workers on site, such as preparing building materials, moving equipment and general housekeeping to keep people safe. They play an important role in the success of any construction project.

7 Trade associations and professional engineering bodies in relation the BSE sector

Regulatory bodies for the building services engineering sector are vital sources of information, guidance and support to maintain and improve industry stansdards for all stakeholders. The tables below outline the main regulatory bodies in the construction industry and other sources of expert advice.

▼ Table 4.3 Trade associations

Regulatory body	Responsibilities
Air Conditioning and Refrigeration Industry Board (ACRIB)	The Air Conditioning and Refrigeration Industry Board (ACRIB) is a UK umbrella organisation that promotes standards, training, and regulatory compliance across the air conditioning, refrigeration, and heat pump sectors.
Association of Plumbing and Heating Contractors (APHC)	APHC represents plumbing and heating contractors while working with key organisations, such as the government and local authorities, to promote best practice, quality work and customer service. Its aim is to provide support for its members to ensure that they can run professional and profitable businesses.
Build Engineering Services Association (BESA)	BESA is a membership association where BSE contractors can access support, guidance and training for the design, installation, maintenance, commissioning, management and control of BSE services and systems.
Electrical Contractors Association (ECA)	The Electrical Contractors Association (ECA) is the UK's leading trade body for electrical and building services professionals, promoting industry standards, technical excellence, and regulatory compliance across the sector.
Federation of Environmental Trade Associations (FETA)	The Federation of Environmental Trade Associations (FETA) is the UK's leading umbrella body representing the interests of manufacturers, suppliers, and contractors in the heating, ventilation, air conditioning, and refrigeration sectors, promoting technical standards, sustainability, and regulatory compliance.
Renewable Energy Association (REA)	The Renewable Energy Association (REA) is the UK's largest trade body for renewable energy and clean technology, working to promote sustainable energy solutions and support the transition to a low-carbon economy through policy advocacy, industry standards, and professional development.

Table 4.4 Professional engineering institutions

Regulatory body	Responsibilities
Chartered Institution of Building Services Engineers (CIBSE)	CIBSE sets standards, publishes guidance and codes of practice, and supports its members in the profession to maintain and enhance their professional excellence. It promotes further and higher education through accredited and approved training courses to achieve industry-recognised professional status.
Chartered Institute of Plumbing and Heating Engineering (CIPHE)	CIPHE is a professional body for members of the heating and plumbing industry. It aims to promote training, health and safety and support emerging new technologies through the publication of research and development papers. It encourages lifelong learning with CPD programmes for its members from apprentices to master plumber certification.
Institute of Engineering and Technology (IET)	The IET is a multidisciplinary engineering global institution that helps to create new technologies with innovative solutions to solve challenges faced in the world today. Its vision is to inspire, influence and inform its members with resources, events, CPD and training courses.
Institute of Lighting Professionals (ILP)	The Institute of Lighting Professionals (ILP) is the UK's leading professional body for lighting engineers and designers, promoting best practice, technical excellence, and sustainable lighting solutions across public and private sectors.
Institute of Refrigeration (IoR).	The Institute of Refrigeration (IoR) is a UK professional body dedicated to advancing refrigeration, air conditioning, and heat pump technologies through education, innovation, and industry standards.

8 The role of continuing professional development (CPD) in developing the knowledge and skills of those working in the sector

8.1 Role of CPD

Construction and building services engineering are fast-moving industries, and the people working within them must keep pace with any new developments.

Professionals and skilled operatives beginning a career in the sector must undertake training and demonstrate competence in order to achieve formal qualifications. These achievements often create employment opportunities, allowing those new to the industry to gain valuable work experience and develop a greater understanding of their job role.

Some professionals and operatives have legal responsibilities to belong to professional, accredited or certified organisations in order to actively continue with their job roles, for example gas fitters have to keep a yearly registration with Gas Safe. This often involves having to maintain professional standards and keep records of **continuing professional development (CPD)**.

Many businesses insist on employees maintaining professional CPD and outline workforce planning in contracts of employment or codes of practice. There is further information on chartered, professional, accreditation and certification bodies in Chapter 7.

Often people working in the construction industry have to undertake CPD to demonstrate the currency of their professional competence. For this reason, they may focus on a particular area of weakness that they have identified or that has been brought to their attention, such as:

- legislation and regulations
- management and supervision
- health, safety and welfare
- digital technology
- conservation and refurbishment
- sustainability
- maintenance
- tools and equipment
- industry standards and best practice.

Key term

Continuing professional development (CPD): the process of maintaining, improving and developing knowledge and skills related to one's profession in order to demonstrate competence

8.2 Types of CPD

CPD does not have to be the completion of onerous formal qualifications at another organisation. Examples of CPD include:

- work experience
- a short bespoke training session taught in-house by an employer or external training provider
- an onsite toolbox talk presented by a site manager on a particular topic of health and safety
- a self-learning online course.

8.3 Benefits of CPD

CPD is a legal requirement for some professionals. However, there are also many benefits for employers and individuals in upskilling, including:

- protecting clients, customers and the public
- keeping up to date with the latest regulation changes, product developments and technological advancements
- developing product knowledge
- working more efficiently
- improving knowledge and skills
- enhancing the company image
- career progression.

Businesses that actively encourage employee CPD will not only have a highly skilled and professional team, but also members of staff that are committed and remain loyal. The more businesses invest time, money and effort in the development of their staff, the safer and more skilled the building industry will become as a whole.

Providers of Continuing Professional Development (CPD) offer structured learning opportunities that help individuals maintain, enhance, and expand their professional skills and knowledge throughout their careers.

Providers of CPD include:

- Professional bodies
- Accreditation bodies
- Certification bodies
- Manufacturers
- In house/toolbox talk.

Research

Research the following organisations and suggest which professionals may belong to them:

- Chartered Institute of Building (CIOB)
- Royal Institution of Chartered Surveyors (RICS)
- Royal Institute of British Architects (RIBA).

Improve your English

Create a template on your computer and input the CPD you have completed over the past year. Reflect on each area and how it may have benefited you in your job role. Ask your tutor to check your spelling, grammar and punctuation.

9 Building Information Modelling (BIM)

Building Information Modelling (BIM) uses **smart technology** to allow effective and efficient collaboration between designers and the construction team at every stage of a building project. It is adaptable to suit the size and complexity of each project and allows technical information to be shared throughout the management and construction teams.

Key term

Smart technology: computers, smartphones and tablets with software that connects to the internet in order to control, report, monitor and analyse devices and appliances remotely

▲ Figure 4.29 Building Information Modelling (BIM)

9.1 BIM process

At the start of a project, the client usually meets with the designer and contractor to discuss the information they want to receive at each stage up until handover, and how this will be shared with them through BIM. This detail is recorded in a document known as the Employer's Information Requirements (EIR), so that the construction team understands the service it needs to provide to the client.

Designers usually start the BIM process by translating information captured from the construction site and the client's drawings into digital 3D models of the building and the infrastructure around it. Clients and other stakeholders can use BIM for virtual-reality tours throughout the building before it has been constructed, helping them to gain a better understanding of the structure.

BIM illustrates every detail of the project in graphical form (drawings) and non-graphical form (written information), recording the relationships between components and how they all fit together. This information can be shared easily with all members of the project team, so that they can analyse every aspect of the design, and where necessary make changes to the model in real time.

BIM brings different professionals together to show how their work fits into the overall project by analysing data and exploring **visualisations** to help understand how a building is to be constructed. Working collaboratively on complex building projects using BIM can eliminate mistakes before they happen, save costs and achieve better project outcomes.

At the design stage, the intelligent model process is used to generate documentation for construction specifications and schedules for the building phase of the project, referred to as the Digital Plan of Work (DPoW). During the building phase, this information is used by the contractor and subcontractors to inform planning and organise resources for an efficient site, and ensure the smooth running of the project.

Another advantage of BIM is that it can be accessed by any member of the project team on the construction site using a mobile device, and it is often used with augmented reality (AR) technology. In AR, digitally generated images are superimposed over real-world images in real time. This technology is useful for building services engineers when planning the arrangement of services, such as air-conditioning ducts and vents.

▲ Figure 4.30 Augmented reality (AR)

Key term

Visualisations: digital or virtual representations of a structure

Research

Find out about the use of augmented reality (AR) in construction. Explain how it can be used by the project team to manage building work.

9.2 BIM levels

The extent to which BIM is used will vary, depending on the nature and size of a project. The government recognises the positive steps the construction industry has taken towards the adoption of BIM and how it has contributed to substantial savings. It also understands the complexity of BIM and how it could be used broadly to describe many different systems.

To bring clarity to how BIM is used in a project, the government has defined a number of levels:

- Level 0 – Designers are using computer-aided design (CAD) to produce 2D drawings and plans, however this information has to be printed to be shared with the project team. There is no digital collaboration between stakeholders using the same platform.
- Level 1 – A Common Data Environment (CDE) is established, usually by the contractor, allowing graphical and non-graphical information to be stored centrally and accessed by the whole project team. CAD is used in the production of 2D drawings, 3D BIM models and other construction

documentation. Clear roles and responsibilities are established in the project team, and common forms of data are used to share information between different parties.

- Level 2 – Any CAD software can be used at this level, but it must be capable of being exported to common file formats. The project team will also be able to work collaboratively using different systems throughout a construction project. This is the minimum level required by the government for all public construction projects.
- Level 3 – This level is in its preliminary stages of development, however it is intended to improve on level 2 BIM with the use of a single server. This will allow an 'open data' standard, so that all stakeholders are able to work simultaneously on the same project, from anywhere in the world.

The original UK BIM framework was structured around BIM Levels 0 to 3, which described the maturity of digital collaboration in construction projects. In contrast, the new BIM standards are based on the BS EN ISO 19650 series, which provide a globally recognised framework for managing information across the entire lifecycle of built assets. Rather than focusing on levels of maturity, ISO 19650 emphasises structured processes, clear roles, and consistent data management. It replaces the PAS 1192 series and supports international collaboration by aligning UK practices with global standards.

BIM Level 0 → ISO 19650 Pre-implementation - This is considered pre-BIM or non-compliant with ISO standards. Projects at this stage lack formal information management and do not meet the minimum requirements of ISO 19650.

BIM Level 1 → ISO 19650 Part 1 (Concepts and Principles) - ISO 19650. Part 1 formalises the principles of structured information management, including the use of a Common Data Environment (CDE), naming conventions, and metadata. It builds on Level 1 by embedding these practices into a consistent framework.

BIM Level 2 → ISO 19650 Part 2 (Project Delivery Phase) - ISO 19650. Part 2 replaces Level 2 with a more robust, internationally aligned process for managing information during design and construction. It defines roles (e.g. Information Manager), workflows, and verification protocols to ensure data quality and collaboration.

BIM Level 3 → ISO 19650 Part 3 (Operational Phase) - ISO 19650. Part 3 supports this vision by managing asset information during operation and maintenance. It enables digital twins, facilities management, and whole-life asset optimisation, aligning with Level 3's goals but through a structured, scalable process.

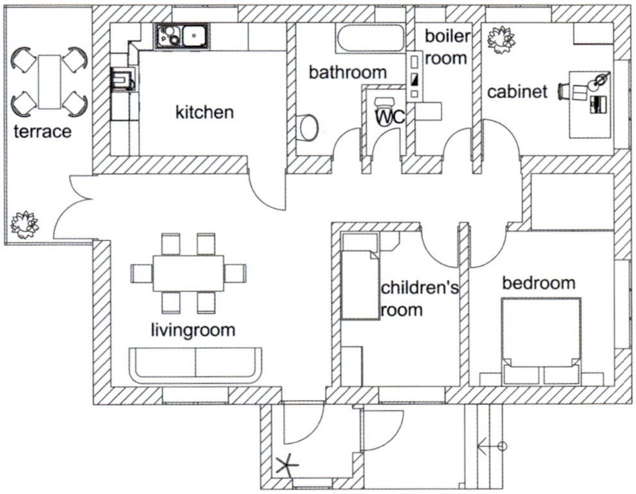

▲ Figure 4.31 BIM: 2D computer-aided design (CAD)

9.3 Data warehouse

Once a building project has been completed, contractors usually measure its success by analysing data and project outcomes. Besides the technical information distributed through BIM, contractors usually want to focus on particular subject areas, such as the procurement of resources or finance. In order to do this successfully, the most recent information is gathered from different sources and fed into a single digital storage system known as a data warehouse.

Data warehouse software systems are designed to meet the needs of each contractor by structuring the information, evaluating it and presenting visual reports. Harnessing the data in this way enables the contractor to make informed business decisions for future construction projects.

Test yourself

Explain the main benefits of using BIM.

10 PESTLE factors

There are many internal and external factors that can affect the way a business is controlled. Some of these factors can be planned for and managed, however there are some that are beyond the control of the business and will inevitably happen.

It is important that businesses develop strategies to prepare for these factors before they happen, by analysing their market from many different angles and the environment as a whole.

Factors that are beyond the control of a business are known by the acronym PESTLE:

- **P**olitical – political situations could affect local and national government spending in the construction sector, for example a change of government.
- **E**conomic –
  - If people are not spending money, the UK economy will slow down. When this happens, the building industry is one of the first to be impacted.
  - Changes in gross domestic product (GDP) value and exchange rates will affect the market value of construction goods and services.
  - When interest rates increase, people are more likely to save rather than spend their money. If interest rates decrease, people will often borrow money to spend on new homes or invest in improvements to their existing properties.
  - Higher taxation also has an impact on businesses and the self-employed, because it results in increased costs of building materials and labour.
- **S**ocial – population demographics and the movement of social and community groups can influence the type of buildings and structures constructed in particular areas to meet the needs of the community.
- **T**echnology – technological innovations could affect how a building or structure is designed and constructed, for example the use of green energy as a source of power.
- **L**egal – legislation made by Parliament can influence the planning, design and construction of buildings. For example, following the Grenfell Tower fire in 2017 there was a review of the Building Regulations 2010 and Approved Document B (Fire safety), effectively banning the use of combustible cladding in external walls of relevant high-rise buildings.
- **E**nvironment – companies have corporate social responsibility (CSR) and protect the environment by lowering carbon emissions produced during manufacturing and construction.

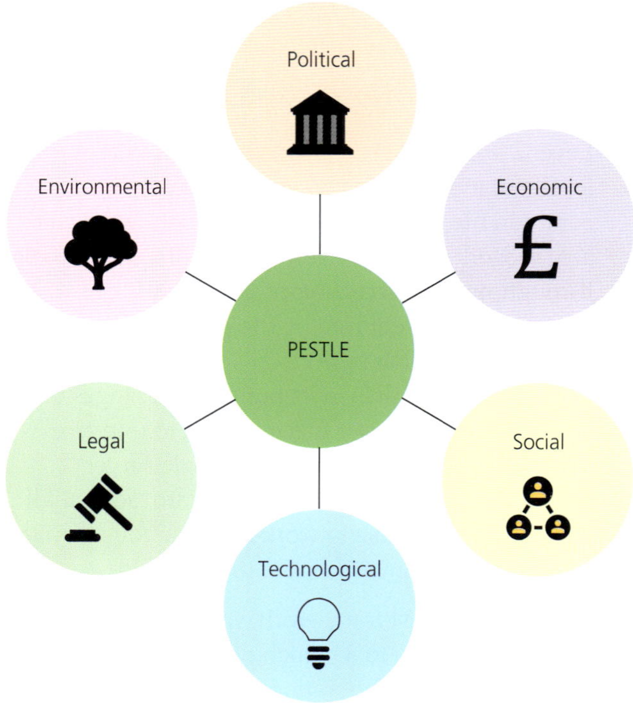

▲ Figure 4.32 PESTLE

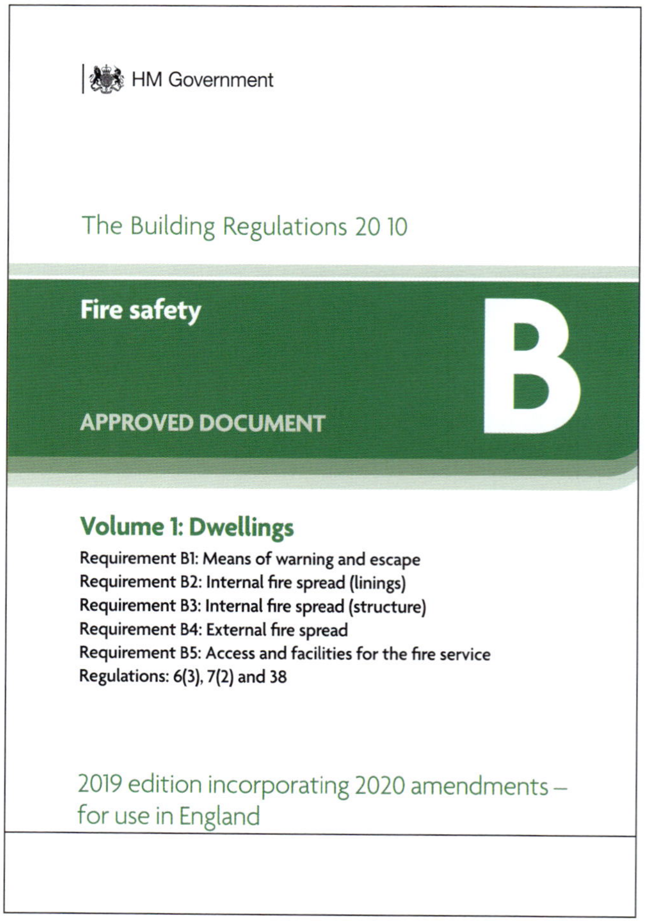

HM Government

The Building Regulations 20 10

Fire safety

APPROVED DOCUMENT

B

Volume 1: Dwellings

Requirement B1: Means of warning and escape
Requirement B2: Internal fire spread (linings)
Requirement B3: Internal fire spread (structure)
Requirement B4: External fire spread
Requirement B5: Access and facilities for the fire service
Regulations: 6(3), 7(2) and 38

2019 edition incorporating 2020 amendments – for use in England

▲ Figure 4.33 A PESTLE factor that can affect construction businesses

Test yourself

List the internal and external factors that can affect a business over which a contractor may have no control.

11 Documentation used in construction projects

At every stage of a construction project, people have to collaborate to share different types of information. If these lines of communication are not efficient, or the information is not clear, it could be detrimental to the project in terms of mistakes and financial implications. While verbal communication is quick and effective in many situations, it can be misunderstood or forgotten, and there is nothing to refer back to at a later date or if a dispute occurs.

In the construction industry, important information is recorded in different types of written documentation. It is advantageous to record information in this way because it is clear, understood by everyone and can easily be duplicated and distributed at various levels in the project team.

Other than health and safety documentation (which is covered in Chapter 1), the following documents are commonly used in construction projects:

- take-off sheets
- contracts
- schedule of rates
- estimates
- quotations
- delivery notes
- purchase orders
- bills of quantities
- wiring diagrams.

11.1 Take-off sheets

To calculate quantities of building materials and labour costs for a construction project, estimators complete take-off sheets (also referred to as material take-off) using information contained in working drawings and specifications.

While this estimation process may not be necessary for small construction jobs, it is always used for major projects to determine their feasibility, as well as being a financial planning tool. Take-off sheets are also used by estimators to produce bills of quantities for the tendering process, and to support bids for future contracts.

Take-off sheets can be completed manually, referring to hard copies of drawings to determine measurements and quantities using a scale rule. However, this is reliant on the methodical approach of an estimator and is at risk from human error. Accurate take-off sheets can be created using computer software, providing estimates that can easily be shared between the office and construction sites.

Requisitioning materials necessary for a construction project from take-off sheets can result in workers getting the resources they need on time, for smoother running of the job.

Test yourself

Explain the purpose of take-off sheets.

11.2 Contracts

Lump-sum and measurement contracts

Once a client has decided which contractor they are going to appoint for a project, either through the tendering process or otherwise, they can formally secure their services with a lump-sum contract. A lump sum is a fixed price for the project based on drawings, specifications and the bill of quantities (see 10.8).

If a lump sum cannot be accurately determined by the contractor before work starts on site, a measurement contract may be used. This type of contract is based on agreed values or rates (prices) for work between the client and the contractor, where the total cost is not finalised until the end of the project.

When neither of these options can be used, a contract can be agreed for work to be completed on the understanding that the client will be charged the **prime cost** of the materials, labour and plant for the project, plus an additional fee to cover the contractor's profit margin and overheads. This is

DOOR & WINDOW TAKE-OFF											
Project:							Sheet # 1 of 1				
Take-off by:							Date:				
DOORS											
How many	No.	Width	Height	Thickness	Type	Material	Glass	Frame type	Frame material	Jamb size	Notes
	1	3′0	6′8	1 ¾″	S/C	Birch	(2) 1′6″ sidelights		Ponderosa pine	4 ¼″	6′0×6′8 opening
	2	6′0	6′8		Sliding	Glass	1/4″ insulated		Aluminium	4 9/16″	Includes screen
	3	2′8	6′8	1 ⅜″	H/C	Birch			Ponderosa pine	″	
	4	3′0	6′8	1 ¾″	S/C	-			″	″	4-1/4″ jamb for door from garage to garden
	5	2′4	6′8	1 ¾″	H/C	-			″	″	
	6	7′0	8′0		Overhead	Wood			″	4 ¼″	Sectional garage door
											** Allow £150 for exterior doors
											*** £190 allowance for all hardware
WINDOWS											
How many	No.	Width	Height	Thickness	Type	Material	Glass	Frame type	Frame material	Jamb size	Notes
	1	6′0	5′0		Casement	Wood					
	2	4′0	3′0		″	″					
	3	3′0	4′0		″	″					
	4	4′0	4′0		″	″					

▲ Figure 4.34 Take-off sheets

known as a cost-reimbursement contract. The extra charges added to the prime cost can be based on a percentage of the overall project value, or a fixed sum of money decided before the contract is agreed.

Where the services of a contractor may be required for design and build, management only or construction management purposes, other forms of contract can be designed to suit these requirements.

Key term

Prime cost: the actual value of goods and services without any additional costs added, for example profit margins

When both parties have agreed to the terms and conditions of a contract, they will sign it in the presence of a witness to complete the legally binding document. This is known as an agreement.

Research

Research the Joint Contracts Tribunal (JCT) for further information on standard forms of building contracts. Explain the factors that may influence the choice of contract agreed between a client and their contractor.

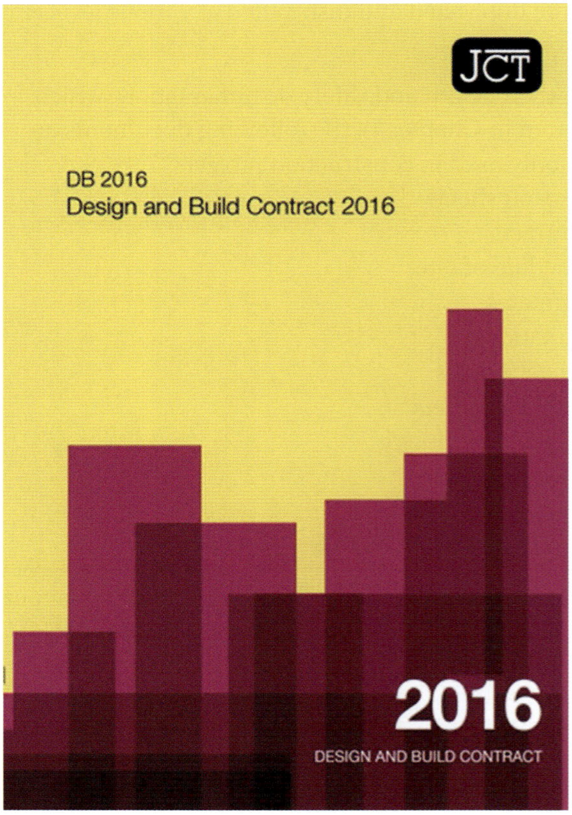

▲ Figure 4.35 Building contract

Payment schedule

An important part of the terms and conditions of any building contract is the payment schedule. This determines how and when the client will pay the contractor for the agreed work. Where minor works are undertaken for a short duration, one payment could be agreed to be made at the end of the project. However, for bigger contracts payments are usually made regularly throughout the project until completion.

Without interim payments, the contractor will have to finance the client's project by paying labour, materials and plant for the duration of the contract, which could cause cash-flow problems. Payments can be agreed to be made in advance of each significant stage of building work, for example:

- the first payment when the contracts have been signed
- the second payment when the foundations are built and inspected
- and so on for each stage of the project.

Most clients prefer to pay contractors interim payments for work completed. This reduces the risk for the client if the contractor is unable to complete the project after payments have been made.

Retention

Clients sometimes withhold a percentage of money due to the contractor at each stage of the building work; this is known as a retention. The exact percentage of money to be withheld has to be agreed between the client and contractor before work starts and is usually 3–5 per cent. The retention acts as financial security for the client, to make sure the contractor finishes the building work and any **snagging** within a reasonable amount of time after completion of the work; this is known as the defects liability period.

If the contractor does not return to complete the work within the period agreed in the contract, the client has reasonable grounds to use the money to instruct other contractors to undertake the outstanding work. On the other hand, if the contractor does complete the project and resolve all of the defects identified within the defects liability period, the client must release the outstanding retention payment without delay.

▲ Figure 4.36 Snagging

Key term

Snagging: corrective work undertaken by a contractor or their subcontractors that has been identified by the client or their representative

Penalty clauses

To ensure a building project is completed by the date stipulated in the contract, a penalty clause is usually included. In the event that the contractor does not finish the agreed work by the completion date, a sum of money will be owed to the client by the contractor. This amount will increase with every day past the original deadline, until the project is completed.

Other documents

There are various other documents included in a contract between a client and a contractor, which may include:

- Building Information Modelling (BIM)
- working drawings
- specifications
- a schedule of work.

11.3 Schedule of rates

The cost of building materials will naturally increase over a long period of time due to inflation. However, it can also quickly rise above this level if there is an increase in transport costs or demand and a shortage of materials.

When prices for materials are not stable, it can cause difficulties for contractors and subcontractors in the period between pricing or tendering for contracts and actually starting the projects when purchase orders are placed.

So that contractors are not caught out by price increases, suppliers often agree a schedule of rates for a set period of time. This allows contractors to determine more accurate quotations for clients, without having to overcompensate for unforeseen price increases after they have submitted their fixed price for work.

11.4 Estimates

An estimate is a prediction of costs for building work provided by contractors for clients, based on all the information provided for a job. Although there has to be a degree of accuracy in an estimate, there are factors that are liable to change after the costs have been finally calculated, which could lead to additional charges once the work has been completed.

When a client is considering the affordability of a project, they may not be able to provide all the information needed by the contractor, for example the final details of finishes. A percentage of money is usually added to the estimate as a **contingency**, to cover any unknown financial risk until more information becomes available.

Key term

Contingency: provision for an unforeseen circumstance; financial contingencies are often considered when planning for a construction project

Industry tip

Never agree to verbal estimates; they often lead to disputes with the client once the work has been completed.

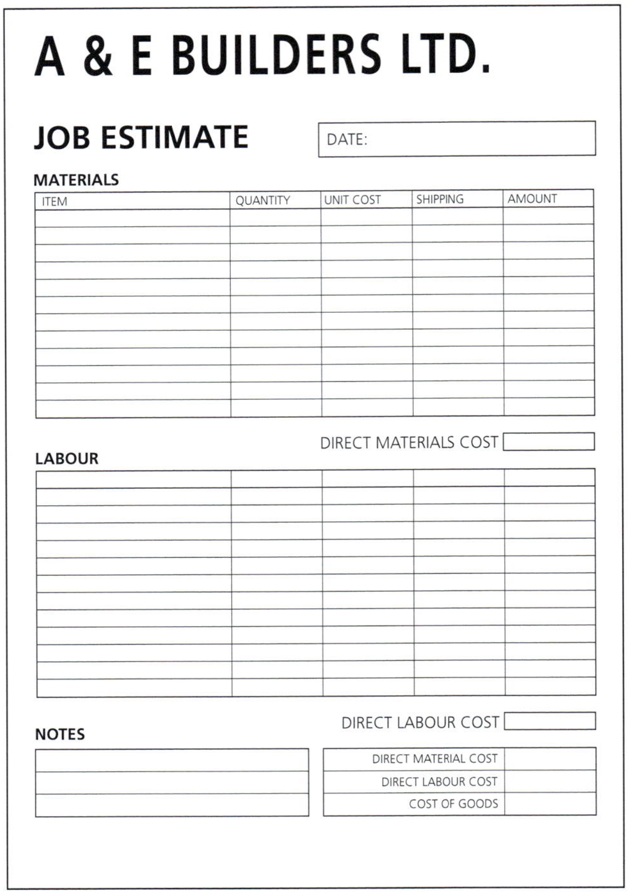

A & E BUILDERS LTD.

JOB ESTIMATE

DATE:

MATERIALS

ITEM	QUANTITY	UNIT COST	SHIPPING	AMOUNT

DIRECT MATERIALS COST

LABOUR

DIRECT LABOUR COST

NOTES

DIRECT MATERIAL COST

DIRECT LABOUR COST

COST OF GOODS

▲ Figure 4.37 Written estimate

Improve your maths

Research the average hourly rate for workers employed in your area in the same trade as you. Consider a routine task that you would undertake and calculate the total cost. This should include labour, materials, consumables, overheads and profit. You are VAT registered, therefore you will also have to include a further 20 per cent tax.

Improve your English

Create a template for an estimate on your computer and include the costs calculated in the 'Improve your maths' task. You could use your parent, carer or guardian as the customer to complete all the details needed for the estimate.

Once you have completed the task, ask your parent, carer or guardian for feedback.

11.5 Quotations

A quotation is different to an estimate; this is a fixed price for goods and services offered by contractors or subcontractors to potential clients. Once a quotation has been agreed in writing between both parties and work has started on site, the price cannot be amended, even if this means that the contractor has underestimated their work. To avoid disputes over money between the client and contractor at the end of the project, it is important that the quotation details exactly what services and goods are being provided.

Most written quotations have an expiry time, for example 30 days from the date provided, after which time the quote is no longer valid. If this date is not stipulated on the document, then the cost of construction materials and overheads may increase due to increased transport costs or demand, which could reduce profit margins for the contractor. Where the client is responsible for a sizable project, they will often request quotations from different contractors as part of the tendering process.

Test yourself

Explain the difference between an estimate and a quotation.

11.6 Delivery notes

A delivery note is a document that accompanies a shipment from a supplier and describes the goods and quantities being delivered.

When goods are delivered to site, the delivery driver will ask for a dated signature on the delivery note from a responsible person, confirming that the correct items have been received in good condition and that they match the original purchase order. Any damaged or missing goods must be recorded on the delivery note before signing it or allowing the driver to leave. Defects or missing goods unreported when a delivery is made are much more difficult to justify to the supplier or put right at a later date.

While the format for delivery notes varies between different suppliers, the information contained on them is basically the same and includes:

- purchase order number or reference
- purchaser's name or company
- delivery address
- order date
- item reference/description
- quantities
- sizes
- signatures of the delivery driver and recipient
- delivery date.

While paper delivery notes are still used by many suppliers of construction materials, some use electronic pads for signatures, with receipts emailed afterwards. This avoids the need to store the documents on site where they could get lost or damaged, as well as providing a digital copy of the delivery note that can easily be stored.

Construction Supplies Ltd
Delivery note

Customer name and address:	Delivery Date / time / Order number
CPF Building Co Penburthy House Falmouth Cornwall	**Delivery Date:** 16/12/14 **Delivery time:** 9am **Order number:** 26213263CPF

Item number	Quantity	Description	Unit Price	Total
X22433	100	75 mm 4 mm gauge countersunk brass screws slotted	30p	£30
YK7334	100	Brass cups to suit	5p	£5
V23879	1 litre	Sadikkens water based clear varnish	£20	£20

Subtotal	£55.00
VAT	20%
Total	£66.00

Discrepancies: ....

Customer Signature: ....

Print name: ....

Date: ....

▲ Figure 4.38 Delivery note

Neither version of delivery notes contains individual prices, nor the total cost for the goods; those details are sent to the purchaser's business address on an invoice. Contractors that have credit accounts with suppliers will receive an invoice after each delivery or collection of goods. At the end of each month, the contractor will receive a statement listing each invoice number and cost, amounting to the total money due for that period.

11.7 Purchase orders (POs)

To keep control of spending and budgets, contractors usually set up a **credit agreement** with their suppliers. Only then can goods be ordered with an official purchase order from the contractor or their nominated employees, for example a buyer. Hard-copy purchase orders can be used, although these are often much slower to process than electronic versions, which can be emailed directly to the supplier without delay.

Before issuing a purchase order, the contractor or their buyer will usually negotiate the best prices for the goods or services with different suppliers, until they are satisfied that they have value for money.

All purchase orders have a unique reference number or code that links them to a particular job or contract. The same references are then used by suppliers on their delivery notes and invoices. This makes it easier for the contractor to assign money actually spent on resources for a job against estimates and quotations; it also prevents unauthorised purchases from suppliers.

Key term

Credit agreement: a legal contract made between a person or party borrowing money and a lender; it states the terms and conditions of the acceptance of credit, including how the debt will be repaid

11.8 Bill of quantities (BoQ)

A bill of quantities is a document usually produced for the client by a quantity surveyor at the planning stage of a building project. It contains a list of all the quantities of materials and resources needed to complete the work, measured in number, length, area, volume or time.

The bill of quantities is part of tender documentation, allowing potential contractors to provide itemised costs for work based on the same specified quantities (rather than taking off quantities from drawings and specifications). The client is then able to compare easily both the overall cost and individual item costs between tenderers.

The bill of quantities is not just used for cost planning during the tendering process, but also later for construction planning, material scheduling and the valuation of interim payments for the contractor.

BILL OF QUANTITIES

Number	Item description	Unit	Quantity	Rate	Amount £	Amount p
	CLASS A: GENERAL ITEMS					
	Specified requirements					
	Testing of materials					
A250	Testing of recycled and secondary aggregates	sum				
	Information to be provided by the contractor					
A290	Production of materials management plan	sum				
	Method related charges					
	Recycling plant/equipment					
A339.01	Mobilise; fixed	sum				
A339.02	Operate; time-related	sum				
A339.03	De-mobilise; fixed	sum				
	CLASS D: DEMOLITION AND SITE CLEARANCE					
	Other structures					
D522.01	Other structures; concrete	sum				
D522.02	Grading/processing of demolition material to produce recycled and secondary aggregates	m^3	70			
D522.03	Disposal of demolition material offsite	m^3	30			
	CLASS E: EARTHWORKS					
	Excavation ancillaries					
E542	Double handling of recycled and secondary aggregates produced from demolition material	m^3	70			
	Filling					
E615	Importing primary aggregates for filling to structures	m^3	15			
E619.1	Importing recycled and secondary aggregates for filling to structures	m^3	15			

▲ Figure 4.39 Bill of quantities (BoQ)

Test yourself

What is the difference between a bill of quantities and a take-off sheet?

11.9 Wiring diagrams

Wiring diagrams are technical drawings provided by the client to building services engineers, such as electricians, for the installation of electrical systems and circuits. They are simplified visual representations of the layout of electrical systems or circuits. They show how electrical wires are interconnected and where fixtures such as switches, sockets and lighting components are positioned, both internally and externally on every level of the building. Symbols, abbreviations and notes are used on the diagrams to indicate the incoming power source, explain how components relate to one another and provide other important information such as voltages and ratings, as well as the size and capacity of the system.

At the end of the project, the wiring diagrams are handed over to the occupants in the health and safety file, and they should be kept for the lifetime of the building. They can be referred to at a later date during any further building work, or during the repair or maintenance of the electrical system.

12 Procedures for handing over projects to clients

12.1 Snagging

After the subcontractors have completed the building work for a project, the site management team will complete a rigorous inspection for defects and record any that they find on a document known as a **snagging list**. The defects identified could be minor, such as a scuff mark on paintwork or a door binding in its frame, or something more serious, for example the drainage system not working.

Once the list has been completed, the contractor usually instructs the relevant subcontractors to rectify the defects at their own expense. Once all the defects have been resolved to a satisfactory standard, the property is ready to hand over to the client.

Key term

Snagging list: a document used to record faults and defects discovered in building work or materials

12.2 Handover package

Besides the simple act of handing over the keys or fobs (an alternative to keys, used for keyless entry) for a property to the client, the contractor is also responsible for the smooth transition from construction to ownership.

To maintain a consistent approach for this process, the site management team usually has a checklist of key points that the client must understand so they can be confident using the building and operating the services within it.

Part of the handover process involves a member of the site management team walking the client through the building, demonstrating the use of essential services and appliances such as heating, hot and cold water and electricity for lighting. The client will also be provided with a handover package containing a number of essential documents:

- the building owner's manual and user guide
- guidance documents on defects reporting and aftercare
- operational and maintenance manuals
- a building regulations completion certificate
- the health and safety file (including construction drawings/BIM)
- the building log book
- testing and commissioning certificates, for example a Building Regulations Compliance Certificate for gas installations
- the building warranty/insurance certificate and policy booklet.

Once the occupant has taken ownership of a property, they may discover small faults with the standard of work or the quality of materials used. In this case, the owner would follow the information in the handover package to compile their own snagging list and report the faults directly to the contractor so that they can put them right.

12.3 Health and safety file

The Construction (Design and Management) (CDM) Regulations 2015 place a legal duty on the designer (or the principal designer, where there is more than one) to produce a health and safety file for projects they have worked on.

The file is prepared during the pre-construction phase of a project and contains relevant health and safety design information that may be used by the contractor in the construction phase, and again once the project is completed, when further work may be carried out on the building, for example servicing, maintenance and alterations.

A copy of the health and safety file should remain with the property owner and be available for reference for the lifetime of the building.

Health and safety

The Construction (Design and Management) (CDM) Regulations 2015 require clients, designers and contractors to plan, manage and control health and safety for construction activities, from start to finish. Everyone involved with construction projects has a part to play in protecting people from harm, and these regulations define the roles of duty holders. Further information can be found in Chapter 1.

12.4 Log book

In order to comply with building regulations and Approved Document L2A, the handover package for a new building (other than a domestic property) given to the client must contain a building log book.

The log book is a management tool that contains information about the services in the building and how they can be operated properly and efficiently. If any changes are made to the original design of the building, the details must be updated in the log book by the facilities manager or others responsible for the management of the building.

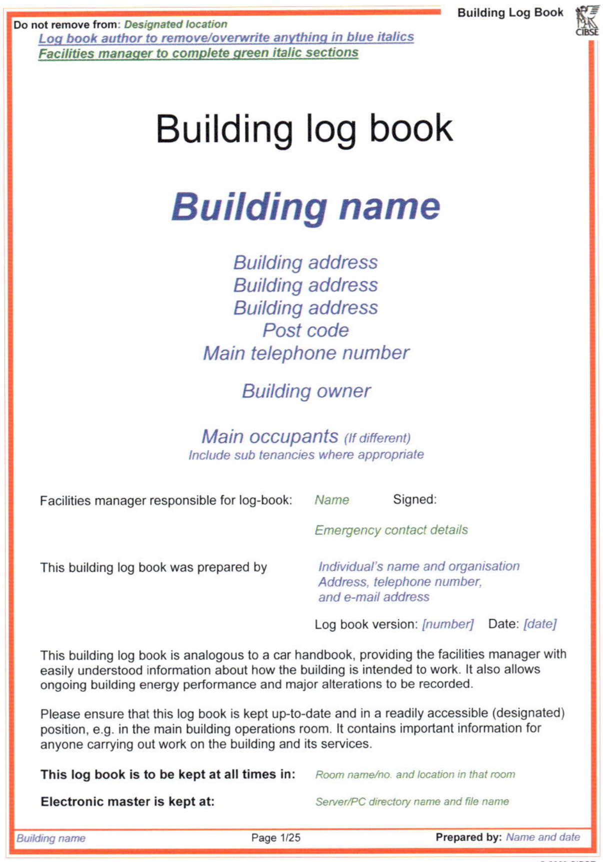
Building Log Book

Do not remove from: *Designated location*
Log book author to remove/overwrite anything in blue italics
Facilities manager to complete green italic sections

Building log book

Building name

Building address
Building address
Building address
Post code
Main telephone number

Building owner

Main occupants (If different)
Include sub tenancies where appropriate

Facilities manager responsible for log-book: *Name* Signed:
Emergency contact details

This building log book was prepared by *Individual's name and organisation*
Address, telephone number,
and e-mail address

Log book version: *[number]* Date: *[date]*

This building log book is analogous to a car handbook, providing the facilities manager with easily understood information about how the building is intended to work. It also allows ongoing building energy performance and major alterations to be recorded.

Please ensure that this log book is kept up-to-date and in a readily accessible (designated) position, e.g. in the main building operations room. It contains important information for anyone carrying out work on the building and its services.

This log book is to be kept at all times in: *Room name/no. and location in that room*

Electronic master is kept at: *Server/PC directory name and file name*

Building name Page 1/25 **Prepared by:** *Name and date*

© 2006 CIBSE

▲ Figure 4.40 Building log book

12.5 Insurance schemes

Most contractors provide the client with a warranty to protect them against the builder becoming insolvent and problems occurring in the building within the first ten years. There are three approved insurance schemes widely recognised and used by contractors in the construction industry:

- Local Authority Building Control (LABC) Warranty
- Premier Guarantee
- National House Building Council (NHBC) Buildmark Warranty.

When one of these policies has been implemented by the contractor, the building work will usually be inspected at various stages by the insurance company to make sure it adheres to standards of work and materials.

The advantage of choosing LABC Warranty is that it works in partnership with Local Authority Building Control. This avoids having more than one body inspect the building work and can be a cheaper option.

Like most building warranties, the Premier Warranty offers a ten-year cover period following completion. This is split into two periods – the defects insurance period, which covers the first two years, and the structural insurance period, which covers years three to ten.

The disadvantage of both these policies is that the client must pay an excess of £1000 for each claim, therefore the majority of new homes are covered by the NHBC Buildmark Warranty. Buildmark provides insurance for new or converted homes from exchange of contracts to a maximum of ten years after legal completion.

Raising Standards. Protecting Homeowners

▲ Figure 4.41 National House Building Council (NHBC) logo

12.6 Storage of documents

Many of the documents included in the handover package must be stored safely by the client for the lifetime of the building, in order to comply with legal requirements. For example, fire safety information must be kept in order to meet regulation 38 of the Regulatory Reform (Fire Safety) Order 2005.

An effective way to do this is to store the documents in an online **repository**, where they can be accessed and managed by the occupiers, Local Authority Building Control or emergency services when they are needed.

Various companies offer to store building documentation online in a digital format, known as a 'building passport'.

Key term

Repository: a central location where something can be stored

Improve your English

Write a sentence about the construction industry and built environment that includes the word 'repository'.

Case study

Theo and his family live in a detached two-storey house in a suburban area of Leicester. The number of family members has grown since buying the property; therefore they would like to extend the back of the house to increase the current size of the kitchen and add a further bedroom upstairs. The property is not listed, nor is it in a conservation area.

- Does Theo need planning permission for the extension?
- List the building regulations that would have to be complied with for the project.
- Explain the process that Theo would have to follow to design and build the proposal.
- Research the professionals that would be involved in this process and explain their roles in the project.

Assessment practice

Short answer

1 When must a business become VAT registered?

2 Identify one type of client that may have construction work undertaken by a contractor.

3 Name one document that would be included in a handover package.

4 Which document is used to record building faults and defects discovered in work or materials in the construction industry?

5 Identify one factor that can have an impact on a business that it is unable to control.

Long answer

6 List the disadvantages of being a self-employed subcontractor.

7 Explain the purpose of Building Information Modelling (BIM).

8 Explain the term 'retention' often found in a building contract.

9 Explain the benefits of continuous professional development (CPD).

10 List the documents that should be included in a tender package.

Project practice

Your employer has noticed that the construction team has not been collaborating well recently, which has reduced its efficiency and caused inaccuracies in information that has been shared.

Your task is to:

- outline the roles and responsibilities of each member of the construction team
- create a hierarchy of job roles within the company on a computer, illustrating the lines of communication throughout the team
- present your work to the team, explaining how it will improve collaborative working.

Chapter 5 Sustainability principles

This chapter explores principles of sustainability in the construction industry, which play an important role when planning and delivering projects. It identifies a range of sustainable solutions, including different materials currently available.

We will then look at responsibilities under environmental legislation and the key measures for environmental performance. In particular, we will focus on the importance of waste management, as well as on how renewable energy and energy conservation systems can be incorporated into buildings.

Learning outcomes

By the end of this chapter, you will understand:

1. sustainability when planning and delivering a construction project
2. types of sustainable solutions
3. environmental legislation
4. environmental performance measures
5. principles of heritage and conservation
6. lean construction
7. waste management legislation
8. waste management
9. energy production and energy use
10. renewable energy and energy conservation
11. digital technologies.

1 Sustainability when planning and delivering a construction project

1.1 Sustainability

When planning and delivering a construction project, sustainability is achieved by:

- using renewable and recyclable resources
- sourcing materials locally
- protecting resources
- reusing and refurbishing materials
- reducing energy consumption and waste
- creating a healthy and eco-friendly environment
- protecting the natural environment.

Research

Choose one of the following stages of project development:

- design
- planning
- delivery.

Find out about the importance of sustainability at this stage of the construction process. Present your findings to a classmate: why is it important to consider sustainability at this point?

1.2 Assessment methods

Various assessment methods are used to determine how well a building performs against environmental, social and economic standards. Buildings are evaluated according to regional, national and global criteria, to assist architects and planners during the design stages of a project.

Building Research Establishment Environmental Assessment Method (BREEAM)

The Building Research Establishment (BRE) has created best practice standards for the environmental performance of buildings through their design, specification, construction and operation. BREEAM sets out benchmarks for standard categories of development and also offers a scheme for non-standard buildings. The assessment method can be applied to both new developments and refurbishment projects.

Research

Look up the BREEAM rating benchmarks and list each one.

BREEAM is an example of **life cycle assessment**, which covers all stages in the life of a construction project, including the transportation of goods, the extraction and manufacture of raw building materials, **demolition** of the building, and how much of the building is reused or repurposed.

Key terms

Life cycle assessment: assessing the total environmental impact of a building, considering all stages of the life of the products and processes used in it

Demolition: when something (such as a building) is torn down and destroyed

Leadership in Energy and Environmental Design (LEED)

Leadership in Energy and Environmental Design (LEED) is an environmental certification scheme, developed by the US Green Building Council. It offers a set of rating systems covering the design, construction, operation and maintenance of green buildings, homes and neighbourhoods, and provides **third-party verification** that a project meets high standards of sustainability. It aims to help building owners and operators be environmentally responsible and use resources efficiently.

LEED-certified buildings are healthier, more productive places that reduce stress on the environment. They are energy and resource efficient, and enjoy increased building value and decreased utility costs.

Key term

Third-party verification: confirmation from an independent party that a project meets standards

Test yourself

List five ways sustainability can be achieved when planning and delivering a construction project.

Research

Visit the following websites:

www.breeam.com/

http://leed.usgbc.org/

Research and compare the strengths and weaknesses of BREEAM and LEED. Present your findings in a PowerPoint presentation.

Timber Research and Development Association (TRADA)

The Timber Research and Development Association (TRADA) is an international membership organisation dedicated to inspiring and informing best practice design, specification and use of wood in the built environment and related fields.

It has a comprehensive online library, which is free to access for its members. Among its resources are wood information sheets, case studies and technical guidance relating to healthy buildings.

TRADA also provides training and events throughout the country to educate designers and engineers.

▲ Figure 5.1 Use of timber in construction

Research

Research the different types of wood information provided on the TRADA website and produce a written report.

WELL Building Standard

The WELL Building Standard measures, certifies and monitors aspects of the built environment that affect human health and wellbeing. It looks at seven different factors:

- Air: achieve optimum air quality by preventing and removing contaminants and purifying the air.
- Water: achieve optimum water quality by using filtration and treatment, and ensure water is accessible by placing water points in optimum positions.
- Nourishment: ensure the availability of and access to healthier food choices, and encourage better eating habits by providing information to users of the building.
- Light: ensure lighting levels are appropriate for the tasks carried out in the building through the design of windows and lighting.
- Fitness: provide opportunities for physical activity for users of the building, to maintain a fitness regime.
- Comfort: create a soothing and comforting environment, for example by ensuring that thermal and acoustic parameters that might cause discomfort can be controlled.
- Mind: support mental and emotional health by providing building users with relaxation spaces and knowledge about their environment.

1.3 Retrofitting to improve energy efficiency

PAS 2035:2019

To ensure the UK can meet its obligations under the Climate Change Act 2008, the energy efficiency of all its existing building stock needs to be improved.

PAS 2035 provides a specification and best practice guidance on retrofitting dwellings (domestic buildings) for improved energy efficiency. It covers how to:

- assess dwellings for retrofit
- identify and evaluate improvement options
- design and specify **energy efficiency measures (EEMs)**
- monitor and evaluate retrofit projects.

Key term

Energy efficiency measures (EEMs): measures intended to reduce energy use, including insulation, glazing upgrades and the installation of green technologies such as heat pumps

The standard requires a whole-house approach. This involves identifying and installing all energy-saving interventions at the same time, to ensure they interact in such a way as to optimise energy efficiency.

PAS 2035 also specifies five new retrofit roles:

- Retrofit assessors undertake retrofit assessments for dwellings in accordance with PAS 2035.
- Retrofit co-ordinators provide a project management role.
- Retrofit advisors provide advice to clients and homeowners on the retrofit process.
- Retrofit designers prepare a safe and effective retrofit design.
- Retrofit evaluators monitor the impact of installed EEMs to ensure they meet the intended outcomes.

PAS 2038:2021

This specification sets out requirements for retrofitting non-domestic buildings for improved energy efficiency. It covers all buildings except those used as private dwellings, including multi-residential buildings in which occupants share some communal facilities (for example hotels and student accommodation).

The PAS defines technically robust and responsible whole-building retrofit processes that support:

> 'improved functionality, usability and durability of buildings
>
> improved comfort, wellbeing, health and safety (including fire safety) and productivity of building occupants and visitors
>
> enabling buildings to use low- or zero-carbon energy supplies
>
> improved energy efficiency, leading to reduced fuel use, fuel costs and pollution (especially greenhouse gas emissions associated with energy use)
>
> reduced environmental impacts of buildings
>
> protection and enhancement of the architectural and cultural heritage as represented by the building stock
>
> avoidance of unintended consequences related to any of the above, and
>
> minimisation of the 'performance gap' that occurs when reductions in fuel use, fuel cost and carbon dioxide emissions are not as large as intended or predicted.'

Source: PAS 2038:2021

Test yourself

List the seven factors that the WELL Building Standard considers.

Explain two new retrofit roles specified by PAS 2035:2019.

Research

Research and give two examples of EEMs that can be retrofitted in an existing terraced property.

2 Types of sustainable solutions

Sustainability involves a commitment to environmental, economic and social objectives:

- Environmental sustainability is about acting responsibly to avoid the depletion of or damage to natural resources. It involves protecting the environment from the impact of emissions, sewage and waste.
- Economic sustainability is about supporting long-term economic growth and ensuring profitability. It involves making more efficient use of resources such as labour, materials, energy and water.
- Social sustainability focuses on wellbeing and quality of life. It involves recognising the needs of everyone impacted by construction projects, from design to demolition. This includes construction workers, local communities, project supply chains and users of the building.

Sustainable construction is guided by these objectives during all stages of a project.

2.1 Sustainable solutions

Prefabricated construction

Prefabricated construction combines pre-engineered units to form major elements of a building. They are manufactured in factories and then transported to site where they are assembled.

Benefits of prefabricated construction include:

- reduced waste (a huge environmental benefit)
- lower build costs
- design and build flexibility
- consistent accuracy and quality of components
- reduced construction time.

See Chapter 3, section 2.1 and Chapter 7, section 1.2 for more details.

Self-healing concrete

Self-healing concrete contains the spores of limestone-producing bacteria and a food source. When cracks occur, moisture in the air causes the spores to germinate. The reactivated bacteria then eat the food source and excrete calcite to heal the crack.

Green roofs

A green roof, also known as a living roof, is an attractive and sustainable roof system that involves installing additional waterproof membranes and drainage mediums, onto which soil is added to allow growth of vegetation.

The benefits of green roofs include:
- reduction of water run-off from roof areas
- extended roof life
- insulation of the building (keeping it warm in winter and cool in summer)
- sound insulation
- providing a habitat for wildlife.

▲ Figure 5.2 Green roof

Smart glass

Smart glass (also known as switchable glass) changes from transparent to translucent (and vice versa) when exposed to specific levels of voltage, heat or light.

Benefits of smart glass include:
- reduced energy costs
- increased privacy
- controlled room temperatures
- improved security.

Grey water

Grey water refers to waste water generated from hand basins, washing machines, showers and baths. Rather than sending it down the drain, it can be reused for watering plants and flushing toilets.

For more information on grey water, see section 10 of this chapter.

Reed beds

Reed beds are artificially constructed wetlands that use natural filtration and biological processes to break down organic matter in waste water and sewage **effluent**.

There are two different types:
- horizontal flow
- vertical flow.

Key terms

Grey water: water that has not been purified for the purpose of drinking, for example recycled water from a sink

Effluent: liquid waste or wastewater

Case study

John and his family live in a remote detached property in the Scottish Highlands. As there is no access to mains drainage, they would like to explore the use of reed beds as a form of sustainable drainage.
- Research the types of reed bed suitable for the property.
- Provide an overview of the working principles of each type of reed bed.
- List the building regulations that would have to be complied with for the project.

Soakaways

Soakaways are an effective way of dealing with surface water. They are essentially large underground holes, filled with coarse stones or purpose-made plastic crates, which allow water to filter through and soak into the ground.

Smart cement

Smart cement contains potassium ions, which allow it to store electricity for long periods of time. If a building was constructed using this cement, and connected to a power source such as photovoltaic panels, it could store power during the day and release it at night.

Research

Find out how the following materials are used in the construction of buildings and roofs:

- recycled bricks
- recycled tiles/slates
- sustainable timber.

Test yourself

Describe two benefits of a green roof.

3 Environmental legislation

3.1 Environmental Protection Act 1990

The Environmental Protection Act 1990 defines legal responsibilities for the management of waste and pollution, and places a duty on local authorities for collecting waste. Under this legislation, businesses have a duty to handle waste safely.

3.2 Climate Change Act 2008

The Climate Change Act 2008 set a target for the UK to reduce its greenhouse gas emissions by 80 per cent by 2050, compared to 1990 levels. This target was updated in 2019 to 100 per cent – net zero.

The main elements of the act are:

- setting emissions reduction targets and carbon budgeting
- establishing a system for annual government reporting on emissions
- creating an independent advisory body (the Committee on Climate Change)
- giving powers to enable the government to introduce trading schemes to lower emissions
- providing a procedure for assessing the risks of climate change
- requiring the government to develop an adaptation programme for sustainable development
- supporting emissions reductions through various policy measures, such as amendments to improve renewable transport fuel obligations, powers to introduce charges for single-use plastic bags and powers to trial incentive schemes for household waste minimisation and recycling.

3.3 Clean Air Act 1993

The Clean Air Act was first enacted in 1956, in response to the Great Smog of London of 1952 – a dense yellow smog that descended on London and contributed to an estimated 4,000 deaths. It was modified by the Clean Air Act 1968, and repealed by the Clean Air Act 1993.

The act covers a range of topics, including:

- prohibition of dark smoke from chimneys
- prohibition of dark smoke from industrial or trade premises
- the requirement for new furnaces to be smokeless (so far as is reasonably practicable)
- the emission of grit and dust from furnaces
- the height of chimneys for furnaces
- declarations of smoke control areas by local authorities.

Under this legislation, local authorities may declare the whole or part of the district of the authority to be a smoke control area, where it is an offence to emit smoke from the chimney of any building, furnace or fixed boiler.

▲ Figure 5.3 It is an offence to emit smoke in a smoke control area

Improve your English

Your client lives in a smoke control area and therefore cannot emit smoke from their chimney unless they use authorised fuels or exempt appliances.

Research either authorised fuels or exempt appliances and write an email to your client explaining the options available to them. What do you recommend?

3.4 Water Act 2014

The main aims of the Water Act 2014 were to:

- reform the water industry to make it more innovative and responsive to customers
- increase the resilience of water supplies to natural hazards such as droughts and floods
- address the availability and affordability of insurance for households at high risk of flooding.

Source: www.gov.uk/government/publications/2010-to-2015-government-policy-water-industry

It includes information on topics such as:

- the water industry:
  - water supply licences and sewerage licences
  - **water and sewerage undertakers**
  - regulation of the water industry
- water resources:
  - water abstraction reform
  - main rivers in England and Wales
  - maps of waterworks
- environmental regulation
- flood insurance:
  - the flood reinsurance scheme
  - flood insurance obligations.

▲ Figure 5.4 Some areas are at higher risk of flooding

Key term

Water undertakers: suppliers of wholesome water and sewerage services to properties

3.5 Building Regulations 2010

Nearly all new construction work and alterations to existing structures have to comply with the Building Regulations 2010. The requirements are set out in Schedule 1 and cover a range of topics:

- Part A: Structure
- Part B: Fire safety
- Part C: Site preparation and resistance to contaminants and moisture
- Part D: Toxic substances
- Part E: Resistance to the passage of sound
- Part F: Ventilation
- Part G: Sanitation, hot water safety and water efficiency
- Part H: Drainage and waste disposal
- Part J: Combustion appliances and fuel storage systems
- Part K: Protection from falling, collision and impact
- Part L: Conservation of fuel and power
- Part M: Access to and use of buildings
- Part N: Glazing – safety in relation to impact, opening and cleaning
- Part O: Overheating
- Part P: Electrical safety
- Part Q: Security in dwellings
- Part R: Infrastructure for high-speed electronic communications networks
- Part S: Infrastructure for the charging of electric vehicles.
- Part T: Toilet accommodation

To help people comply with the regulations, the Ministry of Housing, Communities and Local Government publishes approved documents that offer general guidance on each part of the law (more on these in Chapter 7, section 3.1).

Sustainability is a common theme throughout the approved documents. For example, Approved Document G states that water consumption in a new building should not be greater than 125 litres per person per day. Approved Document L sets minimum appliance efficiencies and control requirements for gas, oil and solid-fuel heating equipment.

The Domestic Building Services Compliance Guide supports Approved Document L and provides guidance for the installation of fixed building services in new and existing dwellings to help compliance with the energy-efficiency requirements of the Building Regulations.

The Domestic Ventilation Compliance Guide supports Approved Document F. It helps architects, planners and installers to comply with the Building Regulations by ensuring the provision of adequate ventilation while minimising energy use and environmental issues.

3.6 Control of Substances Hazardous to Health Regulations 2002

The Control of Substances Hazardous to Health (COSHH) Regulations 2002 are intended to protect people from ill health caused by exposure to hazardous substances. Details of the COSHH regulations can be found in Chapter 1.

Manufacturers and suppliers of hazardous substances produce safety data sheets that contain important information about how products should be transported, used, stored and safely disposed of after use, any special conditions you should be aware of and how to deal with the substance in an emergency.

The COSHH Regulations detail measures that must be taken to ensure substances do not pose a hazard to the environment, for example chemicals being discharged or leaked into water sources.

Industry tip

Manufacturers produce material data sheets, which provide all the required information relating to each chemical.

Test yourself

What is the purpose of the Water Act 2014?

Which of the 2010 Building Regulations provides detail on conservation of fuel and power?

What was the first Clean Air Act (1956) a response to?

3.7 Hazardous Waste (England and Wales) Regulations 2005

Under environmental legislation, waste is considered hazardous if it contains substances or has properties that might make it harmful to either human health or the environment.

The Hazardous Waste (England and Wales) Regulations 2005 restrict the production, movement, receiving and disposal of hazardous waste, such as fluorescent tubes, refrigerators and asbestos. They introduced a registration process for producers of hazardous waste and a new system for controlling, tracking and recording the movement of hazardous waste.

Research

Visit: **www.gov.uk/dispose-hazardous-waste**

Find out the responsibilities of the following with regard to hazardous waste:

- a producer or holder (who produces or stores waste)
- a carrier (who collects and transports waste)
- a consignee (who receives waste for recycling or disposal).

3.8 Control of Pollution (Oil Storage) (England) Regulations 2001

Also known as the Oil Storage Regulations or simply OSR, these regulations were designed to reduce incidents of oil escaping into the environment. They require anyone in England who stores more than 200 litres of oil to provide a more secure containment facility for tanks, drums, bulk containers and **mobile bowsers**.

Key term

Mobile bowser: a wheeled trailer fitted with a tank for carrying oil

All types of oil are covered by the regulations, except for waste mineral oil.

▲ Figure 5.5 Oil storage

3.9 Energy Performance of Buildings Directive

The Energy Performance of Buildings Directive is an EU legislative instrument that aims to reduce the carbon emissions produced by buildings. It requires:

- the production of an energy performance certificate whenever a building is sold, rented out or constructed
- the production of a display energy certificate for large public buildings, which must be displayed in a prominent place
- the regular inspection of air-conditioning systems and boilers.

The requirements of the directive were implemented on a phased basis by the Energy Performance of Buildings (Certificates and Inspections) (England and Wales) Regulations 2007. Later, the Energy Performance (England and Wales) Regulations 2012 consolidated and revoked all previous regulations.

Research

Visit: **www.legislation.gov.uk/uksi/2012/3118/contents/made**

Who is required to display energy certificates? What should the display certificate include?

3.10 Waste Electrical and Electronic Equipment Directive

Details of the Waste Electrical and Electronic Equipment (WEEE) Directive can be found in section 7.1 of this chapter.

4 Environmental performance measures

4.1 Performance measures

Environmental performance can be measured by looking at a range of factors during the design, construction, use and demolition of a building.

Source and use of materials

Sourcing materials locally has a lower environmental impact, as it uses less transportation and therefore causes fewer carbon emissions.

When choosing materials to construct a building, it is also important to consider their:

- recyclability
- maintenance requirements
- thermal properties.

▲ Figure 5.6 Using locally sourced materials will cause fewer carbon emissions

Energy source and consumption

The energy efficiency of a building depends on its use, design, orientation, location and the materials used in its construction.

Choosing a form of heating that is based on renewable energy, such as ground or air source heat pumps, will improve a building's environmental performance. Similarly, environmentally friendly electrical systems can generate power using solar panels or **wind turbines**.

It is also important to ensure a thermally efficient building, that reduces heat losses/gains and so requires minimal heating and cooling. Materials with a lower U-value lose less heat and are more efficient.

Maximising daylight in a building can help to reduce lighting costs, as well as energy consumption due to solar heat gain.

Key term

Wind turbine: a vaned wheel that is rotated by wind to generate electricity

Water source and usage

In order to improve a building's environmental impact, it is important to minimise water usage. This can be achieved by installing:

- grey-water recycling systems
- rainwater harvesting systems
- flow-limiting valves
- dual-flush toilets
- water-saving showers
- infrared taps and non-concussive percussion taps.

Flexibility

The more easily a change of use can be accommodated by a building, the better its sustainability. A building should therefore be designed to allow adaptation, conversion and extension in the future, for example using building components that can be easily disassembled and repositioned for flexibility.

Adapting an established structure for a new purpose can support sustainability, since a substantial part of the existing building fabric will be reused. Depending on the repurposed design, fewer new materials will be required, which translates into less energy being consumed during manufacture. It also avoids the negative impact of demolition on the environment.

Durability and resilience

The environmental impact of a building depends on its:

- resistance to degradation over time (durability)
- ability to adapt and respond to changing conditions while maintaining functionality (resilience).

A durable and resilient building will have a lower environmental impact as it will require less maintenance and repair throughout its life cycle.

Pollution and waste processing

Construction that generates large amounts of waste or causes pollution to land, air or water will have a negative environmental impact. It is important to design waste out of all the stages of a building's life cycle, from construction to use to demolition.

For example, waste can be reduced by:

- using standardised components (for example bricks and blocks)
- storing materials carefully to avoid damage or deterioration (for example cement or lengths of timber)
- recycling, reusing and repurposing materials (for example demolition materials)
- installing energy-efficient and non-polluting systems (for example heating systems).

Transport

Transporting construction materials over long distances will have a negative effect on the environment due to vehicle emissions. Therefore, buildings that use locally sourced materials are considered more sustainable.

Buildings that are easily accessible using local public transport systems will have a lower impact on the environment than those that are only accessible by car.

Demolition

Well-planned management of demolition means that building materials used during construction can be recycled, thereby reducing environmental impact. The more of a building that can be recycled, the better. For example:

- Concrete can be broken down as aggregate.
- Drywall's paper covering can be recycled like most paper products, and the gypsum core is ideal for recycling as garden fertiliser.
- Glass can either be recycled into new glass or broken down as aggregate.
- Steel can easily be melted down and recycled.

4.2 Schemes

Several schemes can be used to certify levels of environmental performance in construction, for example:

- Building Research Establishment Environmental Assessment Method (BREEAM) (see section 1.2 of this chapter)
- Leadership in Energy and Environmental Design (LEED) (see section 1.2 of this chapter)
- **Passivhaus**.

Key term

Passivhaus: ('Passive house' in English) an energy performance standard intended primarily for new buildings, which ensures that buildings are so well constructed, insulated and ventilated that they require little energy for heating or cooling

Test yourself

Explain what a U-value is and why it is important to consider when designing buildings.

5 Principles of heritage and conservation

5.1 Listed buildings

There are three types of listed status for buildings in England and Wales:

- Grade I – buildings of exceptional interest
- Grade II* – particularly important buildings of more than special interest
- Grade II – buildings of special interest, warranting every effort to preserve them.

Buildings are listed in their entirety, even though some parts may be more important than others. The designation regime is set out in the Planning (Listed Buildings and Conservation Areas) Act 1990.

The government document *Principles of Selection for Listed Buildings* sets out the criteria that the Secretary of State applies when assessing whether a building is of special architectural or historic interest:

- To be of special architectural interest, a building must be of importance in its architectural design, decoration or craftsmanship.
- To be of special historic interest, a building must illustrate important aspects of the nation's social, economic, cultural or military history.

Listed building consent is required for all works of demolition, alteration or extension to a listed building that affect its character. This is to ensure the impact of any proposed changes is reviewed before they are approved. This is usually carried out by a conservation officer within the local authority planning department.

Local planning authorities may serve a building preservation notice (BPN) on the owner and occupier of a building which is not listed, but which they consider is of special architectural or historic interest and which is in danger of demolition or of alteration in such a way as to affect its character.

When carrying out work on listed buildings, it is important that you do not use modern repair methods on traditional construction. For example:

- Using cement on older buildings made from materials such as lime mortar can cause irreparable damage.
- Traditional cast-iron soil stacks should be replaced like for like, and not substituted with modern plastic pipework.
- Any original architectural features such as doors, decorative stonework, fireplaces or windows should not be altered.

Any changes and the materials used must be agreed with the conservation officer.

On a listed building, you cannot paint or render stonework, demolish chimney stacks, or add new pipework, flues or alarm boxes on principal elevations of the property.

Research

Review this page: **https://historicengland.org.uk/advice/technical-advice/buildings/building-materials-for-historic-buildings**

What can be used for consolidating weathered limestone? How is this substance applied? Produce a short materials guide.

The Heritage Protection Bill is a legislative and policy framework which protects the historic environment. It requires consents and permissions to protect England's heritage via a balanced, democratic and informed approach to managing changes in historic places. The different heritage assets are protected in different ways and various consents are required when carrying out works to existing buildings and for new development.

5.2 Conservation areas

Local planning authorities have the power to designate any area of special architectural or historic interest as a conservation area, where the character or appearance should be preserved. The special character of these areas is not just made up of buildings but can also be defined by features which contribute to particular scenic views, such as woodland or open spaces.

5.3 Town and Country Planning Act 1990

Under the Town and Country Planning Act 1990, planning permission needs to be sought when carrying out work to:

- build a new property
- increase the size of an existing property
- make significant alterations to an existing property
- change the use of an existing property.

The process of applying for planning permission is straightforward and completed with the local authority planning department.

Test yourself

List and describe the three types of listed status for buildings in England and Wales.

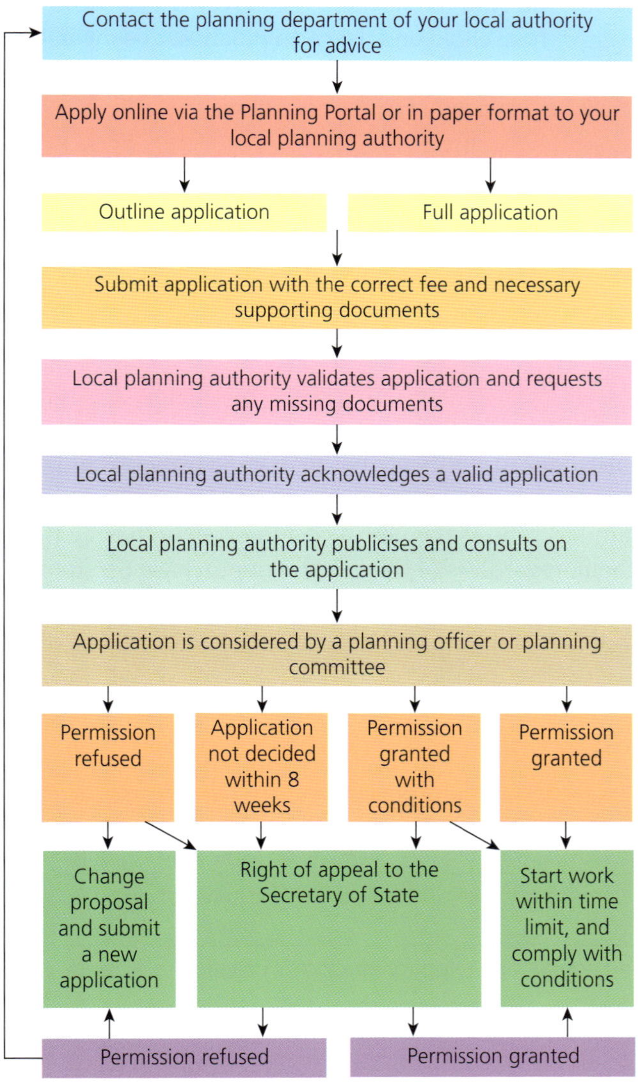

▲ Figure 5.7 Steps in planning permission

Research

Research the local planning permission requirements and application process in your area.

6 Lean construction

6.1 Principles

Lean construction is a construction methodology that aims to minimise waste in terms of costs, materials, time and effort, while maximising productivity and value.

The principles of lean construction are:

- efficiency
- best value
- ensuring the work environment is clean and safe
- improving planning
- continuous review and improvement.

Aims of lean construction:

- eliminating waste and errors through reduction, recycling and repurposing
- improving work planning and forward scheduling
- identifying the processes that deliver best value
- eliminating activities that do not add value
- ensuring the working environment is clean, safe and efficient
- continuous improvement.

6.2 Techniques

A common example of a lean construction technique is just-in-time deliveries. This is a method of providing the required materials for a project in precisely the correct order and quantity at exactly the right time for installation.

The benefits of just-in-time deliveries include:

- elimination of waste
- less storage space needed
- a stabilised work schedule
- reduction in the cost of inventories and inventory requirements.

Another example of a lean construction technique is artificial intelligence (AI). In the construction sector, AI is commonly used to develop safety systems for construction worksites. This includes tracking the interactions of tasks, workers, machinery and objects on a construction site and alerting supervisors to potential safety issues, construction errors and productivity issues, thereby reducing errors and health and safety risks during the construction of buildings.

Recycling is another commonly used method of lean construction. Recycling materials reduces the requirements for the use of new raw materials during the construction stage, eliminating waste removal from site and avoiding over ordering of materials, which can result in additional waste and an increase in the carbon footprint of the building.

Research

Research other methods of lean construction used in the industry today. Explain two to a classmate using a PowerPoint presentation.

7 Waste management legislation

7.1 Waste Electrical and Electronic Equipment Regulations 2013

The Waste Electrical and Electronic Equipment (WEEE) Regulations 2013 implemented in UK law the provisions of the EU WEEE Directive. They aim to support sustainable production and consumption through the collection, reuse, recycling, recovery and treatment of end-of-life electrical and electronic equipment (EEE). This both reduces the need to produce new equipment and minimises the amount of waste sent to landfill, reducing associated environmental impacts.

The WEEE Regulations cover the categories of appliance shown in Table 5.1.

Under the regulations, producers of EEE have obligations with regards to:

- the EEE they sell
- financing the collection, treatment, recovery and environmentally sound disposal of WEEE.

There are specific obligations on producers of non-household equipment when it is discarded as waste by non-household end users. In practice, this means that electrical tools used in construction by contractors must be disposed of in the correct way.

The regulations also require producers to prioritise, where appropriate, the reuse of whole appliances.

Regulations state that if you are a large producer (i.e. have more than 5 tonnes of EEE on the market), you must join a producer compliance scheme (PCS). The PCS takes on your obligations to finance the collection, treatment, recovery and environmentally sound disposal of household WEEE collected in the UK.

Distributors of EEE also have specific responsibilities. They must offer to take back waste of the same (or similar) type as the item a customer is purchasing from them, regardless of the method of purchase (in store, online or via mail order) and irrespective of the brand of the item.

▼ Table 5.1 Examples of appliances covered by the WEEE Regulations

Type of appliance	Examples
Large household appliances	Refrigerators, cookers, microwaves, washing machines, dishwashers, air conditioners
Small household appliances	Vacuum cleaners, irons, toasters, clocks, kettles, fryers, electric knives, body care appliances
IT and telecommunications equipment	Personal computers, copying equipment, telephones, pocket calculators, screens, printers
Consumer equipment and photovoltaic panels	Radios, televisions, hi-fi equipment, camcorders, musical instruments, photovoltaic panels, radios
Lighting equipment	Straight and compact fluorescent tubes, high-intensity discharge lamps
Electrical and electronic tools	Drills, saws, sewing machines, electric lawnmowers, jigsaws and other gardening equipment
Toys, leisure and sports equipment	Electric trains, games consoles, running machines
Medical devices	(Non-infected) dialysis machines, analysers, medical freezers, cardiology equipment
Monitoring and control instruments	Smoke detectors, thermostats, heating regulators, heating controls, alarm components
Automatic dispensers	Hot drinks dispensers, money dispensers, snack dispensers

They can discharge these take-back obligations by:

- joining the Distributor Takeback Scheme (DTS)
- offering in-store take-back
- providing an alternative free take-back service.

Research

Find out what the current UK WEEE targets are. Does data show that the UK is meeting these targets?

7.2 Waste carrier licences

Under section 34 of the Environmental Protection Act 1990, there is a duty of care to ensure that waste is managed and disposed of properly.

If a business transports waste, either for itself or for someone else, it needs to register with the Environment Agency as a waste carrier. If a business' waste is being collected by someone else, it must ensure that the carrier is registered.

There are two types of licence, depending on the type of waste to be transported:

- An upper-tier waste carrier transports other people's waste on a professional basis (for example a waste management company) or its own construction or demolition waste (for example a builder).
- A lower-tier waste carrier was either previously exempt from registration or carries its own (non-construction/non-demolition) waste on a regular and normal basis. This might include carpet fitters taking away offcuts, businesses carrying waste from maintenance work or businesses collecting confidential waste from different locations and transporting it to a centralised office.

▲ Figure 5.8 Waste carriers

7.3 The Fluorinated Greenhouse Gases Regulations (2018)

Fluorinated greenhouse gases (known as F-gases) are synthetic and originate from human activities. The most common type is hydrofluorocarbons, which are used in applications such as refrigeration, air conditioning, fire extinguishers and aerosols.

While these gases are ozone friendly, and therefore often used in place of ozone-depleting substances, they have a high global warming potential (GWP) and therefore require regulation.

Under the Fluorinated Greenhouse Gases Regulations:

- the installation, leak testing, maintenance, repair and disposal of equipment containing F-gases must be performed by trained and certified personnel
- equipment containing F-gases must be tested regularly to check for leaks and any leaks must be repaired as soon as possible
- equipment containing F-gases must be clearly labelled with the type and quantity of the gas (refrigeration and air-conditioning equipment should have an information plate affixed to it)
- records must be kept of:
  - types and quantities of gases in use
  - leaks detected and repairs carried out
  - maintenance history of equipment, including details of the maintenance company or operatives
- F-gases must be recovered at the end of equipment life and managed appropriately.

Further information on the requirements for working with F-gases can be found by searching on **www.gov.uk**.

Industry tip

Operatives who work on systems that contain F-gases, for example heat pumps or air conditioning, must be F-gas qualified.

8 Waste management

The waste hierarchy sets out the order in which actions should be taken to manage waste, from the most to least preferable in terms of environmental impact. There are five actions:

1. Reduce: this is about producing less waste and can be achieved by using fewer materials during design and manufacture, keeping products for longer, choosing products with less packaging and using fewer hazardous materials.

2 Reuse: this involves either using a product multiple times, rather than single use, or repurposing products or their parts at the end of their life by checking, cleaning, repairing or refurbishing them.
3 Recycle: this involves turning waste into new substances or products and, if it meets quality protocols, composting the material.
4 Recover: by using the waste products as fuel to provide heat and power; other methods include **anaerobic digestion**.
5 Landfill: this is the last resort and involves disposal and incineration without energy recovery.

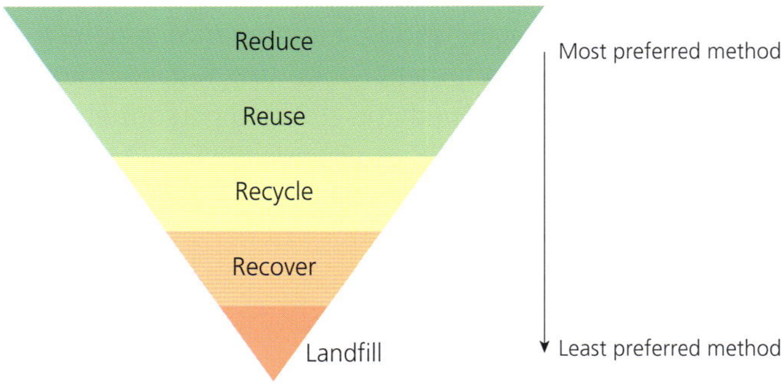

▲ Figure 5.9 Waste hierarchy

> **Key term**
>
> ***Anaerobic digestion:*** a sequence of processes by which microorganisms break down biodegradable material in the absence of oxygen

8.1 Site waste management plans

While no longer a legal requirement, a site waste management plan (SWMP) is an important document for setting out how waste will be managed and disposed of during a construction project. It should be compiled by the design team, contractor and subcontractors and refer to the waste hierarchy to reduce the volume of waste to landfill and increase the use of sustainable materials. This ensures compliance with environment legislation, improves resource efficiency and increases profitability.

A typical SWMP should contain information on:

- who has overall responsibility for the management of waste on site
- which contractors will be involved in the waste management process and in what capacity
- which types of waste will be generated, including hazardous waste
- the expected quantity of waste that will be produced
- how waste will be managed both on site and off site.

Managing waste responsibly can contribute to an improved business reputation, leading to future contracts.

8.2 Waste segregation

Waste segregation means dividing waste into different categories for efficient disposal. Not only does this support the waste hierarchy and protect the environment, by offering opportunities to reuse and recycle waste before resorting to landfill, but it also offers cost savings to construction businesses. Sorted waste is cheaper to dispose of, and some types of waste can even be sold as a source of income.

Hazardous waste in particular should be segregated and disposed of carefully, due to the risk of harm to human health and the environment. There are specific regulations that cover this, namely the Hazardous Waste Regulations 2005 (see section 3.7 in this chapter).

Waste on a construction site is usually segregated into the following streams:

- general (for example insulating materials that do not contain asbestos)
- hazardous (for example asbestos)
- clean fill (material that can be recycled or reused in future construction projects)
- hard fill (for example soil, concrete, bricks and blocks)
- plastic
- metal
- wood
- plasterboard.

It is important that all staff on a construction site are trained on the site's waste policy and the waste hierarchy, to ensure waste segregation is maximised.

8.3 Recycling

Prior to the adoption of environmentally friendly methods of managing waste, such as recycling, reusing and repurposing, most waste was disposed of by burying it in landfill sites or burning it at extremely high temperatures (incineration), depending on the type of material. Some waste is still disposed of in this way.

Recycling or reusing materials can reduce the overall carbon emissions that are produced during manufacturing.

Metals

Often, a building may appear to consist mostly of masonry or concrete. However, surprising quantities of different metals may be present in the structure. Scrap or waste metal can be recovered and reprocessed into new products, for example:

- structural steel columns, beams and lintels
- pipework made from lead (in older buildings) or copper
- cables with copper or aluminium cores.

Metal to be recycled is usually separated into ferrous (containing iron) and non-ferrous before collection.

Plastic

Plastic waste can be very harmful to the environment. When it is disposed of in landfill, it can take hundreds of years to break down, and if it ends up in our oceans, it can damage ecosystems and kill marine life.

It is therefore important to segregate and recycle as much plastic waste as possible. Waste plastic can be reprocessed into many different products, from drinks bottles to car components to building materials.

Timber

Offcuts of timber can be reprocessed into new resources, such as:

- chipboard and medium density fibreboard (MDF)
- paper and cardboard
- mulch
- bedding for animals
- biomass fuel.

Bricks and blocks

Undamaged bricks and blocks can easily be reused in new building projects. Brick and block waste from cutting operations can be crushed and used as aggregate in some types of concrete.

Test yourself

Which new products can offcuts of timber be reprocessed into?

9 Energy production and energy use

9.1 Energy production

In 1831, Michael Faraday discovered that electricity could be generated by moving a bar magnet through a wire coil. This is the principle upon which modern generators work.

In a power station generator, a huge magnet is mounted on a central rotating shaft; this is known as the rotor. Around the rotor is a series of coils known as the stator, which is where the electricity is generated by the rotating magnetic field. As the rotor and stator may weigh several hundred tons, an energy source is required to turn the rotor, for example a turbine.

In the UK, electricity is generated in several different ways, for example by:

- burning fossil fuels, such as coal, gas and oil
- using nuclear fission
- using renewable energy, such as wind, wave, hydro, biomass and solar.

Most power stations are steam powered. The heat created by burning fossil fuels or through nuclear fission is used to produce superheated steam to drive a turbine. This causes the rotor to turn, creating an **alternator** as in Faraday's prototype. The rotor turns at 3,000 revolutions per minute (50 revolutions per second) and produces an alternating current (AC) with a frequency of 50 hertz (cycles per second). The electromotive force causes current to flow.

See Chapter 2, section 5, for more information on electricity generation and transmission.

Key term

Alternator: an electrical generator that converts mechanical energy into electrical energy

Non-renewable energy sources

Fossil fuel is a term used to describe an energy source formed by the decomposition of organic matter beneath the Earth's surface over millions of years:

- coal is formed from dead trees and plants
- crude oil and gas are formed from dead marine plants and animals.

Fossil fuels are predominantly composed of carbon and hydrogen. They are burned in power stations to heat water and provide steam for turbines. However, this process releases carbon dioxide, which can contribute to the greenhouse effect and climate change.

Fossil fuels are a finite resource, which means that they are being used up faster than they can be replaced.

Nuclear power plants use reactors to split atoms, causing a large amount of thermal energy to be released. This process is called nuclear fission. Uranium is the main fuel for nuclear reactors.

- Nuclear reactions generate thermal energy.
- Thermal energy generates water vapour (steam).
- Steam powers a turbine, producing mechanical energy (kinetic energy).
- Mechanical energy actuates an electrical generator, producing electricity.

▼ Table 5.2 Advantages and disadvantages of nuclear power

Advantages	• Does not produce polluting gases • Does not contribute to global warming • Low fuel cost • Low fuel quantity required • Power station has a long life
Disadvantages	• Costs are high for building and decommissioning • Waste is radioactive • Accidents can be catastrophic

Renewable energy sources

Renewable energy sources are obtained from the environment around us and are naturally replenished. As the availability of non-renewable energy diminishes and the effects of global warming intensify, it is important that society considers the use of these technologies as alternatives.

Bioenergy

Bioenergy is created using biomass – organic material such as wood, plants, agricultural crops, food waste or even sewage.

Biomass can be burned in power stations to heat water and produce high-pressure steam. The steam flows over turbine blades, causing them to rotate, which in turn drives a generator and produces electricity.

Biomass can also be converted into liquid fuels, for example biodiesel for use in vehicles.

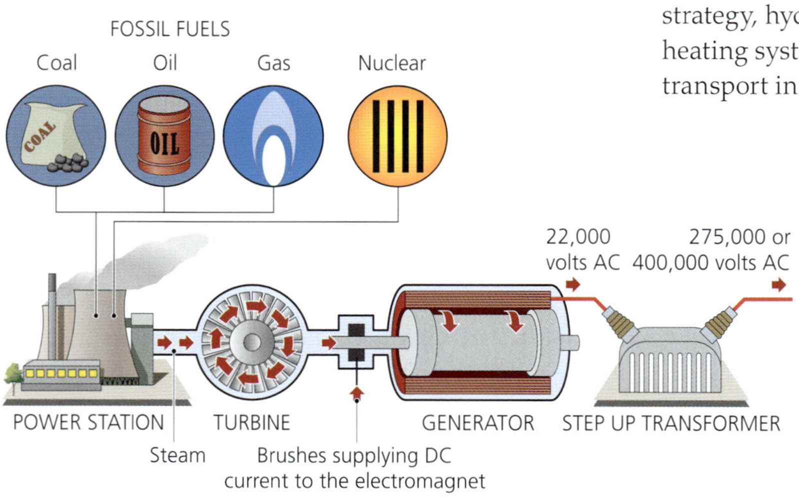

▲ Figure 5.10 How electricity is produced

Bio-methane

Bio-methane is a renewable gas produced by upgrading biogas, which is generated through the anaerobic digestion of organic materials such as food waste, agricultural residues, and sewage. Once purified, bio-methane can be injected into the gas grid or used as a low-carbon fuel for heating, electricity generation, or transport. It helps reduce reliance on fossil fuels and supports circular waste management.

Carbon Capture and Storage (CCS)

CCS is a technology designed to capture carbon dioxide emissions from industrial processes or power generation before they enter the atmosphere. The captured CO_2 is then transported and stored underground in geological formations. CCS plays a critical role in reducing greenhouse gas emissions and enabling low-carbon energy systems, especially in hard-to-decarbonise sectors like cement and steel.

Hydrogen

Hydrogen is a versatile energy carrier that can be used for power generation, heating, transport, and industrial processes. When produced using renewable electricity (known as green hydrogen), it offers a zero-carbon alternative to fossil fuels. Hydrogen can be stored and transported, making it suitable for balancing energy supply and demand, especially in sectors that are hard to electrify—like heavy industry, shipping, and aviation. It can be used in fuel cells to generate electricity with water as the only by-product, or burned directly for heat. As part of the UK's net-zero strategy, hydrogen is being developed to decarbonise heating systems, replace natural gas, and support clean transport infrastructure.

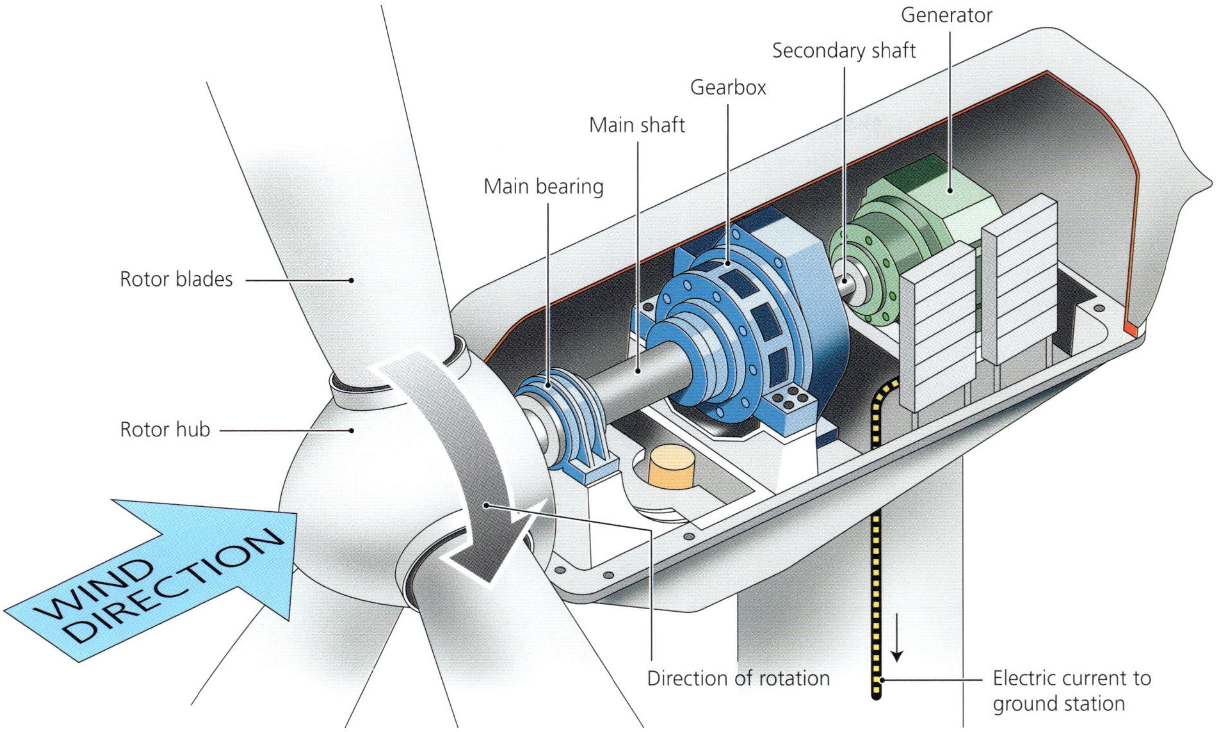

▲ Figure 5.11 How wind turbines generate electricity

Wind energy

The natural **kinetic energy** of the wind can be used to drive an aerodynamic bladed turbine, rotating a generator and producing electricity. Often many wind turbines are grouped together to form a wind farm and provide bulk power to the National Grid.

Turbines can be installed on land or offshore where there is a reliable source of wind.

Industry tip

The greater the wind speed, the faster the rotation of a wind turbine's blades, generating more electricity.

Key term

Kinetic energy: energy derived from motion

Solar energy

Solar energy utilises photovoltaic (PV) panels to convert the sun's radiation into electricity. The panels can be installed on the roofs of buildings or in larger-scale solar farms. Solar farms are capable of delivering bulk power to the National Grid.

In order to maximise the energy generated by PV panels, is it important to install them carefully, for example where there is enough sunlight and facing South.

Hydroelectric energy

Hydroelectric power plants produce energy by channelling running water through a turbine connected to an electrical generator. The electricity is then fed into the National Grid for distribution.

They can use the natural flow of a river as it falls from a greater to a lesser height, or artificial reservoirs and dams, which hold back water and release it as required.

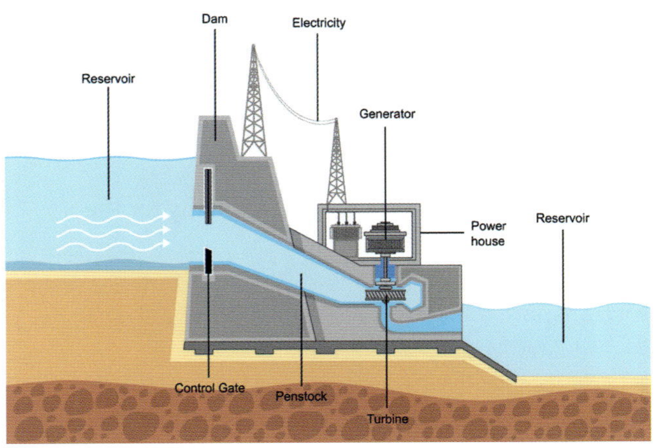

▲ Figure 5.12 Generating electricity through hydroelectric power

Wave and tidal energy

Energy from the movement of waves and tidal flows can be used to generate electricity. Wave and tidal generators are similar to wind generators, except that they use the ocean's current to rotate the turbine instead of wind. This type of energy production is ever-changing, with a wide range of options available to harness the power of waves and tides. These options include absorbers, attenuators, water columns and pendulum devices.

The amount of energy created is determined by the wave's height, speed, wavelength and water density. Wave power is much more predictable than wind power and it increases during the winter, when electricity demand is at its highest. Tidal energy is also predictable and consistent.

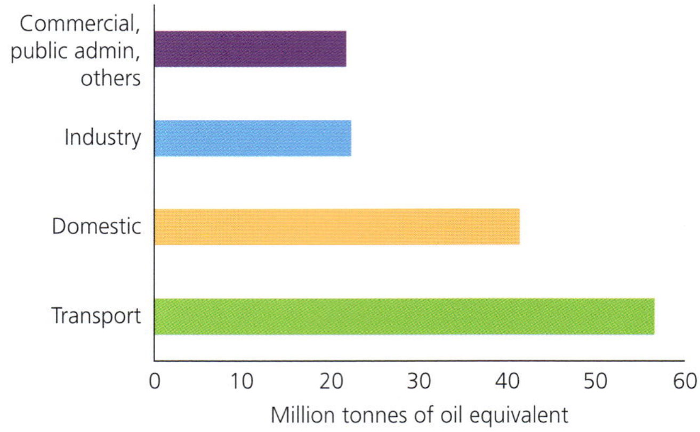

▲ Figure 5.13 Energy consumption in the UK 2019

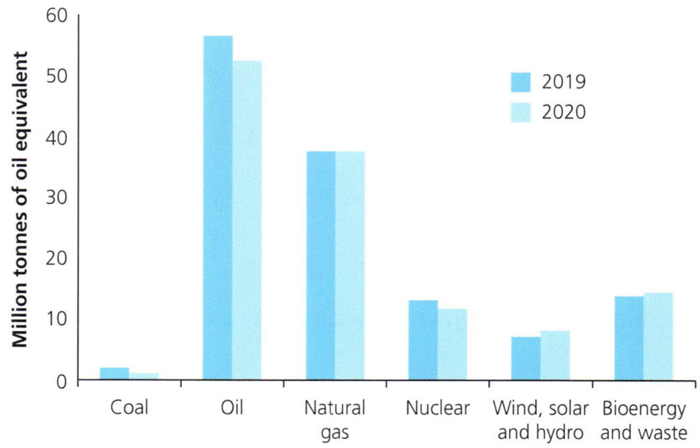

▲ Figure 5.14 Primary energy consumption in the UK in 2019 and 2020 by fuel/technology type

Research

Research the advantages and disadvantages of using non-renewable and renewable energy sources and create a table to show your findings.

Test yourself

Describe how energy can be harnessed from wave and tidal power.

9.2 Energy use

The **Department for Business, Energy and Industrial Strategy (BEIS)** publishes an annual statistical publication titled 'Energy Consumption in the United Kingdom' (ECUK). This collates data from a variety of sources to provide a comprehensive review of UK energy consumption and changes in intensity and output since the 1970s.

Key term

Department for Business, Energy and Industrial Strategy (BEIS): the department of the UK government responsible for the UK's business, energy and industrial strategy

Energy consumption is reported annually for the following sectors:

- domestic
- transport
- industry
- services.

10 Renewable energy and energy conservation

10.1 Uses of renewable energy sources

As mentioned earlier, renewable energy sources are naturally replenished, such as solar, wind, tidal and geothermal. They can be used in a wide range of applications, from small-scale systems such as solar thermal providing hot water for one property to large-scale wind farms producing energy for national use.

Non-renewable energy sources are finite and cannot be replenished as quickly as they are used, for example coal, oil and gas.

Energy sources fall into one of three categories according to their level of carbon emissions: high carbon, low carbon and zero carbon.

▼ Table 5.3 Energy sources categorised according to their carbon emissions

Category	Energy sources
High carbon	• Coal • Oil • Natural gas • Liquified petroleum gas (LPG)
Low carbon	• Solar thermal • Biomass • Heat pumps • Combined heat and power (CHP) • Combined cooling, heat and power (CCHP) • Hydrogen fuel cells
Zero carbon	• Wind • Wave/tidal • Hydroelectric • Solar photovoltaic • Nuclear

Research

Find out how hydrogen fuel cells work and what they are used for.

Solar thermal

Solar thermal systems use either a roof-mounted or ground-mounted collector to harness the sun's energy and heat domestic water. A conventional or 'system' boiler or immersion heater (auxiliary heat source) is used to top up the water temperature to 60°C.

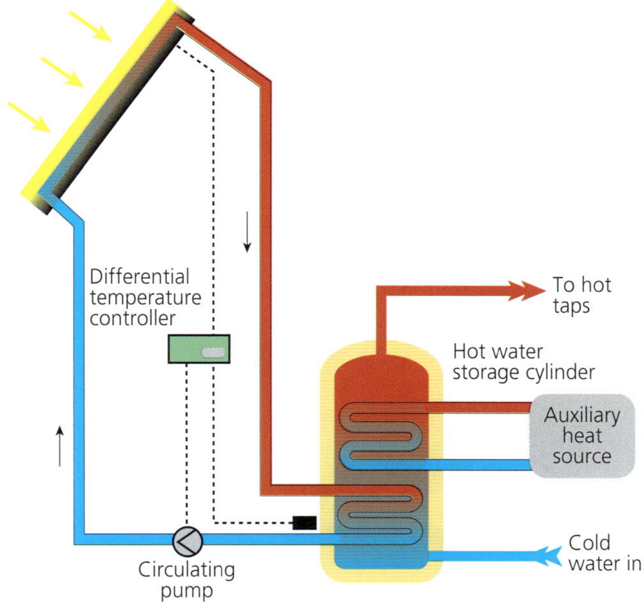

▲ Figure 5.15 Solar thermal system components

Key components of a solar thermal system are as follows:

- A solar collector absorbs heat radiation from the sun. There are two types:
  - a flat plate collector, consisting of a series of flat glass or plastic pipes with a black metal surface, used to absorb the heat
  - an evacuated tube collector, consisting of a set of tubes connected to a header, used to absorb the heat and transfer it to the water contained within it.
- A differential temperature controller (DTC) has sensors connected to the solar collector and the hot water cylinder. Its purpose is to monitor the temperature at both points of the system and switch the circulating pump on and off.
- A circulating pump is controlled by the DTC and circulates the system's heat transfer liquid around the solar hot water circuit.
- A hot water storage cylinder enables the transfer of heat from the solar collector circuit to the stored water for use at terminal fittings.
- An auxiliary heat source is used to provide hot water when solar energy is unavailable.

Industry tip

The best direction for a solar collector to face is due south.

▼ Table 5.4 Advantages and disadvantages of solar thermal systems

Advantages	• Reduced carbon emissions • Lower energy costs • Low maintenance • Improved energy performance certificate rating
Disadvantages	• Not compatible with all existing hot water systems • Less solar energy available in the winter months • High initial installation costs • Require an auxiliary heat source

Research

Research the different types of solar thermal systems in the UK and their uses.

Ground source heat pumps (GSHPs)

Ground source heat pumps (GSHPs) extract heat from the ground using a system of underground pipes filled with a mixture of water and antifreeze known as glycol. A heat exchanger then transfers the heat to the pump itself, ready for distribution to:

- warm-air heating systems (ground to air pumps)
- underfloor heating systems, radiators and hot water cylinders (ground to water pumps).

Key components of a GSHP include:

- a heat collection ground loop, set out in one of two arrays:
  - a horizontal array has flattened, overlapping coils known as slinkies that are spread out in shallow trenches about 1.5–2 m deep
  - a vertical array has pipes installed in holes bored to a depth of 15–60 m, depending on soil conditions
- a circulating pump, which circulates a mixture of water and antifreeze around the pipes
- a heat pump, which extracts heat from the ground and transfers it for domestic use by means of an electric compressor
- a heating system, which is used to heat the property.

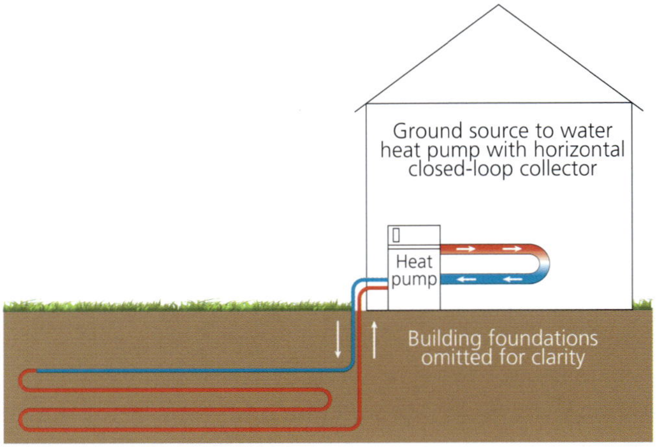

▲ Figure 5.16 Ground source heat pump

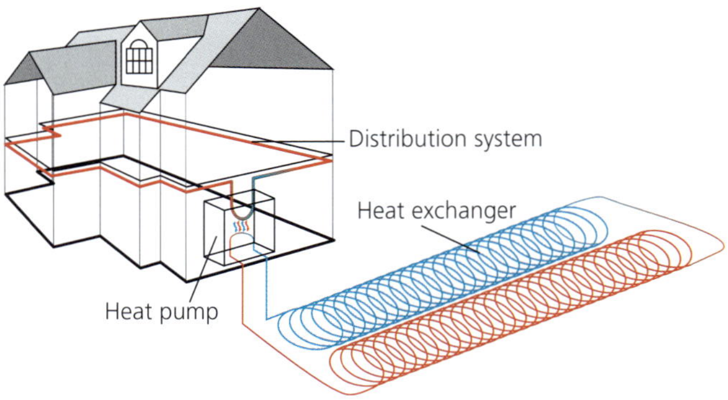

▲ Figure 5.17 Slinkies

Research

Research how a GSHP works in greater detail and produce a poster to showcase this to your classmates.

Industry tip

A 10 m slinky trench will produce around 1 kW of heating load.

Research

Select a GSHP manufacturer and research the installation requirements. Produce a pocket guide for a classmate.

Health and safety

Working in excavations can be dangerous. Ensure that adequate control measures have been implemented such as trench supports and correct access.

Air source heat pumps (ASHPs)

Air source heat pumps (ASHPs) extract heat from outside air. Using a heat exchanger, compressor and expansion valve, they distribute the heat to:

- warm-air convectors (air to air pumps, usually used in commercial buildings as a reverse cycle heat pump that can be used for both heating and cooling)
- underfloor heating systems, radiators, hot water cylinders and even swimming pools (air to water pumps).

ASHPs can get heat from the air even when the temperature is as low as –15°C.

▲ Figure 5.18 Air source heat pump

Water source heat pumps (WSHPs)

Water source heat pumps (WSHPs) operate in a similar way to GSHPs, with the collector located in a body of water (such as a lake) to extract heat.

Heat pumps are not able to provide instant heat, so they work better when they run continuously. Start-stop operations will shorten their lifespan. A buffer tank is incorporated into the system so that when heat is not required, the heat pump can 'dump' the heat into the vessel and keep running. When heat is needed, it can be drawn from the buffer tank.

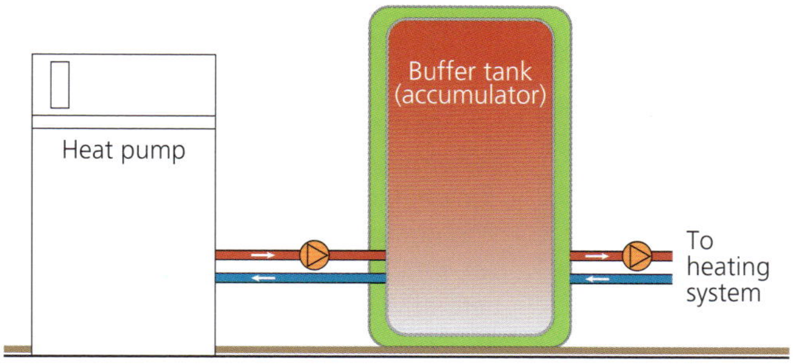

▲ Figure 5.19 Buffer tank

▼ Table 5.5 Advantages and disadvantages of heat pumps

Advantages	• Reduced carbon emissions • Typical efficiencies between 300 and 500 per cent • Low maintenance • Improved energy performance certificate rating
Disadvantages	• Not usually suitable for connection to existing heating systems using panel radiators • High initial installation costs • Air source installations can present a noise issue • Ground source installations require a large ground area or a borehole

Biomass

Biomass is plant or animal material that can be burned to create heat. Unlike fossil fuels, which have taken millions of years to form, biomass material has been sourced recently.

When burned, both fossil fuels and biomass produce carbon dioxide, which is a greenhouse gas linked to global warming. However, biomass material absorbs carbon dioxide as it grows, reducing the amount in the atmosphere, so burning biomass does not lead to a net increase in levels.

Biomass can be used to heat:

- individual rooms using stoves (these may be fitted with a back boiler to heat water too)
- whole properties, using boilers that supply central heating and hot water systems.

▼ Table 5.6 Advantages and disadvantages of biomass systems

Advantages	• Carbon-neutral and inexhaustible energy source • Lower fuel costs • Do not rely on building orientation or weather conditions to operate effectively
Disadvantages	• Require a suitable flue/chimney • High initial installation costs • Require a large space to store fuel • Not suitable for smaller properties

Solar photovoltaics

As mentioned earlier, solar energy uses photovoltaic (PV) panels to convert the sun's radiation into electricity.

PV systems can be divided into two categories:

- off-grid systems, where the PV modules are used to charge batteries
- on-grid systems, where the PV modules are connected to the National Grid via an inverter.

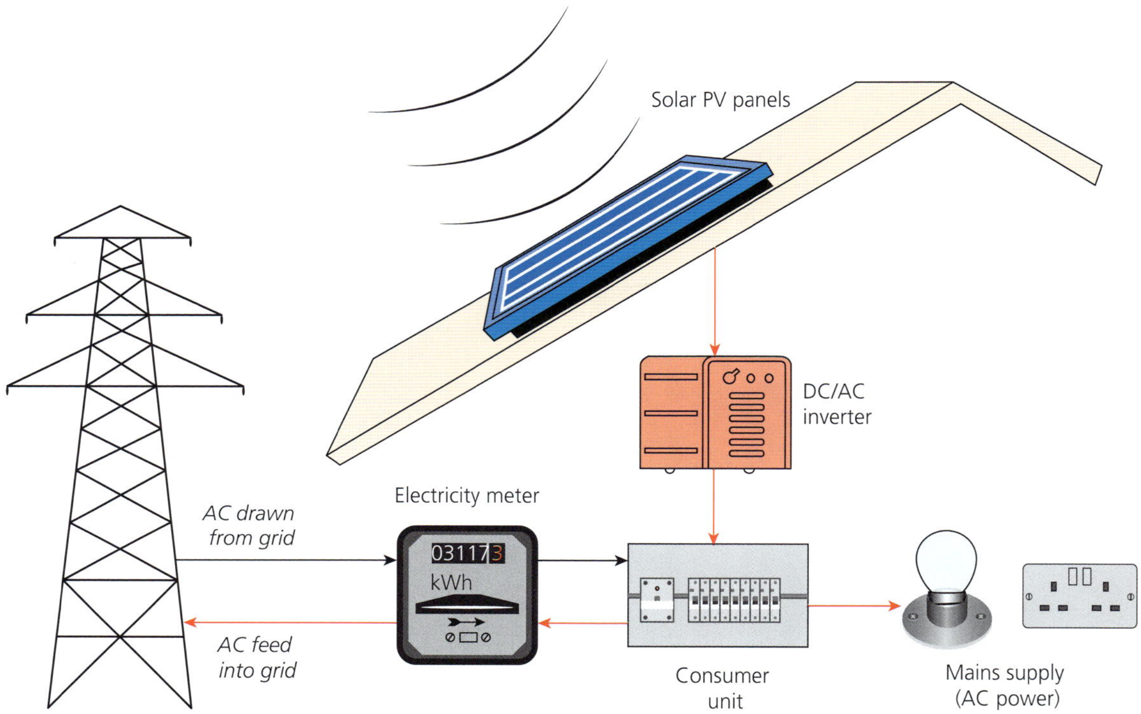

▲ Figure 5.20 Solar photovoltaics

▼ Table 5.7 Key components of off-grid and on-grid PV systems

Off-grid systems	On-grid systems
PV modules	PV modules
PV module mounting system	PV module mounting system
DC cabling	DC cabling
Charge controller	Inverter
Deep-discharge battery bank	AC cabling
Inverter	Metering
	Connection to the National Grid

▲ Figure 5.21 Off-grid system components

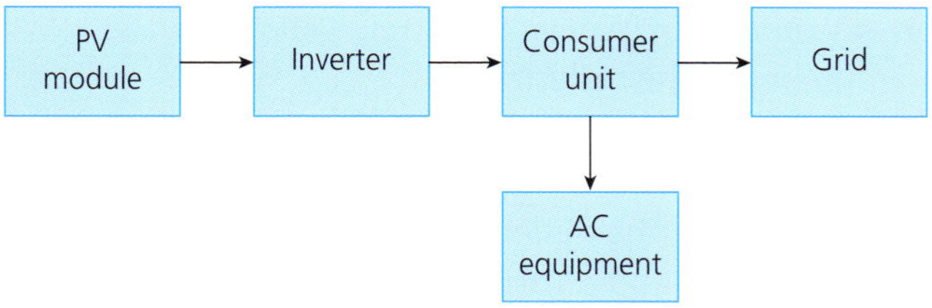

▲ Figure 5.22 On-grid system components

Health and safety

When working on solar PV systems it is important to follow industry safe isolation procedures.

PV modules are available in different efficiencies:

- Monocrystalline modules range in efficiency from 15 to 20 per cent.
- Polycrystalline modules range in efficiency from 13 to 16 per cent.
- Amorphous film ranges in efficiency from 5 to 7 per cent.

Photovoltaic modules can be fitted in different ways:

- On-roof systems comprise aluminium rails, which are fixed to the roof structure by means of roof hooks.
- In-roof systems replace roof tiles with the PV modules.
- Ground-mount and pole-mount systems provide free-standing PV arrays, which can also be installed as computer-controlled motorised mounting systems that rotate and track the sun as it moves across the sky.

An inverter converts the DC input to a 230 V AC 50 Hz output and synchronises it with the frequency of the mains supply.

Research

Research the operating principles of solar photovoltaic (PV) panels and provide an overview of how they convert sunlight to electricity.

Micro-wind

Micro-wind turbines harness energy from the wind and turn it into electricity. The UK is an ideal location for the installation of wind turbines, as about 40 per cent of Europe's wind energy passes over the UK.

The working principles of a wind turbine are as follows:

- The wind passes over the rotor blades of a turbine, causing it to turn.
- The hub is connected by a low-speed shaft to a gearbox.
- The gearbox output is connected to a high-speed shaft that drives a generator which, in turn, produces electricity.

Turbines are available as either horizontal-axis wind turbines (HAWTs) or vertical-axis wind turbines (VAWTs).

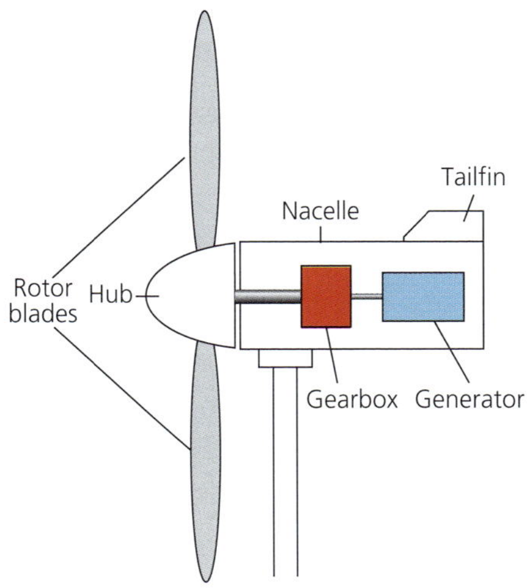

▲ Figure 5.23 Horizontal-axis wind turbine

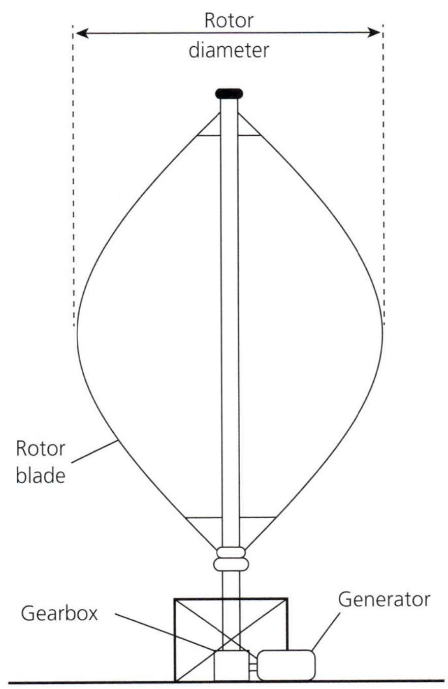

▲ Figure 5.24 Vertical-axis wind turbine

There are two types of micro-wind turbine suitable for domestic installation:

- pole-mounted (free-standing)
- building-mounted (fitted directly to the building, generally smaller than pole-mounted turbines).

As with solar PV, there are several connection options:

- off-grid systems charge batteries in order to store electricity for later use
- on-grid (grid-tied) systems are connected in parallel with the grid supply via an inverter.

▼ Table 5.8 Advantages and disadvantages of micro-wind systems

Advantages	• Good electricity output levels can be achieved • Zero-carbon technology • Effective where no mains electricity is available
Disadvantages	• Require a mounting site away from buildings and obstructions • High initial installation costs • Variable performance according to the availability of wind and location • Turbines can cause noise, vibration and **shadow flicker**

Key term

Shadow flicker: when the rotating blades of a wind turbine create moving shadows

Improve your English

Referring to your local planning requirements, research the location and building requirements for the installation of a micro-wind turbine. Draft an email to a potential customer who is considering one.

Micro-hydro

Micro-hydro is a type of hydroelectric power that harnesses kinetic energy as water moves from a higher to a lower level. The water passes across or through a turbine, which turns a generator and produces electricity.

As with the other micro-generation technologies, there are two possible system arrangements for micro-hydro: on-grid and off-grid systems.

The main components of the watercourse construction are:

- intake (the point where some of the river's water is diverted from the main stream)
- the canal (connecting the intake to the forebay)
- the forebay (a reservoir of water ensuring that the penstock is always pressurised and allowing surges in demand to be catered for)
- the penstock (pipework taking water from the forebay to the turbine)
- the powerhouse (the building housing the turbine and generator)
- the tailrace (the outlet that takes the water exiting the turbine and returns it to the mainstream of the river).

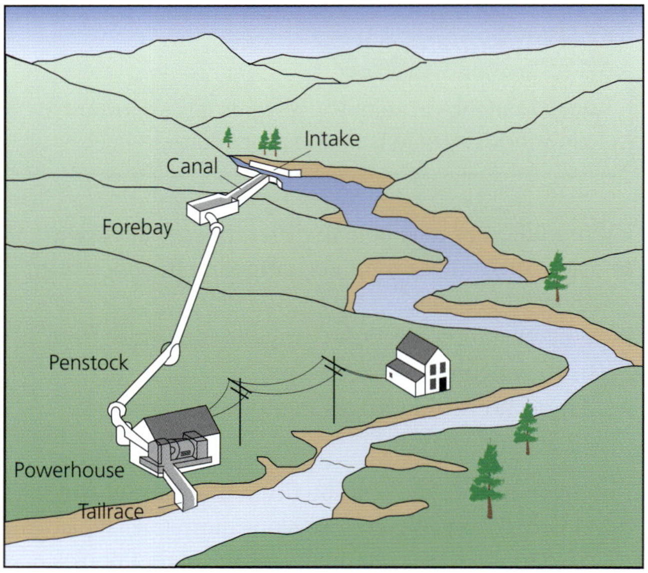

▲ Figure 5.25 Watercourse for a micro-hydro system

To ascertain the suitability of a water source for hydroelectric generation, it is necessary to consider the head (vertical height difference between the proposed inlet position and the proposed outlet) and the amount of water flowing through the watercourse.

▼ Table 5.9 Advantages and disadvantages of micro-hydro systems

Advantages	• Excellent payback potential • Zero-carbon technology • Effective where no mains electricity is available
Disadvantages	• Require a watercourse with suitable head and flow • High initial installation costs • Require planning permission from the Environment Agency

Industry tip

A survey is required prior to the installation of any renewable technology.

Micro combined heat and power (micro CHP)

Micro combined heat and power (micro CHP) systems generate useable heat and electricity for properties at the same time. The fuel source is usually natural gas or liquefied petroleum gas (LPG), but it could also be biomass.

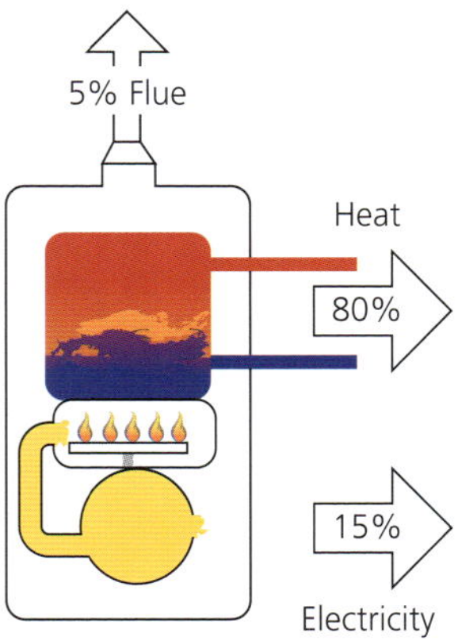

▲ Figure 5.26 Micro CHP boiler

The key components of a micro CHP boiler are the:

- engine burner (1)
- Stirling generator (2)
- supplementary burner (3)
- heat exchanger (4).

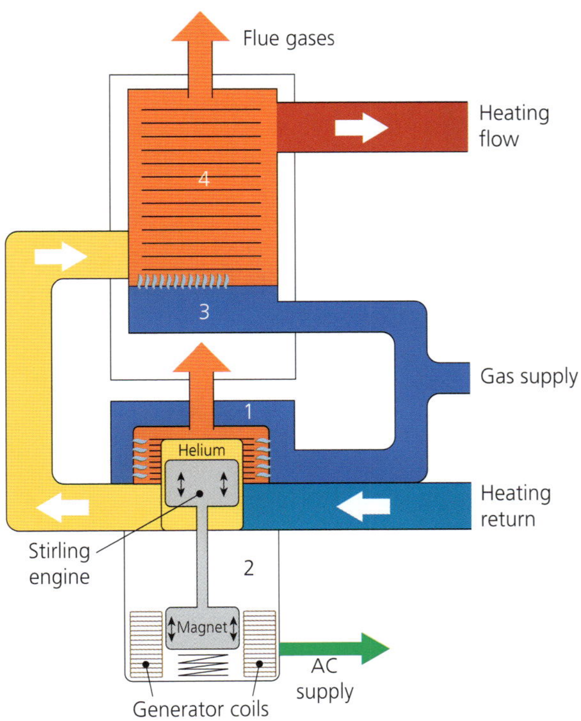

▲ Figure 5.27 Components of a micro CHP boiler

The boiler system works as follows:

- When there is a call for heat, the engine burner fires and starts the Stirling generator. The engine burner produces about 25 per cent of the full heat output. The burner preheats the heating-system return water before it passes to the main heat exchanger.
- Hot flue gases from the engine burner are passed across the heat exchanger to heat the heating-system water even more. If there is more demand than supply, the supplementary burner operates.
- The Stirling generator uses the expansion and contraction of internal gas (often helium) to operate the piston. When the engine burner fires, the gas expands, forcing the piston downwards.
- The return water from the heating system passes across the engine, causing the gas to contract. A spring mechanism in the engine returns the piston to the stop, and the process starts again.

▼ Table 5.10 Advantages and disadvantages of micro CHP systems

Advantages	• Domestic units now similar in size to central-heating boilers • Produce free electricity while generating heat • Do not rely on building orientation or weather conditions to operate effectively
Disadvantages	• More expensive than central-heating boilers • Not suitable for properties with low heat demand • Limited electrical generation capacity • Low-carbon rather than zero-carbon

Combined heat and power (CHP)

Combined heat and power (CHP) systems (also known as cogeneration systems) use waste thermal energy from the generation of electricity for district or industrial heating purposes. The heat is produced centrally at one or more locations and supplied to an unrestricted number of residential and industrial customers, using either steam or high-temperature water.

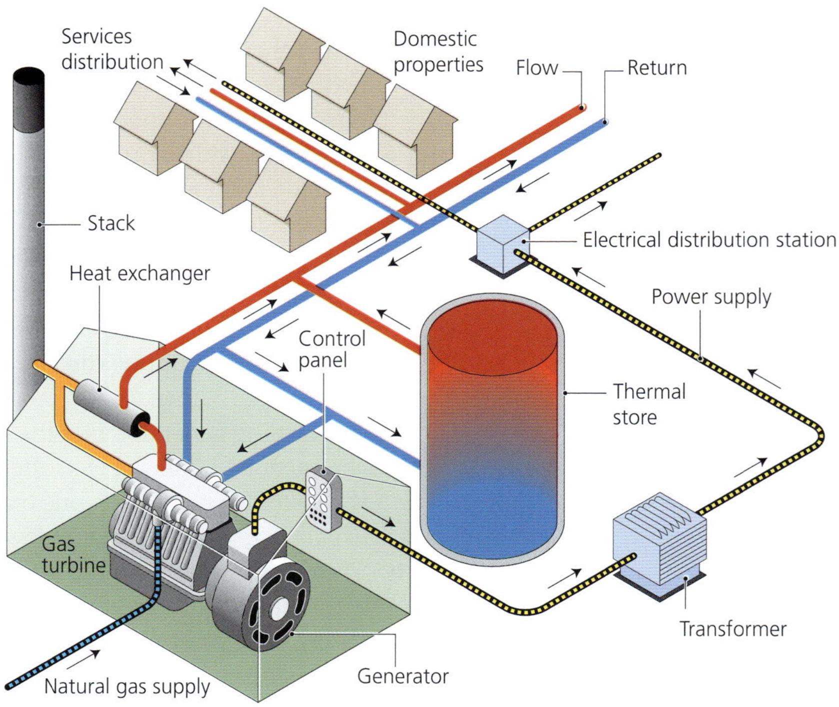

▲ Figure 5.28 Combined heat and power system

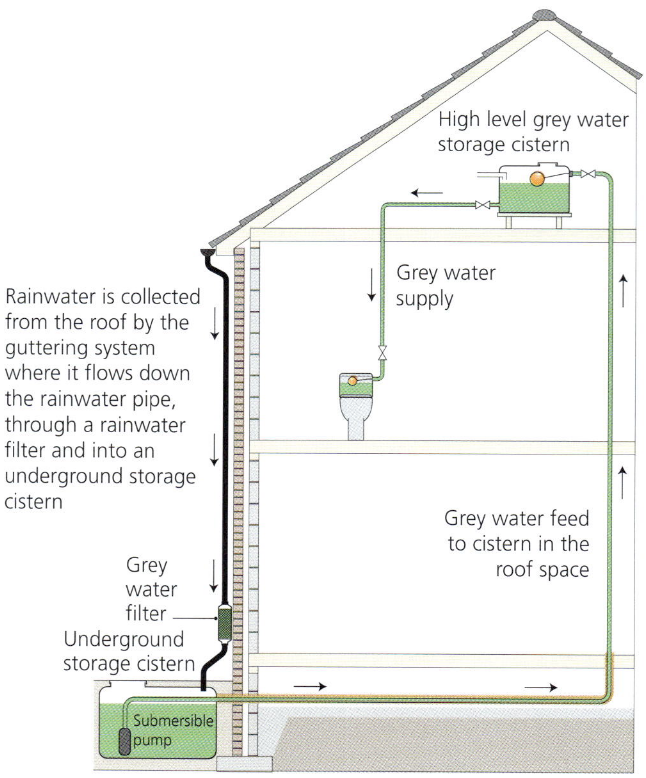

▲ Figure 5.29 Rainwater harvesting

▼ Table 5.11 Advantages and disadvantages of combined heat and power systems

Advantages	• Reduce energy costs • Reduce carbon emissions • Wide choice of fuels available, including coal, biogas, heating oil and bioethanol
Disadvantages	• Not suitable for all sites because of constant and consistent power demands • Can be expensive: initial cost is high due to the technology and payback periods can be very long

10.2 Energy conservation

Rainwater harvesting

Rainwater harvesting involves collecting rainwater from roofs, filtering it and storing it for reuse, thereby saving energy, lowering carbon emissions and reducing mains water consumption.

Systems also include a back-up mains water connection, for when rainfall is limited.

Harvested rainwater is not suitable for:

- drinking
- personal washing, showering and bathing
- preparing food
- dishwashing (by hand or machine).

However, if stored correctly it can be used for:

- flushing toilets
- washing clothes with a washing machine
- watering garden lawns, plants and flowers
- washing cars.

▼ Table 5.12 Advantages and disadvantages of rainwater harvesting

Advantages	• Conserves water • Indirectly reduces energy consumption and carbon emissions • Wide range of system options available • Annual cost of water reduces where a water meter is fitted
Disadvantages	• Potentially long payback periods • Not always straightforward to install in existing buildings • Risk of contamination or cross-connection • Only certain types of outlet and appliance can be supplied

Industry tip

All harvested rainwater pipework systems must be suitably marked to identify their use.

Grey-water recycling

Grey water is waste water that has been used for washing and is generated from hand basins, washing machines, showers and baths. Rather than sending it down the drain, it can be reused for watering plants and flushing toilets, thereby reducing mains water usage.

There are several different types of grey-water system available in the UK, but they all have similar features, including some form of treatment, a storage tank for the treated water, a pump and a distribution system.

Grey water should not be used for:

- drinking
- personal washing, showering and bathing
- food preparation
- dishwashing (by hand or machine).

If grey water is filtered, stored correctly and treated where necessary, it is suitable for:

- flushing toilets
- washing clothes with a washing machine (requires treatment)
- watering garden lawns, plants and flowers
- washing cars.

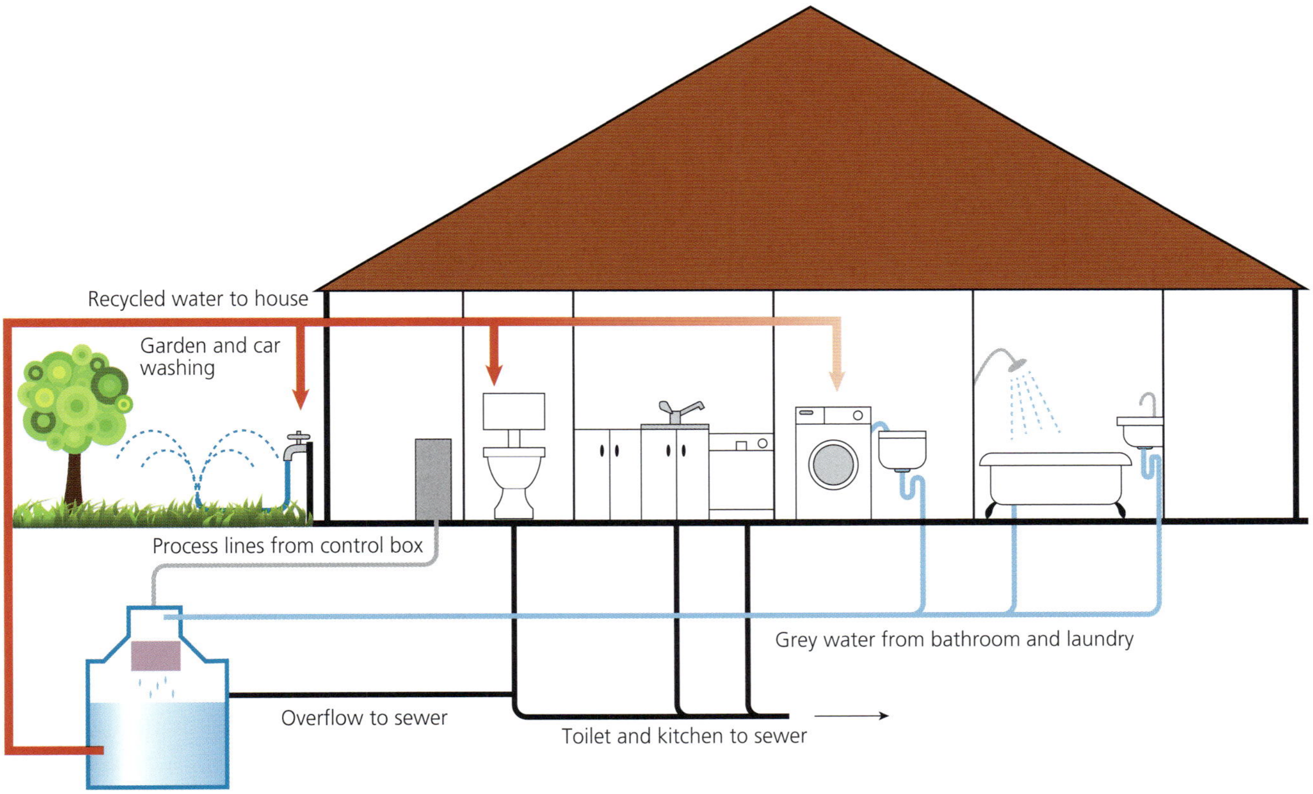

▲ Figure 5.30 Grey-water recycling

▼ Table 5.13 Advantages and disadvantages of grey-water recycling

Advantages	• Conserves water • Indirectly reduces energy consumption and carbon emissions • Wide range of system options available • Annual cost of water reduces where a water meter is fitted
Disadvantages	• Potentially long payback periods • Not straightforward to install in existing buildings • Risk of contamination or cross-connection • Only certain types of outlet and appliance can be supplied

Heat recovery ventilation (HRV) systems

Heat recovery ventilation (HRV) systems reduce the heating and cooling demands of a building. Installed at the top of a building, they extract warm, stale air and pass it through a heat exchanger. The heat is then drawn to incoming fresh air, which is transferred back into the building. They recover between 60 and 95 per cent of the heat in exhaust air and markedly improve a building's energy efficiency.

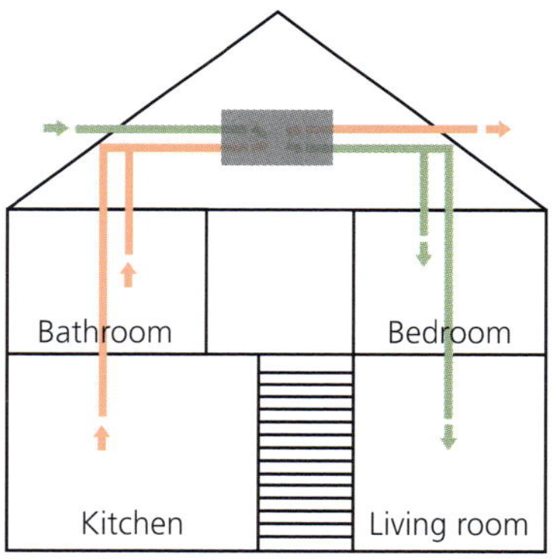

▲ Figure 5.31 Heat recovery

Research

Research one of the following types of heat recovery system:

- counterflow heat exchangers
- crossflow heat exchangers
- rotary wheel heat exchangers.

Explain this system to a classmate.

Test yourself

List three domestic uses for recycled water.

Energy-efficient lighting

Energy-efficient lighting not only lowers electricity bills but it also benefits the environment by reducing carbon emissions.

Incandescent bulbs are inefficient, converting around five per cent of the electricity they use into visible light. Halogen bulbs use the same filament technology as traditional bulbs but run at a higher temperature, making them slightly more efficient. These types of lighting have now been phased out and replaced with more efficient types:

- Compact fluorescent lamps (CFLs) comprise a mercury-filled glass tube with an inner coating of phosphor powder which glows when charged with electricity. They use around 60–80 per cent less electricity than equivalent traditional bulbs and last around ten times longer.
- Light emitting diodes (LEDs) use 90 per cent less energy than traditional bulbs and have begun to replace CFLs. They turn on instantly at full brightness and are available to fit almost any light fitting. There may be a large number of LEDs within a single bulb to create sufficient brightness.

▲ Figure 5.32 Energy-efficient lightbulbs

Electric vehicle charging points

New cars and vans powered wholly by petrol and diesel will be banned in the UK from 2030, resulting in a move towards electric vehicles. These are powered by rechargeable batteries and there are currently three types of public charging point:

- Rapid chargers have outputs of 43 kW and use a Type 2 connector. They are the fastest way to charge an electric vehicle and take between 30 and 80 minutes to reach 80 per cent charge.
- Fast chargers have outputs between 7 and 22 kW and use Type 1 or Type 2 sockets. It typically takes between three and four hours to charge a vehicle.

- Slow chargers have a three-pin plug, like a home charger. It can take up to ten hours to fully recharge a vehicle.

▲ Figure 5.33 Electric car charging point

Research

Visit the following website, which introduces electric vehicle charging: **www.voltimum.co.uk/articles/installation-electric-vehicle-charging**

Find out:

- What type of circuit is required?
- What type of protection is required?
- How is electric shock prevented?

Appliance efficiency ratings

The Energy-related Products (ErP) Directive was introduced in 2009 to help EU countries reduce energy consumption and emissions.

▲ Figure 5.34 Energy efficiency ratings

Energy labels are a common sight on many appliances, such as white goods, grading them according to their energy efficiency with a letter-based rating. These labels were first introduced in 1995 to drive innovation and help consumers make informed decisions about the products they purchase. However, in 2021 a new generation of labels was introduced for refrigerators, freezers, washing machines, washer-dryers, dishwashers, televisions and electronic displays, and light sources.

The energy efficiency rating system for these appliances now runs from A (most efficient) to G (least efficient). The energy labels also include QR codes, clearer guidance on energy consumption and extra information specific to product types, such as noise levels and water usage for washing machines.

Energy labels on other appliances, such as tumble dryers, ovens and air conditioners, still use the old rating system of A+++ to G, although there are plans to update these in 2022. For now, the UK is following EU legislation changes, although post-Brexit it also has the freedom to make its own changes.

Manufacturers are responsible for ensuring their products are accurately labelled.

Since September 2015, boilers, water heaters and heating products have been labelled in the same way as refrigerators and freezers to show how energy efficient they are. Under the ErP Directive, boilers must be labelled with their efficiency level from G (lowest efficiency) to A+++ (highest efficiency). Boiler manufacturers and retailers are responsible for ensuring their products are accurately labelled.

Insulation for pipework and ductwork

All pipes installed in vulnerable or exposed locations inside and outside a building, such as unheated cellars, roof spaces, garages and outbuildings, must be insulated to improve energy efficiency, prevent freezing and retain heat energy.

Types of pipework insulation include:

- PVC foam
- expanded polystyrene
- extruded polystyrene
- cross-linked polyethylene foam
- expanded nitrile rubber
- expanded synthetic rubber.

Ductwork should be insulated in order to:

- keep air within the ducts at the required temperature
- prevent ducts from leaking air.

Ductwork carrying warm air should be insulated to conserve energy. Ductwork carrying cold air should be insulated to conserve energy and control condensation on the outer surface.

Various thermal insulation products are available for ducting, including rigid preformed slabs or boards and flexible rolls, blankets, mats and mattresses.

Building insulation

Insulation is needed to reduce heat loss from a building during cold weather (see Figure 5.36). This in turn reduces the energy required to maintain a comfortable temperature for the occupants.

Insulation should also reduce the heat entering a building in warm weather. This saves energy by reducing the need for air-conditioning systems.

There are many different ways to insulate a building, making it more energy efficient and lowering running costs.

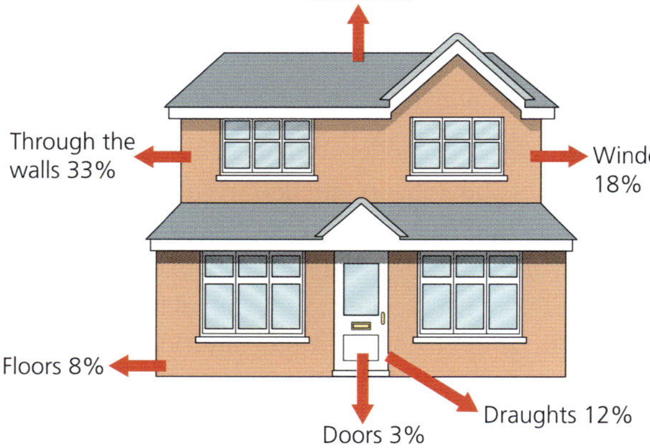

▲ Figure 5.35 Heat loss from a building

Cavity wall insulation

Cavity walls are constructed from two leaves of masonry, separated by a nominal 50 mm gap. The outer leaf is usually facing brickwork, and the inner leaf is brickwork or blockwork. Insulating an existing cavity wall involves drilling a series of holes through the outer leaf and blowing or injecting the insulation material into the cavity.

Cavity wall insulation materials include:

- blown mineral fibre
- polystyrene beads (EPS)
- rigid board insulation (installed during construction)
- urea formaldehyde foam.

External and internal wall insulation

External wall insulation involves fixing an insulating layer over the existing wall, either mechanically or adhesively, and covering it with a **render**.

Where external insulation is not an option, for example in conservation areas, internal wall insulation can be installed. This consists of thermally engineered insulated studs and insulation slabs. The use of internal wall insulation will reduce the internal space.

> **Key term**
>
> ***Render:*** a pre-mixed layer of sand and cement, similar to mortar, used to make masonry walls flat and prepare them for top coats of finishing plaster

Loft insulation

Installing loft insulation, or upgrading existing loft insulation, is the quickest and most efficient method of improving the thermal performance of an existing building. Loft insulation materials include:

- blown insulation material
- insulation boards
- roll loft insulation.

Current building regulations state that insulation materials should be laid to a minimum thickness of 270 mm.

Floor insulation

Around 15 per cent of heat is lost through an uninsulated floor. Insulating a ground floor will keep the building warmer and reduce energy bills. This can be done using rigid board insulation.

Upstairs floors do not need to be insulated, as they are within the thermal envelope of the building.

> **Research**
>
> Research the purpose, limitations and benefits of insulation materials. Copy and complete the following table:

Insulation method	Purpose	Benefits	Limitations
Foam board			
Loose fill			
Insulation roll			
Sprayed foam			
Reflective system			
Rigid fibrous			

Test yourself

Describe three methods that can be used to insulate a property.

11 Digital technologies

11.1 Internet of things

The term 'internet of things' (IoT) refers to objects connected via a digital network. The potential for integrating the IoT into building services engineering systems is vast, and this is covered in more detail in Chapter 10.

11.2 Control and monitoring systems

Building management systems

A building management system (BMS), sometimes referred to as a building automation system (BAS), is a computer-based system that monitors and controls a wide range of building services, for example:

- heating, ventilation and air conditioning (HVAC)
- lighting
- energy
- fire systems (smoke detection and alarms)
- security systems (CCTV, motion detectors and access controls)
- ICT (information and communications technology) systems
- lifts.

Successful implementation of a BMS will help building managers to understand how a building is operating in real time and allow them to control and adjust the system to ensure optimum performance and energy efficiency.

A BMS can provide:

- a visualisation of real-life building data with a graphic user interface, allowing comparisons between buildings and benchmark data
- a method for generating building usage reports
- time scheduling of building operations function, such as lighting, heating and security controls
- notification of faults and failures through a set of alarms and alerts.

Automated controls

Automated controls can be used to maximise energy efficiency in a building.

Movement sensors can switch building services on and off when required, for example lighting. As a person enters a room, a **passive infrared (PIR) sensor** detects them and switches on the light; when the person leaves, the sensor is no longer able to detect them and turns off the light.

Key term

Passive infrared (PIR) sensor: an electronic sensor that measures infrared (IR) light radiating from objects in its field of view

Infrared sensors are increasingly being used to operate taps, toilets, urinals and shower outlets, particularly in public buildings and hotels. The sensors operate solenoid valves that allow water to flow when the infrared beam is interrupted. These automated facilities not only save water, but they are also more hygienic because they are touch-free.

Smart controls

Smart controls allow users to control and monitor a range of building services from a mobile device such as a tablet or phone, for example:

- lighting
- heating
- security
- air conditioning
- ventilation.

Smart sensors connect to consumer Wi-Fi and have a companion app to control their functionality and set parameters, for example time and temperature.

▲ Figure 5.36 Smart controls

Smart meters

Smart meters are replacing traditional gas and electricity meters. They offer a number of benefits:

- Meter readings are sent automatically to energy suppliers.
- Energy bills are more accurate, as they do not rely on estimates.
- Consumers can see their energy usage in real time and track costs via an in-home display. They can look at how much energy has been used in the past day, week, month and year.

The smart meter equipment installed by energy suppliers usually comprises:

- a smart electricity or gas meter
- a communications hub
- an in-home display.

▲ Figure 5.37 A smart meter

Case study

Sian and her family live on the south coast of England. They are conscious of their current energy usage and would like to explore the use of smart controls and renewable energy sources to produce electricity. Their property is not listed, nor is it in a conservation area.

- What options could be considered for generating electricity?
- For each option, discuss the building regulations that would need to be complied with.
- Explain the smart controls that could be used to reduce electricity consumption.

Assessment practice

Short answer

1 State two common sustainability assessment methods.

2 List three control and monitoring systems for BSE systems.

3 List three sustainable solutions for construction.

4 Which document provides guidance for the installation of fixed building services in new and existing dwellings to help compliance with the energy-efficiency requirements of the Building Regulations?

5 State the three types of listed status for buildings in England and Wales.

Long answer

6 Explain how sustainability is achieved when planning and delivering a construction project.

7 Describe five retrofit roles.

8 Explain the term 'social sustainability'.

9 Explain what a passive house is.

10 Describe the operating principles of grey-water recycling and rainwater harvesting systems.

Project practice

Services PLC is an established building contractor, with branches throughout the UK. Each branch works on different types of development, including new-build residential and commercial retail.

Services PLC is preparing a tender submission for the building of a new school. You are part of the team working together to respond to the tender.

- Suggest a range of renewable technologies that could be included in the build.
- Summarise the purpose of each system, including advantages and disadvantages.
- For each system, explain the components and their purpose.
- Give examples of sustainable solutions that could be incorporated into the building, for example automatic lighting.
- Give examples of how waste will be managed on the construction project, including completion of a basic site waste management plan.

The information should be included in a project plan, which will be discussed in a group as part of the tender process.

Chapter 6 Measurement principles

This chapter looks at the importance of accurate and appropriate measurement, identifying the standard units of measurement and appropriate measurement techniques used in construction.

By applying tried and tested construction measurement principles, it is possible to ensure accuracy both in calculating the quantities of materials needed for a project and in setting out and building a structure. We will also look at the possible consequences of inaccurate measurements for both client and contractor.

The chapter concludes by looking at scales and tolerances used in construction.

Learning outcomes

By the end of this chapter, you will understand:

1. accurate and appropriate measurement
2. standard units of measurement and measurement techniques
3. measurement standards, guidance and practice.

1 Accurate and appropriate measurement

The methods of measurement used for construction activities vary depending on the work task and the operational phase of the construction project.

Well before construction work starts on site, a range of personnel will have taken many accurate measurements in order to design, plan and prepare for a project. Measurements will continue to be taken with accuracy during the subsequent construction phase.

▲ Figure 6.1 Appropriate measurements are taken when planning a project

Improve your English

Search online for definitions of the word 'measurement'. Select three definitions that you think could be applied to construction activities and combine them into one definition in your own words.

1.1 The benefits of accurate measurement

Accurate measurement ensures a project meets the requirements of the client and runs smoothly through to completion for the contractor.

Benefits to the client

When a client commissions a project, they have certain expectations:

- The work should be completed within an agreed budget and timescale.
- The **design brief** should be wholly fulfilled.

Key term

Design brief: a working document which specifies what a client wants; it makes clear all the design requirements of a project

These outcomes can only be achieved if accurate and appropriate methods of measurement are used during all phases of construction.

During the planning stage, accurate measurement of quantities of materials and components required for the job will mean the right amounts are delivered to site. This avoids a shortage of materials, which can cause production delays, or a surplus of materials, which can cause waste. Resulting cost savings can be passed on to the client.

▲ Figure 6.2 Accurate calculation of materials for delivery to site avoids production delays and waste

Industry tip

Costing a project of any size is a balance between making an acceptable profit for the contractor and offering a fair price to the client. Accurate costings at the beginning of a project may allow a contractor to pass savings on to the client while still maintaining healthy profit margins.

Using established principles when calculating required quantities of materials allows for assessment of the time needed to use and install those materials, contributing to the creation of a realistic work programme. Completion targets can then be set to support efficiency and productivity and satisfy the client's requirements for timely handover of the building.

Creation of a work programme founded on accurate calculations will also allow workforce requirements for the project to be established, which will sustain productivity and avoid potential delays due to shortfalls in workforce numbers. The client benefits because the project is more likely to be delivered on time.

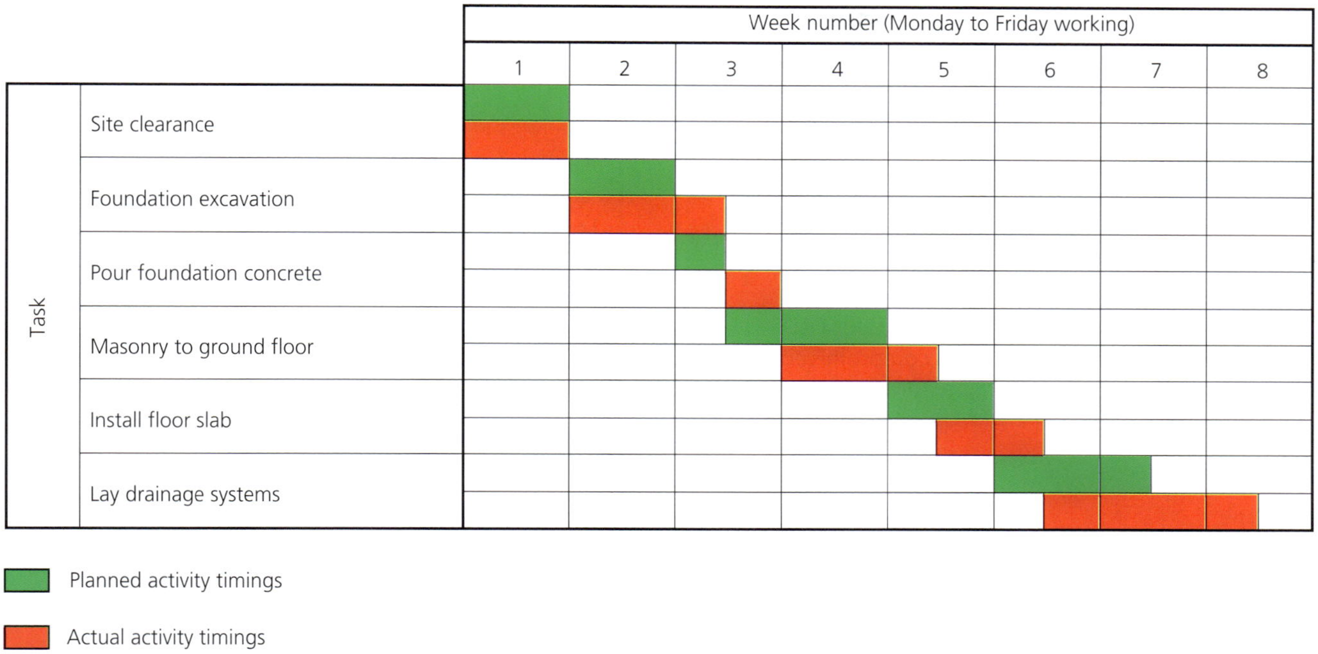

▲ Figure 6.3 Accurate calculation of quantities of materials contributes to creating a realistic work programme

Research

Once all the quantities of materials for a project have been accurately calculated, an evaluation of costs can be carried out, and the amount of time needed to complete specific tasks can be estimated.

Examine the room you are in carefully. Make a list of all the materials you think are contained in the room, for example timber for the floor and cables for electrics. Then for each material in your list, assign a tradesperson or operative who might use it. The list will vary depending on the use of the room, so repeat the exercise with a partner in a second room and discuss your findings. How many materials have you identified?

▲ Figure 6.4 A new building must be constructed using accurate measurements

In order to construct a building exactly as shown in the drawings and in accordance with specifications, it is vital to use accurate and appropriate methods of measurement.

If measurement principles are not applied correctly, it can cause adverse consequences:

- If a building that is not dimensionally accurate or not **built to square** is handed over to the client, significant problems can emerge when the installation of items such as floor tiles is attempted, or fittings and **modular components** such as kitchen units are positioned.
- Using unspecified materials or components may mean the building is not suitable for its intended purpose, for example reducing the width of timbers used in floor joists.

Key terms

Built to square: where all the corners of a square or rectangular building are accurately set at 90°

Modular components: standardised units that fit together in a number of ways

Health and safety

If a building is not constructed in accordance with drawings and specifications because of carelessness in measurement, the structural integrity of the design could be compromised, leading to dangerous structural failure.

Test yourself

State two benefits to the client of accurate measurement.

Benefits to the contractor

Accurate measurement allows for accurate costing of a project, which in turn can:

- help a contractor to manage profitability and cash flow, to maintain business activity and ensure company growth
- allow a contractor to negotiate with suppliers for favourable prices on bulk materials and specified components, to improve company profit margins.

▲ Figure 6.5 Costing a project must be done carefully

Careful use of accurate and appropriate methods of measurement during the construction of a building prevents errors, which may be costly to remedy. It would be especially problematic if errors were discovered once the building was occupied after handover to the client. Avoiding errors means the job runs smoothly and is more likely to be completed on time.

If a company consistently completes projects on time and to a high standard, by accurately constructing buildings to specification, it will over time build up a good reputation and ensure continuity of work. Carelessness and inaccurate work can quickly destroy a hard-earned and valuable reputation.

▲ Figure 6.6 Accurate work can result in a good company reputation

Test yourself

State one benefit to a contractor of accurate measurement.

1.2 Choosing appropriate methods of measurement

Throughout the stages of a construction project, different work activities require different methods of measurement. These should always be applied accurately.

▼ Table 6.1 Appropriate methods of measurement and the benefits of accuracy

Construction activity	Method of measurement	How the method is used	Benefits of accuracy
Site survey  ▲ A survey of a new site must be undertaken	Calculation of area	A building must fit on a proposed site, with allowances for roadways, paths, parking and landscaped areas. There should be consideration for adequate space between the new building and existing structures.	Accurate calculation of the area of the site is needed to confirm that the proposed project is feasible. Early confirmation that the site is suitable allows for an appropriate design concept to be created.
Locating services on a site  ▲ Existing services must be accurately located	**Linear measurement**	One of the first tasks on a new construction project is to locate the position of any existing services on the project site. Records regarding the position and route of services are used to accurately measure from stated positions on site. Once located, the services will be removed or rerouted.	Finding the precise location of services by accurate measurement is essential so that subsequent site work can proceed safely. Accidental damage to existing services could also cause delays and create added costs.
Groundworks including removal of topsoil and excavations  ▲ Large volumes of material may need to be moved	Calculation of volume	Before a building can be constructed, the soft topsoil must be removed from the area on which the structure will be positioned. To measure the amount to be moved, the volume of material is calculated in cubic metres (m^3). Excavations for foundations and underground drainage systems also require calculation of the volume of material to be moved.	Moving large volumes of soil or other similar material involves the use of heavy machinery, which is expensive to hire or buy. Accurate measurement of the quantity of material to be moved means the number of machines and the time needed can be established and the cost confirmed.
Correctly locating the new structure on the plot  ▲ Accurate measurements are needed when positioning a building on a plot	Linear measurement	Project drawings show where a building will be located on a site. Accurate measurement from given reference points is needed to position the building in the correct location.	If a structure is wrongly positioned, it may interfere with other elements of the development, or its position may contravene planning requirements. In extreme cases, a wrongly located building has had to be demolished and rebuilt in the correct position – at great expense.

▼ Table 6.1 Appropriate methods of measurement and the benefits of accuracy

Construction activity	Method of measurement	How the method is used	Benefits of accuracy
Building the new structure  ▲ Accurate measurements are needed throughout the course of construction	Linear measurement, calculation of area, calculation of volume	Accurate linear measurement is needed to set out the outline of a building, the position of internal walls, doors and windows, vertical dimensions to establish floor and roof heights and much more. Accurate calculation of area will confirm quantities of materials required for activities such as plastering, bricklaying and roofing. Calculation of volume identifies requirements in cubic metres for materials such as concrete or **screed** required for solid floors.	Accurate measurement of the outline and features of a building is essential if the building is to be constructed exactly as designed. For example, wrongly positioned internal walls may interfere with other design elements. Measurements to calculate quantities of materials must be accurate to avoid shortfalls or excess materials, affecting productivity or adding to costs.

Key terms

Linear measurement: the distance between two given points along a line

Screed: a levelled layer of material (often sand and cement) applied to a floor or other surface

Health and safety

It is essential to accurately locate existing services in order to allow subsequent work to be conducted safely:

- Damage to water services can cause flooding and disrupt supply to others in the locality.
- Accidental contact with electricity services can cause death or injury to workers or damage to machinery.
- Gas services carry a risk of explosion if pipes are damaged or breached.

Test yourself

For what type of work task would linear measurement be used? Give two examples.

Improve your English

Write a clear explanation of the difference between area and volume.

1.3 Costing techniques for different projects

Construction projects vary greatly in size and complexity. A project may be relatively simple, such as building a boundary wall, or highly complex, such as building a multi-storey office block.

Calculating the cost of such a wide variety of project types requires the use of different techniques to arrive at figures that are as accurate as possible. Let us look at some frequently used costing techniques.

▲ Figure 6.7 A complex construction project in London

Job costing

This technique is usually applied to specific client requirements for a distinct project or 'job'. It involves analysing the job in detail, breaking down the costs of each element and tracking and recording them as the job proceeds.

The cost elements are broken down into labour, materials and overheads. This costing technique is sometimes confused with process costing, where the steps to complete a job are identified and an average cost is applied to each step based on past experience.

Industry tip

When a contractor prices a job, the calculated labour and materials are sometimes referred to as 'direct labour' and 'direct materials', since they are sourced and paid for by the contractor.

Job costing sometimes includes a 'cost reimbursement' condition, where the client agrees to meet the cost of job elements that cannot be accurately calculated at the start of a project, for example if the design choice of certain items such as windows has not been finalised.

▲ Figure 6.8 Various techniques can be used to cost a project

Batch costing

This can be viewed as an extension of job costing, since a batch (like a job) is broken down into component materials to be costed individually. For example, concrete is manufactured from sand, aggregate and cement. The quantities of the three component materials can be costed separately and those costs totalled as a batch, usually expressed in cubic metres (m^3) in the case of concrete.

Batch costing is often used where the manufacturing process takes the form of a continuous flow of mass-produced construction materials, such as plasterboards or bricks.

Activity costing

This technique assigns costs to specific activities rather than whole jobs or processes. Identifying the many activities required to produce a building provides costing data that can feed into the overall costing process.

For example, the activity of building a brick wall can be costed by using specific information:

- The dimensions of the wall allow the number of bricks required to be accurately calculated.
- The quantity of sand and cement for mortar can be calculated.
- The hourly rate of pay for a bricklayer can be precisely set.
- The number of bricks laid per hour can be established from past work experience.

By identifying quantifiable activities, the associated costs for materials, labour, equipment and overheads can inform the costing process – both for the current project and also for future projects with similar characteristics.

Life cycle cost analysis

This technique is used to estimate the overall costs of a building throughout its entire life cycle. It allows for comparison of alternative building concepts at the design stage, so that an evaluation can be made of which design is most economically and environmentally beneficial over time. The analysis includes the cost of everything throughout the life cycle of the building, including:

- land acquisition
- construction
- energy
- water
- operation
- maintenance and repair
- replacement.

Some professionals consider life cycle analysis to be little more than a 'best guess' of costings for a building that could potentially be in use for 100 years or more. Over such an extended time period, circumstances may change unpredictably.

However, with increasingly demanding environmental regulations impacting on building design and operation, refinement of this technique could provide an important tool for costing new buildings that are designed to be more environmentally friendly.

Research

Search online for the 'Designing Buildings Wiki' website. Enter 'costing' into the search bar and research the differences between life cycle costing and whole-life costing.

Test yourself

Look at the costing techniques discussed in this section. Which technique would you use to cost the replacement of slates on a roof?

▲ Figure 6.9 A slate roof covering

Improve your English

There are other types of costing technique that mainly apply to manufacturing processes but can have an application in construction. Find out about target costing and write down your opinion on whether this technique could be used in a construction setting.

2 Standard units of measurement and measurement techniques

The UK construction industry uses **metric units** as standard. These are easy to use because they are based on a **decimal system** which provides a consistent relationship between different measurement units. The value of a measured number can be changed simply by moving the decimal point (see Table 6.2 later in this chapter).

▲ Figure 6.10 Metric units of measurement are easy to work with

Industry tip

You may hear **imperial units of measurement** being discussed on site. Material sizes are often referred to in imperial units, even though they are now sold in metric units. An example of this is 8ft × 4ft sheets of plywood or plasterboard, where the correct size is 2400mm × 1200mm.

Key terms

Metric units: decimal units of measurement based on the metre and the kilogram

Decimal system: a number system based on the number ten, tenth parts and powers of ten

Imperial units of measurement: units of measurement of the British Imperial System, used until 1965 when the metric system was adopted

2.1 Units of measurement

Some standard metric units of measurement and their abbreviations are as follows:

- Length
  - millimetres (mm)
  - centimetres (cm)
  - metres (m)
  - kilometres (km)

- Weight or mass
  - gram (g)
  - kilogram (kg)
  - tonne (t)
- Liquid
  - millilitre (ml)
  - litre (L or l).

Table 6.2 shows the relationships between metric units of measurement and gives examples of how they are applied.

▼ Table 6.2 Standard metric units of measurement

Measurement	Relationship between units	Example
Length	10 mm = 1 cm	1 mm × 10 = 1 cm
	100 cm = 1 m	1 cm × 100 = 1 m
	1000 m = 1 km	1 m × 1000 = 1 km
Moving the decimal point changes the value: • 6250 mm can be shown as 6.250 m • 6250 m can be shown as 6.250 km.		
Weight or mass	1000 g = 1 kg	1 g × 1000 = 1 kg
	1000 kg = 1 t	1 kg × 1000 = 1 t
Liquid	1000 ml = 1 l	1 ml × 1000 = 1 l

Research

Research the units of measurement for length and weight used in the Imperial system. Create a table showing how they relate to each other, for example how many ounces in a pound. Can you see how much easier it is to use the metric system?

Improve your maths

Convert 12.6 cm into millimetres. Work with someone else to think up some other examples to convert.

Units of area and volume are derived from units of length:

- area is measured in square units, for example square metres (m^2)
- volume is measured in cubic units, for example cubic metres (m^3).

Every building must be designed to withstand the loads and pressures created by:

- the weight of the materials it is constructed from
- the weight of equipment and occupants within it
- external forces, such as wind or the weight of snow.

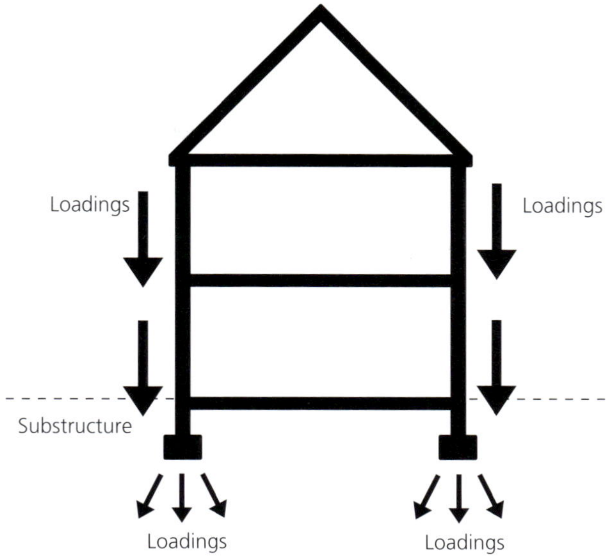

▲ Figure 6.11 The measurement unit for loadings is the newton (N)

These forces are measured in newtons (N). For example, the force exerted by pressure on concrete can be measured as newtons per square millimetre (N/mm^2). This measurement is used to classify the amount of crushing force that concrete can withstand.

Industry tip

Structural engineers must have a clear understanding of the definition of the newton: 'The force which gives a mass of 1 kilogram an acceleration of 1 metre per second, per second (1 kg m/s^2)'. This enables them to produce accurate and reliable structural calculations for a building.

Research

Research the origin of the newton as a unit of measurement.

Even time has a standard unit of measurement – the second (s), which is the basis for time management in every aspect of human activity, including construction.

Many of these units of measurement we take for granted, using them every day without thinking too much about how they fit together or have an influence in our lives.

In construction, these standard units of measurement are used in specific ways using appropriate equipment. Let us look now at some measurement techniques.

2.2 Measurement techniques

Measuring equipment used in construction work can be relatively simple or technically complex.

Simple equipment such as a tape measure can give accurate results when used carefully to measure:

- height
- length
- distance.

In any structure, it is important to measure heights accurately, for example of floors, ceilings, door and window openings, and roofs. Any inaccuracies could have serious consequences:

- If the height of a ceiling is not measured accurately during the building process, it may clash with the height measurement of the tops of doors or windows in the room.
- If the height of floors is not measured accurately during installation, a staircase will be placed at the wrong angle of slope, and it may not sit properly where the top of the stairs meets the upper floor.

▲ Figure 6.12 Getting a height measurement wrong could have serious consequences

Accurate measurement of length is also important if the design is to function as intended. Accuracy is vital when measuring the outline of the building, the position of internal walls and the position of doors and windows along an **elevation** of a building.

Test yourself

Give two reasons why accurate measurement of height is important.

Key term

Elevation: a view of the front, back or sides of a building

Technically complex equipment can be used to measure height, length or distance to high levels of accuracy. Laser instruments can be used to measure differences in height and 3D laser scanning can be used to give accurate data about the height and length of existing structures during a survey.

▲ Figure 6.13 A 3D laser scanner

Traditionally, an instrument called a theodolite has been used to measure distance and height, using mathematical methods of trigonometry and triangulation to establish accurate angles and measurement details for setting out new buildings or surveying existing structures.

Improve your English

Search online for 'how to use a theodolite'. Write a short report on the steps for setting up the instrument.

The optical theodolite has evolved into an instrument known as a total station. This modern surveying instrument combines an electronic version of the theodolite (which can record location data using GPS) with an electronic distance meter to speed up the measurement process. Some versions can even be operated remotely by a single operator and the gathered data can be streamed to an office location for reliable storage.

▲ Figure 6.14 A total station surveying instrument

Test yourself

What is the name of the instrument that has been developed from the theodolite?

To ensure efficiency and productivity throughout the construction process, accurate measurements are important for calculating the quantities of materials required. Calculations of area and volume are used to establish the amount of plasterboard needed for walls and ceilings, bricks and blocks needed for walls, and concrete needed for foundations and solid floors.

Industry tip

Although accurate measurement is important when calculating quantities of materials, sometimes to speed up a specific job it might be acceptable to use approximation. For example, you may see someone pacing out the length of an excavated trench to establish an approximate measurement in metres, to quickly calculate the volume of concrete required for a foundation.

Accurate measurement is also important when establishing the weight or mass of materials and the capability of lifting equipment. There is usually no means of weighing things on site, but the weight of items should be labelled on packaging, and equipment for lifting and transporting materials should have a safe working load (SWL) clearly and permanently indicated on it.

Industry tip

While scales and weighing equipment are not routinely installed on a site, if large volumes of material such as topsoil are to be removed it may be considered economical to install a weighbridge to monitor and record the weight of materials being transported off site.

Legal requirements to make new structures more environmentally friendly have led to the use of specialised measurement techniques for calculating heat transfer and energy use. Energy used in the construction and operation of a building produces carbon. This combines with oxygen in the atmosphere to produce carbon dioxide (CO_2) gas, which contributes to climate change.

To calculate CO_2 emissions, the thermal transmittance of the building's structure must be established. This is known as the U-value, which expresses the rate of transfer of heat through a structure (or more correctly through one square metre of a structure), taking account of the difference between internal and external temperatures. It is expressed in watts per metre squared kelvin (W/m^2K).

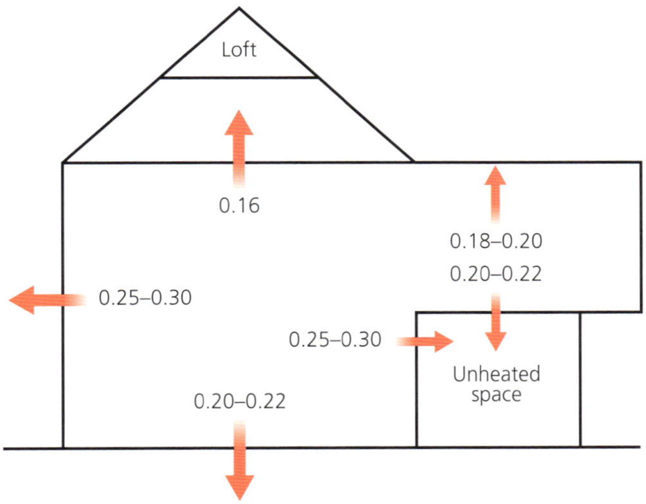

▲ Figure 6.15 Heat loss expressed as U-values in a typical house

To control heat transfer in a building, insulation must be used. Measured resistance to heat transfer is expressed as the R-value. The higher the R-value, the more resistance a material provides to heat transfer.

▲ Figure 6.16 Insulation must be installed to control heat transfer

The complex process of calculating U-values and R-values has been simplified with the introduction of software that analyses all relevant data in order to calculate the values to a high degree of accuracy.

Industry tip

It is pointless calculating the required level of insulation if operatives do not use appropriate techniques when installing insulation materials. Measures to limit heat transfer are only effective when insulation materials are properly installed.

The standard units of measurement and many of the measuring techniques discussed in this section can be linked to the use of data sources such as Building Information Modelling (BIM). This is a system for digital storage, handling and distribution of information and data. Chapter 8 covers BIM in greater detail.

More traditional data sources, such as drawings, are commonly used to 'take off' information for calculating quantities of materials and establishing timescales for work programmes. Let us look at some points regarding standards and practice when using this type of data source.

Improve your English

Describe the difference between U-values and R-values.

3 Measurement standards, guidance and practice

Construction drawings are produced to a set of **conventions**, which include particular units of measurement, views and sheet sizes. Conventions in measurement standards include the consistent use and application of **scale** when producing drawings as information sources.

▲ Figure 6.17 Construction drawings

Key terms

Conventions: agreed, consistent standards and rules

Scale: when accurate sizes of an object are reduced or enlarged by a stated amount

An architectural technician or a draughtsperson produces drawings of a building to scale. This means that large structures can be represented with accurate proportions on a document that is much smaller, so that it is easy to work with. It would obviously be difficult to work with a drawing of a house that was produced at full size.

Scale is shown using a **ratio**, such as 1 to 10. This would usually be written in the form '1:10'. Using this scale, a drawing of a feature that is 1 m (or 1000 mm) long in real life would be drawn 100 mm long on the paper. This is because 100 mm is a tenth of 1000 mm.

Key term

Ratio: the amount or proportion of one thing compared to another

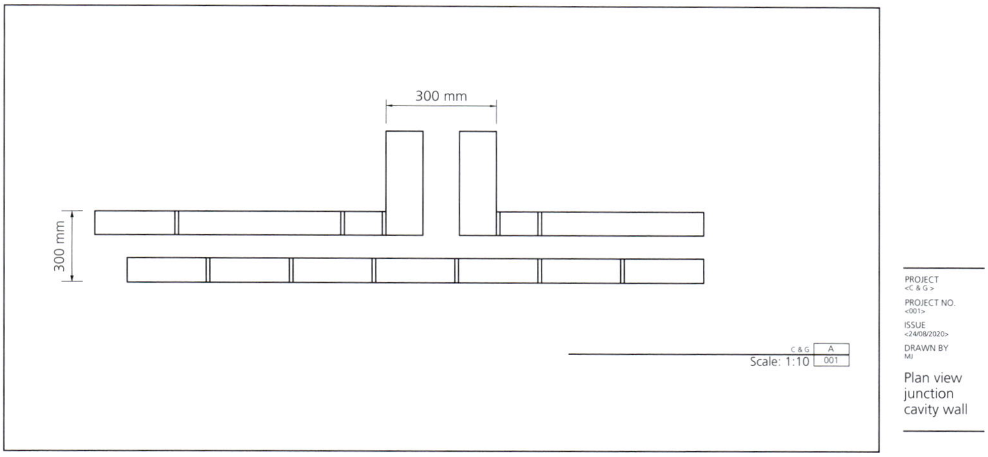

▲ Figure 6.18 A drawing showing a scale of 1:10

Test yourself

What name is given to a set of agreed, consistent standards and rules?

In some cases, more than one scale will be used on the same drawing, with different scales being used for different views, perhaps to allow for greater detail to be shown or a wider view to be given in context. A scale drawing may also use more than one unit of measurement, for example to show air-flow velocity in a ventilation system an arrow 1 cm long could represent a flow rate of one metre per second (1 cm = 1 m/s), regardless of the scale of the main drawing.

3.1 Common scales

Table 6.3 shows the scales commonly used for different types of drawing.

▼ Table 6.3 Types of drawing and common scales

Type of drawing	Description	Scales commonly used
Detail drawing	This shows accurate, large-scale details of the construction of a particular item, such as a timber-frame structural corner or the makeup of a suspended concrete floor.	1:1, 1:2 and 1:5
Floor plan	This shows the layout of internal walls, doors and stairs in a building. In a dwelling, it also shows the arrangement of special-use rooms, such as bathrooms and kitchens.	1:50 and 1:100
Elevation drawing	This shows the external appearance of each face of the building, with features such as slope of the land, doors, windows and the roof arrangement.	1:50 and 1:100
Sectional drawing	This is a slice or cut through of a structure to give a clear view of details that would otherwise be hidden. For example, on a working drawing for a house, a sectional drawing could allow us to clearly see the layout of the stairs within the building.	1:50 and 1:100
General arrangement drawing	Sometimes referred to as a location drawing, this can be used to show a single building element and what it should contain. It can also be used to show the main elements of a structure, such as the external walls, internal or partition walls, floor details and stairs.	1:50, 1:100 and 1:200, depending on the level of detail required
Site plan	This shows the proposed development in relation to the property boundary. It also shows the positions of drainage and other services and access roads and drives. It may show the position of trees and shrubs if they are required as part of the planning details.	1:200 or 1:500
Block plan	This shows the proposed development in relation to surrounding properties. It must be based on an up-to-date map and drawn at an identified standard metric scale. It usually shows individual plots and road layouts on the site as a simple outline with few dimensions.	1:1250 or 1:2500

Improve your maths

Show your understanding of how scale works by filling in the gaps in this table:

Scale size on drawing	Scale	Actual size (mm)
10 mm	1:10	100 mm
25 mm	1:20	
	1:50	300 mm
50 mm	1:200	

In order to be used reliably during both the planning and construction stages of a project, drawings should comply with British Standards.

BS EN ISO 19650 sets out a comprehensive code of practice for managing the production, distribution and quality of construction information, including drawings. It allows for the co-ordination of a range of essential construction data and information.

To produce drawings that are consistent and can be formatted to suit particular applications, they are laid out on standard-sized sheets of paper. Since different sizes of paper can be used and drawings can be printed and formatted in different ways, it is important that the scale used for drawings is clearly stated. Larger standard-sized paper can be used to produce drawings to a scale that shows greater detail, if required.

Table 6.4 shows the dimensions of standard paper sizes. Note that the higher the number, the smaller the paper.

▼ Table 6.4 Standard paper sizes

Title	Size (mm)
A0	1189 × 841
A1	841 × 594
A2	594 × 420
A3	420 × 297
A4	297 × 210

Research

Table 6.4 shows standard paper sizes from A0 to A4. Research additional paper sizes and list their measurements.

When using a drawing to build a structure or install components, written dimensions and measurements on the drawing should be used to ensure the work is completed exactly as designed. Using a scale rule on a drawing to establish dimensions for a work task can lead to errors. However, a scale rule can be used to 'take off' measurements from a drawing when calculating quantities, if written dimensions are not shown.

▲ Figure 6.19 A scale rule

Research

Using a scale rule to determine measurements that are not shown on a drawing can cause problems. This is because it is difficult to interpret the scaled thickness of drawn lines, especially for scales that greatly reduce the size of the object, such as 1:500. A relatively thin line on a drawing could be interpreted as having a thickness much greater than in real life.

Obtain a scale rule (borrow one if you do not own one). On a blank piece of paper, use a pen to draw two parallel lines about 100 mm apart. Find the 1:500 markings on the scale rule and read the distance between the two lines you have drawn. Note the measurement and ask someone else to do the same. Are both your measurements the same?

Costing a project from drawings and other data sources must be done in accordance with 'rules of measurement'. The rules most commonly used are the New Rules of Measurement (NRM), produced by the Royal Institution of Chartered Surveyors (RICS). This is a standard set of measurement rules and essential guidance for the cost management of construction projects and maintenance works. Their use helps to prevent disputes and maintain efficiency and productivity.

3.2 Tolerances

As stated a number of times in this chapter, accuracy when using and interpreting information from a range of data sources is essential if a building is to be completed exactly as designed. However, a building will often be constructed in conditions that are not perfect. In addition, many materials have characteristics that make it difficult to produce a flawless result.

For example, a concrete foundation or solid floor slab may easily be shown on a drawing with straight lines to indicate its depth and position, but the properties of the material do not allow it to be installed with a perfectly straight profile or edges.

▲ Figure 6.20 It is difficult to install concrete with a perfectly straight profile or edges

Allowances can be made for variations in the installation of certain materials and components in the form of **tolerances**, expressed as plus or minus (±) a stated amount. These tolerances will vary, depending on the nature of the material. For example, a double-glazed unit has a very limited range of tolerance – if it is larger than the window frame to be glazed by only 1 mm, it simply will not fit.

Key term

Tolerances: allowable variations between specified measurements and actual measurements

By contrast, when setting out a brick wall, it may be necessary to adjust the size of the mortar joints to fit whole brick sizes into the overall length of the wall, to avoid cutting bricks. In this case, a tolerance of ± 3 mm is allowable. This means that a 10 mm mortar joint can be enlarged by 3 mm to 13 mm or reduced by 3 mm to 7 mm to accommodate whatever is needed.

Tolerances may also be used to allow for acceptable variations in the strength of materials, the performance of a heating or ventilation system, temperature ranges in which materials can be used and many other situations.

Test yourself

Give one reason why tolerances are necessary when using certain materials.

Industry tip

Manufacturers often provide guidance on tolerances for the installation of components and materials they make. Building regulations and British Standards are another source of information on tolerances that are acceptable for a range of materials and associated construction activities.

Exceeding allowable tolerances for a given construction task will mean that the work does not meet the specification. The function of the structure could be compromised and other elements of the building may not be able to be installed properly.

Assessment practice

Short answer

1 What units are used when calculating the quantity of topsoil to be removed from a site?
2 What two expectations for a project must a contractor fulfil to satisfy the client?
3 Write a definition of the decimal measuring system.
4 Name two work activities when constructing a building that use linear measurements.
5 What does NRM stand for in relation to costing a project?

Long answer

6 Explain how accurately calculating quantities of materials and components assists in the creation of a feasible work programme.
7 Explain the reason for the use of tolerances and the implications of not meeting them.
8 Explain the possible consequences to the contractor and the client if a building is constructed out of square due to carelessness in measurement.
9 Describe how the installation of a staircase could be affected if floor levels in a building are not accurately set at the designed height measurement.
10 Explain how heat transfer related to construction activities is measured.

Project practice

A new college is to be constructed on a site located on a hillside. The site has an old farmhouse and a number of farm buildings on it, which have to be demolished and removed.

Adequate parking must be provided for staff and students, and a sports and recreation area is included in the plans.

Working in a small group, list the sequence of work from start to finish and select a suitable costing technique for each stage of the work.

You will need to consider the work needed to:

- carry out demolition
- manage the flow of rainwater down sloping ground
- move soil and shape the landscape
- construct the parking area, college building and recreation facilities.

Chapter 7 Building technology principles

The construction industry is evolving rapidly, with innovations in building technology to meet the demand for more affordable housing while reducing the impact of construction work and manufacturing on the environment.

In this chapter, we will look at the history of domestic buildings, forms of construction and different building materials. We will also compare traditional onsite construction methods with modern off-site manufacturing, and look at sources of building information and guidance.

Learning outcomes

By the end of this chapter, you will understand:

1. construction methods
2. forms of construction
3. key content and required notifications of UK building regulations and approved documents
4. building standards
5. regulatory bodies and guidance on technical safety and legislative aspects
6. manufacturers' instructions
7. building structure and fabric.

1 Construction methods

1.1 Onsite construction

In the early years of construction, all buildings in the UK were made on site, with materials often sourced locally from the surrounding land. This led to the development of low-rise dwellings with walls made from heavy timbers, **cob** or natural stone bonded together with lime mortar, or a combination of these materials. The walls were solid and often built directly on the ground with little or no foundations to support them. Roofs were often pitched with timber rafters and covered with slate or thatch to protect the occupants from the elements.

Over the years, construction methods have evolved, due to advances in construction design and technology, alongside the ability to use different building materials which are manufactured and transported to site from further away.

Table 7.1 illustrates how traditional and modern construction methods have transformed the way we build today.

Key terms

Cob: a blend of subsoil (clay or earth), sand and straw mixed with water to make an organic material historically used to construct walls for homes and agricultural buildings in Devon, Cornwall and Wales

Wattle and daub: a building method whereby a woven matrix of wooden strips (wattle) is covered with a mud-based daub; this technique was used for the construction of non-load-bearing walls and infill panels in buildings for thousands of years, but since the development of plasterboard it has become obsolete and is only ever used in restoration of historical or eco buildings

Mullioned windows: windows divided by vertical members, usually made of stone or wood

Pebbledash: a coarse surface finish for external walls consisting of small shells, gravel or pebbles applied to cement render while it is still wet

▼ Table 7.1 Traditional and modern construction methods

Period	Build method
Tudor (1485–1603)	• Thatched or slate roofs • Exposed timber frame with **wattle and daub** walls
Stuart (1603–1714)	• Single-skin stone and brick walls, sometimes rendered • Granite-framed and **mullioned windows** with coloured glass • Slate floors
Queen Anne (1702–14)	• Terracotta tiles and panels • Single-skin red-brick walls • Windows with glazing bars
Georgian (1714–1837)	• Single-skin stone or handmade brick walls • Buildings constructed either directly off firm ground, or on stone, brick or concrete foundations • Vertical sliding wooden sash windows • Architecture, symmetrical facades
Victorian (1837–1901)	• Single-skin brick solid walls, built on concrete, hydraulic lime or stepped brick foundations • In some areas of the country, cavity walls were used for the first time in late Victorian dwellings • First use of a damp-proof course (DPC), made of materials such as lead or copper • Pointed arches constructed over doors and wooden or metal-framed windows (Note: most of the population still lived in small houses or cottages)
Edwardian (1901–10)	• External cavity walls used in some buildings • Introduction of electric lighting • Some timber framing • Hanging wall tiles • **Pebbledash** walls • Timber porches and balconies

Period	Build method
Addison Act (1919)	• Introduction of affordable council housing after the First World War • Commonly 3 to 4 bedrooms, with indoor toilets, baths and hot running water • Often constructed with brick, block or concrete walls • Window openings designed for lots of natural light
Semi-detached (1930s)	• Hipped roofs and pebbledash walls • Many homes still constructed with single-skin brick walls bonded with lime mortar • Recessed porches • Mock timber framing • Timber bay windows
Art Deco (1920s–40s)	• Flat roofs • Plain white walls • Metal-framed windows (known as 'Crittall' windows) • Open interiors with Egyptian influences used in the design • Other forms of foundations introduced, including piles and raft
Prefabs (1940s) after the Second World War	• Mass produced in factories and assembled on site • Precast concrete columns and metal tubing • Small windows
Terraced (1960s–70s)	• Integral garages • Clad with hanging wall tiles or weatherboarding • Single-glazed aluminium windows and doors • Polythene damp-proof membranes (DPM) used as a barrier at ground level for the first time to prevent moisture entering the building
New build (1990s)	• uPVC double glazing • Insulated roofs, floors and cavity walls with a rendered finish • Period features
New build (current day)	• Energy efficient, eco-friendly building materials • Open plan • Good use of glass for solar gain

▲ Figure 7.1 Victorian house

▲ Figure 7.2 Art Deco house

▲ Figure 7.3 A dwelling constructed from cob

▲ Figure 7.4 Wattle and daub

There are many different forms of construction used to build new dwellings.

Onsite construction using traditional methods such as brick and block cavity walls or timber frame can be slow, and progress can often be delayed by adverse weather.

In the pursuit of building sustainable homes more quickly, with less of an impact on the environment, alternative methods of construction are sometimes used. These might include the use of straw-bale walls or, less conventionally, shipping containers.

Although the idea of building with straw bales ticks a lot of sustainable-construction boxes, this method is unlikely to be used for large housing developments because of its limitations, for example it can only be used for low-rise buildings, as well as having a limited lifespan and often rustic appearance. On the other hand, the idea of using shipping containers to build is not as extreme as it may first appear, especially if they are converted into habitable spaces in factories and delivered to site in modular form.

▲ Figure 7.5 Modular container house

Besides the construction of most new dwellings on site, the services and fabric of existing buildings still need to be maintained, and at some point upgraded, while in situ. Cosmetic and structural renovation and refurbishment of buildings are also jobs that are often only able to be completed on site because of the nature of the work.

The use of 3D printing and robotics as methods of onsite construction are covered later in this chapter.

1.2 Modular construction

Modular construction combines pre-engineered units (or modules) to form major elements of a building. These modules are manufactured in factories and then transported to site where they are assembled.

After the Second World War, there was a housing shortage in the UK. In order to provide homes quickly, many temporary buildings known as **prefabs** were made using precast concrete sections. However, modern prefabricated buildings are far superior to historical examples, and modular construction is now used to build permanent structures that are comfortable for the occupants, attractive to look at and sustainable.

Key term

Prefabs: buildings manufactured using factory-made components or units that are transported and assembled on site

Approximately 13 per cent of all new houses built in the UK use modular construction, as it offers many benefits:

- Automated systems improve accuracy and quality control.
- Large numbers of components can be manufactured efficiently with reduced waste.

- By carrying out a large proportion of the construction work in a controlled factory environment, the delays associated with adverse weather conditions on site are avoided.
- Up to 95 per cent of a building can be constructed in a factory, therefore reducing the number of trades needed on site during the assembly and final finishing of the project.

Due to the cost savings from mass production in factories, standard houses using modular construction can be produced more cheaply and efficiently, with better quality control, than using traditional bricks and mortar. It is also possible to design and construct unique houses in modular form, although these are usually more expensive, depending on the specification.

Most modular houses are constructed using structural insulated panels (SIPs) or timber frame, although heavier and less-sustainable precast concrete can also be used. SIPs are made using two layers of oriented strand board (OSB) bonded to each side of a polyurethane insulation core to provide a strong, rigid and highly insulated panel.

There are several types of modular building, each providing a different level of finish:

- turnkey (pre-assembled) – fully finished with plastered walls, fixtures and fittings, such as bathrooms and kitchens
- shell only (pre-assembled) – walls, floors and roof assembled, with no fixtures and fittings or internal finishes
- panel system (flat pack) – the basic elements of the structure provided, ready to be assembled on site.

Further information on modular construction can be found in Chapter 3.

▲ Figure 7.6 Pre-fab dwelling

Test yourself

Explain the advantages of using modular construction.

1.3 First fix

The construction of a new building is broadly divided into three areas of work, known as structural, first fix and second fix. As the name suggests, structural work includes the main load-bearing elements of a building, such as the foundations, basement (where there is one), walls, precast concrete staircases, floors and roof.

First-fix building work has a different meaning for each trade, although it is generally considered to be the phase of work completed after the structure has been erected and before plastering commences.

Table 7.2 outlines some of the first-fix tasks completed by various trades.

▼ Table 7.2 Work undertaken at first fix

Construction operative/ trade	Range of work undertaken
Plumber	Installation of pipes through the building for hot- and cold-water systems
Electrician	Installation of cables and back boxes for switches, electrical sockets, alarms, smoke detectors, CCTV, data-networking cables and television
Gas fitter/heating and ventilation fitter	Installation of pipes for gas distribution for boilers, underfloor heating, and ducting for heating, ventilation and air-conditioning systems
Carpenter	Installation of flooring, door frames, linings, staircases, stud partition walls and windows **Encasing services**
Dryliner/plasterer	**Dryling** ceilings and walls in preparation for plastering, or taping and filling the joints between the plasterboard sheets

Key terms

Encasing services: creating a framework of materials to cover pipes, cables and other services in a building; also referred to as boxing-in

Drylining: fixing plasterboard to ceiling and wall surfaces, such as metal suspended ceilings, timber and metal stud partition walls, and masonry walls

▲ Figure 7.7 First-fix plumbing

The first-fix stage of construction needs to be well planned and executed, especially when installing pipes, cables and ducting. If mistakes made at this stage are identified later in the building process, surface finishes on the ceilings or walls may need to be damaged to allow access to the hidden services.

1.4 Second fix

Once the various trades have completed their first-fix work and the plastering/drylining is finished, second-fix work can commence. Second fixing often involves the installation of items and equipment that could have easily been damaged or affected by earlier stages of construction work.

Table 7.3 outlines some of the second-fix tasks completed by various trades.

Items installed during second fix are still vulnerable to damage from other construction workers and equipment on site, for example tilers, painters and decorators, therefore where necessary adequate measures should be taken to protect them until completion of the project.

Finishing trades such as tilers and decorators are usually scheduled to work on site at the end of the second-fix stage. However, in some cases they can work after the first-fix stage. For example, tiling a bathroom before the shower, basin and toilet are installed. It is important that they co-ordinate the completion of their work with the other trades, to avoid getting in each other's way and to prevent unnecessary damage to their work.

Further information on first-fix and second-fix tasks undertaken by different trades can be found in Chapter 4.

▼ Table 7.3 Work undertaken at second fix

Construction operative/trade	Range of work undertaken
Plumber	Installation of fixtures and fittings, for example sinks, toilets, showers, baths, towel rails and radiators
Electrician	Installation of electrical fixtures and fittings, for example switches, sockets, fuse boards (power distribution units and circuit protection) and light fittings Testing and commissioning of new electrical systems
Gas fitter/heating and ventilation fitter	Installation, testing and commissioning of gas boilers and other heating, air-conditioning and ventilation systems
Carpenter and joiner	Fixing skirting, architraves, doors, kitchens and fitted wardrobes/cupboards

1.5 Self-driving vehicles

Fully **autonomous vehicles** can be found on construction sites today. Due to the unpredictable nature of the environment, for example the movement of workers and plant, current deployments of autonomous vehicles typically have the machines operating within a separate, fenced-off area away from people. Semi-autonomous vehicles, which depend on a small amount of human intervention, are already being used in some areas of construction in the UK, for example earth-moving plant used to prepare sites ahead of building work starting.

Key term

Autonomous vehicles: vehicles able to operate without human intervention

In other countries, semi-autonomous and fully autonomous excavators, trucks, diggers and cranes are also being used successfully for simple but time-consuming and otherwise dangerous construction tasks, such as digging, trenching, drilling and excavating the ground on site.

Using autonomous vehicles for construction projects can reduce costs by improving productivity, machine utilisation and site progress, because they are able to operate 24 hours a day without a driver. They also produce fewer carbon emissions than vehicles driven by humans, due to less erratic movements and more efficient use.

▲ Figure 7.8 Self-driving (autonomous) vehicle

1.6 Computer-controlled manufacturing robots

While manual labour will likely always be a huge factor in building projects, the development of robots to automate construction tasks has led to improvements in productivity. They have been designed to undertake repetitive and labour-intensive tasks, such as demolition, laying bricks and blocks and plastering.

Advantages of using robots include:

- protecting workers from hazards and reducing workplace injuries
- reducing operating costs and waste
- increasing production rates and efficiency
- ensuring accuracy.

▲ Figure 7.9 Computer-controlled manufacturing robot

1.7 Large-scale 3D printers

Large-scale 3D printers can produce perfectly formed walls for large buildings on demand. They function by controlling a mobile robotic arm on site to 'print' cement-based mortar walls, layer by layer, until the full height of the walls is reached.

The advantages of using 3D printers on site are similar to those for using robots, however this method also produces seamless walls.

▲ Figure 7.10 Large-scale 3D printer

1.8 Drones

Recent innovations in drone technology have provided the construction industry with a resource that can be used to plan, manage, report and communicate efficiently through a number of digital platforms, for example Building Information Modelling (BIM), Auto CAD and mobile apps on site.

With the right software, drones can reduce costs and labour needs, and improve the efficiency of construction projects. Typical uses of drones include:

- surveying land using laser scanning (topography, volume and measuring distances)
- producing 2D and 3D maps
- identifying improvements that can be made to infrastructure
- pre-construction site planning, for example traffic-management systems and welfare facilities
- performing maintenance inspections that may otherwise be difficult to complete without the risk of harm, working at height or the use of costly access equipment, for example inspecting a roof
- producing thermal imaging to identify hot spots in a building which could be potential hazards, or cold spots resulting in heat loss and poor energy efficiency
- streaming live footage to the project team with the use of virtual reality (VR) glasses
- inspecting work for quality control
- monitoring progress on site in real time and sharing updates with the project team
- identifying hazards and improving site security.

The law states that drones used for business purposes must be flown responsibly, following Civil Aviation Authority (CAA) regulations. From 2021, drone operators must also have several licences to fly them safely and confidently. To obtain these, operators must successfully complete both of the following qualifications:

- A2 Certificate of Competency (A2 CofC)
- General Visual Line of Sight Certificate (GVC).

▲ Figure 7.11 Drone in use on a construction site

Health and safety

Drones can be used for mitigating health and safety risks, by reducing the exposure of workers to hazards. They can also be used during site inductions and training sessions, to navigate the construction site without risk of harm.

Test yourself

Describe a situation in which a drone could be used by a building services engineer.

2 Key content and required notifications of UK building regulations and approved documents

2.1 Approved documents

Virtually all new construction work and alterations to existing structures must comply with the Building Regulations 2010. These regulations are developed by the government and approved by parliament. Their purpose is to define the minimum standards of design, building materials and work in the UK.

To help people understand and comply with the building regulations, the Ministry of Housing, Communities and Local Government publishes general guidance on each part of the law in a range of approved documents. These documents provide practical solutions and examples of ways to comply with the building regulations in some common situations, for example water efficiency in a dwelling.

Table 7.4 contains a list of the approved documents and an outline of their key content.

Test yourself

Explain the purpose of approved documents.

Research

Find out about the Building Regulations 2010 by visiting **www.legislation.gov.uk/uksi/2010/2214/contents/made**

Explain the type of construction work that is exempt from legislation.

▼ Table 7.4 Approved documents

Approved document	Heading	Key content
A	Structure	This document covers the **loadings** on a building and the construction of structural elements such as foundations, walls, floors, roofs and chimneys.
B	Fire safety	This document covers fire safety precautions in new and existing residential buildings, flats, offices, schools and colleges.
C	Site preparation and resistance to contaminants and moisture	This document covers the removal or treatment of site waste materials, subsoil drainage and resistance to contaminants. It also provides details about the resistance to moisture of roofs, floors and walls, with damp-proofing and ventilation solutions.
D	Toxic substances	This document provides health and safety guidance on the use of toxic substances during building works, for example the prevention of harmful fumes produced by urea-formaldehyde in cavity-wall insulation from entering a building.
E	Resistance to the passage of sound	This document covers resistance to the passage of sound in new domestic buildings, flats and schools, and buildings that have been converted. It also provides information on particular areas of concern in structures, for example between connecting buildings.
F	Ventilation	This document provides guidance on the design, installation, inspection, testing and commissioning of ventilation systems in new dwellings and other types of building. It also covers air quality and the provision of adequate ventilation to prevent condensation.
G	Sanitation, hot water safety and water efficiency	This document sets out standards for the supply of softened wholesome water in any area of a building for the purpose of washing, for example bathrooms, kitchens, showers, sanitary conveniences and any sinks provided in food-preparation areas. Its purpose is to ensure water is sanitary, safe and used efficiently. It also explains where alternative sources of water can be supplied to sanitary conveniences for flushing, for example grey water and rain water.
H	Drainage and waste disposal	This document provides guidance on surface- and foul-water drainage above and below ground and includes information on: • pipe sizes • protection of pipes • treatment and disposal of waste • sewage infrastructure and maintenance • refuse storage • hygienic pipework, discharge and cesspools • pollution prevention.

▼ Table 7.4 Approved documents

Approved document	Heading	Key content
J	Combustion appliances and fuel storage systems	This document covers the supply of air, discharge of combustion products and protection of buildings for oil, gas and solid-fuel appliances. It also provides guidance on the safe installation of hearths, fireplaces, chimneys and flues.
K	Protection from falling, collision and impact	This document includes guidance on designing and fitting staircases, ladders, ramps, guarding and vehicle barriers in and around all types of building, in order to avoid falls, collisions and impacts. Note: Approved Document N (Glazing) has now been combined with Approved Document K.
L	Conservation of fuel and power	This document provides guidance on the energy efficiency of different structures. It is updated regularly to reflect developments in building materials and new technologies, therefore people must ensure they are using the most up-to-date version. Approved Document L is divided into four parts, with each providing specific details on a particular building type: L1A new dwellings, L1B existing dwellings, L2A new buildings other than dwellings, L2B existing buildings other than dwellings.
		The documents cover the energy efficiency of heating, ventilation and air-conditioning systems, insulation, boiler productivity, lighting and hot-water storage. Further information on Part L can be found in Chapter 3.
M	Access to and use of buildings	This document looks at dwellings in three different categories: • Category 1 – Visitable dwellings • Category 2 – Accessible and adaptable dwellings • Category 3 – Wheelchair user dwellings. It provides information about ease of access to, and use of, buildings, including facilities for disabled visitors or occupants, and the ability to move through a building easily. This includes accessible stairs, corridors and lifts.
O	Overheating	This document addresses the risk of overheating in new residential buildings. It sets out design strategies to limit solar gains and ensure adequate ventilation, helping maintain safe indoor temperatures during hot weather. The guidance applies to dwellings, care homes, and student accommodation, and supports climate resilience and occupant health.
P	Electrical safety	This document states that reasonable provision must be made in the design and installation of electrical equipment to protect people operating, maintaining or altering the installations from fire or injury. Only competent people should complete electrical work in dwellings, or on land where the supply comes from the same source. This document also provides information on the range of work that is considered notifiable under the building regulations.
Q	Security – dwellings	This document details the standards required for the design, installation and fixing of doors/**door sets** and windows in new dwellings and flats, to resist a physical attack by a 'casual or opportunist burglar'. To achieve this aim, doors and windows should be sufficiently robust and fitted with appropriate **ironmongery**.

▼ Table 7.4 Approved documents

Approved document	Heading	Key content
R	Physical infrastructure for high-speed electronic communications networks	This document provides guidance on the design and installation (physical infrastructure) of high-speed electronic communications networks within a new building or one that is undergoing major renovation works. Copper, fibreoptic or wireless devices must be capable of delivering broadband speeds of no less than 30 mbps. Note: some buildings are exempt, for example in isolated areas where it is too remote to connect to a high-speed network.
S	Infrastructure for the charging of electric vehicles	This document requires new homes and buildings undergoing major renovation to include infrastructure for electric vehicle (EV) charging. It outlines technical standards for installing charge points or cable routes, supporting the UK's transition to low-emission transport and aligning with net-zero goals.
T	Toilet accommodation	The document provides guidance on the provision of toilet facilities in non-domestic buildings. It covers quantity, accessibility, and layout requirements for different building types, ensuring sanitary provision meets the needs of users, including disabled access and gender inclusivity.

Key terms

Loadings: the application of a mechanical load or force on a structure

Door set: a combination of a door, frame and any associated ironmongery

Ironmongery: metalwork such as locks, latches and handles

Research

In the event that electrical work is notifiable, who needs to be informed?

3 Building standards

The British Standards Institute (BSI) produces agreed standards across a wide variety of industry sectors. In construction, these relate to structures, materials and sustainability.

British Standards (BS) are used as common minimum standards for **public-sector projects**, covering areas such as BIM, fire safety and waste management. Their purpose is to provide general and specific guidance for the construction industry to improve working practices.

Key term

Public-sector projects: projects funded by the government

The BSI also develops Publicly Available Specifications (PAS). These are fast-tracked standardisation documents produced to meet an urgent market need.

The International Organization for Standardization (ISO) is an independent, non-governmental organisation that develops and publishes international standards.

Table 7.5 outlines some of the standards used in construction and renovation.

▼ Table 7.5 Standards used in construction and renovation

Building standards	
BS 1192-4:2014	Collaborative production of information – Fulfilling employer's information exchange requirements using COBie. Code of practice
BS 7000-4:2013	Design management systems – Guide to managing design in construction
BS 7913:2013	Guide to the conservation of historic buildings
BS 8536-1:2015	Briefing for design and construction – Code of practice for facilities management (Buildings infrastructure)
BS 8541	Library objects for architecture, engineering and construction
BS 9999:2017	Fire safety in the design, management and use of buildings. Code of practice
BS ISO 55000:2014	Asset management – Overview, principles and terminology
ISO 16739-1:2020	Industry Foundation Classes (IFC) for data sharing in the construction and facility management industries – Data schema
ISO 14001:2015	Environmental management systems – Requirements with guidance for use
ISO 19650	Organization and digitization of information about buildings and civil engineering works, including Building Information Modelling (BIM) — Information management using Building Information Modelling
ISO 9001:2015	Quality management systems – Requirements
ISO 50001	Energy management systems
PAS 91:2013+A1:2017	Construction prequalification questionnaires
PAS 180:2014	Smart cities. Vocabulary
PAS 2035/2030:2019	Retrofitting dwellings for improved energy efficiency. Specification and guidance.
PAS 2038:2021	Retrofitting non-domestic buildings for improved energy efficiency. Specification
PAS 2080:2016	Carbon management in infrastructure
PAS 8811:2017	Temporary works. Major infrastructure client procedures. Code of practice
PD 7503:2003	Introduction to knowledge management in construction

4 Manufacturers' instructions

Any product that has been designed and manufactured in the UK, or imported into the UK, must conform to section 6 of the Health and Safety at Work etc. Act 1974 (HASAWA) – 'General duties of manufacturers etc. as regards articles and substances for use at work'. The legislation states that:

- products must be designed and constructed to eliminate risks, so that they are safe while being used, cleaned and maintained
- where necessary, products must be calibrated, tested, inspected and certified to meet with the relevant product regulations
- adequate information must be provided for the safe maintenance, operation and disposal of products
- as far as is reasonably practicable, revisions of information must be supplied if a serious risk to health or safety concerning a product has been discovered.

Further information on HASAWA can be found in Chapter 1.

Manufacturers of new products, importers of products and suppliers of significantly refurbished products must provide adequate information concerning safety and the absence of risks to health, and other issues such as the protection of the environment.

Product information is usually supplied with goods in the form of printed manufacturer's instructions, with further copies available on the manufacturer's website. This information will reflect the type of product, article or substance. For example, if the product is a new energy-efficient boiler, manuals should be provided to explain how it must be installed, operated, serviced and maintained to meet the minimum standards laid out in building regulations.

Industry tip

QR codes linking to manufacturer's instructions can be found on the body of some tools and equipment. This avoids having to keep paper copies of the instructions and can allow people to find specific product information much more quickly.

Test yourself

Under what regulation do manufacturers have to provide information on their products?

5 Building structure and fabric

5.1 Structure

Working in building services engineering, it is important to recognise different building materials and fabrics, their uses and implications.

Forms of construction

In the construction industry, the structure of a building is looked at in two parts:

- The substructure is the section of a building extending below the horizontal damp-proof course (DPC), including **basements** and **retaining walls**.
- The superstructure is the section of a building from the level of the DPC upwards.

Key terms

Basements: habitable rooms or spaces constructed below ground level

Retainer walls: structural walls designed to bear the weight of lateral loads imposed from one side; they are built below or above ground to hold back earth, and are often **battered** in form

Battered: a type of retaining wall designed with a sloped/leaning surface to provide increased structural support

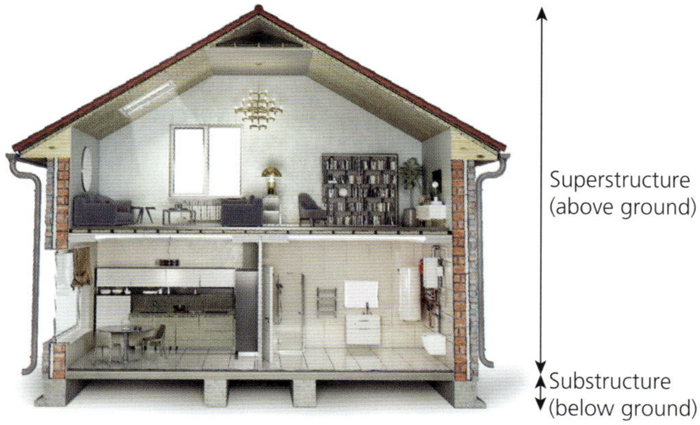

▲ Figure 7.12 Substructure and superstructure of a building

Earlier in this chapter, we looked at methods used for the structures of different types of building and some of the materials used, including:

- timber frame
- steel frame
- masonry
- concrete.

Timber frame

Timber used for structural purposes such as walls, floors and roofs is referred to as carcassing. All load-bearing carcassing timber must be tested and stress graded to ensure it will support any imposed loads. The most commonly used strength classes of softwoods are C16 and C24, however TR26 is also used to manufacture roof trusses and open web joists. Hardwoods range in strength classifications from D18 to D70, although these are rarely used.

Some hardwood and softwood species are naturally stronger than others because of their grain and cell structure. However, not all of these timbers are sustainable or suitable for the structures of buildings because of the time they take to grow and their cost.

- Fast-grown commercial softwoods such as redwood, fir and spruce (also known as whitewood) imported from Scandinavia are mainly used for carcassing work. These can be supplied with rough sawn edges or planed straight and smooth.
- Carcassing timber that is planed to uniformed dimensions is referred to as regularised; this is often easier to work with than sawn timber.

- Most timber used for structural purposes has been dried to a moisture content of 16–18 per cent and planed smooth with eased edges (removed with a small rounded edge). This type of timber originated in Canada and is commonly referred to as CLS (Canadian lumber stock) or ALS (American lumber stock).
- Scant is also a type of timber used for the structure of buildings. It is very similar to CLS, but the finished (planed) sizes are usually smaller.
- Sheathing is the timber-based sheet material used to cover one face of a timber frame to provide strength to the structure. Oriented strand board (OSB) is the most commonly used type of sheathing due to its strength, durability and relatively low cost.
- Breathable membrane is the moisture resistant material used to protect the timber frame from water ingress, while allowing any moisture in the structure of the wall to escape through it.
- A vapour barrier is a plastic sheet that is usually fixed between a timber frame and plasterboard on the inside face of the wall. The impermeable vapour barrier prevents moisture passing through the wall. Vapour barriers may also be used on floors and ceilings in timber frame buildings.

▲ Figure 7.13 Timber frame

▼ Table 7.6 Timber frame

Timber frame	
Fixings	Nails, screws, bolts, builders' metalwork, e.g. brackets, truss clips, restraint straps and joist hangers
Implications	Timber frame is at risk from dry rot, wet rot, insect attack and shrinkage. There may be defects in the timber, for example knots and splitting.
Hazards	Cutting and drilling create airborne dust.

Steel frame

Steel-framed buildings are often made from deep section I beams (also referred to as H beams) and columns bolted together and sometimes welded at the joints to form rigid structures resistant to lateral movement.

Most steel frames (also known as portal frames) are made from structural steel beams, protected with primer paint. Where the frame may be exposed to the weather, further coats of specialist paint have to be applied, such as epoxy or fire-resistant paint. Alternatively, the frame could be galvanised or made from aluminium.

▼ Table 7.7 Steel frame

Steel frame	
Fixings	Welded joints, bolts, brackets, plates
Implications	Steel frame is difficult to cut, drill and secure fixings to. It is also heavy and prone to corrosion.
Hazards	Grinding, welding and drilling activities may result in fire, fumes, projectile debris, hot metal and razor-sharp edges. Cutting and welding expose workers to a very bright light source.

Masonry and in-situ and pre-cast concrete

Masonry is a term used to describe bricks, blocks and stone bonded together with cement or lime mortar. Bricks made to construct walls for dwellings are often manufactured from clay, concrete or calcium silicate (sand lime). The colour, texture, strength and **porosity** of bricks can vary, depending on the choice of materials used and how they are manufactured.

Concrete is made from a mixture of Portland cement, fine aggregate (sand), coarse aggregate (gravel) and water. British Standard BS 8500 specifies the ratios and size of the aggregates for concrete designed for different uses, for example foundations, paving, reinforced concrete and general purpose.

There are two types of concrete block:

- Aggregate blocks are manufactured from cement, water and natural or man-made aggregates.
- Aircrete blocks are made from cement, lime, sand, **pulverised fuel ash**, aluminium powder and water.

Both types of concrete block are made in a range of different densities and compressive strengths, designed to provide good thermal, acoustic and fire-resistant properties. In general, aggregate blocks are stronger and aircrete blocks are lighter and easier to handle with better thermal-insulation properties.

▼ Table 7.8 Masonry and concrete

Masonry and concrete	
Fixings	Masonry screws/bolts, screws and plugs, chemical fixings, masonry nails, cartridge fixings
Implications	Drilling large holes with core drills can be slow, therefore hole positions should be planned for in advance. Retrospectively drilling into concrete that is reinforced with steel can cause difficulties.
Hazards	Cutting and drilling create silica dust and projectile debris.

Key terms

Porosity: the measure of a substance's ability to hold water or allow water to pass through it

Pulverised fuel ash: a fine powder by-product produced in power stations from the burning of coal to produce electricity

5.2 Substructure

The substructure of a building is designed to bear (support) the weight of the superstructure by transferring the imposed loads down to the ground below. It includes a building's foundations. Designers, such as structural engineers and architects, usually determine the type and size of foundations needed by considering the following factors:

- type of building (industrial, commercial or domestic)
- size and weight of the superstructure
- build method, for example timber frame or brick and block
- number of levels
- location, for example exposed position where there are high winds or an area that is prone to flooding
- topography of the land
- load-bearing capacity of the ground on site
- impact on the environment and sustainability
- construction costs.

The weight of materials used to construct a building is referred to as the dead load. This remains constant and can easily be calculated for the design of building foundations. As well as this weight, designers also have to consider variable imposed loads, known as live loads, such as people, furniture, snow and wind.

If the foundation of a building is inadequate or has been poorly constructed, it will often fail and result in subsidence. If this occurs, the building will partially sink into the ground, either evenly or unevenly:

- A small amount of even subsidence is known as settlement, and this sometimes happens over a short period of time after a building has been newly constructed. Providing this type of subsidence is only short lived, it should not cause any structural damage to the building.
- Where uneven subsidence occurs, it often causes structural damage, which may result in cracking of hard building materials, such as brick and block walls. Cracks in masonry walls usually occur between the openings for doors and windows, where they follow the line of the mortar joints from the ground to the top of the building.

Health and safety

Any subsidence identified in a building should be inspected and carefully monitored by a building surveyor to ensure it does not get worse over time. Subsidence can lead to the collapse of part or all of a structure.

It is sometimes possible to undertake retrospective structural work, known as 'underpinning', on existing foundations when they are unable to support current loads or proposed additional construction work, for example an extension or loft conversion. This involves digging underneath the foundations, one section at a time, to enable deeper concrete foundations to be created. These are often reinforced with steel.

▲ Figure 7.14 Retaining wall

Improve your English

Look up the meaning of the word 'lateral' and include it in a sentence about part of a building other than a retaining wall.

The term 'footings' was traditionally used to describe a shallow foundation of a wall constructed of stepped brickwork. It can now be used to mean any building foundation below DPC level.

Foundations are broadly divided into two categories, determined by their depth:

- shallow foundations (less than 2 metres)
- deep foundations (2 metres or more).

Shallow foundations

There are several types of shallow foundation, used to construct a variety of buildings:

- narrow strip and trench fill
- wide strip
- raft
- pad.

Narrow strip and trench fill foundations

Narrow strip foundations (also referred to as 'strip foundations') are the most common type of foundation for **low-rise buildings**. They are constructed by digging a narrow trench approximately 600 mm wide (the width of a digger bucket, although the width will depend on the thickness of the wall) and the length of the building's load-bearing walls. Approved Document A provides the minimum widths of strip foundations determined by the type of ground. The foundations have to be deep enough so that frost does not affect them, usually at least 1 metre deep or until load-bearing soil is reached, and approved by a building control officer.

Traditional strip foundations are built with a layer of concrete filling the bottom of the trench, to provide a strong base for the blocks laid below ground level. Trench fill foundations are very similar to traditional strip foundations, but they have a deeper strip of concrete poured in the trench, just a couple of brick courses below ground level. Trench fill foundations are often cheaper and safer to build than traditional strip foundations, because they are quicker to construct and avoid the risk of the sides of the trench collapsing while bricklayers are working below ground level.

When the concrete has cured, brickwork is laid directly on top of the foundations up to the damp-proof course (DPC) 150 mm above finished ground level.

Where buildings are constructed on sloping sites, the narrow foundations are usually stepped up the gradient. This avoids unnecessary deeper foundations on one side of the building and reduces the volume of concrete needed.

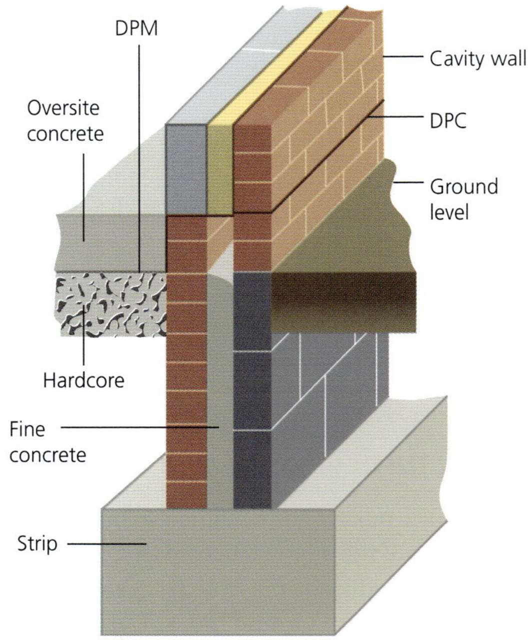

▲ Figure 7.15 Narrow strip foundation

Key term

Low-rise buildings: buildings with up to four storeys

Health and safety

Nobody should enter or work in a deep excavation for a foundation unless it has been adequately supported using timbering or trench supports and risk assessed, with satisfactory control measures in place to protect people from harm. Further information on excavations and confined spaces is covered in Chapter 1.

Industry tip

Wet concrete usually contains small pockets of trapped air that can weaken the foundation if they are not removed before it cures. There are several methods to remove air from wet concrete, including:

- tamping – tapping the surface of the wet concrete with a timber batten to release the trapped air to the surface; this method is quite basic and often does not remove all of the air trapped at the bottom of deeper concrete foundations, therefore it is only suitable for floor slabs

▲ Figure 7.16 Tamping

- using a vibrating concrete poker – a mechanical tool with a vibrating head; by partially inserting the poker into the wet concrete, the vibration from the tool head causes trapped air to be released deep within the concrete.

▲ Figure 7.17 Vibrating concrete poker

Health and safety

Extended use of a vibrating concrete poker could be harmful to health, for example causing vibration white finger. An appropriate risk assessment must be carried out and the operator must be trained.

Improve your maths

Calculate the volume of concrete needed for a narrow strip foundation measuring 7.5 metres × 4.25 metres, with a depth of 1 metre and a width of 0.6 metres.

Wide strip foundations

Wide strip foundations are similar in construction to narrow strip foundations. However, they are normally at least 1.5 metres deep and much wider. This provides enough space for bricklayers to work safely in the trenches to build their brick and block walls to the DPC level. They are used to support the superstructure of buildings with heavier loads or where the soil has a lower bearing capacity, for example soft sandy clays, by distributing the weight through the reinforced concrete foundation over a wider area.

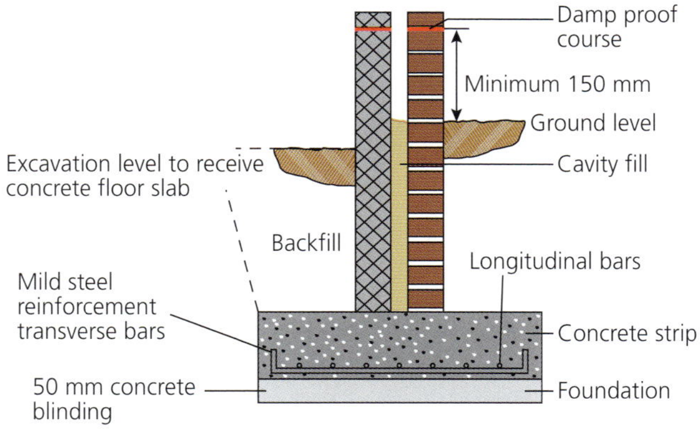

▲ Figure 7.18 Wide strip foundations

Raft foundations

Sometimes it is impractical or uneconomical to keep digging into the ground until firm soil with load-bearing capacity is found for strip foundations, therefore other types of foundation may need to be considered.

Raft foundations are often used for small low-rise domestic buildings, because they are relatively quick to construct and use less concrete compared with other methods.

A raft foundation consists of a slab of concrete reinforced with steel frame, constructed under the entire footprint of a building. The concrete slab acts like a raft would on water, by distributing any imposed loads over a large area to the ground below. Some areas of the raft will have additional loads imposed directly on them from the load-bearing walls above, for example the edge (also known as the toe). At these points, the depth of the foundation is increased to create a ring beam in the slab, which improves the load-bearing capacity of the raft and prevents cracking.

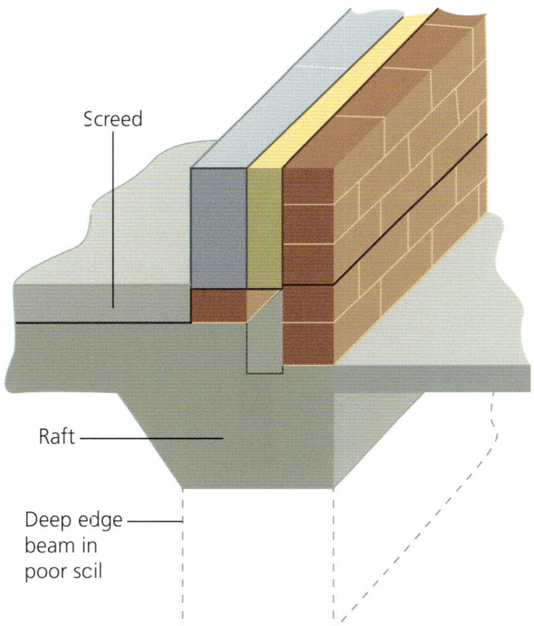

▲ Figure 7.19 Raft foundations

Pad foundations

The methods used to construct industrial units and commercial buildings are very different to those used for domestic houses. This is because they are designed to be erected quickly and efficiently, by maximising the use of large open spaces for industrial and commercial purposes rather than as places to live. These types of building are usually constructed around a steel **portal frame**, with suspended external cladding and low-height walls. Most of the weight of these buildings is transferred from the superstructure down through the steel frame to single point loads at ground level. At these positions, square or circular concrete pad foundations are built to support the structure by distributing the imposed loads down to the ground below.

Pad foundations can also be used to support ground or ring beams. These connect the pad foundations together, so that the weight of the structure can be distributed along the beams and down to the pads in the ground below.

Key term

Portal frame: a large structural frame made from load-bearing timber and steel beams and columns

▲ Figure 7.20 Portal frame

Deep foundations

In ground conditions where the surface layer of soil is unable to support the superstructure of a low-rise building with shallow foundations or the weight of a high-rise building, deep pile foundations have to be constructed. This involves driving long, **pre-cast** cylindrical steel and concrete piles deep into the ground until they reach rock or strong load-bearing soil; these are referred to as driven piles. This method displaces the soil as the piles are driven into the ground to provide a strong foundation resistant to downwards and lateral forces.

Piles that rely on the firm soil around them to support any imposed loads are referred to as friction piles, whereas piles that are supported from below by bedrock or another solid surface are known as end bearing piles.

Piles are often designed to be evenly spaced at strategic positions around the footprint of the structure, where they are cut off to the same level and connected at the surface with a concrete and steel pile cap, also known as a ring beam.

Key term

Pre-cast: formed into a shape in a factory before being delivered for use on site

Driving deep piles into the ground with heavy plant is an extremely noisy process, which could be a nuisance to the public. Furthermore, the amount of vibration caused in the ground could cause damage to the structure of nearby buildings. To overcome these issues, piles can be cast **in situ** by boring holes into the ground with long augers (drill bits) and then filling these holes with wet concrete, reinforced with steel; this method of construction is referred to as replacement.

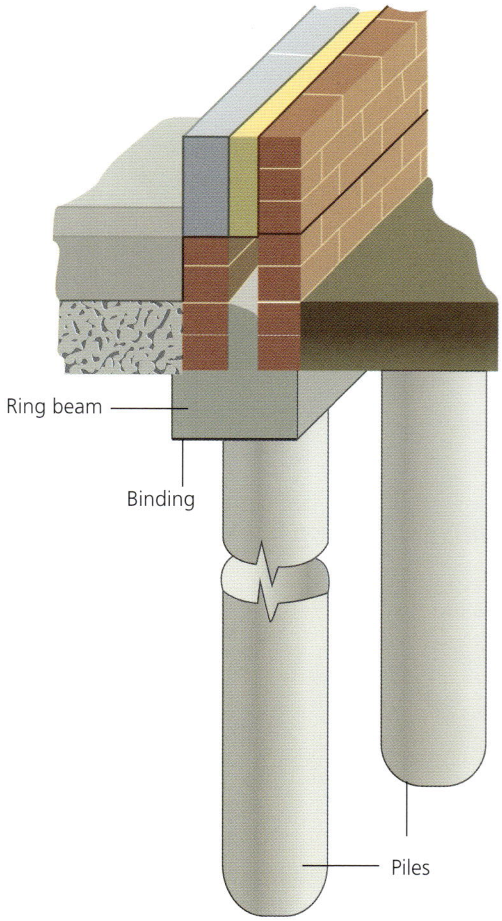

▲ Figure 7.21 Pile foundation

Key term

In situ: in position

Health and safety

Steel reinforcement bars are often left exposed at the top of pile foundations until the pile cap has been cast. These can be extremely hazardous to construction workers, so bar protection caps must always be fitted over them to reduce the risk of injury.

Test yourself

List the four main types of building foundations.

Research

Research screw pile foundations and explain when and why they may be used.

Research

Research alternative types of building foundations and explain why they might be preferred over conventional types.

5.3 Superstructure

As previously mentioned in this chapter, the superstructure of a building is everything above the DPC level on the foundations, including:

- floors
- walls
- roofs
- windows
- frames and doors.

Roofs

Roofs protect occupants of a building (and their possessions) from the elements.

Pitched roof

A basic pitched roof on a domestic building is traditionally constructed with pairs of **common rafters**, connected to a spine (known as a ridge board) at the **apex**. Towards the bottom (foot) of each rafter, a bird's mouth joint is cut to fit over the wall plates, which are secured to the top of the inner skin of the external walls. These joints provide secure fixing points for the rafters, so that the imposed loads from the roof can be transferred down through the load-bearing walls to the foundations below.

> **Key terms**
>
> ***Common rafters:*** rafters that run from a ridge board to the wall plate at 90 degrees on plan
>
> ***Apex:*** the top of a pitched roof

The sectional sizes of the rafters are determined by the distance they have to span and the weight of the roof covering; guidance can be found in Building Regulations Approved Document A: Structure or the Timber Research and Development Association (TRADA) tables. If the sections of the rafters become too large, it may be more economical to reduce them in size and support them from below, mid span, with a heavy structural beam known as a purlin. These days, metal purlins (and sometimes ridges) are preferred rather than traditional timber ones, because they are stronger, lighter and more durable.

The process for constructing a basic pitched roof is referred to as a 'cut roof' or 'cut and pitch'. This is because all the timber components are marked out and cut (sawn) on site by highly skilled carpenters. This method has now been largely replaced with prefabricated **roof trusses** manufactured in factories and delivered to site where they can be erected much more quickly than cut roofs.

> **Key term**
>
> ***Roof trusses:*** prefabricated roof sections, held together with gang nail plates at the intersections (nodes)

> **Improve your maths**
>
> Use a formula or scaled drawing to calculate the true length of a rafter with a rise of 1.8 metres and a run (half the span of the roof) of 2.4 metres.

Roof trusses

Each member of a trussed roof is specifically designed and calculated to suit its intended use, for example roof shape, pitch, span and weight of the roof covering. There are many advantages of using trussed roofs, including:

- less waste
- greater energy efficiency
- greater accuracy
- reduced material costs, building times and onsite labour costs
- less-skilled labour required
- lighter and more sustainable materials.

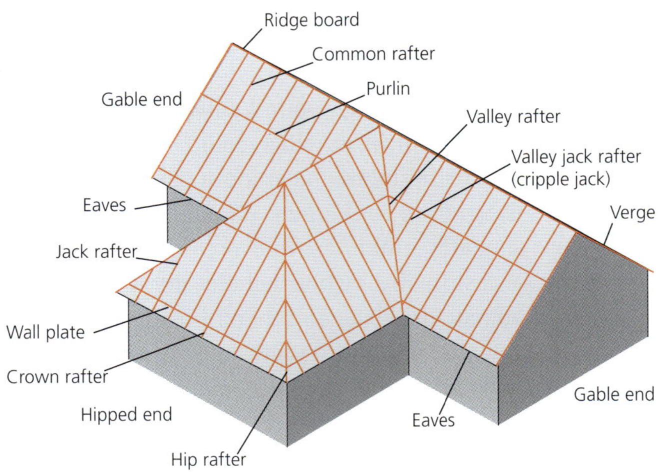

▲ Figure 7.22 Traditional cut roof

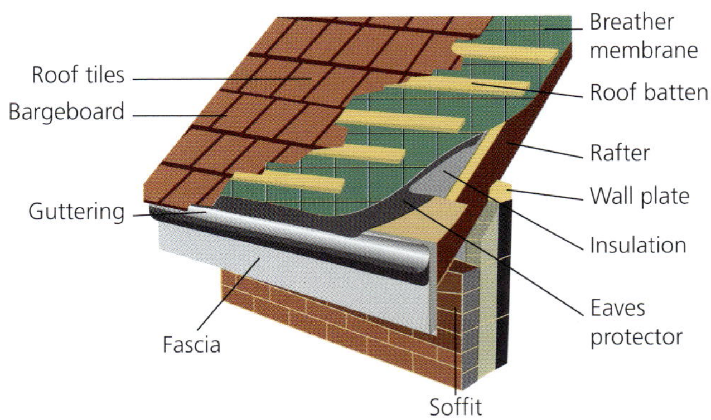

▲ Figure 7.23 Typical section through a pitched roof

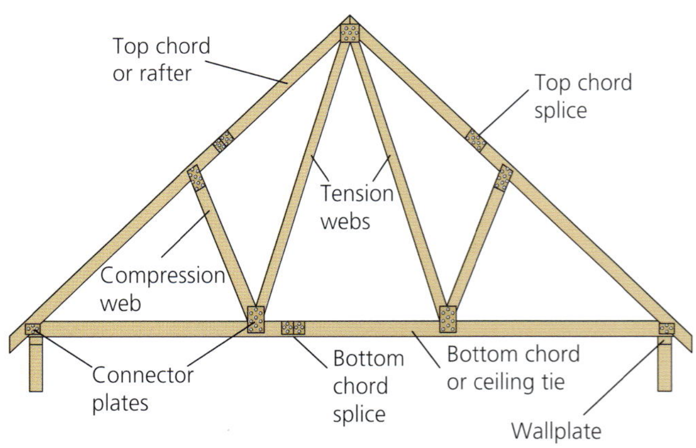

▲ Figure 7.24 Roof truss

▲ Figure 7.25 Nail plate

Loft space

The loft space created in a domestic pitched roof can be utilised as living areas, such as bedrooms, bathrooms and home offices. These are either designed into the building when it is initially constructed or completed retrospectively at a later stage. Constructing a loft conversion in a cut roof is generally a simple process, because the arrangement of the rafters and supporting roof members leaves open spaces for the additional rooms, which will require additional steelwork to support a subfloor. However, converting a trussed roof is much more difficult, because the supporting braces have to be removed and the smaller sections of timber trusses result in a loss of structural integrity.

Where a loft has been designed for a living space, it may be restricted in headroom in some areas because of the triangular shape of the roof. To overcome this issue and to create a source of natural light and ventilation in the loft, the roof may be extended to the back to create dormers. When it is not possible to use dormer windows because of planning restrictions or a tight budget, natural light and extra headroom can be created with standard roof windows, known as skylights, or balcony windows.

▲ Figure 7.26 Dormer

▲ Figure 7.27 Roof window

▲ Figure 7.28 Balcony window

Research

Research the planning permission and building regulations needed for a loft conversion in a domestic property.

Flat roofs

When a cheaper form of roof construction is needed or has been designed for all types of structures, the pitch of a roof can be lowered to less than 10 degrees to form a flat roof.

Flat roofs are constructed with a slight fall (pitch or angle) to allow water to run towards the guttering on one side of the roof and the drainage system at ground level. If a flat roof does not have the correct fall, it leads to rainwater sitting on the roof (referred to as 'ponding') for long periods of time which could promote weed and moss growth.

Due to the low angle of a flat roof, some roof coverings, for example slates and tiles, are unsuitable because driving wind will cause them to lift and rainwater could be blown underneath.

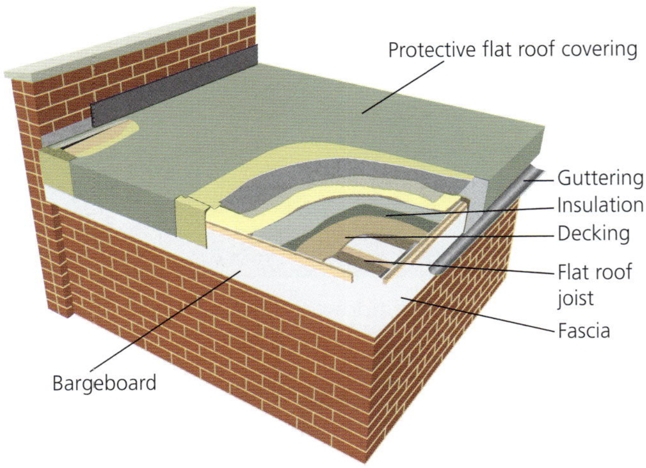

▲ Figure 7.29 Typical section through a flat roof

Figure 7.30 shows a range of basic roof shapes used in the construction of domestic houses.

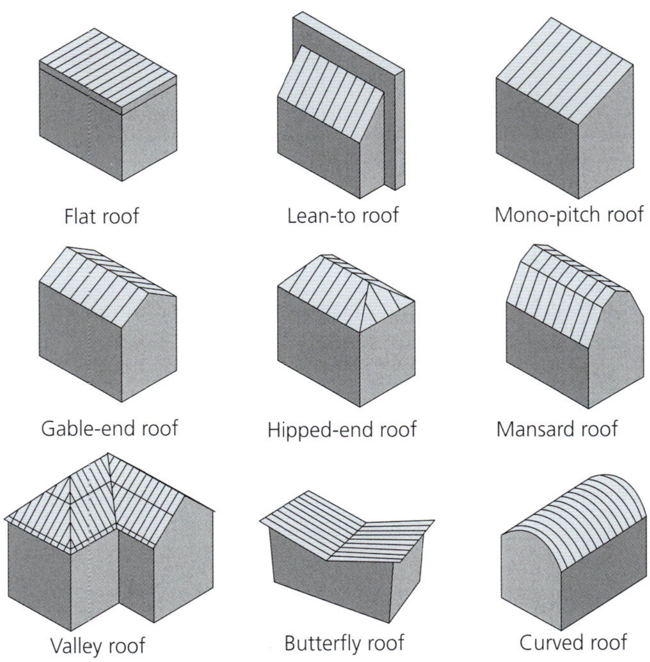

▲ Figure 7.30 Roof shapes

Protecting roofs

The timber framework for a pitched domestic roof is usually protected with a layer of breathable membrane to prevent water entering the building; it also allows any moist air to be ventilated from the loft space. If moist air does become trapped, it will increase the moisture content in the roof timbers above the 20% threshold, leaving it susceptible to an attack of dry rot and possible collapse of the structure.

A range of products and materials can be used to provide a waterproof covering to protect different-shaped roofs from the elements. For pitched roofs, these include:

- clay and concrete tiles
- natural and synthetic slates
- metal, for example lead, zinc and copper
- fibreglass
- glass
- asphalt shingles
- timber shingles
- thatch
- solar tiles.

For flat roofs, the following might be considered:

- metal, for example lead, zinc and copper
- fibreglass
- glass
- rubber
- green roofs.

Warm and cold roofs

To maintain the energy efficiency of a building and prevent heat loss, all types of roof must be highly insulated to conform to building regulations. The location of insulation in a roof is important, because it determines the dew point. This is the place where warm moist air rising in a building meets cold air from outside to create condensation.

Pitched and flat roofs insulated between ceiling joists/rafters are referred to as cold roofs. They must be adequately ventilated along the **eaves** on both sides of the roof or the ridge where the insulation is between the rafters, to allow moisture to escape. If the moisture does not escape, the dew point may be within the structure of the roof. This is known as interstitial condensation of the roof and can cause rot in the timber rafters and joists.

The thickness of insulation required in a cold roof is usually determined by the type of insulation; however, where it is positioned between the joists/rafters, a gap of at least 50 mm has to be maintained to provide adequate ventilation.

Key term

Eaves: the part of a roof that overhangs the internal skin of the external walls

Alternatively, a warm roof can be used. This is where reflective foil rigid insulation is placed over the top of the rafters or flat-roof joists with an external grade **decking** fixed either below or on top, depending on the specification. Warm roofs do not need to be ventilated because the dew point occurs outside the building, and although they are initially more expensive to construct, they are more energy efficient compared to cold roofs.

Key term

Decking: timber boards or sheet materials such as chipboard or plywood used to cover structural joists

Test yourself

Explain the difference between a traditional cut roof and a trussed roof.

External walls (load-bearing)

Solid walls

Internal and external walls are classified as either load-bearing or non-load-bearing, depending on the method of construction and materials used. Traditional load-bearing external walls were often made of natural resources, such as stone and cob, because these materials were readily available on the land where they were being used. Although the thermal mass of solid walls kept buildings relatively warm in the winter and cool in the summer, they did not achieve the same levels of energy efficiency as the methods we use now.

▲ Figure 7.31 Solid brick wall

Cavity masonry walls

A cavity masonry wall actually consists of two walls (referred to as leaves or skins) constructed with a space between them, known as a cavity. The external skin of a cavity wall is usually built with attractive clay facing bricks **bonded** with 10 mm mortar joints. The internal walls are constructed with cheaper concrete blocks, because they are often covered with plaster or plasterboard on one face, to provide a smooth surface on the inside of the building.

Key term

Bonded: the arrangement of staggered joints

The internal and external skins of a cavity wall are connected with metal wall ties, built across the cavity into the mortar joints as the walls are constructed. Wall ties improve the load-bearing capacity of the walls by preventing the skins from moving apart and possibly collapsing under the weight from the load above.

During the 1990s, building regulations changed, making it compulsory to include insulation in all cavity walls to reduce heat loss and improve the energy efficiency of buildings. There are several types of insulation that can be used in a cavity wall; some partially fill the cavity and leave a compulsory minimum 50 mm void to comply with building regulations, while others completely fill it. Where an existing cavity wall does not contain insulation, loose-fill can be retrofitted by blowing it into the cavity through holes drilled into the external wall. Blown-in insulation uses small particles of recycled materials such as fibreglass, mineral wool and cellulose (made from old newspapers), which has less of an impact on the environment than alternative types of insulation.

If insulation can be built into the walls as they are being constructed, full-fill **fibreglass batts** could be used in the cavity. These provide low-cost insulation with excellent sound resistance. However, over time the fibreglass batts can settle in the cavity resulting in cold air pockets in the wall, which could result in heat loss. To maintain the air gap in a cavity wall and avoid moisture **bridging** from the external skin to the inner skin, rigid reflective insulation is often preferred. As a cavity wall is built, the reflective foil sheets are positioned in the cavity and held tight against the internal wall with plastic retaining clips fixed over the wall ties. Although this type of insulation is slightly more expensive, the reflective foil faces are highly effective at preventing heat loss from a building.

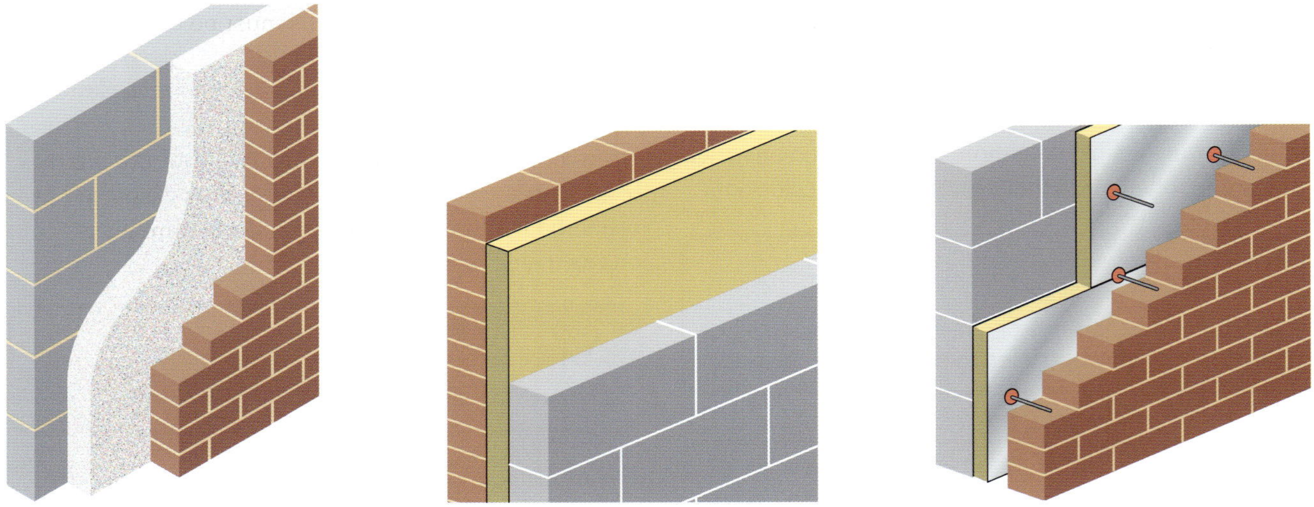

▲ Figure 7.32 Cavity walls with loose fill, fibreglass batt and reflective foil insulation

Key terms

Fibreglass batts: pieces of fibreglass insulation made from sand and recycled glass; besides providing good heat insulation, they are also fire resistant, sound absorbent and water repellent

Bridging: the action of water travelling across materials

Industry tip

Bricklayers should avoid building cavity masonry walls any higher than 1.5 metres in one day to avoid overstressing the mortar joints and making the wall unstable.

Industry tip

Frog bricks should be laid with the frog facing up, so that it can be filled with mortar to improve the strength of the wall and make it more resistant to sound transmission.

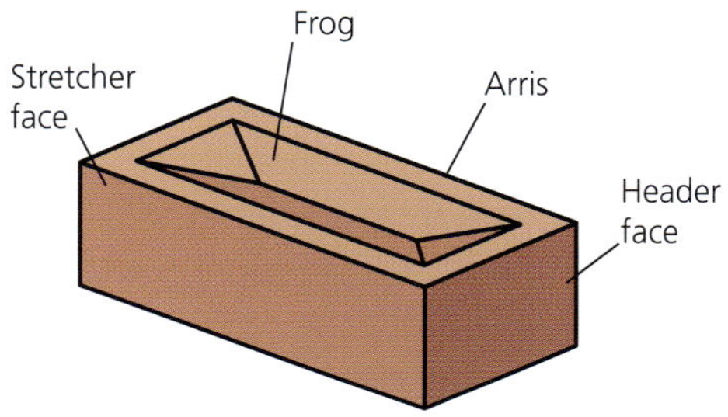

▲ Figure 7.33 Parts of a brick

Industry tip

The internal and external skins in a cavity wall should be constructed at the same time to keep the mortar joints (bed) correctly aligned and prevent weakening a single skin if left unsupported.

Lintels

Wherever an opening is created in a masonry wall for a window or door, the brickwork over it has to be supported either with a lintel or by forming a self-supporting brick arch. Lintels are structural beams manufactured from steel, concrete or reinforced brickwork. Alternatively, when a feature is required, natural stone is often used, although this is generally more expensive. Traditionally, heavy timber beams were also used for lintels but these are not as durable as metal, stone and concrete and are therefore rarely specified for external masonry walls on new builds unless they are protected from the weather.

The sectional size of a lintel and the amount of bearing that it has on the masonry wall either side of an opening will determine its load-bearing capacity. Lintel sizes are usually calculated and specified by the project designer. However, where they span distances up to 1.2 metres, they must have at least 100 mm bearing on the wall each side of the opening and 150 mm where the opening is greater than 1.2 metres.

Health and safety

Lintels can be extremely heavy, therefore mechanical lifting equipment should always be used whenever possible to prevent manual handling injuries.

▲ Figure 7.34 Metal lintel

▲ Figure 7.35 Natural stone lintel

Weep holes

Moisture can sometimes penetrate porous brickwork or build up in a cavity wall. To prevent this happening, weep holes are built into the external skin above window and door openings and towards the bottom of the walls, just above ground level. Weep holes are plastic vents, positioned in the vertical mortar joints (known as perps, short for perpendicular) in the external skin of a wall. They allow any build-up of water to escape the cavity to prevent water ingress.

Closing-off

Building regulations state that all cavity edges in external masonry cavity walls must be closed off where they are exposed, for example around window and door openings, around vents and at the top of the wall. Closing the cavity at these points improves the energy efficiency of the building by reducing heat loss. It also prevents smoke, dangerous gases and flames travelling up the cavity like a chimney in the event of a fire.

There are several methods that can be used to close off a cavity:

- Cavity closers can be built into the brickwork around window and door openings. They are available in a range of sizes to suit the width of standard cavity openings.
- The blockwork on the internal skin of the cavity wall can be returned, to reduce the width of the cavity. A gap of at least 50 mm must still be maintained between the internal and external skins to prevent 'cold bridging'. This joint is then filled with an insulated vertical damp-proof course (DPC), to stop water ingress around the opening in the wall and improve its thermal efficiency.
- Cavity barriers, also known as cavity stop socks, are often used to seal the cavity along the top of the wall to provide fire resistance.

Note: building regulations will determine the exact requirements for cavity closers and cavity barriers in a cavity wall.

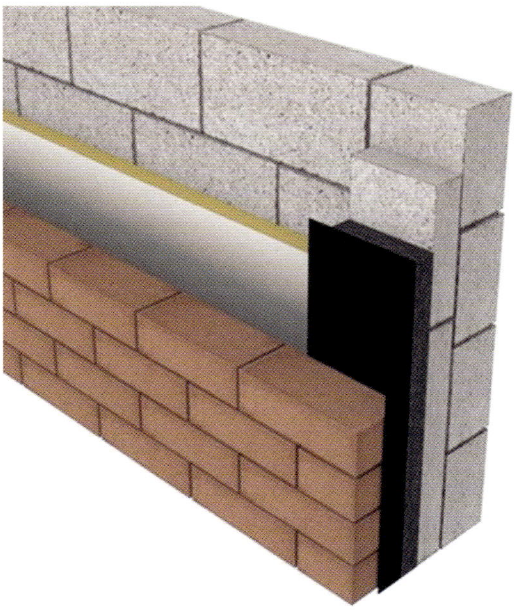

▲ Figure 7.36 Vertical DPC in a cavity wall

Internal walls

Concrete block

An internal wall used to divide a space is known as a partition. These can either be load-bearing or non-load-bearing, depending on their method of construction.

Solid concrete blocks are often used to build load-bearing internal walls, because they are cheaper and quicker to build than bricks and also provide excellent sound insulation and fire resistance. The exposed faces of a concrete block wall are traditionally covered with cement render as a base coat and finished with gypsum plaster to provide a flat, smooth surface ready for decoration.

One disadvantage of this method is the time it takes for the cement render to set, therefore faster-drying backing plasters are now preferred to reduce drying times. Alternatively, the masonry of solid internal walls can be covered with sheets of plasterboard bonded to the wall with **drywall** adhesive, in a process known as dot and dab. The long edges on the plasterboard sheets used for this process are slightly tapered. This allows for the joints between the boards to be covered with jointing tape and made flush with drywall joint filler.

Key term

Drywall: another name for plasterboard

Test yourself

Explain the purpose of cavity walls compared to solid walls.

Timber partition walls

Non-load-bearing timber partition walls are more commonly known as stud partitions, because they consist of vertical members known as studs. Studs are usually spaced equally with 400 to 600 mm between the centres of the uprights to suit the size of plasterboard wall covering and to provide support for fixing abutting edges, with extra studs being used to form corners and junctions.

Each stud in a partition is fixed into a head plate at the top of the wall and along the floor into a sole plate, with two wire nails in each butt joint. The studs are strengthened further with short noggins fixed between them, in positions determined by the height of the wall or where the wall is most vulnerable, for example waist height in a school or hospital. Further noggins may also be built into a partition before it is covered with plasterboard, to provide additional fixing points for heavy items, such as kitchen units, radiators or electrical.

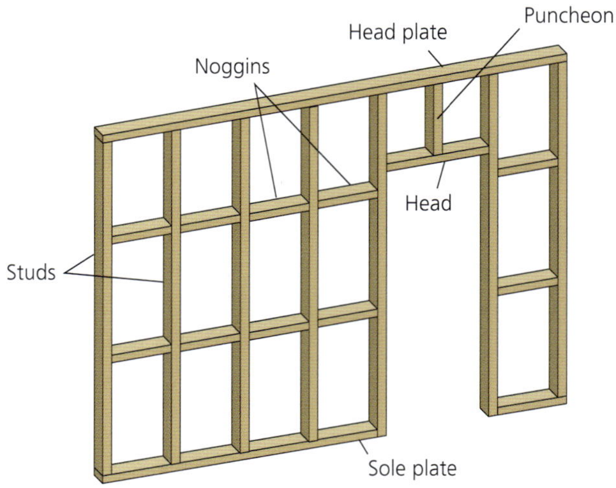

▲ Figure 7.37 Non-load-bearing stud partition wall

The arrangement of timbers in load-bearing partition walls is similar to non-load-bearing walls. However, the timbers used must be **stress-graded**. Two head plates must be fixed in these types of walls to spread the weight of any imposed loads and there must be additional structural timbers (lintels) around any openings for doors. Further timber studs should also be used where there are specific heavy loads, for example directly above load-bearing walls.

Key term

Stress-graded: timber that has been visually or machine-assessed for its strength and stiffness, and certified according to its structural classification to British Standard rules

Research

Research the methods for stress grading timber and the grading system used for each class.

Identify which stress-graded timber classifications are most commonly used and explain why you think this is the case.

Industry tip

Masonry walls should not be supported by timber floors or partition walls, because any movement in the timber will crack the masonry.

Metal stud partition walls

Metal stud partitions are an alternative solution for non-load-bearing walls. They are often used to divide spaces in domestic and commercial buildings.

The low-cost, lightweight metal sections of this system are easy to cut and quick to install. In positions where additional support is needed, such as doorways or hatches, the hollow sections of the metal studs are filled with timber to strengthen the wall. The timber inserts also provide strong fixing points for any door or hatch linings that may be fitted after the wall has been constructed.

Each metal stud section is designed and manufactured with elongated holes in it. These are used to route services through the walls and prevent the need for tradespeople to create holes and notches in the metal studs themselves, as they would have to in timber.

▲ Figure 7.38 Metal stud partition wall

Health and safety

You should always wear gloves when handling metal stud-wall components to protect your hands from the razor-sharp edges.

Improve your English

Non-load-bearing walls are not limited to timber and metal stud partitions. Research proprietary partition walls and write a definition in your own words.

Floors

Floors are designed to provide a level surface for the occupants to live and work, while preventing moisture rising from the ground and weed growth. Modern floors are also designed and constructed to:

- prevent heat loss
- prevent the transfer of sound from one level to another
- slow the spread of flames in the event of a fire.

In domestic buildings, the methods used to construct ground floors are often different to those used for upper floors, mainly due to the increased risk of moisture affecting materials in floors closest to ground level.

There are two types of floor construction:

- solid floors
- suspended floors.

Solid floors

Solid floors are used at ground level. They are constructed using different layers to make a solid concrete base:

- The first layer is compacted hardcore; this consists of materials that are **chemically inert** and not affected by water. Clean, broken bricks and aggregates or similar **unbound** materials can be used as hardcore, because they easily drain water and are not affected by the chemicals in other materials in the floor.

Key terms

Chemically inert: will not react with chemicals

Unbound: not bonded together with cement

- A thin layer of sand blinding is used for the next layer, to protect the polythene damp-proof membrane (DPM). The DPM is a protective barrier to prevent moisture penetrating the floor, which could lead to rising damp in other areas of the building such as the walls.

- Concrete is poured over the DPM to form the floor slab. Also referred to as oversite concrete, this layer provides most of the structural integrity in a solid floor and should not be confused with the finished floor level.
- Sheets of rigid reflective foil insulation are laid over the concrete slab to prevent heat loss through the floor. If an underfloor heating system is being used in the building, it will be fixed either in or on top of the insulation at this level.
- The floor insulation is protected with a thin-polythene sheet, before the final layer of cement screed is laid.

Test yourself

What type of timber should be used in load-bearing partition walls?

Research

Research dry and liquid screeds used in floor construction and explain the benefits of each system.

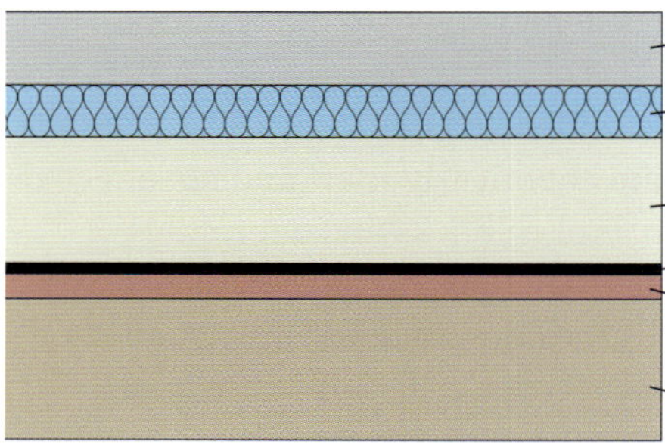

▲ Figure 7.39 Solid floor

Suspended ground floors

Traditionally, ground floors were built using timber joists suspended between load-bearing walls, with low-height sleeper walls supporting them mid span. Solid timber floor boards were laid across the joists at right angles and fixed in position with nails. The void underneath the floor was ventilated with air bricks built into the external walls; this created airflow around the timber floor and reduced the risk of rot.

Suspended timber floors at ground-floor level are not commonly used in modern construction, because the ends of the joists built into the external walls would often suffer with decay from moisture penetration over a period of time, unless they were supported with joist hangers. These days, a durable system known as block and beam is preferred. This involves using reinforced concrete beams infilled with blocks. The floor can be used as soon as the blocks and beams have been positioned on the DPC on the inner skin of a cavity wall, therefore construction work can continue without delay. The exposed block and beam floor is completed with insulation and screed in the same way that a solid floor would be finished.

Suspended upper floors

The block and beam system used for ground floors can also be used for upper floors in domestic buildings. However, it is more expensive than suspended timber floors and the additional weight would have to be supported with deeper foundations, which would increase construction costs. For these reasons, suspended timber floors are often chosen by designers for use in low-rise domestic buildings.

Besides traditional solid timber beams, engineered joists have now been developed to span longer distances unsupported and without the natural defects sometimes found in solid timber, for example twisting, springing and large knots. These joists are wider and lighter than solid joists, making them easier to handle, and stronger joints are formed with the chipboard floor covering. They are also designed so that services such as pipes and cables can easily be laid through them, without weakening them by drilling holes and notches.

▲ Figure 7.40 Suspended timber upper floor with solid joists

▲ Figure 7.41 Suspended timber upper floor with Eco-joists

▲ Figure 7.42 Suspended timber upper floor with I-beam joists

Where timber suspended floors are unsuitable, for example in multi-storey buildings, **pre-stressed concrete** slabs manufactured off site can be used. This type of floor is quick to install with the use of a crane. It also has fewer joints than the block and beam system, meaning it has better fire resistance.

Alternatively, a floor constructed from a concrete slab reinforced with steel can be cast in-situ using formwork (a temporary mould), which avoids the need for expensive cranes. Suspended floors can also be cast in-situ using a lightweight concrete floor system. Rather than formwork, this method uses a steel deck which forms a permanent part of the floor.

Key term

Pre-stressed concrete: a type of concrete that has been compressed during production to improve its strength; this is normally achieved by 'tensioning' (stretching) high-tensile steel wires in the concrete

▲ Figure 7.43 Pre-stressed concrete slabs

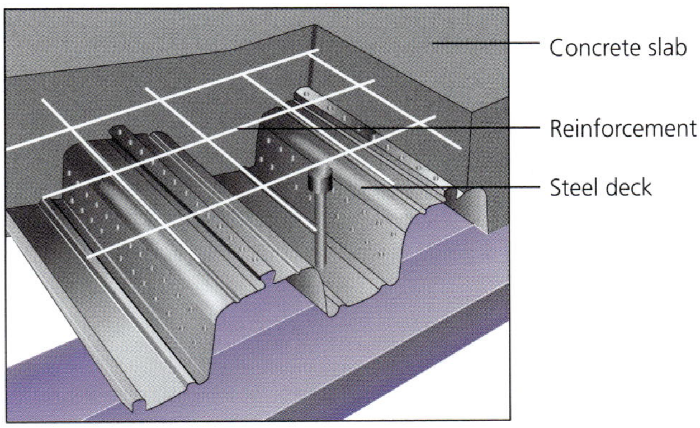

▲ Figure 7.44 Lightweight concrete floor system

Windows, doors and frames

Windows, doors and frames can allow heat to escape a building if they are not correctly designed and installed, hence they are subject to planning permission and inspection by the building control officer after they have been fitted.

High-performance, energy-efficient doors, windows and frames can be made from hardwood, softwood, unplasticised polyvinyl chloride (uPVC) and aluminium. Although timber frames are not used as often as the other examples, because they have to be maintained more regularly, they are sustainable and therefore have less of an impact on the environment compared with other materials.

The type of glass used is just as important as the frames. Double and triple glazing is mainly specified for new construction projects, because the high-performance, **low-emissivity (low-E) glass** and **argon gas** filled units act as insulators to reflect heat back into a building to reduce energy bills.

Key terms

Low-emissivity (low-E) glass: a type of thermally efficient glass that is covered with a microscopic coating on its surface to minimise the amount of infrared and ultraviolet light that can pass through it to keep a building warm in the winter and cool in the summer

Argon gas: a colourless and odourless gas with excellent thermal properties; often used to fill the voids between the panes of double and triple glazing to improve the energy efficiency of a building

Test yourself

Describe where a DPC is used in the superstructure of a building.

5.4 Infrastructure

Chapter 3 looks at how well-designed infrastructure, including roads, sewage systems, railways and bridges, can contribute to the community and the success of construction projects.

5.5 External work

External construction work could involve the following:

- paving
- boundaries
- drainage
- parking.

The infrastructure listed in section 2.3 is subject to planning restrictions and building regulations. Many of the main sewer systems in the UK were initially built in the late 1800s, since when the population has grown substantially, meaning that some of them no longer work efficiently.

Besides raw sewage, surface water from paving and driveways was directed into the main sewer system until 2008, when new regulations were introduced to prevent flooding and other environmental problems in populated urban areas. The law states that planning permission is not needed if a new or replacement driveway at the front of a property is made from porous materials (sub-base materials and surface finishes) such as permeable concrete, block paving or resin-bound gravel. These regulations do not apply to other parts of a property. Work in or around listed buildings may also need consent.

Where a kerb lining a path is raised across a new driveway, permission will have to be granted before it can be lowered. The local authority is responsible for public roads and paths and should therefore undertake this task, to ensure that no damage occurs to services under the paving surface and that the lowered kerb is adequately supported.

Walls that form a boundary between two or more properties are known as party walls. The Party Wall etc. Act 1996 is a framework of legislation that aims to prevent and resolve disputes over boundaries. The law states that you must inform your neighbour if you are intending to build on the boundary, work on an existing party wall or excavate near and below the level of their foundations. Further information on party wall agreements can be found on the following website: **www.gov.uk/party-walls-building-works**

5.6 Fabric

Timber

Natural timber and manufactured timber products are commonly used in the fabric of buildings for stairs, windows, doors, kitchens, frames, skirtings and architraves. Carpenters and joiners are usually responsible for cutting and fixing timber products.

▼ Table 7.9 Timber

Timber	
Fixings	Nails, screws, bolts, brackets, adhesive
Implications	Timber is at risk from dry rot, wet rot, insect attack, splitting, shrinkage and expansion.
Hazards	Cutting and drilling creates airborne dust.

Cladding

Cladding is the term used to describe panelling that covers or protects surfaces. It can be made of various materials.

External wall cladding (also referred to as weatherboarding) is usually supplied in narrow widths and long lengths shaped to overlap each board or interlock with tongue and groove (T&G) joints to conceal any movement in the materials and to protect the building from the elements. Timber cladding can be used externally, but it must be treated with preservative, chemically modified (for example Accoya wood) or naturally resistant to water (such as cedar).

Composite cladding is a durable alternative to timber. It is manufactured from recycled plastic and wood in a range of solid colours designed with woodgrain or smooth finishes. Composite cladding can be cut and fixed just like timber, but it does not need to be painted or stained and will not rot.

Fibre cement planks are also used for external wall cladding. These are manufactured from sand, cement, cellulose, synthetic fibres and water and designed with a woodgrain or smooth finish. Fibre cement cladding is extremely durable and does not need to be maintained like other products.

Ceramic and porcelain tiles have been used in kitchens, bathrooms and shower rooms for many years. However, PVC and acrylic resin wall panels are now being used as well, because they are easy to install and they do not have the issues associated with grouted joints.

Research

Research Accoya wood. Write a brief description to explain it properties, uses and benefits.

▲ Figure 7.45 Composite cladding

▼ Table 7.10 Cladding

Cladding	
Fixings	Screws, nails, brackets, adhesive
Implications	Bare timber may have to be painted or treated if it has been cut or drilled. Composite and timber cladding will expand and contract in different temperatures.
Hazards	Cutting and fixing create airborne dust.

Granite, glass, marble, stone and concrete

Granite, glass, marble, stone and concrete are used to provide attractive, durable and hardwearing wall, floor and work surfaces in buildings.

▼ Table 7.11 Granite, glass, marble, stone and concrete

Granite, glass, marble, stone and concrete	
Fixings	Adhesive, mortar, grout
Implications	Some masonry surfaces are difficult to drill and fix into.
Hazards	Silica dust can be created when cutting, drilling or grinding.

Brick

Bricks are often used to build the outer skin of external walls because they are strong, durable and attractive. Common bricks, also referred to as facing bricks, are made from fired clay and are available in a variety of colours and textures to suit the building or structure design. Old clay bricks varied in size and shape; however the standard size of a common brick is now 215 mm long, 102.5 mm wide and 65 mm high.

Bricks are bonded together with 10 mm thick mortar joints and laid so that the vertical joints are staggered, to increase the strength of the wall. The arrangement of the bricks is referred to as the bond. The most common types of brick bonds used in the UK are

▲ Figure 7.46a English bond (left) and Stretcher bond (right)

▲ Figure 7.46b Flemish bond (left) and Header bond (right)

▼ Table 7.12 Brick

Brick	
Fixings	Screws and plugs, masonry screws, resin fixings, frame and hammer fixings
Implications	Fixings should be positioned in the centre of bricks or mortar joints to provide the best fixing points. If a fixing is positioned on the edge of a soft brick it may cause the brick to crack. If the mortar bond between the bricks is not good, it may weaken the fixing; for example, the joints between the bricks are not completely filled with mortar, therefore creating a void and weakness in the wall.
Hazards	Silica dust can be created when cutting, drilling or grinding.

Fenestration

Fenestration is a term used to describe the design and arrangement of windows in a building. Window types were covered earlier in this chapter.

▼ Table 7.13 Fenestration

Fenestration	
Fixings	Masonry screws, frame anchors, brackets, frame fixings, expanding foam
Implications	If windows are not designed or installed properly, they could cause air leakage and condensation and reduce the energy efficiency of the building.
Hazards	Drilling into different surfaces to secure the frames could create airborne dust. The work is often undertaken at height.

Industry tip

Always try to conceal the fixings used to secure doors and windows by either hiding them in the **rebates** or using fixing brackets.

Key term

Rebate: a profile often used in timber products such as doors and windows

Plasterboard

The studwork frames of timber and metal stud partitions are usually clad with plasterboard fixed with drywall screws, before being plastered or taped and filled along the joints.

Where the studs are spaced less than 450 mm apart, 9.5 mm plasterboard can be used, otherwise 12.5 mm plasterboard is required.

There are various types of plasterboard that can be used to clad a partition wall, each with different characteristics. The exact type to be used will be specified by the designers, however it could include one or more of the following:

- standard plasterboard (ivory)
- fire-rated (pink paper)
- thermal (white)
- moisture-resistant (green paper)
- acoustic (blue)
- fixable (enhanced properties to support increased fixing weight)
- impact resistant (ivory).

▼ Table 7.14 Plasterboard

Plasterboard	
Fixings	Drywall screws Plasterboard fixings, for example: • spring toggles • plasterboard plugs • self-drilling plasterboard fixings
Implications	It can be difficult to fix heavy items to partition plasterboard walls unless they are fixed directly into the studs or noggins. Standard plasterboard is unsuitable for rooms with high humidity such as bathrooms.
Hazards	Cutting and drilling create airborne dust.

Industry tip

Plasterboard must be stored flat and in a dry location. If it gets damp and distorted, it will be difficult to cut neatly during installation.

6 Approved documents and guidance for penetrating building structure and fabric

When installing services such as plumbing, electrical, HVAC, or data cabling, penetration through walls, floors, ceilings, and other structural elements must be carefully planned and executed to maintain the building's integrity, safety, and performance. This includes:

- pre-assessment of structural drawings and service routes
- coordination with structural engineers and designers
- use of appropriate drilling, cutting, or core boring methods
- sealing and firestopping measures post-installation
- inspection and documentation for regulatory compliance.

Approved document A – Structure

- Ensure penetrations do not compromise load-bearing elements or structural stability.
- Avoid weakening beams, joists, or load paths without engineered solutions.
- Reinforce or reroute services where necessary to maintain structural integrity.

Approved document B – Fire safety

- Maintain compartmentation by using fire-rated sealants, collars, or sleeves.
- Penetrations through fire-resisting walls, floors, or ceilings must be fire-stopped to the same rating.
- Follow manufacturer-tested systems and ensure third-party certification where applicable.

Approved document C – Site preparation and resistance to contaminants and moisture

- Prevent moisture ingress by sealing penetrations in external walls, foundations, and damp-proof courses.
- Use waterproof grommets, sleeves, or membranes to protect against water and ground contaminants.
- Maintain integrity of radon barriers or gas membranes where present.

Case study

Mia is a self-employed gas engineer and is registered Gas Safe. She has been asked to install a new energy-efficient boiler in a three-bedroom semi-detached dwelling for a domestic customer. A hole needs to be bored in the fabric and structure of the building in an external wall for a new boiler flue.

- Explain the potential issues that could be caused for the occupants of the dwelling as a result of Mia drilling a hole for a flue in an external wall.
- Explain the processes and procedures that should be followed to complete the task.
- List the approved documents and guidance that Mia should refer to before installing the new boiler.

Assessment practice

Short answer

1 Which approved document provides advice and guidance on fire safety?
2 Which regulatory body must gas engineers be registered members of?
3 What type of foundation is often used to support metal portal frame structures?
4 What is the minimum gap permitted by building regulations between the skins in a cavity wall?
5 Name one type of fixing that can be used in a masonry wall.

Long answer

6 Explain the purpose and benefits of off-site construction.
7 Explain the difference between first fix and second fix.
8 List the benefits of using autonomous vehicles on construction sites.
9 List the materials used to construct a solid floor and explain the purpose of each component.
10 Describe the implications of not following manufacturer's instructions.

Project practice

Your employer has asked you to design a free-standing modern garden room for one of his customers.

The building should be made from as many sustainable materials as possible, with a single pitched roof and composite cladding. It can be constructed under permitted development rights and therefore does not need planning permission.

- Design a garden room to meet the customer's requirements using computer-aided design (CAD) software such as SketchUp.
- Explain the type of foundations you would recommend and why.
- Calculate the total cost of materials needed for the project and present them in a suitable digital spreadsheet.

Chapter 8 Information and data principles

This chapter explores the meaning of the term 'data' and how data is used in planning, costing, constructing and operating a building. We will discuss how data can be processed and organised into useful and valuable information using a range of methods.

In the construction industry, expert analysis and evaluation of data can allow informed decisions to be made when bringing a project to a successful completion. Projections can be made about efficient operation of a building throughout its life cycle.

Being able to interpret information gained from data sources is a valuable skill for those employed in the modern construction workplace.

Learning outcomes

By the end of this chapter, you will understand:

1. data
2. sources of information
3. data management and confidentiality
4. drawings, circuit diagrams and schematics
5. programming and set up of digital systems using IT resources.

1 Data

Humans have been gathering and processing data for thousands of years to refine their understanding of the universe. In modern times, working with data has developed into a science that follows defined methods to perform specific tasks.

Data and information are closely linked. Data is a collection of facts, such as numbers, measurements, words or descriptions, that can be analysed and processed so that meaningful information is created.

Research

Watch an introductory explanation of what data science can be used for (for example, search for 'Intro to data science' on YouTube).

1.1 Key elements of data

Every day, vast quantities of data are generated, often referred to as 'big data'. However, simply collecting masses of data is not the key to creating useful information. Data must be correctly stored, 'cleaned' to remove inaccurate or imprecise records, and identified to assign it for use.

▲ Figure 8.1 Vast quantities of data are generated every day

Managing data in this way can involve many complex data-processing systems working together, requiring them to be operationally compatible. This is known as **data interoperability**.

Complex processing of data can involve one system describing or analysing data in another system, for example to inform an architect regarding multiple intricate design elements of a building.

When a set of data is used to describe or analyse other data, this is referred to as **metadata**.

Key terms

Data interoperability: the ability of data systems to exchange and use information

Metadata: a set of data that describes or analyses other data

Test yourself

How must data be processed to make it usable?

The level of detail in a set of data (or dataset) must be regulated or even limited, so that it can clearly inform users according to the required purpose of the information.

Different data types that can be analysed include:

- numerical – consisting of numbers, such as dimensions and measurements
- categorical – consisting of options, such as 'type A, B or C' or 'up, down, left, right'
- ordinal – consisting of steps or sequencing (ordering) of information.

To allow the extraction of useful information from large datasets, a process known as **generalisation** can be used. This is essentially a process of 'pulling back' from the mass of raw data to give a broader, more general view and reduce the extent of analysis required. This makes information extraction more manageable.

For example, rather than stating the floor area of dozens of individual rooms in a large multi-storey building, the data could be arranged in bands to categorise room volumes, such as 7–10 m^3, 11–15 m^3 and so on. This reduces the volume of data while still providing useful information, to allow analysis of requirements such as heating and ventilation systems or optimal occupancy levels in the completed building.

Key term

Generalisation: (in data processing) creating layers of summarised information from mass data

1.2 Different sources of data

Data is generated from many different sources during the life cycle of a building.

Data generated during design and construction

When a building design is conceived, decisions must be made about the shape and size of the structure. Therefore, early in the design process, measurements and dimensions must be established to ensure the design will fulfil its purpose and the completed building will function well for its occupants.

Quantities of data multiply rapidly as floor areas and volumes of rooms in the structure are calculated, energy requirements are assessed, and thermal performance of the structure is analysed. Calculations will be made to establish structural strength and the forces generated by loadings, to make sure the building will be structurally sound.

Improve your English

Using data to calculate loadings is the job of a structural engineer. Research and write a short report on the role of a structural engineer working on a construction project.

This means that as soon as the design process begins, data is being generated which can be used by other members of the construction team, such as quantity surveyors preparing costings, contractors and subcontractors preparing tenders, and systems designers planning building services requirements.

Industry tip

While a large project is costed by professionals using large volumes of complex data, successfully costing even a simple project involving a small number of workers will rely on careful use of reliable data sources, such as up-to-date lists of material prices.

▲ Figure 8.2 Data is generated right from the beginning of the design process

Building services systems require extensive data calculations to establish **dynamic** elements, such as:

- air flow for ventilation systems
- gas supply rates for heating systems
- water flow rates and pressure levels for plumbing systems
- voltage and current ratings for electrical systems.

▲ Figure 8.3 Data is used to calculate dynamic elements in building services systems

Key term

Dynamic: characterised by frequent change or motion

Test yourself

What kinds of data can be generated during the design stage of a building project?

As the construction process proceeds on site, data regarding productivity and meeting deadlines is constantly scrutinised to manage efficient interactive working between site personnel.

Data generated and recorded during a project is valuable. Information about factors such as speed of operations and the effects of delays caused through bad weather or supply issues can be used in designing and building future projects more efficiently and economically.

A system that is increasingly helpful in managing large volumes of data through the complete life cycle of a building is Building Information Modelling (BIM), which integrates many data sources in a digital format. The system allows authorised users to access a range of important information during the design, construction, occupation and even demolition of a structure. There is more information on BIM later in this chapter, and in Chapter 4, section 8.

Test yourself

During which phases of a building's life cycle can BIM be used?

Industry tip

BIM is used mainly for large-scale complex projects. However, the BIM principle of having a central source of accessible and accurate data and information can be applied to good effect even during a project with just a few workers.

Structures within the built environment must have provisions for access by pedestrians and vehicles, along with reliable connections to services and telecommunications. These requirements mean that extensive transport systems and services delivery infrastructure must be designed, constructed and maintained.

Useful data can be generated by analysing vehicle and pedestrian movements, metering services provision and examining telecommunications activity. Information harvested from careful analysis can point to emerging trends and help in planning the future infrastructure needs of society.

▲ Figure 8.4 Analysis of infrastructure data can help in planning for the future

Data generated on completion and handover of a building

When a building is handed over from the contractor to the client for occupation and brought into use, a great deal can be learned from the analysis of data generated by a number of considerations:

- How successful was the delivery of the project by the contractor?
- Was the project delivered and handed over on time? If not, why not?
- Does the building meet its design brief fully now it is in use?
- What changes could be made to improve performance?

Gathering data to answer questions such as these is referred to as post-occupancy evaluation (POE). It may be undertaken by an independent consultant or by the client's own team and can provide valuable data to contractors who build repeat structures or specialise in specific project types.

POE is an increasingly important tool in analysing data to support the construction industry in meeting more stringent environmental targets, and more rigorous regulation connected with energy use. The design of systems delivering utilities such as electricity, gas and water, along with waste-disposal systems, can be refined and improved by analysis of performance data and metering of usage in completed buildings.

In most buildings, there are systems that automatically control various functions, such as heating thermostats or alarm motion sensors. In large structures, sophisticated computer-based building management systems can be installed to automatically control and monitor aspects such as general energy consumption and distribution, lighting levels, and water management and consumption rates. There is more about digital control systems later in this chapter.

▲ Figure 8.5 Computer-based systems monitor and control a range of functions

Data generated by these systems can be used by estates and buildings managers to understand how buildings are operating, adjust and control systems to optimise their performance, and create reports or set alarms to give an alert when operational **parameters** are exceeded.

In the case of large estate facilities, such as a hospital complex or university campus, specialised computer **software** referred to as enterprise asset management (EAM) can be employed to manage the flow of data. This can inform accounting and purchasing decisions to sustain economic day-to-day operations and support planning for efficient use of individual buildings within the estate.

Key terms

Parameters: limits which define the conditions of operation for a system or process

Software: a sequence of digital instructions designed to operate a computer and perform specific tasks

Research

Find out the key difference between enterprise asset management (EAM) and a computerised maintenance management system (CMMS).

Data generated during maintenance

EAM can be used to collect data about efficiency, reliability and repair costs, allowing planning and work scheduling for maintenance. As inspections take place, data can be logged in real time, and work orders can be sent to personnel already on site to prioritise maintenance or repair tasks.

▲ Figure 8.6 EAM can create 'intelligent data' to manage the maintenance of a building

The real-time capture of data using systems like EAM can be linked with BIM systems to create so-called 'intelligent data' that can lead to improved efficiency throughout a building's life cycle. Benefits of this type of interoperability include:

- performance projections for long-term planning
- reliability of building services
- operational cost monitoring.

Data generated, processed and stored using information and communications technology (ICT) that is managed using equipment in multiple locations allows information-rich activities to be conducted in a collaborative way. Data is a key part of the construction process from design to demolition, used by both personnel in construction company offices and workers on construction sites.

Test yourself

What are the benefits of using 'intelligent data'?

1.3 How data can be used

Table 8.1 identifies how data can be used to complete construction projects successfully.

Key term

Procurement: the process of agreeing business terms and acquiring goods, products and services from suppliers

▼ Table 8.1 How data can be used in construction projects

Data use	Applications
Understanding behaviour	Data can be used to analyse and understand the behaviour of personnel working on a project. Efficient deployment of skilled workers and the creation of workforce motivation to achieve operational efficiency are valuable management skills. Data regarding personnel numbers on site can be referenced to output over time to establish optimum levels of staffing, enabling managers to support efficiency in office environments and on site. In terms of data use, the reference to 'behaviour' could be extended to factors that might impact on project success, such as the behaviour of certain materials and equipment in variable conditions. With the emergence of new materials and the progressive adoption of new construction methods, data about how these materials and methods function is of real value.
Performance assessment	Key performance indicators (KPIs) can be linked to appropriate data to measure performance in areas such as: • monitoring project costs • tracking project progress over time • identifying company strengths and weaknesses • confirming client satisfaction. Uses for performance data include: • tracking profitability by comparing cost with budget • highlighting trends in complaints from clients • evaluating quantities of waste with a view to improving recycling.
Improving market competitiveness	When companies bid for contracts, they need to be well informed about workforce availability, materials and components **procurement**, current regulatory requirements and fluctuating economic conditions. Accurate and current data is critical in providing reliable information, so that a company can improve and maintain its success in competition with other contractors. Fully understanding the area of construction activity that a company works in through efficient data analysis is key to making decisions on future company development. Decisions can be made regarding investment in training and equipment needs. Assessment of which types of construction project match the capabilities and experience of a company will help to maintain financial profitability and a reputation for high-quality work.
Allocation of resources	A company may work on a number of projects simultaneously. This means that assets and resources will need to be actively managed to make best use of them across the range of work being progressed. Data on factors such as personnel placement, work activities, plant and equipment usage, and temporary accommodation on site allows analysis of resource needs and efficient allocation across the range of project commitments.

Case study

After analysing footfall data in a town-centre pedestrianised area, the local authority decided that improvements were required to footpaths, landscaped areas and lighting. The works involved the installation of gravel and paved paths, granite kerbs, stone steps, play and sports equipment and street furniture.

During a site survey as part of project planning, some challenges were identified, including:

- a lack of site access, restricting movement of materials to the work location from delivery points
- a lack of onsite storage.

Management discussions took place and research was undertaken into available options. Based on data made available by local employment agencies and advice from local plant hire companies, the decision was made to:

- temporarily increase the workforce
- hire materials conveyers to move materials from the restricted delivery bays.

Would you agree that these are effective strategies?

If you agree, state why.

If you disagree, state your alternative proposals.

Digital data networks that transfer information efficiently between users have become integral to our increasingly technology-based communities and now have an important role in the successful completion of construction projects.

▲ Figure 8.7 Efficient digital data networks have an important role in everyday life as well as construction

Improve your English

Make a list of the ways that data is transferred using digital networks. Write a short description of each way that you identify. An example you could start with is streaming films over the internet.

Test yourself

How can data be used to improve a company's market competitiveness?

2 Sources of information

So far in this chapter we have considered how data is the raw material that can be processed to provide information. The resulting information can then be presented in various ways and used by construction personnel in a range of activities.

In this section, we will consider different sources of information and how they can be interpreted for use in the workplace.

2.1 Interpreting data sources

Specific data sources may provide information for workers with designated responsibilities.

For example, electricians working on the installation of electrical services would not require information about the brick types used in the building to complete their work successfully. However, they would require information about which size and type of cables to use to satisfy voltage ratings, or about switching units or electrical junctions that must be situated in specific locations.

We can identify specific sources of information and which personnel groups or individuals could use them.

Product data

Product data provides information or instructions about how to use or install a product correctly. For example, a manufacturer has to supply information about how to handle and store a product safely in accordance with health and safety regulations, such as the Control of Substances Hazardous to Health (COSHH) Regulations 2002. There is more on these regulations in Chapter 1.

▲ Figure 8.8 COSHH symbols that could appear on a product label

The correct and safe installation or use of a product could be made clear by using illustrations, diagrams, performance charts, or clear written instructions.

Health and safety

Many materials and substances used on site have the potential to cause harm or injury. Always check for product data labels and examine them carefully.

Product data is closely tied to manufacturers' specifications. These provide very specific information on performance data, such as the temperature range within which installation work must take place or the type of surrounding environment in which the product will operate successfully.

Manufacturers' specifications may include key data and information on how a product should be assembled, dismantled, calibrated, adjusted, maintained, repaired, examined or inspected.

Product data can be used by building designers working in an office environment or by onsite personnel installing a range of systems. Trade workers might use product data to confirm that the materials they are working with are in accordance with specifications for the work task.

Test yourself

What key information can be provided by a manufacturer's specification?

Client specification

In the Royal Institute of British Architects' (RIBA) plan of work 2020, the client specification has been defined in part as 'a statement or document that defines the project outcomes and sets out what the client is trying to achieve'.

The client's needs will obviously depend on the type and use of the required structure, for example:

- A client requiring construction of a dwelling may specify the floor areas needed to satisfy the living requirements of the intended number of occupants.
- A client who has commissioned the construction of an office block may specify the levels of light within the building to allow comfortable work in office spaces.
- The specification for a retail development may centre on the project being attractive and convenient for customers to use.

Information in the client specification must be interpreted carefully in order to produce a design brief that meets the objectives as fully as possible. Professional personnel such as architects, quantity surveyors and engineers will give careful attention to the client specification when performing their work.

Building Information Modelling (BIM)

As mentioned earlier in this chapter, BIM is a structured system that allows authorised users to access a range of important information at all stages of construction and beyond. It uses digitally processed information to analyse design elements of a building, including 3D modelling.

Using BIM, complex design ideas can be transformed into a medium that is easier for all personnel to work with. It allows collaboration between all the designers, engineers and contractors working on a project, providing comprehensive information about each role's workflow. This enhances the process of design and construction and allows exploration of alternative design possibilities before work on site begins.

BIM allows digital data describing internal building engineering services to be presented and analysed visually, so that the way they interact with each other can be seen. Clashes in systems can then be identified at an early stage in the design process; this is known as 'clash detection'.

An example could be drainage system pipework within a building interfering with the route of heating ducts. In the past, this could require extensive redrawing of plans and cause expensive delays if the 'clash' was discovered after work had commenced.

▲ Figure 8.9 BIM allows possible clashes in the systems of a building to be detected before work begins

Research

Find out the history of BIM. When was it developed and what did the first BIM system contain?

The Common Data Environment (CDE) is a single central source of information used within the BIM system. Relevant documents and data are brought together in a shared digital environment that can be accessed by all authorised personnel collaborating on the project.

The CDE can contain different types of digital information, including schedules, contracts, registers, reports and 3D models, forming the foundation of shared information on which collaboration can take place. Large amounts of digital data and information can flow in a controlled way through the CDE during the development of a project, making it possible to reduce mistakes and avoid duplication.

Test yourself

What function does the CDE perform in a BIM system?

Work program planning

To ensure work can be completed on schedule and within budget, careful prior planning of the construction process is required. Planning a programme of work is often undertaken using charts that provide data and information about the sequence of activities.

There are two main documentary methods of planning the sequence of work in construction – Gantt charts and critical path analysis (CPA) (see Chapter 3, section 4.2).

Improve your English

Search online for the origins of the Gantt chart and write a short report.

Research

Look online for examples of critical path analysis. Once you understand how they work, create your own example for a simple activity, such as making a cup of tea. Ask one of your peers to do the same and compare your results.

Commissioning and certification

Commissioning is the process of confirming that all building systems are installed, tested, operated and maintained according to predetermined operational requirements. Data and information linked to engineering techniques and procedures are used to check the correct installation and function of systems such as heating, ventilation and air conditioning.

Systems must be confirmed as fit for purpose before complete handover of the building to the client. However, as part of the contract arrangements, there are circumstances where performance testing of systems may continue after the client has accepted handover. For example, some systems are dependent on different weather conditions, and commissioning inspections and adjustments may take place over the first year of occupation.

Commissioning data and information form a checking and testing format to bring systems into operation and verify that they are in good working order. The testing format confirms specific details and responsibilities, such as:

- the party or parties responsible for each aspect of commissioning
- whether the commissioning process and outcomes must be witnessed (this is a legal requirement in some cases)
- the standards that should be adhered to
- the specific documentation that is required, which can include a record of operations, maintenance and future works.

A certificate is issued to confirm that installation and operational standards have been met, which must comply with technical benchmarks and criteria set by appropriate organisations. The standards could be set by Building Regulations, the British Board of Agrément, the manufacturer or another recognised certification body.

Research

Visit the British Board of Agrément (BBA) website at **www.bbacerts.co.uk**

Search the site and identify the five key areas that the organisation works in. Find a case study and note how certification benefited the company working under BBA guidance.

Even after a building's systems have been commissioned and certified, ongoing inspection and testing are required throughout the life cycle of the building. A schedule for this will include test data that needs to be recorded to ensure optimal operational efficiency is maintained.

▲ Figure 8.10 Testing an electrical system

Records of test data from inspections are a valuable source of information for new owners, if the building changes ownership. This is especially true in the case of a large complex structure that may be owned by **commercial** investors. A building is an expensive asset that must be properly maintained to retain its value.

In the case of a smaller **domestic** property, a condition report can be prepared for a new owner or homebuyer. This takes the form of a visual inspection to generate data and information about factors such as the condition of the building's structure, heating and lighting systems, exterior roof surfaces, attic roof spaces, floor spaces and underfloor spaces.

Efficient commissioning and reputable certification of a building's systems, coupled with planned and effective maintenance, will contribute to greater efficiency throughout the life cycle of a building. Operational efficiency can translate into reduced energy needs, which is an increasingly important factor in the continuing drive to reduce carbon emissions. Accurate and appropriate data and information are essential to the success of energy-saving strategies.

Key terms

Commercial: relating to buying and selling

Domestic: relating to a dwelling or home

Improve your English

Using the definitions of 'commercial' and 'domestic' supplied, write a summary of the ways that data can be used for each category of building in terms of commissioning and certification.

Research

Investigate how much carbon is produced annually by the construction industry. According to the scientific data you find, what targets need to be met to reduce industry carbon emissions to net zero by 2050? What changes in the materials used and established work practices are being made to tackle the issue?

2.2 Using data to calculate outcomes or cost

The term 'calculation' can be defined in different ways, according to the context. It can be used to describe the process of:

- using information to make a judgement or plan something carefully and intentionally, for example to calculate outcomes
- determining something by mathematical or logical methods, for example to calculate costs.

Data can be used to provide results that meet both of these definitions.

Calculating outcomes

When calculating outcomes, data can be used to analyse past or current activities to model what could happen in the future. A familiar example of this is forecasting the weather. Attempting this requires the analysis of past data related to a huge number of variables which, as we see daily, makes a reliable prediction difficult.

Something more straightforward might be calculating the outcome of using timber of specific dimensions to produce a load-bearing element of a building, such as a suspended floor. Consistent data has been recorded over many years which can be applied to produce reliable structural calculations, ensuring that the proposed design is safe and capable of bearing the required loads.

Test yourself

What are the two definitions of the term 'calculation'?

Building services engineering elements of a structure produce a wealth of data that has been harvested in differing types of building under varying conditions. This has led to the compilation of charts and tables that can be referred to when designing similar structures. Data from past experience can be used to calculate acceptable operational tolerances in new designs in order to avoid 'over-engineering', which could be wasteful and add to costs.

▲ Figure 8.11 Analysing data from past projects supports efficient design in new projects

Calculating costs

Many databases are used to help cost projects of all types and sizes. They draw on historical costing data and are often designed to track current prices of materials and components, almost in real time. Using up-to-date data for costing a project is vital to avoid cost overruns and accurately assess financial risks in a competitive construction environment.

Data from across the sector is collated, analysed, modelled and interpreted by the Building Cost Information Service (BCIS) of the Royal Institution of Chartered Surveyors (RICS). This valuable source of data is used by many contractors for projects of different sizes. There are also many other companies and institutions that specialise in providing data for costing a project, such as PropertyData and Bestdata.

A major project with complex design elements needs accurate data to produce costings that will result in profitability for the contractor and a fair price for the client. Of course, profitability is key to the success and financial health of any size of construction company, and small and medium-sized enterprises (SMEs) also rely on accurate data for success. A small contractor or even a sole trader benefits from keeping accurate project records to provide data for costing future work.

▲ Figure 8.12 A small contractor or sole trader can benefit from keeping accurate data from past projects

3 Data management and confidentiality

Data and information sources used in the construction industry are of great value, both operationally and financially. This means they must be stored securely and protected.

Improve your English

Write a short account explaining why data and information sources are valuable operationally and financially.

3.1 Data storage

Physical storage

In the past, data tended to be stored as paper documents and physical drawings. While hard-copy documentation is still used extensively, digital data storage is increasingly used.

There can also be a crossover between physical and digital data storage, if paper documents are scanned and converted to a digital medium for easier transmission to others in different locations. This can speed up communication of useful data that might be needed urgently.

Industry tip

Everyone involved in using and storing physical data must be conscientious in keeping documents and drawings secure and in good condition. Even if they do not contain sensitive details, they are valuable sources of information that should be taken care of.

Virtual storage

Virtual storage is the storage of data in a digital format, for example on a computer hard drive, on a portable flash drive or in a location that is remote from the user.

Remote storage is often referred to as 'cloud storage' or 'in the cloud'. Data is transmitted digitally through a network and then stored on a server. These storage facilities are often owned by specialist companies who use multiple data centres around the world.

Data files are stored remotely in a number of different ways:

- Public cloud: data is stored in the service provider's data centres across multiple regions or continents. Users pay the provider based on the amount of data stored.
- Private cloud: data is stored in an in-house storage facility, forming a dedicated environment that is protected from outside access by a **firewall**.
- Hybrid cloud: this is a mix of public and private cloud storage, offering businesses flexibility and more ways that data can be managed and made available to authorised users.

Data storage companies must provide safe and secure storage and have facilities in place to back up data to prevent loss.

Key term

Firewall: a protective software program or hardware device that monitors, filters and may block data entering and leaving a network

▲ Figure 8.13 A server room in a data storage centre

Test yourself

What is meant by storing data 'in the cloud'?

3.2 Confidentiality

Some data must be kept confidential. For example, data for a construction project commissioned by the government or a bank may contain sensitive details which could threaten security if accessed by unauthorised parties.

Confidential data stored in a physical format must be locked away and protected in secure facilities. It may be necessary to consider protection from fire damage, in addition to prevention of unauthorised access.

▲ Figure 8.14 Confidential physical data should be stored securely

Measures to maintain the security of digital data include:

- using anti-virus software and keeping it up to date
- using a system of strong passwords so that access to data is restricted
- using **encryption** to convert data from a readable format into an encoded format
- installing **firmware** and software updates as soon as they become available (these often contain security patches and new security features)
- logging off (or turning off) computers and locking them if possible.

The term 'hacking' has become increasingly familiar, referring to theft or manipulation of critical data, usually with malicious intent. Hackers are focused on bypassing the security measures designed to keep them out of data processing and storage systems. Security measures must therefore be rigorously designed, carefully adhered to, and updated when necessary, in order to meet the growing challenge of these attacks. Increasingly sophisticated systems are being devised to maintain the security of digital data.

Key terms

Encryption: the process of converting data or information into a code to prevent unauthorised access

Firmware: software that has been permanently installed in a device to make it work as the manufacturer intended

Improve your maths

Find out about the mathematical processes used to encrypt data.

The news often refers to 'cyberattacks' when describing the work of individuals or organisations who steal data or install damaging software known as **malware**. Malware can behave like a virus, in that it can easily be passed from computer to computer, or system to system, so that the 'infection' spreads.

A Trojan horse or Trojan is a type of malware that is often disguised as legitimate software, misleading the user as to its true intent. It tricks systems operators into allowing unauthorised access.

Key term

Malware: software that is specifically designed to disrupt, damage or allow unauthorised access to a computer system

Research

Search online for the names of some common Trojans.

Test yourself

Why is malware sometimes referred to as a Trojan?

Loss of data can be very damaging to the operational efficiency of construction companies that are managing complex project commitments. Recovery of lost or stolen data has developed into an important activity that uses specialist techniques to beat the abilities of hackers and data thieves.

Legal requirements

Data protection in the UK is governed by the UK General Data Protection Regulation (UK GDPR), which came into effect on 1 January 2021. It should be considered alongside the Data Protection Act 2018. All companies that sell goods or services need to comply with these legal texts.

The aim of this legislation is to ensure that personal data is gathered legally, and that those who collect and hold data protect it from misuse and exploitation. A construction company that stores personal data about its workers should always ensure confidentiality.

▲ Figure 8.15 GDPR protects the confidentiality of personal information

Industry tip

GDPR is not simply guidance – it is law. As such, there are legal penalties if it is not adhered to. If your role in construction involves dealing with anyone's personal information, never treat it casually.

Improve your English

Search online for the seven principles of GDPR and write a summary.

4 Drawings, circuit diagrams and schematics

Whether based in a design office or on a construction site, workers rely on accurate information to perform their work tasks efficiently. Graphical documents, in the form of drawings, circuit diagrams or **schematics**, are an important means of communicating data and information, and they can be presented in paper or digital format.

Key term

Schematics: a diagram representing the elements of a system using graphic symbols

4.1 Drawings

Drawings are documents required at every stage of building work. They are an efficient way of providing a great deal of clear data and information, without the need for lots of potentially confusing text.

When producing drawings, a technician or draughtsperson will draw the details and features to scale. This means that large components can be represented with accurate proportions on a document that is much smaller, which is more manageable than if they were drawn full size.

Improve your English

Search online for definitions of 'scale drawing', then write a definition in your own words.

Drawings for construction and installation purposes are produced using conventions. This means that the format and layout of drawings follow agreed standards, allowing the data and information they contain to be consistently understood.

Standardised symbols are used on drawings as a means of giving information in a simpler way. If all the components and installation details on a drawing were labelled in writing, it would soon become very crowded with text, which could be confusing.

Where symbols are used on a drawing, they must conform to an agreed standard, so that everyone using the drawing can interpret the information consistently and accurately.

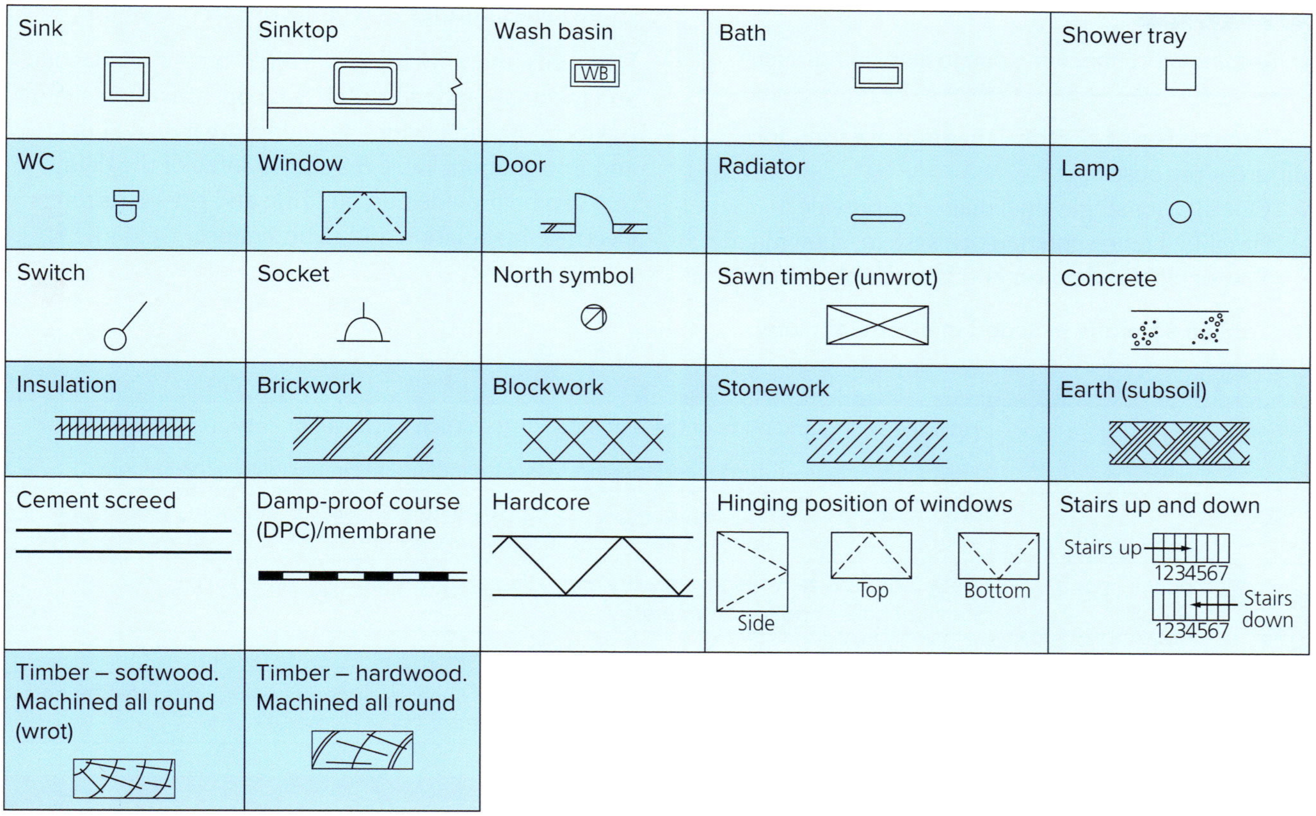

▲ Figure 8.16 Basic drawing symbols

Symbols and abbreviations for construction and technical drawings have been defined by the British Standards Institute. Until 2019, BS 1192 was used to set out methods for managing the production and quality of construction information. This has now been replaced by BS EN 19650 and further amends will be made due to the UK leaving the EU. There are many other British Standards that apply to specific types of construction and technical drawing.

Test yourself

What is an advantage of presenting technical data in the form of drawings?

4.2 Circuit diagrams and schematics

Circuit diagrams are technical drawings that provide visual representations of electrical circuits or systems. They are used to explain a design to electricians and technicians, who will use them during installation, maintenance or repair work. The complexity of drawings will vary, depending on the intended purpose of the electrical system and who will carry out the work.

Electrical circuits are shown on a circuit diagram using lines to indicate electrical connection routes within the system, with related components shown as symbols. As mentioned previously, diagrams using symbols are often referred to as schematics. Pictorial diagrams differ in that they represent elements of an electrical system using graphic drawings or realistic pictures to provide relevant information.

Wiring diagrams show a simplified representation of an electrical circuit, usually giving more information about the relative placement of devices and components and their **terminal** positions.

> **Key term**
>
> ***Terminal:*** a connection point to an external circuit

Different types of electrical diagram are used for different purposes:

- One-line (or single-line) diagrams provide a simplified representation of a system, showing the flow of electrical power and how the electrical circuit is connected. They show the major components, such as switches and transformers, using standardised symbols. The representation does not depict the physical size or position of the components.
- Three-line diagrams are more complex, with separate lines drawn to represent live, neutral and earth (or ground) electrical current. They provide a detailed visual guide for cabling arrangements, switches, fittings and protective device connections.

Diagrams and drawings for a building provide layout and installation details for use by operatives on site in the form of plans, which include a great deal of data and information. Table 8.2 shows some of the details that can be provided on building and site plans for electrical installation and maintenance.

▼ Table 8.2 Details included in electrical plans

Details provided	What it shows
Wiring	How the wiring and parts of the electrical system such as power outlets are interconnected How switching is arranged and located to isolate individual circuits within the system
Fixtures	Where fixtures such as light fittings or electric heating units are located in the electrical system, along with details of the electrical loadings they create
Incoming power lines	Details of the size, voltage, capacity and rating of power-delivery items such as cables, junction boxes and transformers
Fuses and circuit breakers	Where they are located within the electrical system and what is protected
Equipment	The location within the electrical system of items such as air-conditioning units, solar panels and generators, as well as their voltage and size

Remember, drawings and diagrams produced for use on site typically include data in the form of standardised symbols, and in some cases annotations, to clearly explain design and installation details of the building's electrical systems.

> **Industry tip**
>
> If you work with electrical drawings and diagrams, you will quickly become familiar with the basic symbols and abbreviations. However, the range of symbols is extensive – make a conscious effort over time to expand your knowledge of them.

> **Health and safety**
>
> Since electricity has the potential to cause injury or even death when not managed correctly, the ability to accurately interpret data from drawings and diagrams is vital in order to keep the installers and users of electrical systems safe.

5 Programming and set up of digital systems using IT resources

The term 'programming' applies to:

- manually entering commands into a control unit to set up operation routines
- writing lines of code to create a software program that is part of the internal operational command structure of a control unit.

Information technology (IT) used in the creation of software programs is found in many aspects of daily life. For example, smart phones and smart televisions are familiar items that make use of IT software programming to process and present information.

The development of software programming skills requires an understanding of an appropriate programming language. There are many programming languages that are designed for specific purposes, such as C++ and Python.

The International Electrotechnical Commission (IEC) has devised a standardised approach to programming digital controllers referred to as IEC 61131, which aims to improve the efficiency and speed of implementing automation solutions in a range of applications.

Programmable controllers can be found all around us, controlling and automating many important functions.

Industry tip

Keep in mind that standards formulated by recognised organisations are updated or amended periodically. It is important to check regularly that you are working to the latest relevant standards, to make sure your technical understanding and work practices are current.

5.1 Digital systems

Digital controllers are used to control systems such as heating, ventilation and air conditioning. These can use appropriately positioned sensors in a large building to transmit data to the control unit. Changes can be made automatically or manually to adjust temperature, humidity or the amount of ventilation, in order to maintain comfort levels for the occupants.

An example of a familiar digital controller is a domestic heating system control panel, which allows programming of 'time of day' on–off timings and can regulate temperature according to data from a room thermostat set at the level desired by the occupants.

▲ Figure 8.17 Manually entering commands into a heating system controller

Using digital technology (as opposed to mechanical controllers and timers) means that specific software can be integrated into the controller design for individual devices. This allows direct control and programming at the controller location (sometimes referred to as manual operation mode). The controller may be separate from or attached to the device being controlled.

Chapter 10 covers digital systems used in construction in more detail.

Research

Digital systems use binary code. Research what this means and how it works in a digital system.

Test yourself

What type of systems in a building are digital controllers used for?

5.2 IT resources

Computer-based systems make it possible to control devices remotely through digital communication networks. These systems may be designed to allow a single operator to control a number of devices from a central location (sometimes referred to as supervisory control). This allows the gathering of performance data, which can be used to create maintenance schedules and provide information for troubleshooting faults with equipment.

▲ Figure 8.18 A single operator can control many processes through digital IT systems

'Smart' controls are designed to allow remote operation by persons who may not even be in the building. Smartphones can be used to input control data, allowing convenient operation and adjustment of digital systems from any location with mobile telecoms network access.

For more information on smart controls, see Chapter 5.

Improve your English

Visit an app store and search for 'smart controls'. Choose an app and write a description of what it can do.

Data harvested from digital control systems in existing occupied buildings can be integrated into BIM systems during the design stage of a new building, to model the new building's digital control systems. The potential efficiency and economy of important systems in the proposed development can be tested through digital simulations using suitable IT resources such as Simflow.

This type of 'intelligent' digital data management takes the use of computer-aided design (CAD) beyond simply producing 3D models of a structure. This is a highly sophisticated use of data to support the design and construction of environmentally friendly buildings that are economical to build and operate, providing high levels of user comfort and convenience.

Assessment practice

Short answer

1 In what different forms can data be produced?
2 Write a definition of data interoperability.
3 Who can make use of data generated at the start of the building design process?
4 What do the letters EAM stand for?
5 Why might commissioning and certification of systems in a new building continue over the first year of occupancy?

Long answer

6 Describe a work task where metadata would contribute to efficiency when designing a building.
7 Explain how generalisation can make producing information from data analysis easier and more manageable.
8 Explain how BIM is structured to manage large volumes of data to allow collaboration between project personnel.
9 Explain the purpose and list the benefits of using data for post-occupancy evaluation (POE).
10 Explain how data can be used to calculate outcomes and costs. In your explanation, suggest what sources of data can be used.

Project practice

A detached house has an integral garage which the owner wishes to convert into living space. The proposal is to remove the existing wall between the lounge and the garage to create a larger lounge area.

A new gas heater will be installed in the existing garage area to heat the enlarged lounge.

Attached to the wall that must be removed are the gas and electricity meters for the dwelling.

- Write a report that outlines the sequence of tasks needed to maintain services to the dwelling during the alteration works.
- Copy the drawing in Figure 8.19 and create a diagram showing suggested new positions for the gas and electricity meters.
- Using the correct symbols (research them online), show suggested positions on your diagram for four power sockets, two light fittings and a light switch.
- Using an electrical supplier website as a data source, produce a costing for the electrical fittings used in the project.
- Add to the diagram a suggested suitable position for the new gas heater and indicate the route of the gas supply pipe from the gas meter.

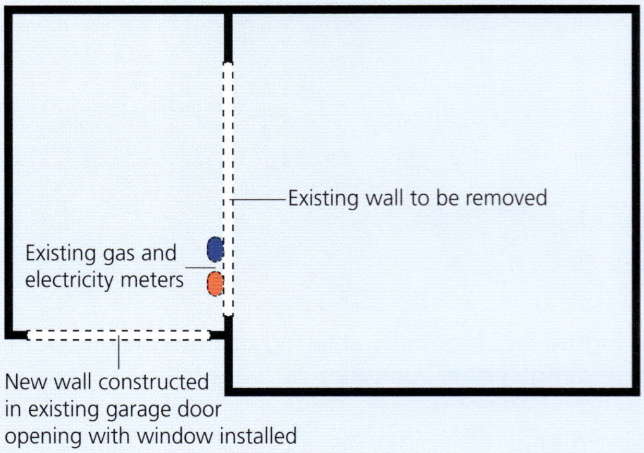

▲ **Figure 8.19** Working drawing

Chapter 9 Relationship management in construction

This chapter looks at the benefits to the construction industry of positive interpersonal relationships, good customer service and team working. It identifies legal and moral responsibilities with regards to equality and diversity, and examines how a strong representation of society in the workplace can improve business productivity and performance.

You will learn about the benefits of different communication methods and styles, as well as how to negotiate and resolve disputes. To close the chapter, we will explore employment rights and responsibilities, and the importance of ethical behaviour to the reputation and growth of an organisation.

Learning outcomes

By the end of this chapter, you will understand:

1. stakeholders
2. roles, expectations and interrelationships
3. the importance of collaborative working to project delivery and reporting
4. customer service principles
5. the importance of team work to team and project performance
6. team dynamics
7. equality, diversity and representation
8. negotiation techniques
9. conflict-management techniques
10. methods and styles of communication
11. employment rights and responsibilities
12. ethics and ethical behaviour
13. sources of information.

1 Stakeholders

Any group or individual that has an invested interest in the long-term success of an organisation is referred to as a stakeholder. Stakeholders can affect and be affected by the achievement of an organisation's business objectives. Therefore, it is important that they can work together.

Internal stakeholders support or have concern for an organisation and benefit from their direct relationship with it, for example employers and employees. If an organisation did not have any internal stakeholders, it would simply cease to exist.

There is a risk that some internal stakeholders will have conflicting interests. For example, employees want to earn a good wage, whereas their employer may want to reduce outgoings on staff remuneration. To avoid this problem, successful construction businesses keep the number of stakeholders to a minimum and ensure they have a good balance between all of their interests in order to prevent conflict.

External stakeholders do not have a direct relationship with an organisation and are not employed by it, yet they can still have an indirect effect on it. For example, a building control officer enforcing building regulations will determine the standards that a contractor has to adhere to.

External stakeholders that do not benefit from a direct interest in an organisation are also known as secondary stakeholders.

Table 9.1 gives examples of internal and external stakeholders.

▼ Table 9.1 Types of stakeholders

Internal stakeholders	External (secondary) stakeholders
• Clients • Architects • Investors • Managing directors • Site managers and supervisors • Structural and civil engineers • Buyers • Contracts managers • Estimators • Trades forepersons • Quantity surveyors • Project managers • Subcontractors	• Suppliers • Government • Trade associations • Emergency services • Building control officers (BCOs) • Regulatory authorities • Communities and local residents • Health and Safety Executive (HSE) • Local councils • Lobby groups and activists, for example **environmental activists** • Trade unions • Professional bodies • End users in a construction project

Key term

Environmental activists: people who campaign for the protection of the natural environment

▲ Figure 9.1 A site manager is an internal stakeholder

▲ Figure 9.2 The Health and Safety Executive (HSE) is an external stakeholder

Test yourself

Give a definition of 'stakeholder' in the construction industry.

Stakeholders will always be present in construction projects, and their influence plays a part in determining success or failure. For this reason, whenever possible, internal and external stakeholders should be selected to suit specific construction projects because of their skills, knowledge and experience working on similar jobs.

Research

Find out what shareholders are. Describe the main differences between shareholders and stakeholders.

Test yourself

How could an external stakeholder, such as a community, indirectly affect an organisation?

2 Roles, expectations and interrelationships

The roles, expectations and interrelationships of stakeholders throughout all stages of a construction project, from design to construction to handover and use, are covered in detail throughout this book:

- hierarchy of project management (see Chapter 3)
- promoting good relationships across the project (see this chapter)
- cost-control measures (see Chapter 3)
- time-management methods (see Chapter 3)
- handover processes (see Chapter 4)
- public relations, including the behaviour of employees outside of work hours (see this chapter)
- follow-up and review (see Chapter 3).

Test yourself

Explain the benefit of using a hierarchy to structure an organisation in the construction industry.

3 The importance of collaborative working to project delivery and reporting

A **collaborative** approach to project delivery and reporting is essential to ensure work is completed on schedule, within budget and to the client's specification.

Reporting any foreseeable problems before they occur, and working with the project team to resolve them, is essential for meeting project aims and objectives. Collaborative working also ensures projects are completed to the minimum industry standards and building regulations.

Project teams and other stakeholders that communicate effectively at every stage of construction work are more likely to provide a safe working environment for employees and others that may be affected by work activities.

Figure 9.3 identifies various stages of construction work when members of a project team need to collaborate in order to achieve a successful project outcome.

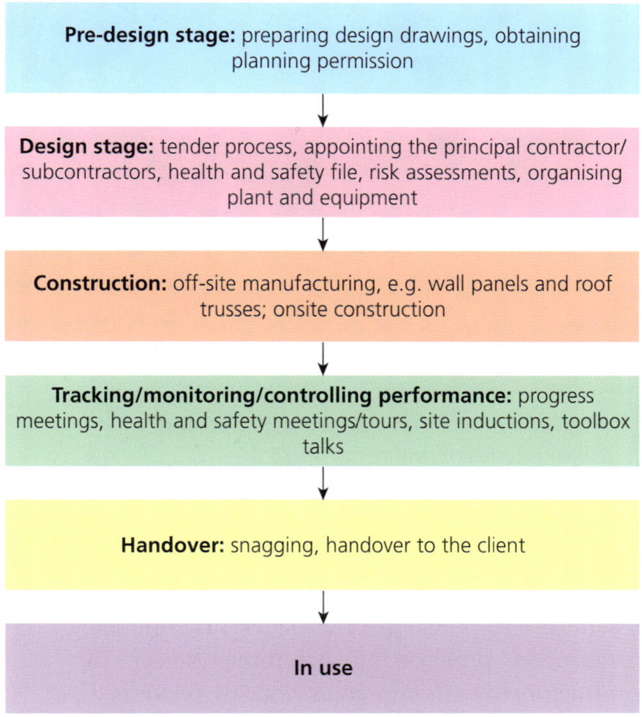

▲ Figure 9.3 Stages of construction work

In Chapter 4, we looked at Building Information Modelling (BIM) and how this can be used to collaborate at various stages of a construction project, by identifying design problems and communicating efficiently with the project team. The construction management team can also use workflow software packages to plan and organise work. Information can then be shared easily with others and worked on as a **living document** to bring a project in on schedule.

Later in this chapter, we will look at the benefits for project delivery of members of the project team collaborating using other communication methods, including face-to-face meetings.

Key terms

Collaborative: involving people or groups of people working together with a single common interest or aim

Living document: a document that can be changed, amended or updated when necessary for the benefit of the user

▲ Figure 9.4 A project management team discussing project progress

Test yourself

Why is a collaborative approach important for project delivery and reporting?

4 Customer service principles

Clients are responsible for initiating and financing construction projects. It is therefore important for contractors to maintain good working relationships with their clients, in order to keep them happy throughout a project and increase the chances of repeat business.

Contractors need to create a good first impression. This can be achieved through a portfolio of successfully completed contracts and positive **testimonials** from previous clients. Demonstrating good product knowledge is also extremely important, as the client is unlikely to be an expert and will be looking for guidance to steer them through the project and achieve their vision.

Key term

Testimonials: statements of recommendation produced by satisfied customers or clients that confirm the quality of a product or service

Improve your English

Write a testimonial on a supplier's website for a product you have recently purchased.

▲ Figure 9.5 A contractor meeting with a client

When a contractor works with a client for the first time, it is essential to establish a good level of trust. This can start during the planning stage of construction by maintaining good lines of communication with the client, such as responding promptly to any messages and involving the client in decision making. Listening to the views of clients at every stage of a project and treating them with empathy is also important in building their confidence.

Total trust between parties can take time to establish, but just a moment to ruin. However, if a contractor can demonstrate their ability to manage the client's project efficiently by meeting agreed timescales and working with honesty and integrity, then their professional bond will grow.

Dealing with clients and employees is just as important as managing a project itself, especially if they have a complaint. The way a contractor responds to negative feedback is important; they need to demonstrate that they are continuously looking for ways to improve, in order to keep clients satisfied and maintain a good reputation.

▲ Figure 9.6 A developer handing over the keys to a new house

Improve your English

Write a sentence to explain the term 'empathy'.

Test yourself

Describe the customer service principles that should be applied at various stages of a construction project.

5 The importance of team work to team and project performance

When employees collaborate towards common goals without friction or conflict, they create a healthy work environment that improves staff morale. Employees who feel part of a team and are offered the chance to be creative and learn new skills develop positive 'can-do' attitudes that contribute to the long-term success of an organisation.

Teams perform best when individual members are motivated by satisfaction with their roles and everyone has accountability for the part they play.

▲ Figure 9.7 Employees working together

Team members will work well together if they can communicate openly, without fear of reprisal. Individuals should be comfortable both asking their peers for help when needed and providing feedback to the team when the opportunities arise. This will create an environment that improves trust between team members and ultimately leads to improvements in efficiency and productivity.

Conversely, conflict, tension, low engagement and lack of trust in a team will have a negative impact on project performance.

Test yourself

Explain the advantages of team work for team and project performance.

6 Team dynamics

Team dynamics refers to psychological processes and behaviours occurring in a team that influence its direction and performance.

People's personalities and behaviour are often uncontrollable and unpredictable, and they can have either a positive or negative influence on team dynamics. With the right strategies in place to address any issues early on, a good team leader can manage the relationships between team members, in order to keep lines of communication open and ensure active participation and co-operation.

When all members of a team work collectively, listen to and support each other, they are more likely to resolve challenges and achieve objectives. Signs of positive team dynamics are when members work together without conflict and trust the expertise, knowledge and abilities of individuals within the team.

▲ Figure 9.8 Construction professionals demonstrating good team dynamics

Businesses rely on positive team dynamics to generate new ideas and improve performance. It is therefore important to identify poor team dynamics as soon as possible. Signs of negative team dynamics include:

- disruptive work behaviour
- failing to achieve positive results
- not being accountable
- poor decision making.

Poor team dynamics can cause individuals to feel vulnerable, making them less reliable and less able to adapt to the changing situations that may arise during building projects.

7 Equality, diversity and representation

Employers that embrace **equality**, diversity and inclusion develop an organisational culture that helps people to achieve their full potential based on their talent, not on factors such as their age, background, race, religion, gender, disability or sexual orientation. Employing a talented, diverse representation of society allows different views and ways of thinking to be shared within the organisation, to reflect the local culture and community.

Organisations that promote the values of equality, diversity and inclusion are often more sociable, productive and efficient, which improves profitability for their shareholders.

Key term

Equality: a state where all members of a society or group have the same status, rights and opportunities

▲ Figure 9.9 A female construction worker

▲ Figure 9.10 People with disabilities should be supported to work in the construction industry

▲ Figure 9.11 The construction industry must aim for a truly inclusive and diverse working environment

To demonstrate its commitment to equality, diversity and inclusivity, an organisation usually has a written equality and diversity policy, which should be read and signed by all employees as part of their induction.

The policy should include:

- values
- legal duties

- processes which employees should follow to challenge discrimination or if they have a **grievance**, for example **whistleblowing**.

The policy also helps to identify any staff training needs and how to support them.

Key terms

Whistleblowing: the act of reporting information about wrongdoing

Grievance: a feeling of having been treated unfairly

Employees who work for organisations that have a robust equality and diversity policy and recognise equal opportunities will feel more valued.

Improve your English

Read the equality and diversity policies for your employer and your training provider. Summarise the main differences in a short paragraph.

Discrimination is the unfair treatment of someone because of their characteristics. An employer could openly discriminate against certain groups or individuals while recruiting new employees or promoting existing employees within their organisation. For example, they should not advertise for specific age groups or genders, unless they are specifically needed to fulfil the job role.

Besides the moral reasons for equality and diversity, businesses also have legal duties as outlined below.

7.1 Equality Act 2010

This is the main piece of equality and diversity legislation that protects people from discrimination at work and in wider society. It outlines the different ways that it is unlawful to treat someone, and it strengthens protection in some situations.

Under the act, the following are protected characteristics:

- age
- disability
- gender reassignment
- marriage and civil partnership
- pregnancy and maternity
- race
- religion or belief
- sex
- sexual orientation.

Improve your maths

A construction site has 267 workers, with age ranges as follows:

- 18–25 years – 25 workers
- 26–35 years – 116 workers
- 36–45 years – 53 workers
- 46 years and over – 73 workers.

What percentage of the workers are aged between 18 and 25 years?

7.2 Employment Rights Act 1996

This act legally protects the personal rights of employees and **workers**. It covers areas such as:

- contracts of employment
- protection of wages
- zero-hours workers
- Sunday working
- flexible working
- rest breaks
- study and training
- unfair dismissal
- maternity and parental leave
- redundancy payments.

▲ Figure 9.12 Rest break on a construction site

Key term

Workers: people who do not have a permanent contract of employment with an employer but are contracted for work or services

7.3 Employment Act 2008

This act covers:

- the procedure for the resolution of employment disputes between employers and staff
- compensation for financial loss in cases of unlawful underpayment or non-payment of staff

- the enforcement of minimum wages
- the enforcement of offences under the Employment Agencies Act 1973
- the right of trade unions to expel or exclude members on the grounds of membership of a political party.

7.4 Human Rights Act 1998

The Human Rights Act 1998 incorporates the European Convention on Human Rights into UK law. It sets out fundamental rights and freedoms that everyone in the UK is entitled to, including:

- the right to life
- freedom from torture or inhuman or degrading treatment
- freedom from slavery
- the right to liberty and security
- the right to a fair trial in a court of law
- the right to respect for private and family life, home and correspondence
- freedom of thought, conscience and religion
- freedom of expression
- freedom of peaceful assembly and association
- the right to marry freely
- the right to an effective remedy in a national court
- freedom from discrimination
- the right to education
- the right to vote.

Under this act, people who are not treated fairly, equally and with dignity and respect can take legal action to defend their rights.

Case study

Ashley is employed as an electrical engineer by a small construction company with a workforce of 12 full-time employees. She is the only female member of staff. Since completing her apprenticeship when she was 19 years old, she has worked for her current employer for six months.

Ashley has started to notice that she is being treated differently to the other employees, for example being given less-challenging jobs and being paid slightly less than another electrical engineer who started at the same time as her.

- What employment rights does Ashley have?
- Working in a small group, discuss Ashley's situation and explain the steps she should take to achieve a satisfactory outcome.
- Do you think this is a case of workplace discrimination? Why?

Test yourself

Explain the benefits of equality and diversity in the workplace.

8 Negotiation techniques

At various stages of construction projects, different parties need to negotiate money, assets or resources in the best interests of the organisation or as an individual.

Negotiation skills are important in the construction industry, for example when buyers acquire land. Negotiations between buyers and landowners must be managed carefully due to the large sums of money involved. A poorly negotiated deal to purchase land will impact on profit margins when projects are completed and sold.

Once land has been acquired, planning permission may be needed before building work can start. This involves the project planning team working closely with the local planning department, to develop and negotiate initial concepts into a mutually agreeable proposal. If the initial planning application is refused, there is usually an opportunity to appeal the decision at a hearing, where negotiation skills will be tested once again.

▲ Figure 9.13 Local authority planning committee

In situations where a client already owns a building plot, they will usually select a principal contractor through the tender process and award the contract after carefully negotiating the terms. For more information about principal contractors and the tender process, see Chapter 4.

Different negotiation techniques are needed during the construction phase of building work, for example when changes occur to the original contract agreement (change orders) or time extensions may be needed to avoid penalty clauses or to resolve disputes with subcontractors.

8.1 Distributive negotiation

Distributive negotiation is used to **haggle** over a common single interest at stake, known as a fixed sum.

A fixed sum is best described as a pie that parties are battling over for a bigger slice, with exchange offers back and forth. If one party gains more of the pie through the distributive negotiation, then the other party loses a percentage.

When preparing a bargaining strategy for a negotiation, both parties should have a preconceived goal. This is the point at which they would walk away without a deal, known as the **reservation point**.

Due to the nature of this method of negotiation, it is not possible for both parties to have a whole pie each. However, a mutual agreement can be reached where one party has a smaller slice of the pie, providing they have accepted a deal no less than the reservation point.

Key terms

Haggle: to negotiate for the best terms of an agreement or financial arrangement

Reservation point: the highest price a buyer is willing to pay for an item and the lowest price a seller will accept for the item

Improve your maths

A developer negotiates to buy a plot of land for £470,000. The plot measures 23 m × 29 m. The developer then decides to sell a part of the plot that measures 29 m × 9 m, and adds a further 15 per cent to the rate at which he bought it.

How much will the developer sell the plot for?

8.2 Win–lose approach

This is probably the most common negotiation method used to settle disputes between two parties. However, an agreement is more difficult to reach, because one side has to compromise in order for the other to experience a positive outcome.

If the win–lose approach is used to negotiate the distribution of resources, bitter disputes could cause conflict between the parties and may damage their future relationship, especially if negotiations break down and one party walks away.

8.3 Lose–lose approach

Sometimes during business negotiations, all concerned parties end up worse off and not achieving their desired result, for example unavoidable financial cuts within an organisation due to low annual profits.

In these situations, all participants should try to minimise their losses as much as possible, and to make sure they are fair. Even though participants are no better off using the lose–lose approach, their relationship remains intact because the loss has been evenly distributed.

8.4 Compromise approach

Negotiations between professionals to find an acceptable middle ground can be time consuming. When negotiators are unable to reach a mutual agreement and have nothing more to negotiate with, they often make **concessions** in order to meet the needs of the other party or to get something else they want.

The compromise approach is used to settle disputes quickly by one party settling for less than it may have hoped for, in order to reduce strained negotiations and maintain or fix a relationship. The risk is that one party gives up much more than it should, resulting in financial loss.

Key term

Concessions: something granted in response to a demand

8.5 Integrative negotiation (integrative bargaining)

This type of negotiation takes place between parties with common interests, in order to collaborate in finding a mutually beneficial solution. It is used when multiple issues have to be agreed.

Negotiators often make trade-offs across the issues being discussed and add further issues into the negotiations to create additional value for all parties, so that a successful outcome can be achieved.

8.6 Win–win approach

In this approach, negotiators with shared interests work together to find resolutions they are both satisfied with, rather than seeking to fulfil self-interests. This avoids disagreements and helps to maintain strong relationships between parties, while achieving a fair outcome for both sides.

9 Conflict-management techniques

Working in a fast-paced industry such as construction can be very rewarding. However, there can sometimes be conflict between the management team, subcontractors, suppliers and clients.

Disputes can occur for several reasons, such as:

- ambiguous contract terms
- breaches of contract conditions
- late supply of building materials, resources or equipment
- breaches of site rules, for example failing to wear mandatory PPE
- programme delays.

People working at all levels in the construction industry will experience the pressure of having to meet tight deadlines while working long hours. This can have an impact on their physical, physiological and mental health, and is also a contributing factor when conflicts occur.

Initial grievances may be verbal exchanges, although these could escalate to formal written correspondence, for example emails and letters. Wherever possible, conflict should be avoided, because litigation can be a time-consuming and costly process; it may also damage professional working relationships between the development team and the client.

If a dispute occurs, each side will compete for their best interests using one of a number of conflict-management techniques, in order to force the other side to concede or reach an amicable agreement.

There are no benefits to conflict, so it should be avoided to prevent friction between staff, a negative working environment and delays to a project. Building Information Modelling (BIM) (see Chapter 4) can be used as a preventative measure by identifying issues that could provoke conflict between parties before they become a problem. When changes or alterations occur to a project design or specification, BIM can be used to update all concerned parties.

A dispute can quickly heighten emotions, unless a conflict-management technique is used. One way to resolve a problem is for parties to have an informal discussion in a neutral environment, where both sides have an opportunity to air their views without fear of backlash. If an amicable agreement cannot be reached however, the dispute will have to be resolved using mediation, conciliation or arbitration.

9.1 Mediation

This process involves appointing a mediator – an impartial third party who uses negotiation and communication techniques to encourage participation from both sides to find common ground and resolve their dispute.

One side cannot force another to use mediation; however, agreeing to follow this route can avoid the financial expense of going to court.

9.2 Conciliation

This is similar to mediation. An unbiased conciliator is appointed to meet with the parties separately and together, to weigh up both sides of an argument. They will then make a proposal based on the relative merits of each side, to find an amicable solution to the matter and bring it to a close as quickly as possible.

9.3 Arbitration

During this process, an appointed arbitrator (or tribunal hearing) resolves disputes between parties. When both sides of an argument have been heard, the arbitrator makes a legally binding decision, known as the award. If either party disagrees with the decision, they can take the case to court.

Test yourself

Why is it important to be able to use different negotiation techniques when working in the construction industry?

Improve your English

For the last two months, a supplier has been late in supplying building materials to your construction projects. Write a formal letter to the supplier to try to resolve the matter.

10 Methods and styles of communication

Communication is the process of exchanging information from one place, person or group to another.

The success of any project depends on accurate and effective lines of communication throughout the hierarchy of the construction team. Inaccurate, ambiguous, incomplete or confusing information can lead to delays, missed deadlines, initiated penalty clauses and additional project costs.

10.1 Methods of communication

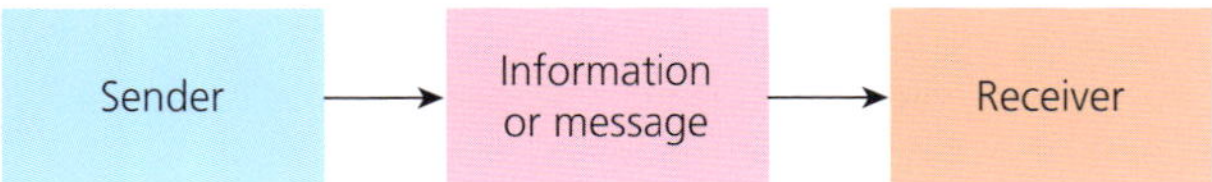

▲ Figure 9.14 Three-part communication process

Information can be exchanged verbally, non-verbally or in the form of visualisations (graphics). While there are advantages and disadvantages to each of these methods, information should always be clear, comprehensible and in a format that is easy to understand.

▲ Figure 9.15 Graphical communication – a safety sign

Complex information received by two or more people can sometimes be interpreted slightly differently by each recipient; just because information has been sent and received does not mean it has been understood as the sender intended. To confirm the recipient has fully understood the information, feedback needs to be given to the sender. However, this is not always possible using some forms of written or graphical communication.

The number of methods used to communicate in the construction industry has grown substantially in recent years, with the development of new digital platforms and technology.

Health and safety

Under the Construction (Design and Management) (CDM) Regulations 2015, employers have a legal duty to ensure information is easy to understand to protect the health, safety and welfare of people from work activities. (See Chapter 1 for more information on these regulations.)

In this section, we will look at some of the most commonly used methods of communication, their advantages, barriers to their effective use and problems that can occur.

Verbal communication

Face-to-face communication and indirect communication, such as talking on the phone or using walkie-talkies, are the simplest, quickest and probably most frequently used methods of transferring information from one person to another. Working in the construction industry, you will interact with your colleagues each day to build good working relationships, trust and confidence.

When team members engage positively and have respect for one another, they are generally happier, which in turn creates a better working environment. **Top-down** positive communication is a sign of a strong management team, which can result in employees having increased confidence in their employers, being more productive and having less desire to leave their jobs.

Negative forms of communication, for example in terms of pitch and tone of voice, cause conflict in the workplace, damage staff morale and increase stress levels. They often create **passive-aggressive behaviour** in individuals and a toxic culture, impacting on the business' performance and causing an increase in staff turnover.

▲ Figure 9.16 Construction workers communicating verbally

Key terms

Top-down: proceeding from the most senior to the least senior

Passive-aggressive behaviour: indirectly or subtly expressing negative feelings

Test yourself

Explain the term 'arbitration' and describe a situation in the workplace when this may be used.

Table 9.2 outlines strengths and weaknesses of using verbal communication.

▼ Table 9.2 Strengths and weaknesses of using verbal communication

Strengths	Weaknesses
• It is quick and simple. • There is no reliance on technology, for example an internet connection. • Two-way conversation is possible. • It is personal. • It is direct. • It allows the recipient to confirm that they have understood the information (give feedback). • Non-verbal impressions can also be used with verbal communication, for example raised voice, gestures or body language. • The sender can use both open and closed questioning (open questioning allows someone to give a free-form answer; closed questioning requires a response from a limited set of answers such as 'yes' or 'no').	• There is no written record. • Information can be forgotten by the recipient. • Information can be misunderstood if the sender has a strong accent or regional dialect. • Language barriers may exist between people who speak different languages. • Information can be misheard with background noises. • The recipient may have a hearing impairment. • The message could be **ambiguous**.

Key term

Ambiguous: unclear and difficult to understand

Industry tip

Listening is as important as speaking, as this is one way that we learn. While undertaking your industry placement as part of your T Level course, your employer will assess both of these skills.

Health and safety

The Health and Safety Executive (HSE) reported in 2021 that one in four people living in the UK has had mental health problems. Work-related stress can aggravate pre-existing mental health issues, such as anxiety and depression. A report by the Office for National Statistics (ONS) stated that more than 1,400 construction workers took their own lives between 2011 and 2015, over three times the national average.

Talking with co-workers informally has many benefits, including providing support and improving wellbeing and general mental health.

Non-verbal communication

People can communicate with each other without the need to speak, using written information, body language, gestures and even the way they dress.

Wherever possible, formal communication in a work environment is written down, because it provides a permanent record and is easier to distribute to others without misinterpretation.

Written information in some forms is a legal requirement, for example building contracts and policies. When information is recorded, it can be referred to or even used as evidence if a dispute or grievance occurs between employers, employees or clients.

▲ Figure 9.17 Non-verbal communication – a contract of employment

▼ Table 9.3 Strengths and weaknesses of written communication

Strengths	Weaknesses
• It can be referred back to. • There is a permanent record of the communication. • The same information can be distributed easily without diluting it. • The sender does not have to meet the recipient to pass on the information. • It can be used to communicate if the recipient has a hearing impairment.	• There may be no opportunity to feed back quickly. • It is not possible to clarify immediately whether information has been received and understood. • It is impersonal. • It is indirect. • A recipient may have dyslexia or other reading difficulties. • There may be language barriers. • It takes time to write. • It is slow to distribute. • A recipient may have impaired vision. • The message could be ambiguous.

Besides written communication, non-verbal communication can involve body language and behaviours such as eye contact and facial expressions. However, these methods can be easily misinterpreted and there is no permanent record.

Industry tip

Good communication skills are essential in most job roles in the construction industry, and they are a desirable quality for employers. The first communication you have with an employer could be a job application, so it is vital that it is well written without any spelling or grammatical errors, in order to create a good impression. Before submitting your CV and job application to an employer, always get someone to **proofread** them.

Key term

Proofread: to check a piece of written communication for errors in spelling, grammar, punctuation and accuracy before it is shared or published

Industry tip

The way you dress, act and speak during an interview is vital in creating a positive first impression with a potential employer. Even though you may be applying for a job working in a particularly dirty environment, you should still dress appropriately for an interview. Remember, you only get one chance to make a first impression!

▲ Figure 9.18 A job interview

Visualisations (graphics)

Information displayed in a graphical form, such as posters, safety signs and notices, is often used in the construction industry to reinforce information already communicated by some other method. For example, during a site induction you may have been told to wear a range of personal protective equipment (PPE); mandatory safety signage displayed around the site may also illustrate this requirement.

Technical information is often communicated using detailed visualisations, either in hard-copy format or using digital technology.

Industry tip

Posters and information boards should be updated regularly to keep them current, and to ensure they are not disregarded by the intended recipients.

▼ Table 9.4 Strengths and weaknesses of using visualisations

Strengths	Weaknesses
• The message can be repeated to a wide audience. • There is no language barrier. • The message is clear and consistent. • It is quick to interpret a simple message. • There is a written record. • They are eye-catching. • They are jargon-free. • Symbols and pictograms are often standardised in the UK construction industry.	• The recipient may have a visual impairment. • There is no confirmation that they have been acknowledged. • They are easily dismissed. • The recipient may have to receive training to understand the symbols, pictures or pictograms. • They can be lost, removed or defaced if on display. • The recipient cannot ask questions. • There is no immediate feedback. • Only simple messages can be conveyed.

Table 9.5 outlines various methods of communication and how they might be used in a typical construction project.

▼ Table 9.5 Methods of communication

Type of communication	Use in a typical construction project
Face to face	• Job interviews • Site inductions • Toolbox talks • Informal discussions with work colleagues • Site meetings • **Appraisals**
Email	• Communications with clients and other stakeholders • Quotations and estimates • Exchanging of building contracts
Letter	• Formal exchanges of information between contractors, duty holders and clients, for example building contracts • Business letters • Formal exchanges of information with the local authority, HSE and professionals involved in a construction project • Notes • Memorandums (memos)
Telephone (land line)	• Internal communication between staff within a business: • Better signal strength compared with mobile phones in remote locations • Less likely to run out of battery charge • Gives the impression of a professional business when communicating with clients
Mobile phone	• Communication between members of the site management team on a construction site (providing they have a good network signal) • Accessing the internet for emails, livestream video calls and text messaging • Recording videos or taking photographs to communicate information quickly from the construction site to the office or other stakeholders • Contacting the emergency services

Walkie-talkies	• Shortwave radio contact between members of the construction team on building sites:
	  • durable and moisture resistant
	  • quick and direct
	  • ideal for use in remote locations where phone signals may be weak

Drawn information	• Communicating information from an office to construction sites, for example as site plans, elevations, section drawings or detail drawings • Scheduling and planning • Sharing technical information with clients, the planning department and subcontractors • Forming part of a tender package • Supporting oral communication on construction sites in the form of sketches

Type of communication	Use in a typical construction project

Tablets, notebooks and laptops	• Video-conferencing site meetings
	• Site inductions
	• Letter writing
	• Snagging lists
	• Surveying
	• Requisitions (Orders)
	• Auditing
	• Scheduling
	• Digital project management:
	  • augmented reality (AR)
	  • virtual reality (VR)
	  • Building Information Modelling (BIM)
	  • controlling drones to broadcast livestreams in real time

Signs and notices	• Safety signage displayed around the workplace, for example fire escape • Labelling on packaging • Site notice boards • Improvement and prohibition notices served by the HSE
Videos	• Site inductions • Toolbox talks • Marketing • Site inspections • Monitoring progress and health and safety standards

Key term

Appraisals: scheduled routine meetings between an employee and their employer to review their work performance against their job description

▲ Figure 9.19 Walkie-talkies are ideal for use in remote locations where phone signals may be weak

10.2 Communication styles

Communication styles can be either formal or informal:

- Formal communication is used by organisations to pass information through prescribed official channels, following an organisational structure.
- Informal communication is much quicker, because there are no rules restricting which direction or lines of communication have to be used; therefore, it is 'free flowing'.

The number of non-English-speaking workers, and those who speak English as a second language, is on the increase in the UK construction industry. All employers have a duty of care to communicate effectively with everyone under their control, especially when it concerns their health and safety. To overcome these problems, employers may use a translator or a **bilingual** supervisor or co-worker to interpret key information. Where this is not possible, they may have to translate written information to use for site inductions, toolbox talks and other training materials. In situations where language is a barrier, simple pictorial illustrations are also a useful tool to support employers communicating with their employers.

Improve your English

Make a short toolbox talk video on a topic of your choice, to develop your speaking and presenting skills.

Key term

Bilingual: fluent in two languages

Case study

Lucas has been a self-employed roofer for the past ten years. Last summer, he was subcontracted to tile the roofs on 15 new houses on a development. Unfortunately, he had an accident while working on one of the roofs, and sustained a back injury that has prevented him from continuing to trade as a roofer.

During the investigation into the accident, Lucas said that he attended a site induction, but he had difficulties reading and understanding the RAMS (risk assessments and method statements).

- Whose fault is it that Lucas could not understand the RAMS?
- Work in a small group to consider an alternative site induction for Lucas.

11 Employment rights and responsibilities

Every employee in the UK has legal rights and responsibilities under the Employment Rights Act 1996. Employment law controls employees' rights at work and relationships between employees, employers, trade unions and the government.

As soon as someone accepts a job offer from an employer, they technically have a contract of employment, so before starting work it is important that they understand what they are entitled to and what to expect from their employer. A contract of employment does not have to be in writing, although most employers prefer to record the following terms of the agreement to avoid confusion or disputes:

- employment rights
- employee's duties
- responsibilities – employer to employee
- responsibilities – employee to employer
- employment conditions.

11.1 Employment rights

Wages

Minimum wage

Almost every person with a legal right to work in the UK has employment rights, such as being entitled to the national minimum wage and living wage. The minimum wage a person should receive depends on their age and whether they are an apprentice:

- Workers of at least school-leaving age are entitled to the national minimum wage. In England, you can leave school on the last Friday in June, if you will be 16 by the end of the summer holidays. You will have the following options until you are 18:
  - Spend 20 hours or more a week working or volunteering, while in part-time training or education.
  - Stay in full-time education.
  - Start a traineeship or apprenticeship.
- Workers over the age of 23 are entitled to the national living wage.

▲ Figure 9.20 An apprentice tradesperson

The rates of the National Minimum Wage and National Living Wage are reviewed on 1 April every year. To check the current rates, visit **www.gov.uk** and search for national minimum wage or national living wage.

Apprentices who have completed the first year of their apprenticeship and are aged 19 or older are entitled to the minimum wage for their age.

If an employee is absent from work for at least four days in a row and has earned at least £120 per week, their employer must pay **statutory sick pay** (SSP) for up to 28 weeks. Employers can pay more than the minimum SSP; this is usually agreed in the contract of employment before starting work.

The law states that employers must give their employees **payslips** as proof of their earnings, tax paid and any other deductions, for example pension contributions.

Key terms

Statutory sick pay: payment made by employers to employees when they are too ill to work

Payslips: written documents provided by an employer that contain personal information about the employee and specific details about the amount of money that they have earned for a given period (for example a week or month), as well as their total earnings, tax deductions, National Insurance contributions and pension payments made for that financial year

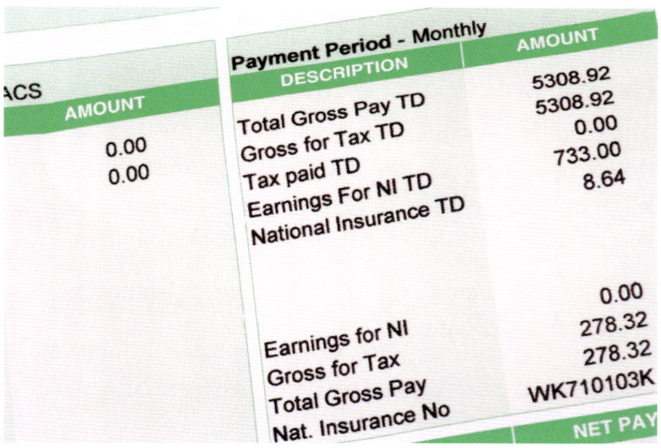

▲ Figure 9.21 A payslip

Research

Find out who is not entitled to the national minimum wage or national living wage.

Pension

Everyone who has made National Insurance contributions through their earnings is entitled to a government **state pension**, which can be claimed on reaching state pension age (although this is currently under review). The amount a person receives depends on their National Insurance record. A state pension alone is unlikely to provide enough money to live on when someone retires, therefore they will also have either a workplace pension (set up by their employer) or a private pension (set up by the individual).

Since 2012, all employers must automatically enrol their employees into a workplace pension scheme and make contributions towards it, provided they are aged 22 or above and earn over the low-income threshold set by the government.

All eligible employees make their pension contributions through deductions taken directly from their income. If an employee prefers not to have a workplace pension, they can opt out of the scheme. However, this is best done within the first month of employment to prevent payments going into a pension fund that can only be accessed when they retire.

Key term

State pension: a regular sum of money paid by the state to people of or above official retirement age

Time off

Almost all employees have the right to paid leave for public duties and responsibilities, such as jury service, under the Working Time (Amendment) Regulations 2007. People who work a five-day week are also entitled to at least 28 days' paid leave each year, which is equivalent to 5.6 weeks of holiday. Part-time employees have the same entitlement as full-time staff, although this is dependent on the number of days that they work, for example four days per week amounts to 4×5.6 weeks = 22.4 days per year.

Employees who are parents of children under the age of 18 may also be eligible for unpaid leave to look after their child's welfare, providing they have been with a company for more than one year.

Employees over the age of 18 are entitled to three different types of unpaid rest breaks:

- rest breaks at work – one uninterrupted 20-minute break if they work more than six hours a day
- daily rest – at least 11 hours between shifts
- weekly rest – 24 hours' uninterrupted break each week, or 48 hours each fortnight.

Employees are entitled to 52 weeks' statutory maternity leave, made up of 26 weeks' ordinary maternity leave and 26 weeks' additional maternity leave. Statutory maternity pay (SMP) is paid for up to 39 weeks and includes:

- 90 per cent of the employee's average weekly earnings (before tax) for the first six weeks

- either a fixed sum set by the government or 90 per cent of average weekly earnings (whichever is lower) for the next 33 weeks.

When an employee's partner is having a baby, adopting a child or having a baby through surrogacy, they might be eligible for one or two weeks' leave or paternity pay. The weekly rate for paternity pay is 90 per cent of average weekly earnings. Employees may not be entitled to both paternity leave and pay, but they may be eligible for shared parental leave and pay.

There are different eligibility criteria for birth, surrogate and adoptive parents to get shared parental leave and pay; further information can be found at **www.gov.uk/shared-parental-leave-and-pay**.

▲ Figure 9.22 Parental leave

Improve your maths

A 19-year-old carpenter earns the national minimum wage:

- One week, he worked 47 hours in total.
- His employer pays him 'time and a half' for every hour after 39 standard-time hours.
- 37 per cent of his total pay is deducted for tax, National Insurance and pension.
- The threshold for PAYE (pay as you earn) is £242 per week, or £1,048 per month.

How much has the carpenter earned after deductions?

Research

Find out the employee leave and pay entitlements for paternity and adoption.

Research

Find out which type of employees would not be permitted time off work for jury service.

Health and safety

Rest breaks from the workplace are extremely important to protect people from physical and psychological harm.

Equal rights

Under the Equality Act 2010, people at work and in wider society are legally protected from direct or indirect discrimination, victimisation and harassment. See section 7.1 of this chapter for details of protected characteristics.

Health, safety and welfare

The Health and Safety at Work etc. Act 1974 (HASAWA) protects the health, safety and welfare of:

- people at work
- those who could be affected by work activities.

An employer's legal duties under this legislation are explained in Chapter 1.

Access to representation in times of grievance

Occasionally, an employee will raise a problem, complaint or concern with their employer that cannot be resolved amicably without intervention.

A grievance is usually raised with an employer following internal procedures laid out in the contract of employment. This often involves notifying an employer in writing at first, and then following this up with a meeting between the employer, a representative from the organisation's **human resources department** (if appropriate), the employee and a companion of the employee's choosing. Employees can represent themselves, although they may exercise their legal right to take a representative from their trade union or a colleague for support.

Key term

Human resources department: department of an organisation that deals with recruiting, administrating and training staff

Redundancy

Employers and employees must adhere to the terms of the employment contract until it ends, when the employee is given notice or dismissed.

If an employee who has worked for the same employer for two or more years is made redundant, they are normally entitled to statutory redundancy pay. The amount of pay received depends on the employee's age at the time of leaving and their length of service (although this is capped at 20 years).

11.2 Employment responsibilities

Employers have responsibilities to their employees to make sure they are treated fairly, by maintaining good practices and complying with employment law.

Employers' responsibilities towards their employees include:

- protecting their health, safety and welfare, for example by providing personal protective equipment, toilets and rest facilities
- providing a contract of employment
- informing and consulting when necessary
- being an inclusive employer and not discriminating while recruiting, employing or promoting staff
- paying the minimum wage, sick pay, maternity pay, holiday pay and other entitlements
- allowing staff to return to the same job after a leave of absence, for example after maternity leave or a **sabbatical**
- following the Working Time Regulations 1998
- abiding by the terms and conditions of the contract of employment
- considering requests from staff for **flexible working**
- auto-enrolling employees into a workplace pension.

Key terms

Sabbatical: an extended period of unpaid leave from work, taken in agreement with an employer, often used for holidays, travelling or pursuing interests

Flexible working: a working arrangement that allows an employee to choose when and where they work, in order to improve their work–life balance, reduce their stress levels and provide better job satisfaction

Employees also have responsibilities to their employer, including working to the terms and conditions agreed in their employment contract and company handbook.

A company handbook, also known as a company policy, provides employees with job-related information, for example working conditions, health and safety, standards, procedures and reasonable behaviour expectations. It also provides employees with details about the rules that everyone responsible for using personal data has to follow, in order to protect the confidentiality of people at work and to comply with the UK General Data Protection Regulation (UK GDPR) and the Data Protection Act 2018 (see section 3.2 of Chapter 8).

Company handbooks often contain a social media policy and code of conduct. These documents explain rules which employees should follow outside of their working hours, to ensure their behaviour does not impact negatively on the employer or damage their brand or reputation. Improper or illegal behaviour by an employee can sometimes prevent them performing their job role as outlined in their job description, and may lead to disciplinary action or dismissal by their employer. Furthermore, they could be prosecuted by external enforcement bodies, for example the police, if they have committed a criminal offence.

▲ Figure 9.23 A company policy document

Test yourself

Describe the most likely outcome for an employee following a minor breach and a serious breach of their contract of employment.

Test yourself

List the employment rights of an employee.

12 Ethics and ethical behaviour

In principle, people should treat others the way they would like to be treated themselves. This is essential to the long-term success of a business and the route to making a better environment in our workplaces and the world around us.

When you think about **ethics**, you should consider how the actions you take could affect others. For example, a business may choose not to provide PPE for employees as a way to increase its profits; however, this is considered **unethical** because of the risk to people's health.

Larger businesses often develop a system of policies and practices to ensure they operate in a fair, moral and legal way; this is known as a code of ethics. Businesses with a consistent set of values and robust moral code develop a strong ethical culture, which means they are less likely to face the consequences of unethical behaviour. Unethical behaviour in business is damaging and can have a negative impact on an organisation's future prospects.

▲ Figure 9.24 As well as being a legal responsibility, providing PPE is ethical behaviour

Key terms

Ethics: moral values that govern a person's behaviour towards others

Unethical: morally wrong or unacceptable

Defamation: the act of damaging someone's good reputation through a false written or verbal statement, also known as libel (written) or slander (spoken)

Table 9.6 provides examples of ethical and unethical behaviour in business.

▼ Table 9.6 Ethical and unethical behaviour in business

Ethical behaviour	Unethical behaviour
• Adherence to health and safety legislation and guidance • Equality and diversity • Fulfilling the terms of building and employment contracts • Honesty, integrity, commitment and loyalty • Corporate social responsibility • Environmental protection • Accurate business statements • Protection of personal or sensitive data • Fair employment rights and responsibilities • Adherence to laws • Respect • Accountability for actions • Submission of accurate and true annual accounts to HMRC	• Ignoring health and safety legislation and guidance • Discrimination • Failure to honour commitments • Corruption • Harassment, for example sexual harassment • Disregard for corporate social responsibility • Wilful damage to the environment, for example inappropriate disposal of waste or destruction of protected trees • False business claims • Disclosure of sensitive personal information • Mistreatment/exploitation of employees, for example paying below the minimum living wage or offering poor working conditions • Sabotage • Disrespect • **Defamation** of a competitor • Tax avoidance and manipulating accounts • Theft • Bribery • Money laundering • Deception

Businesses that demonstrate ethical traits and encourage and reward good attitudes are often more productive and successful, as they develop trust between the management, the employees and the public. When a business with power is seen to be acting responsibly and unselfishly, it can also improve its reputation.

Test yourself

Describe a situation where ethical behaviour is applied in the workplace.

13 Sources of information

Organisations need to create opportunities to meet new customers and develop relationships with contacts and partners for new business ventures.

Networking is a low-cost process used by many organisations to make initial introductions with likeminded people, share information and form long-lasting business relationships. Construction businesses can network on construction sites, attend trade events or exhibitions, join networking groups or use social media. Distributing business cards is another way to expand a networking community, although this method is often slower and restricted to a small demographic compared with the use of digital media.

The internet is used by most construction businesses in the UK to market their product or service, raise their profile and develop people's interest for future sales opportunities. A professionally made and managed website is essential as a platform to advertise and promote a business to the public, customers and other stakeholders.

Key term

Networking: an activity where businesses and people with a common interest meet to share information and develop contacts

Businesses can use social media to engage quickly with a wide demographic. It allows them to share ideas, knowledge and good practice with other businesses, through posts, blogs, podcasts, articles and videos.

Businesses that use social media often provide customers with the opportunity to write reviews, as a way of giving unfiltered honest feedback, although it may not always be positive or constructive. Some businesses also use social media to their advantage by redirecting traffic from posts to their main website, as a marketing strategy to increase sales.

Social media must be controlled carefully by an organisation. It can quickly and easily damage both reputation and brand if errors are made or negative feedback is received and publicised. Due to the fast pace of social media, businesses have to continually monitor and update their content to ensure it remains current and has not been abused with unwanted or inappropriate posts. There is also a risk that confidential information could be accessed by hackers or leaked online, unless a good cyber security system is used to protect it.

Assessment practice

Short answer

1 What term describes a group or individual that has an invested interest in the long-term success of an organisation?

2 Why is it important to maintain a good working relationship with customers?

3 What is the main piece of legislation that prevents discrimination in the workplace?

4 Where would an employee find information on working conditions, health and safety, standards, procedures and reasonable behaviour expectations?

5 What must all employers automatically enrol their employees into and make contributions towards, providing the employees are aged 22 or above and earn over the low-income threshold?

Long answer

6 Describe the negotiation techniques used in the construction industry.

7 Describe a situation involving unethical behaviour in a business.

8 Explain the term 'networking'.

9 Describe a situation when arbitration may be appropriate to use in the workplace.

10 Explain the advantages of non-verbal communication.

Project practice

Liam has just started a new job as a solar panel installer. In the first week, he has been asked by his employer to install six photovoltaic (PV) panels with two other employees.

Liam and his work colleagues attended a site induction on the new development under construction, where they were informed about the site rules and mandatory PPE that has to be worn at all times on the site.

While installing the first PV panel, Liam was challenged by the site manager for not wearing his hard hat. Liam said that it was uncomfortable, and too hot to be wearing on a roof in the middle of the summer. Later that day, Liam was caught again by the site manager not wearing his hard hat as instructed and was asked to leave the site. During an investigation into the matter by his employer, Liam claimed he was being discriminated against.

- Describe the conflict-management technique used by the site manager when he challenged Liam on the second occasion.
- Work in a small group to research Liam's employment rights and responsibilities, and write a brief report to explain his behaviour on the day and whether this was acceptable.
- In your opinion, was Liam discriminated against by the site manager?

Chapter 10 Digital technology in construction

The expanding and evolving use of digital technology in the construction industry has led to improvements in management and production methods, for example:

- better efficiency in project planning
- a more streamlined design approach
- enhanced collaboration between the building team and stakeholders
- greater innovation in construction methods and design.

While Chapter 8 introduced the term 'digital' and outlined a limited number of applications for digital technology, this chapter considers the use of digital technology in construction in greater detail, identifying digital systems and processes that are employed to achieve specific results.

Learning outcomes

By the end of this chapter, you will understand:

1. the internet of things (IoT)
2. digital engineering techniques
3. opportunities for the use of technology.

1 Internet of things (IoT)

The term 'internet of things' (IoT) refers to the system of applying unique digital identifiers to physical objects, such as buildings, which can be connected to a digital network. Communication data can flow between the connected objects, allowing interactivity and collaboration to achieve specific goals.

Objects connected through a digital network can be referred to as 'smart' objects, when the processing of information and data happens without human intervention. Communication can take place directly from machine to machine (M2M). This means that systems management and decision making can be automated in operational areas of a smart building, such as:

- monitoring and optimising systems performance
- detecting systems inefficiencies
- alerting remote operators to potential problems
- diagnosing possible causes of problems.

▲ Figure 10.1 In the IoT, communication can take place from machine to machine

The potential for using the IoT in construction applications is vast. Designing buildings and infrastructure elements to incorporate interconnected and unified sensors and monitoring equipment allows data analysis, which can translate into:

- reduction in energy use
- improvements in manufacturing efficiency
- improvements in safety
- streamlining of materials delivery and supply-chain activity.

Smart technology can help building designers to identify optimal construction materials and methods for future projects, leading to the construction of buildings that rely less on highly skilled workers, use fewer resources and produce less carbon.

The interconnected nature of the IoT allows data analysis which can provide performance information on individual components and materials in a range of structures. As a result, manufacturers can improve and refine manufactured resources.

▲ Figure 10.2 The interconnected nature of the IoT allows for many applications

Smart technology is evolving quickly, and the term 'artificial intelligence' (AI) is becoming more familiar. AI describes digital systems that go beyond simply following programmed instructions. They use methods of analysis that identify patterns and correlations, in order to draw appropriate conclusions more quickly than humans are able to.

Industry tip

AI is used in many devices we take for granted. For example, face and fingerprint recognition software is often used in mobile phones. This application of AI is also often used to prevent unauthorised access to construction sites, using a **biometric** turnstile.

Key term

Biometric: involving the detection and analysis of individuals' unique physical characteristics as a means of verifying identity

▲ Figure 10.3 Controlled access to a construction site using a fingerprint reader

Data analysis using AI holds the promise of real-time **autonomous** decision making when operating buildings systems. It also offers the ability to compare different features, options, costings and performance factors during the design of a new building.

Smart learning that uses digital technology and the networking capability of the IoT is likely to become an integral feature of future design processes and operating systems in the construction industry.

Key term

Autonomous: functioning independently without external control

Test yourself

How can AI be used to control who enters a construction site?

1.1 Uses of digital technology

The development and implementation of digital technology, in both the construction industry and society generally, has led to new methods of working, with technological tools being refined constantly.

Let us look at some uses of digital technology in the operation of completed buildings and during the construction phase.

Digital technology used in completed buildings

Completed buildings that are designed to integrate with the IoT can produce large quantities of useful data.

Sensor technology and data management systems embedded in a structure can record and archive data, as well as produce live data streams, for analysis and evaluation. This allows a clear understanding of the performance of an occupied building and the environment it provides for its occupants.

▲ Figure 10.4 Sensor technology embedded in a structure can provide valuable data

Occupier comfort

In buildings such as offices, smart control systems can be used to monitor and adjust levels of relative humidity, air temperature, ventilation, dust and even air pressure within individual rooms, in order to ensure a comfortable and productive working environment.

Research

Search online for companies that offer control systems for buildings. Write down the range of features in a building that can be controlled by digital technology.

Analysis of data over time can be used to refine the operational **parameters** of control systems, in turn creating and maintaining consistently comfortable working and living conditions.

Experts in AI and smart technology suggest that control systems in the future may be able to sense in real time whether a building's occupants are tired, too hot or too cold, and automatically make adjustments to heating and ventilation to enhance wellbeing.

Key term

Parameters: limits which define the scope of a system, process or activity

▲ Figure 10.5 Analysis of data can be used to set operational parameters of control systems

Improve your English

Increasing numbers of smart products are being marketed for home use. Find out what is available and write an account of how a smart home could be set up for the comfort and convenience of the occupants.

Test yourself

How can the analysis of appropriate data help to make a building comfortable to work in?

Power management

Digital technology can support energy efficiency through the management of electrical power. Demand for electrical power in an occupied building will fluctuate depending on the number of occupants active at certain times. Intelligent systems can gather and compare information over time to detect changes in energy use and potentially predict periodic peaks and troughs in demand.

Through the IoT, energy suppliers could request that consumers command their internal systems to reduce energy consumption when increased general demand puts stress on supply capacity. This model of real-time management of demand is likely to play an increasingly important part in improving efficient energy usage.

Sustainable electricity generation for individual properties, for example in the form of solar panels and wind turbines, is increasing. However, these energy sources may not provide sufficient amounts of energy for larger buildings, which may still need to rely on external energy supplies.

Renewable energy sources such as solar and wind power provide variable energy outputs dependent on weather conditions, so it is necessary to balance the use of onsite and off-site energy sources in a building according to the demand from occupants.

▲ Figure 10.6 Digital technology allows sophisticated control of electrical power

Systems using AI and the IoT have the potential to make these decisions in real time, using a constant feed of up-to-date information. In order to optimise energy use and minimise the production of carbon, sophisticated control systems could make decisions to use onsite energy generation, store energy in battery facilities, export energy to the National Grid, use only external energy sources, or combine some of these options.

Research

By 2030, all new cars in the UK will be fully electric (or at least heavily electrified) instead of powered by petrol and diesel.

Find out how many electric car charging points there are in the UK. Could AI be used to manage this part of the infrastructure?

Maintenance

Benefits of digital technology include the vast quantities of data that can be processed and the speed of data transfer. AI systems can compare a huge range of scenarios based on live data streams, from which warning signs of equipment failure can be identified. This allows predictive maintenance to be a reality.

A building is an asset which must be managed and maintained efficiently in order to preserve its value. The development of AI systems that can detect potential faults and anomalies in equipment means that maintenance interventions can be made before that equipment fails.

Equipment failures cause disruption that can be costly and even dangerous to the building's occupants, so digital technology is a valuable tool in asset management and maintenance of the built environment.

Test yourself

How can digital technology be used to maintain a building effectively?

Digital technology used during the construction of buildings

While some types of digital technology have been used in construction for many years, especially during the design process, new technologies are constantly being developed to improve site safety and efficiency and make construction more sustainable.

Design

Computer-aided design (CAD) has become a standard design tool when developing the concept for a construction project (see later in this chapter for further details).

Research

Research the range of CAD software available online. How many different programs can you find? Write down the cost of two CAD software packages.

Health and safety

Using digital technology can mean spending a long time focused on a screen and sitting in one position. Be aware of the possible damage to your health – poor posture can cause back and neck problems, and excessive screen time can affect your eyesight.

Digital systems can speed up the design process considerably, leading to greater productivity. Building design ideas created using specialist software can be presented as a three-dimensional (3D) model, allowing examination and refinement of the design concept.

▲ Figure 10.7 Digital tools can produce 3D models of a structure

Digital tools are able to generate accurate materials and components lists from 3D models or 2D drawings. These can be linked digitally to materials costs databases, in order to produce an up-to-date costing for a project.

Speeding up the process of designing and costing a project, coupled with the ability to update or amend project details quickly, streamlines development and improves productivity.

On site

Once project design details have been finalised and costings completed, digital technology has a significant role to play on the construction site. Smart equipment, such as surveying tools that use satellite positioning data for setting out buildings, can link with remotely stored project data through the IoT. This type of equipment is becoming increasingly sophisticated and can speed up site preparation considerably.

A similar smart approach can be used to avoid damage to existing buried pipes and cables when excavating a new site. Mechanical excavators can be equipped with satellite positioning equipment in order to locate pipes and cables, and digital depth controls can assist operators in limiting the depth of excavations to safe levels.

Industry tip

Experienced site supervisors will judge the safest way to work around existing services on site. Sometimes the safest way to expose pipes and cables is to dig by hand.

▲ Figure 10.8 It is vital to avoid damage to buried cables and pipes on site

Smart technology can also support the efficient operation of machines on site. For example, removing too much material when shaping the ground will require some material to be replaced later, which is uneconomical. Smart equipment in earth-moving and excavation machines can ensure pinpoint accuracy in levelling the ground and creating gradients and slopes when contouring the terrain.

Digital tracking systems can also be installed in machinery. **Geofencing** uses digital technology such as RFID or GPS to pinpoint the location of construction equipment and create a virtual boundary, ensuring that it remains within a specified area.

Key term

Geofencing: using technology to pinpoint the location of equipment and create a virtual boundary

Improve your English

Search online for 'AI excavators'. Write a report about how construction machines are being developed to operate without drivers. (Hint: Japan is leading the way.)

Materials and resources

Delivery of materials and components to site must be reliable and consistent, in order to avoid project delays and increased costs. However, storage of materials on site is often limited due to congestion and a lack of suitable storage space.

A strategy developed in the car manufacturing industry known as 'just in time' (JIT) can be used in construction activities to improve efficiency, reduce waste and eliminate the need for extensive onsite storage facilities.

Improve your English

Find out the history of the JIT strategy. Write a short report about how the idea came about and where it was first applied.

A JIT strategy requires delivery of the right materials, in the right order, in the right amount, at the right time. For example, concrete is usually mixed off site to the required specification. It must therefore be delivered at the right time and in the right quantity to allow it to be poured at the required location before setting occurs.

There are many software packages designed to facilitate JIT systems.

▲ Figure 10.9 JIT systems can be used in congested locations where access is difficult

Industry tip

The timing and sequencing of deliveries is critical for a JIT system. Goods must therefore be handled carefully to avoid damage. Damaged goods may not be replaced quickly, leading to disruption of production schedules.

Digital technology can be used to track and monitor materials and components during manufacture, transport and delivery. This is useful during complex projects, where a large catalogue of materials and components can be controlled using processing systems.

Test yourself

What is a JIT system?

2 Digital engineering techniques

Digital engineering uses a digital skillset to create and edit data as part of a design process. Engineering principles are applied in a virtual environment, allowing designers to explore a range of design possibilities and develop innovative solutions to design problems.

Improve your English

Find out what is meant by the term 'engineering principles' and write a definition in your own words.

In this section, we will discuss three main techniques applied in the field of digital engineering:

- simulation
- animation
- modelling.

2.1 Simulation

Simulation uses digital methods to mimic the behaviour of real systems and processes. It helps to answer important questions about a proposed construction project, such as:

- Will the building design be energy efficient?
- Which construction methods are best suited to the design concept?
- What is a realistic timescale for the construction of the building?

By using a simulation model, 'what if' analysis becomes possible, so that the design and selected construction processes can be checked for feasibility. For example, it can be confirmed that the intended building plot will accommodate the design, allowing access to deliver and store materials without causing undue disruption to surrounding properties and their occupants.

The sequencing of project stages can be simulated, so that construction processes follow a smooth path without disruptive clashes. Any necessary changes can be made before work commences on site, avoiding errors and expensive delays.

▲ Figure 10.10 Simulation software can assist in planning and design

Simulation can also be used for structural analysis. Performance data harvested from past projects can be used to model the behaviour of materials and components in the planned project, to ensure the proposed design is capable of carrying the stresses and loadings that will be imposed on it.

Test yourself

Suggest some questions about a new building project that simulation can help to answer.

When retrofit or restoration work is planned, simulation can be used to ensure the existing structure can accommodate new materials or components without overloading or stressing load-bearing elements.

▲ Figure 10.11 Digital simulation can be used to confirm structural stability during restoration work

Research

Search online for 'before and after' images of construction restoration projects. Copy and paste the images into a Word document and write a summary of each project.

2.2 Animation

Visualising a completed building and imagining how it will function can be difficult when simply viewing a 2D drawing or diagram. While a 3D digital model can deliver an impressive view of a project, the viewer's experience can be greatly enhanced if that model is animated to help them focus on key elements.

Digital animation is a powerful tool that can be used for refining conceptual details of a project. It can provide an engaging and lifelike view of a building, often by creating a walk-through to show room areas and features. This has many benefits, such as the ability to:

- visually assess the effects of lighting and the proportions of room sizes
- gauge the number of occupants who can use the internal spaces
- evaluate the access to and movement of occupants in shared spaces and corridors to ensure safety and comfort.

▲ Figure 10.12 Digital technology can be used to create a walk-through of a proposed building to assess lighting effects and room proportions

Improve your English

Search online for an example of a construction animation. Watch the whole animation then write a detailed description of what it showed you. (If necessary, you can watch the animation a number of times.)

Test yourself

What does the term 'walk-through' mean in the context of animation?

Case study

Following the Grenfell Tower fire in 2017, a contract has been awarded to replace dangerous external cladding on a group of multi-storey residential buildings. The buildings will be occupied throughout the project, and traffic in the streets around the apartment blocks must be disrupted as little as possible. Storage for the replacement cladding panels is limited, so a JIT delivery system will be used, with delivery vehicles routed to specific offloading points around the area.

In pairs or small groups, discuss how creating an animation of the organisational requirements for the project would be beneficial.

Think about:

- traffic flow
- personnel movements
- equipment movements
- materials handling
- sequence of work
- protection of occupants.

2.3 Modelling

So far, this section has considered how simulation and animation use digital modelling to assist in designing buildings and planning construction work. However, digital modelling can also be used throughout the life cycle of buildings.

For example, sophisticated digital modelling is used to survey existing buildings, in order to collect data that can be used for:

- assessing maintenance and repair requirements
- planning restoration projects
- carrying out alterations and extensions
- making decisions on demolition methods at the end of a building's useful life.

A laser scanner can be used to create 3D imagery of a building's complex geometry or survey surface areas that are not easily accessible. From its setup position, it can digitally record the precise distance of densely grouped points at rapid speed. This process is often referred to as a 'point cloud survey'.

Industry tip

Point cloud surveying may also be referred to as 'light detection and ranging' or LIDAR – a combination of the words light and radar.

▲ Figure 10.13 A laser scanner

The point cloud data generated by laser scanning can be integrated into Building Information Modelling (BIM) software or CAD systems to digitally create detailed 3D models of a structure or features of the built environment for a range of uses. Laser scanners can be mounted on drones to capture the exact contours of landscapes, road layouts, railway routes and even entire towns.

▲ Figure 10.14 Laser scanners can be mounted on drones for surveying

A major advantage of using digital technology in this way is that 2D or 3D digital models and representations of structures can be easily amended and instantly shared by being transmitted to other users anywhere in the world.

Test yourself

State one use for point cloud surveying.

The scanned data can be utilised in **immersive** technologies, such as virtual reality (VR). A VR user is visually completely shut off from the outside world by wearing a head-mounted display (HMD). Whatever input the user sees through the HMD becomes their 'reality', allowing them to experience a digitally generated scene as if they were part of it.

Key term

Immersive: creating a 3D image which appears to surround the viewer

The HMD senses user movement, so that when the user looks up, down or from side to side, the scene they see tracks their range of movement over 360 degrees, to give the sense of looking around a space within the virtual building.

Health and safety

There are obvious safety considerations if a user cannot see their real-life surroundings while moving through a virtual environment. If you use a VR system, plan the way you use it with this in mind.

Research

VR systems have become quite common. Research and make a list of construction and non-construction uses for VR systems.

Multiple users can enter this virtual world to collaborate on design decisions, making the process more creative and efficient. Often, a client will have definite ideas about how they want their project to be completed but they may not be able to express their ideas clearly to designers. Using immersive technology, the designer can accompany the client using the virtual representation of the building to confirm that the proposed design fully matches the client's expectations.

▲ Figure 10.15 An architect using an MR system

Industry tip

Other types of immersive technology are often mistakenly grouped under the title 'virtual reality'. Augmented reality (AR) and mixed reality (MR) are different systems that can be applied in various construction-related contexts. For more on AR, see Chapter 4.

As well as producing dynamic visual representations of buildings and other structures, the data in digital models can also be used to create static illustrations or artistic impressions of a project for use in conventional non-digital documentation, such as a client brochure or report.

2.4 Blockchain technology in construction

Blockchain technology is increasingly being explored in construction to enhance transparency, trust, and operational efficiency across complex project environments. By using decentralised, tamper-proof ledgers, blockchain enables secure tracking of materials, automated execution of smart contracts, and immutable documentation of design changes, inspections, and approvals. It supports supply chain integrity by verifying the origin and certification of components, and helps ensure regulatory compliance through auditable records.

Blockchain also facilitates faster payments and reduces disputes by linking deliverables to smart contract conditions. Additionally, it strengthens workforce management by securely storing and verifying digital credentials, licenses, and training records. Though adoption is still emerging, pilot projects in infrastructure and public sector contracts are demonstrating its potential to reduce risk, improve accountability, and support sustainability goals.

2.5 Generative AI in digital engineering for construction

Generative AI offers significant benefits to digital engineering in construction by streamlining design processes, enhancing collaboration, and improving overall project efficiency. It accelerates design cycles by automatically generating multiple viable solutions based on predefined constraints, reducing manual drafting and revision time. By integrating inputs from various disciplines—architecture, structural engineering, and MEP—it fosters better coordination and early clash detection, minimizing costly rework.

Generative AI also supports sustainability by optimizing material usage and energy performance, while predictive modeling helps anticipate lifecycle issues and environmental impacts. Ultimately, it enables scalable innovation, reduces costs, and empowers teams to make data-driven decisions with greater speed and confidence.

Clash Detection & Coordination: AI can preemptively identify and resolve conflicts between architectural, structural, electrical, and plumbing systems, reducing rework and delays.

3 Opportunities for the use of technology

Digital technology is widely used in many industries to achieve specific user-defined results. Manufacturing, research, transport, medicine and a host of other sectors routinely make use of digital technology in order to operate and grow. Systems and processes using digital data are often taken for granted in our daily life, for example in items such as mobile phones and computers.

Table 10.1 details some of the benefits of using digital technology in construction, especially in relation to planning and organising a project and monitoring the construction phase.

▼ Table 10.1 Benefits of technology in construction

Benefit	Application
Accuracy	Digital technology allows large volumes of data to be processed with high degrees of accuracy. Data concerning measurements, costings, projections and evaluations can be cross-referenced and linked in complex ways to produce rich and valuable assessment data. This can inform reliable decision making and forecasting during the planning of a project and the monitoring of progress during the construction phase.
Accessibility	Digital data can be stored and retrieved quickly and easily. Data can be accessed by anyone who is authorised to use relevant networking systems, regardless of their location. This encourages collaboration and allows data to be instantly updated or modified.
Efficiency	Digital networks can be permanently open to authorised users, allowing efficient working in both local teams and global partnerships across different time zones. Whether working in teams or in isolation, working simultaneously or sequentially, effective work patterns can be created and refined to respond to current needs in maintaining efficiency.
Risk reduction	Digital simulation of construction operations can be used to identify areas of operational risk. Analysis of accident and injury data allows the identification of emerging trends, possibly due to inappropriate behaviour or incorrect work practices. These can be mitigated by the rapid introduction of safer work methods – risk assessments and method statements can be matched more closely with actual working patterns and conditions.

Test yourself

What are the benefits of using technology in construction?

There will no doubt be intriguing opportunities in the future to take advantage of the power of digital equipment, computer modelling and smart technologies.

3.1 Robotics

Industries such as car manufacturing use robotics with digital control systems to assemble components repetitively.

There are some repetitive activities in construction, where industrial robots are already being used or are being developed, for example in off-site construction. This involves the manufacture of parts or sections of a building away from the site location. The manufactured items are then transported to site, where they are assembled to complete the building. Off-site construction is often referred to as 'prefabrication' and may involve modular construction methods. For more information on off-site construction, see Chapter 7.

▲ Figure 10.16 Robotics can be used for repetitive construction operations

When manufacturing sections of a building, the use of robots has many benefits, including:

- consistent quality standards
- a regulated work rate that allows for reliable scheduling of materials deliveries to the factory
- accuracy and efficiency (so less waste)
- enhanced safety standards, as there is reduced human involvement.

▲ Figure 10.17 Factory-produced modules being installed on site

Research

Research off-site construction companies and find out the types of projects modular construction has been used for.

Other types of robots that could be used in the construction industry include:

- inspection robots that can reach areas that are difficult or dangerous to access
- maintenance robots that can clean and repair buildings safely
- 3D-printing robots that 'print' components or whole buildings
- demolition robots that safely dismantle unstable or fragile structures.

▲ Figure 10.18 3D-printing robots can 'print' components or entire buildings

Test yourself

How can robotics be used in modular construction?

3.2 CAD/CAM

CAD/CAM refers to software that is a combination of two processes – computer-aided design (CAD) and computer-aided manufacturing (CAM).

As mentioned above, CAD is a digital tool used to create 2D and 3D representations and drawings during the design process, which allows simulation, testing and refining of ideas to arrive at a satisfactory finished design.

CAM uses digital **geometric** data to control manufacturing machinery, such as computer numerical control (CNC) machines which have motorised tool positioning and manoeuvring capabilities controlled by pre-programmed computers.

Key term

Geometric: consisting of defined angles, patterns and shapes

The combination of CAD and CAM results in a faster production process that provides dimensional control and ensures product consistency. By using automated systems that can work around the clock, the need for skilled production workers is reduced. However, these systems still require input from trained personnel who set up and program the machines and monitor and intervene when problems arise.

▲ Figure 10.19 A CNC cutting machine being set up to perform repetitive tasks

Digital processes such as CAD, CAM and BIM can be integrated, in order to save time and achieve complex outcomes without the possibility of human error in interpreting drawings and written instructions.

For example, a 2D or 3D CAD drawing of a building consists simply of lines to represent its shape and the component parts within it. If the CAD data is linked to BIM, the combinations of simple lines can be allocated grouped geometric patterns, features and dimensions.

This can allow automatic generation of a range of accurate views of a component, along with details of the materials it is made from, so that a CAM machine can manufacture it.

Assessment practice

Short answer

1 What does M2M stand for?
2 How can the IoT improve energy efficiency in a building?
3 State two benefits (besides energy efficiency) that can result from integrating buildings into the IoT.
4 What alternative name is sometimes given to point cloud surveying?
5 How can robotics be used in a construction setting?

Long answer

6 Describe how AI systems go beyond simply following programmed instructions.
7 Explain how AI systems could be used to improve worker productivity in an office building.
8 Describe the range of benefits achievable by using CAD when designing buildings.
9 Explain how an item of equipment or machinery used on site can be described as 'smart'.
10 Describe the difference between simulation and animation when used for analysis of buildings.

Project practice

An old office block in a city centre is to be extended to allow for company expansion. It is a familiar landmark in the area because of its attractive period architectural features. A railway line and a river run close to the building.

The existing parking area next to the building will become the location of the office extension. Parking for existing and new office staff will in future be provided by a new underground car park below the extension.

You have been commissioned to conduct a survey of the building plot and the existing structure to establish the feasibility of the proposed extension project.

- Write a report to compare the advantages and disadvantages of conducting the survey visually on foot or using a drone to create a point cloud survey.
- Make a list of potential problems that must be considered regarding the proposed underground car park. (Hint: think about the railway and river nearby.)
- Consider technological methods that could be used to record details of the architectural features of the existing landmark building, with a view to matching them in the new extension.

Chapter 11 Construction commercial/business principles

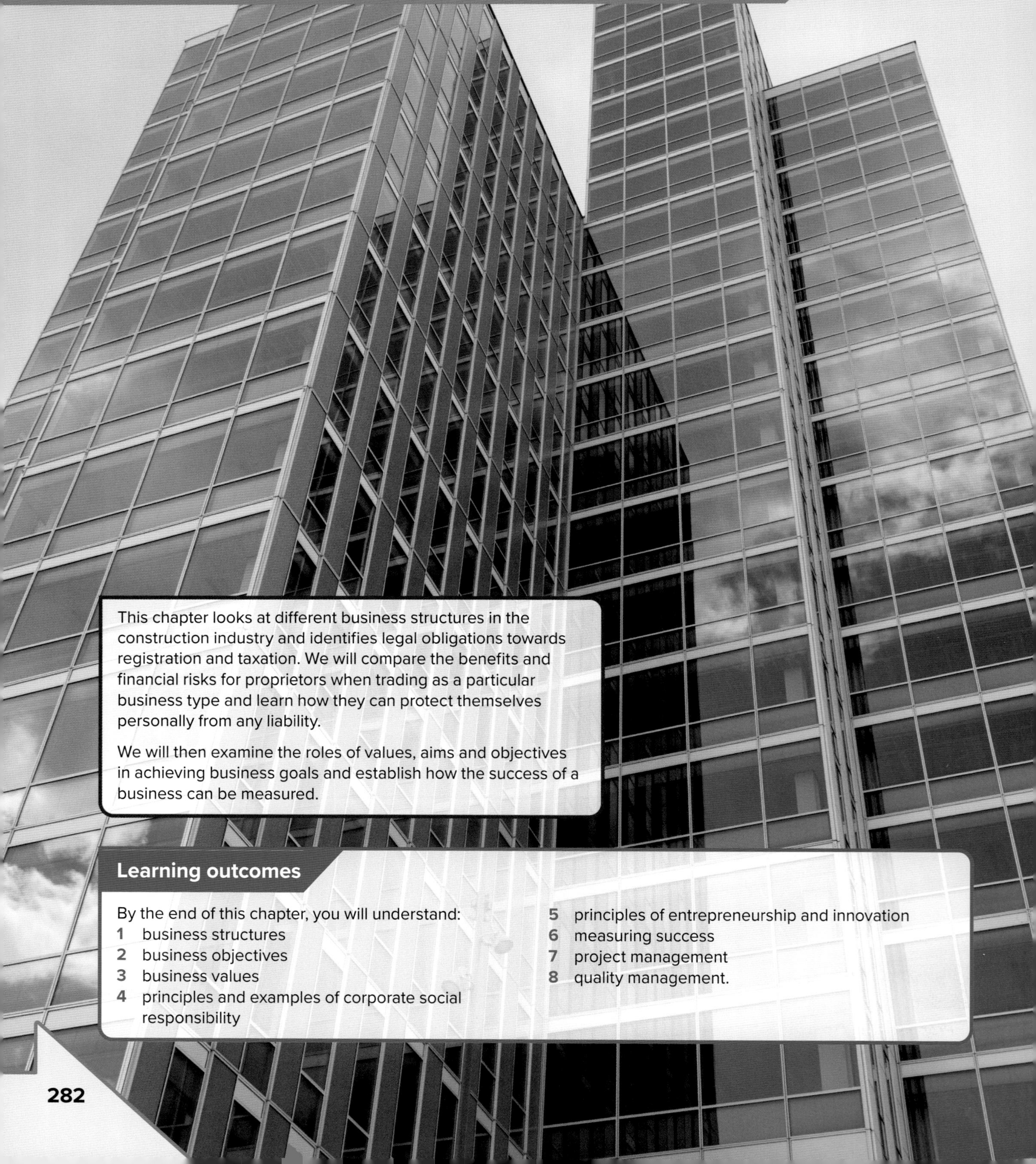

This chapter looks at different business structures in the construction industry and identifies legal obligations towards registration and taxation. We will compare the benefits and financial risks for proprietors when trading as a particular business type and learn how they can protect themselves personally from any liability.

We will then examine the roles of values, aims and objectives in achieving business goals and establish how the success of a business can be measured.

Learning outcomes

By the end of this chapter, you will understand:

1. business structures
2. business objectives
3. business values
4. principles and examples of corporate social responsibility
5. principles of entrepreneurship and innovation
6. measuring success
7. project management
8. quality management.

1 Business structures

1.1 Business types

There are many different types of business that operate in the UK construction industry, ranging from individuals trading on their own to national public limited companies employing thousands of people across the country.

Some business structures are not designed for everyone because of the size of the organisation and its legal status. However, as a business grows it may need to change to a more appropriate business model.

The advantages and disadvantages of each business type have to be considered carefully, because there could be serious legal implications if trading unlawfully.

In this section, we will look at the following types of business:

- sole traders, for example a self-employed tradesperson
- partnerships, for example an architects' practice
- limited companies (private, Ltd, or public, PLC), for example construction companies
- small and medium-sized enterprises (SMEs), for example construction companies
- not-for-profit organisations, for example Her Majesty's Revenue and Customs (HMRC) or the Health and Safety Executive (HSE)
- community interest companies (CICs), for example local community development trusts
- franchises, for example building maintenance businesses.

Sole traders

As soon as someone registers with HMRC as self-employed, they are effectively a sole trader. As a **proprietor**, they own and have full control of their entire business or enterprise, although they can employ other people to work for them.

A sole-trader business can operate under the name of its owner. However, if the proprietor would prefer to use a different business name, they must register it at companies house to prevent another company trading with the same name.

Key term

Proprietor: an individual who owns a business

▲ Figure 11.1 A sole trader

Operating as a sole trader is one of the simplest business structures to run, because there are fewer tax responsibilities (such as filing accounts with Companies House) and therefore less administration. However, a sole trader is still responsible for the day-to-day management of their company and:

- accurately recording expenses, sales and profits
- running payroll
- filing annual self-assessment tax returns with HMRC
- paying National Insurance contributions
- paying value added tax (VAT) if annual turnover exceeds, or is likely to exceed, £85,000.

Research

Find out about value added tax (VAT) and explain the advantages and disadvantages for a sole trader.

Improve your maths

An electrician who is employed as a sole trader has an annual turnover of £92,000 (inclusive of VAT) and is paid a salary of £42,000.

Calculate how much tax they will have to pay to HMRC (excluding any tax savings) if the rates of tax are:

- VAT 20%
- tax and National Insurance 27%.

Note: VAT is paid on whole turnover, but tax and National Insurance are paid on salary.

To help manage the financial aspects of running a business, many sole traders appoint an accountant. However, the sole trader is still legally responsible for the information submitted to HMRC and Companies House each year.

A sole trader is personally liable for any losses or debts (for example to the bank, HMRC or suppliers) that the company may incur if things go wrong. **Creditors** can legally recover any money owed by the **debtor** from their business or their personal assets such as their house. If the debt is considerable and the individual is unable to pay, they could be made personally bankrupt. This could result in further difficulties, for example being unable to obtain any type of credit such as a mortgage, credit card or finance on a vehicle because they are deemed to be a high risk.

Key terms

Creditors: individuals or organisations that are owed money because they have provided goods, services or a monetary loan; HMRC would also be described as a creditor where a tax duty is owed

Debtor: an individual or organisation that owes money

Partnerships

Partnerships are businesses owned by two or more individuals. Partners share the costs, duties and risks of managing a business together, although they may delegate certain responsibilities to their employees.

Each partner's share of the profits is based on the **partnership agreement** between all stakeholders and can be determined by looking at the amount of money they invested in the business and/or the level of involvement they have in the day-to-day running of the business. It is possible to have 'sleeping partners' who have no involvement in managing the business but who contribute financially.

Each partner is personally responsible for paying tax and National Insurance contributions, based on their share of the profits, and must therefore register for self-assessment tax returns with HMRC.

A nominated person must be chosen to manage the partnership's business accounts and file its annual tax returns with HMRC.

Key term

Partnership agreement: a legally binding contract that sets out terms and conditions for each partner in the business

▲ Figure 11.2 A nominated person filing a tax return

There are three different types of business partnership:

- ordinary partnership
- limited partnership
- limited liability partnership (LLP).

There are no special responsibilities for an ordinary partnership other than those described above.

Limited partnerships have to be registered with Companies House. They must have at least one general partner and one **limited partner**:

- General partners are responsible for managing and controlling the business and are liable for any debts that the business cannot repay, in the same way as a sole trader. They can apply for the limited partnership to act as an authorised contractual scheme (ACS), which means that the partners co-own the assets but only pay tax on their share of the profits and are exempt from paying **corporation tax**.
- Limited partners are not responsible for the management of the business and are only liable for any debts up to the amount they have contributed.

A partner cannot be both general and limited at the same time.

Key terms

Limited partner: a part-owner of a business whose financial liabilities cannot exceed their investment

Corporation tax: a duty paid annually to HMRC based on a company's profits

Improve your English

Explain the term 'sleeping partner'.

Research

How long does a business legally have to keep tax records? What are the consequences of not keeping tax records in order?

A limited liability partnership (LLP) can be created (**incorporated**) with two or more partners. While it can have any number of ordinary (non-designated) members, it must have at least two designated members.

Designated members have responsibilities to manage the business, for example:

- registering the business for self-assessment with HMRC (individual LLP members must also register separately)
- registering the business for VAT, if annual turnover is expected to be over £85,000
- keeping accounting records
- completing self-assessment and VAT returns
- preparing and signing annual accounts and filing them with Companies House
- sending an annual confirmation statement to Companies House.

A formal LLP agreement should be written to outline:

- how members can join or leave the LLP
- the roles and responsibilities of members
- how decisions are made
- how profits are shared among members.

Having a legally binding agreement in place when the business is first set up avoids potential disputes or complications between partners at a later stage of trading.

LLPs are sometimes preferred to other types of business because they clearly define their partners as individuals when it comes to earnings, unlike corporations. They also protect the personal assets of partners as a separate legal entity from the liabilities of the business.

As with any partnership, if there are only two members and one leaves, the LLP may have to be dissolved.

Test yourself

Why are LLPs sometimes preferred to other types of business?

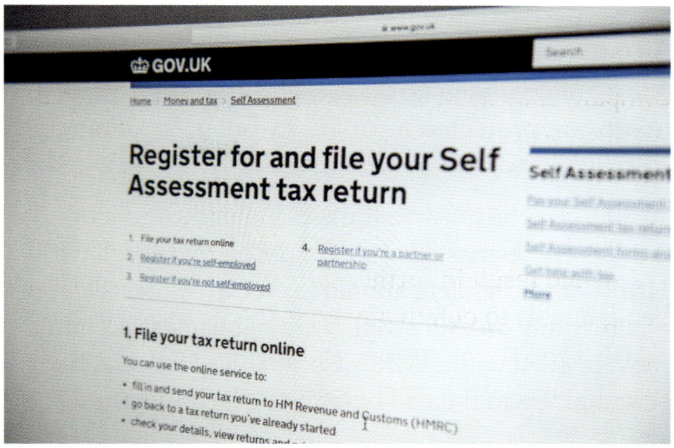

▲ Figure 11.3 Self-assessment tax return

Test yourself

What factors could cause a business partnership to fail? Explain how these could be avoided before starting an enterprise.

Key term

Incorporated: legally registered as a limited company

Limited companies

Businesses often change their legal status to limited when they become more established, because it gives clients more confidence that the business is a credible organisation. Other benefits of limited companies include:

- protection of owners' personal assets from any business debts
- owners given complete control of their business
- no requirement to pay National Insurance
- lower rates of tax
- more tax relief than sole traders.

A disadvantage of being a limited company is that information about the directors, shareholders, registered office and yearly financial statements is on public record. These details are visible not only to the general public, but also to the business' potential competitors.

Private limited companies (Ltd)

A private limited company (Ltd) is a business that is legally registered with Companies House. It is referred to as incorporated, meaning that it has been made into a corporation that is a separate legal entity, with its own rights and obligations. The main benefit of a limited company is that the owners and shareholders have limited liability, so their personal assets are protected against any business debts up to the value of the money that they have invested.

As with any business, a limited company and its owners (known as directors) must be registered with HMRC and complete annual tax returns, regardless of whether any **duty** is owed. Annual business accounts must be filed with HMRC, Companies House and shareholders at the end of each **financial year**.

Directors of the business are paid in the same way as its employees, through the pay as you earn (PAYE) system, and they do not have the flexibility to withdraw money from the company without restrictions or tax consequences. This means that tax and National Insurance are calculated and deducted from salaries, rather than the directors and employees having to complete annual self-assessment returns with HMRC.

At the end of each financial year, the board of directors considers whether to distribute any profits among the shareholders or reinvest the money in the business. Payments made to shareholders are known as dividends; these can vary in size, depending on the amount of investment shares that have been purchased.

The limited company is responsible for paying PAYE duty to HMRC either monthly or quarterly, and any corporation tax annually. There are strict deadlines for filing tax returns and making any payments due, with potential penalties or fines if they are not met. Therefore, most businesses employ an accountant to complete their accounts, or a larger company may have an in-house accounts team.

▲ Figure 11.4 A limited company

Key terms

Duty: a type of tax charged on the value of goods and services

Financial year: a 12-month period over which a business' accounts are calculated, as opposed to a calendar year (January to December); also known as a fiscal year

Research

Research the purpose of National Insurance contributions made through personal income.

Public limited companies (PLC)

A public limited company (PLC) is similar to a private limited company. The main difference is that money can be raised for the business through investors (for example the public) buying shares on the stock exchange.

This type of business benefits from many tax and National Insurance advantages up to a certain value of its profits, including:

- paying less corporation tax
- better dividend payments for shareholders
- pensions
- reinvesting money back into the business.

For a company to change its legal status to PLC, it must have at least two shareholders and a minimum of £50,000 worth of shares to be issued. As with a private limited company, a PLC must be registered with HMRC and Companies House and file tax returns.

Small and medium-sized enterprises (SMEs)

In the UK, small and medium-sized enterprises (SMEs) are defined as follows:

- small – employs on average no more than 50 people and has an annual turnover of £10.2 million or less
- medium-sized – employs on average no more than 250 people and has an annual turnover of £36 million or less.

In 2020, 99.9 per cent of all businesses in the UK were SMEs, which represents the employment of the majority of the working population. These businesses play an important role in the country's economy because of the jobs and tax revenue that they generate. To encourage further growth and innovation, and to help keep these enterprises in business, the government regularly offers various incentives, such as better access to loans and favourable tax treatment.

Not-for-profit organisations

As the name suggests, these are charitable businesses that do not make a financial profit. They often seek to provide a public service or social benefits for individuals or communities in need.

These types of business are usually run by a board of directors that use similar management techniques to for-profit organisations. However, any money earned is used to cover operational costs and reinvested back into the organisation to pursue its objectives or goals.

The type of management structure adopted often depends on whether the organisation is limited, a legal charity or an unincorporated association with voting members.

Research

Find out what unincorporated associations are and how they operate:

- What is their usual purpose?
- Do they have to register with Companies House?
- How much does it cost to set them up?
- What happens if they make a profit?

A not-for-profit organisation can be started by anyone without any money, although they are often funded through charitable donations, grants or self-funding. In many cases, they also have tax-exemption status. However, they are still obligated to deduct tax from their employees' wages and complete annual company tax returns with HMRC.

A successful not-for-profit organisation will demonstrate ethical standards and practices, such as trustworthiness, openness and accountability. This instils public and government confidence in the organisation, resulting in greater financial investments in the future.

Research

Research not-for-profit construction organisations in the UK and explain the motives for running their businesses.

Community interest companies (CICs)

The aim of community interest companies (CICs) is to provide a benefit for the community or trade with a social purpose. Unlike not-for-profit organisations, they are expected to make a profit, which is predominantly reinvested in the company or used for the community they are set up to serve. However, returns to company owners and investors are allowed, as long as they are balanced and reasonable, and a dividend cap must be put in place. A CIC is primarily about benefiting the community and not making a private profit.

Franchises

Successful and profitable businesses may look to grow further by creating a franchise.

A franchise allows a business to expand quickly with lower **capital** outlay by selling the rights to the business name, logo or trademark and business model to self-employed **entrepreneurs**, referred to as franchisees. By operating under the banner of an established business with a proven track record, a franchisee can seek to replicate the success of the bigger brand.

It is sometimes more beneficial to buy a long-term franchise than start a business from scratch with a new name, because of the business system (processes, suppliers and other resources), existing customer loyalty and support provided by the **franchisor**. New businesses often take years to establish themselves with a strong customer base before making a profit, whereas they are more likely to be profitable from the outset if they have invested in a franchise.

Business support provided for a franchisee often includes:

- marketing (promoting the business)
- recruitment of personnel (human resources)
- provision of training and equipment
- provision of resources, for example machinery or vans
- assistance from head office
- collaborative networking with other franchisees
- dealing with suppliers
- financial planning
- accounting.

▲ Figure 11.5 Business support training

Key terms

Capital: the amount of funds or **liquid assets** owned by a business

Liquid assets: things that a business owns that can be sold quickly and easily for cash without any loss in value

Entrepreneurs: individuals who start up their own businesses, taking on financial risks in the hope of making a profit

Franchisor: the original owner of a franchise

Once an agreement has been reached and the contract has been signed, the franchisee must pay any initial fee for the rights to the business. In addition, the franchisor will be paid a regular share of the profits by the franchisee, known as royalties. Royalties can be calculated from a percentage of the business turnover, a mark-up on goods supplied or a flat fee.

For a franchise business to work properly, it has to maintain its standards and brand control. The business model must therefore be transferable, regardless of geographical location. The business activity should also be something that another business can do or be trained to do successfully.

Evidence suggests that franchises often perform well because of the personal financial investment made by their owners and the support network provided. Once established, the franchisee has the option to buy further investments or sell their business to a buyer authorised by the franchisor.

A disadvantage of this type of business is that the franchisee does not have any independence compared with running their own business. They have to adhere to the franchise business plan, operational plan for day-to-day running, systems, policies and procedures. Regardless of how well a franchise is managed, there is always a risk that the organisation can be tarnished by other franchisees, which could impact on future success.

Test yourself

Which term describes an individual who starts up their own business, taking on financial risks in the hope of making a profit?

2 Business objectives

Objectives can be defined as the incremental steps a business needs to take in order to achieve its overall aims, which are often closely aligned to its **business plan**.

Businesses working in different sectors vary in size and scale, therefore their aims and objectives must be individually tailored. For example, an entrepreneur who has recently started a new business may initially have an aim to financially survive the first couple of years of trading, whereas an established organisation may have an ambition to increase its annual turnover or profit.

Key term

Business plan: a written document that defines a business' goals and the strategies and timeframes to achieve them

Improve your English

You have decided to start your own business as a sole trader in your chosen profession and you are looking for investment to get you started, for example to buy a van and tools.

Design headed paper with your business details and write a letter to a bank, requesting a loan and outlining your business plan.

Objectives should be challenging, to give a business purpose and direction. Once they have been achieved, they should be updated to improve the business and move it one step closer to achieving its main goal.

The main categories of business objective used to measure performance of an organisation in the construction industry are:

- financial
- social
- organisational culture
- innovation
- quality
- sustainability
- compliance.

2.1 Financial objectives

Entrepreneurs often prefer the independence associated with having control of their own organisation, rather than being employed by someone else. The freedom to run a business and make important decisions with the potential of determining successful outcomes can be personally satisfying, especially when the owner is challenged beyond their comfort zone.

The financial objective for private organisations is usually to develop their business with innovative ideas, products or services in order to maximise profits. As the business grows, other opportunities can be created to adjust or set new goals in order to achieve a bigger **market share** in a particular geographical location.

As mentioned above, the aims and objectives of businesses can be very different. For example, the financial objective of not-for-profit organisations is not to create wealth but to make enough money to:

- cover their costs
- establish reserves
- fund activities that benefit the community
- reinvest in the business.

Key term

Market share: the percentage of total sales or output that a business has in a specified market; sometimes referred to as market leadership

2.2 Social objectives

Private organisations have social objectives to provide quality products and services that are useful to customers.

Most private organisations serve society by employing competent people (skilled and unskilled) at different levels. Organisations with strategic social objectives to treat their employees fairly and inclusively will create meaningful jobs, which enables their businesses to grow and in turn generate further employment opportunities. These types of job often provide opportunities for continuous professional development (CPD) and promotion, as well as offering other staff benefits. Besides the fair wages that employees should receive from their employers, they may also benefit from a workplace pension or welfare schemes such as private healthcare.

Not-for-profit organisations, such as charities and social enterprises, work for the benefit of society by providing a public service (such as housing, healthcare and education) or helping people, in order to create positive social change.

▲ Figure 11.6 Private organisations seek to provide quality products and services that are useful to customers

2.3 Organisational culture objectives

Organisational culture can be described as a collection of core beliefs, ethical values, expectations and behaviours that define the corporate personality of an organisation. It is embodied in the way an organisation conducts business, treats its customers, manages **workflow** and interacts with its employees. It is essential that all members of an organisation understand its shared cultural objectives and promote them while performing their roles and responsibilities.

Creating a strong organisational culture is vital for the success of a company. Not only does it define a company's internal and external identity, but it also helps to build a strong, unified team of employees who enjoy greater wellbeing. In turn, this will reduce staff turnover and assist with recruiting high-calibre employees.

An organisation that recognises and rewards the success of its employees is likely to transform them into advocates for the business, so that they continue to promote and contribute to its consistent cultural behaviours.

Key term

Workflow: a sequence of activities needed to complete a work task

Test yourself

Why is it important to have a clearly defined organisational culture?

2.4 Innovation objectives

Innovation objectives involve greater levels of uncertainty and risk than general business objectives. They should be challenging but obtainable, by agreeing steps to achieve them with all the business' shareholders.

The primary innovation goals for business improvement can be defined by:

- setting aims and developing a strategy for how these goals can be achieved
- understanding the primary challenges that may prevent targets being met
- understanding data, costs and risk
- being able to implement innovative new ideas in the organisation
- increasing employee retention
- engaging with employees and promoting innovation and new ideas
- identifying and rewarding talent.

A business' innovation objective may be to launch a new product or service, outperform its competitors or become a market leader. However, in order to accomplish these long-term goals, they may first have to achieve the following typical objectives of:

- optimising existing processes or procedures
- increasing business productivity and efficiency
- improving the quality of service or existing products
- being more competitive
- meeting customers' needs
- improving the skills and knowledge of the workforce
- developing performance
- refining sustainability within the organisation.

Innovative construction methods and technology, such as self-driving vehicles, virtual reality (VR), computer-controlled manufacturing robots, 3D printers and drones, are explained in Chapter 7.

▲ Figure 11.7 Virtual reality is a good example of innovation in construction design and development

2.5 Quality objectives

The minimum requirements for behaviour, work and materials in the UK construction industry are specified in health and safety legislation, British standards and building regulations. However, businesses often aim to exceed these standards, in order to ensure the welfare of their employees and offer an improved service to their clients.

Construction businesses can subscribe to a number of voluntary assessment and certification schemes, to ensure best-practice standards that often far exceed those required by legislation.

The Building Research Establishment's Home Quality Mark (HQM) is an independently assessed certification scheme for new homes that focuses on the expectations and needs of occupants and protection of the environment. It assesses the quality and sustainability of the home itself, its surroundings and the construction or renovation in the following key areas, set out in the 'Home Quality Mark ONE' technical manual:

- 'Our surroundings', for example recreational space, local amenities and sustainable transport options
- 'My home', for example sound insulation, ventilation, water efficiency, energy and cost
- 'Delivery', for example construction energy and water use, commissioning and testing, and aftercare.

Where the HQM scheme recognises that the standards of a new home are significantly higher than the minimum requirements, it awards a star rating from 1 to 5. A 1-star rating means that a home meets key baselines beyond minimum standards; a 5-star rating means that a home is outstanding and far exceeds minimum standards.

Attaining the HQM instils confidence in buyers, tenants, investors, developers and insurers that higher levels of build quality have been met and there are significant benefits to living in the home, for example a lower risk of defects usually identified in new-build properties.

The International Organization for Standardization (ISO) is an international standard-setting body with headquarters in Geneva, Switzerland. Its membership comprises 165 national standards bodies from around the world. They work together to develop and publish market-relevant commercial, industrial and technical standards.

The main goal of the ISO is to facilitate trade. It offers solutions to global challenges and supports innovation by providing guidelines to streamline processes and improve quality and safety across a range of businesses and products.

There are over 23,000 ISO standards that have been voluntarily developed by industry experts by consensus, covering all aspects of manufacturing and technology. ISO identifies its certified standards with a unique reference number, for example:

- ISO 45001 – Occupational health and safety management systems
- ISO 9001 – Quality management systems
- ISO 14001 – Environment management systems.

Access to general ISO information can be obtained free online. However, the full guidelines must be purchased from the organisation.

The ISO is not a governing body and does not have authority to enforce any regulations and laws. It is also not involved in the certification of its standards; this is carried out by external certification bodies. Therefore, businesses can only become certified to specific standards, for example 'ISO 9001:2015 certified' and not 'ISO certified'. To become certified, a business will need to adhere to the requirements in the standards and may therefore use these as quality objectives.

Improve your English

Write a short sentence to explain the meaning of the word 'consensus'.

Test yourself

Explain the purpose of the ISO.

2.6 Sustainability and compliance objectives

Construction organisations often embed sustainability into their business objectives for moral reasons, for example to protect the environment by maintaining **ecological balance**, conserving natural resources and preventing pollution. The way we design and construct buildings plays an important part in achieving these aims, from energy-efficient construction to eco-friendly use of building materials.

Building regulations and other external regulatory control measures determine mandatory obligations, for example Part L of the Building Regulations 2010 outlines requirements for conservation of fuel and power. However, businesses may create their own internal control measures beyond these standards and build them into their business objectives.

Besides the moral reasons mentioned above, businesses may set further goals to improve their sustainability in order to reduce waste, increase profits and attract the interest of future clients with similar interests in protecting the environment.

▲ Figure 11.8 Protecting the natural environment from development activities

Key term

Ecological balance: where living organisms, such as plants, animals and humans, co-exist in a sustainable environment

3 Business values

Earlier in this chapter, we acknowledged that organisational culture stems from a company's beliefs and values, sometimes referred to as its **philosophy**.

Transparent business values play an essential role in any organisation. They can:

- create a sense of purpose and commitment
- improve **cohesion** of the workforce
- drive an organisation forward, helping it to reach the goals set by the management team
- help motivate employees by building trust and security
- help develop relationships with partners, stakeholders and customers
- influence sales, customer service and marketing strategies
- demonstrate the business' culture outside of the organisation, which may attract new talent.

It is important that potential new employees share the same values as the organisation; this will determine whether they enjoy their job, engage with the company and work productively.

Key terms

Philosophy: values and beliefs that act as guiding principles for behaviour

Cohesion: a state of working together in unity

The fundamental business values for construction organisations are:

- financial stability
- customer service
- care for life
- ethics and transparency
- codes of conduct
- collaborative working.

3.1 Financial stability

Financial stability means a business having sufficient funds to pay overheads, repay any loans and still make a profit to prepare for risk management in times of potential **economic downturn**.

A business with steady revenue that does not rely too much on debt should be making a healthy profit through sales while its expenses remain the same. This should enable growth of the business.

Key term

Economic downturn: when the economy has stopped growing and is on the decline, resulting in reduced financial turnover

3.2 Customer service

When a customer first approaches a construction business with a concept for a project, they often have expectations of what the business can provide and the level of service it should offer. To build and maintain good customer service, the business must:

- pay close attention to the customer at all times
- be helpful – make the customer feel at ease exchanging information and ideas
- be polite and courteous when communicating verbally or in written form
- listen actively to the customer by responding to all of their needs
- solve problems – offer practical solutions with design, product and scheduling ideas
- provide support to meet the customer's needs and expectations
- be timely – provide the product or service in a mutually acceptable timeframe
- be proactive – anticipate problems before they occur
- demonstrate skills and knowledge – have a strong and up-to-date understanding of the industry and be able to produce work to the highest standards
- know the business product or service offered – understand the business' limitations

- ensure the customer's needs are met – measure the success of customer service by making sure the customer's initial expectations have been met or exceeded
- offer value for money – be able to provide a satisfactory product or service for a reasonable price
- personalise support for each customer.

Providing good customer service is vital to the success of all construction businesses, to ensure a good reputation. Customers who experience good customer service are more likely to provide repeat business and recommend a company to other potential customers.

3.3 Care for life

The health, safety and welfare of all employees and others affected by work activities, for example the general public, should be one of the core values of all businesses.

Besides the legal obligations that businesses have to adhere to, they should also consider the mental and physical wellbeing of their staff and the impact that their job roles may have on them. Businesses that value their employees will make sure they have a good work-life balance and actively promote health and wellbeing through a number of different benefits, such as a tax-deductible cycle-to-work scheme or mentoring.

▲ Figure 11.9 Cycle-to-work scheme

3.4 Ethics and transparency

Business ethics can be described as business practices and policies when faced with arguably controversial subjects, for example corporate social responsibility. Some business ethics are embedded in legislation, for example employers have a duty to treat employees fairly and with respect and not to discriminate. However, ethics often goes beyond legal responsibilities to include a moral code for the business and its employees.

Businesses with positive values and behaviours gain public trust and approval, which will enable them to grow. There are negative implications for businesses that do not have business ethics, such as damage to reputation, difficulties retaining employees and loss of work.

Business transparency is about open and honest communication across all levels within an organisation and the sharing of information both internally and externally.

For a business to be completely transparent, it must provide access to all the information needed, not just what it is willing to share. Some businesses may not want information such as salaries, revenue, future ambitions and hiring policies to be available to the public, although this level of transparency can provide potential investors and customers with an insight into the organisation and the level of service they can expect.

When employees feel financially, physically and emotionally safe within an organisation, they are more likely to communicate openly and honestly with their employers, without fear of losing their job. In turn, when employees feel they are being listened to, they are more motivated and productive.

▲ Figure 11.10 Employees who can communicate openly and honestly with their employers are more motivated and productive

3.5 Codes of conduct

A code of conduct is essentially a set of rules written by an employer for their employees, in order to protect their business and its reputation. It should explain an organisation's values and principles and link them with standards of professional behaviour.

A code of conduct typically includes standards of behaviour regarding equality and diversity, protection of sensitive data, and disclosure of any criminal convictions that an employee may have prior to employment or during their service.

All employees must read and understand their employer's code of conduct before starting work; they must also appreciate the consequences of not following the rules. Any breaches of the code of conduct could result in disciplinary action being taken, or in some cases termination of employment.

3.6 Collaborative working

Internal collaborative working enables a diverse range of employees with specific skills or traits to work together to problem solve and achieve business objectives.

Collaborative working with another organisation is referred to as networking. Organisations of different sizes can collaborate in a non-hierarchical way either in person or remotely using a variety of different technologies, for example video conferencing and BIM.

Besides being more competitive, there are several other advantages of working with another organisation, including:

- sharing responsibilities
- sharing knowledge and expertise
- sharing materials, resources and facilities
- developing skills and innovation
- improving motivation
- increasing productivity and sales
- accelerating business growth
- being able to tender for larger contracts
- achieving aims that may have been out of reach if working independently.

Provided that employees understand and share the same business values, and their employer honours them when making important decisions, a strong business can be created with a positive company image.

Test yourself

List the benefits of networking with another organisation.

Research

Research your industry placement's key business values and explain how it is working to achieve these.

Case study

You and your business partner have successfully been trading for a couple of years in the construction industry and have decided to grow the enterprise further by employing some staff.

Complete the following tasks in preparation for your new employees:

- Outline your business values.
- Write a brief code of conduct.

4 Principles and examples of corporate social responsibility

Construction and maintenance of the built environment can have a negative impact on the natural environment and local communities, through pollution and disruption. Employees in the construction industry can also experience short- and long-term physical and mental health issues due to a number of factors, for example working to time constraints in a competitive market.

Corporate social responsibility (CSR) is the commitment of an organisation to carry out its business activities in a socially and environmentally responsible way.

▲ Figure 11.11 Disruption to a community due to construction activities

CSR is not a legal requirement. However, developing a CSR strategy that will have a positive impact on the community and wider society, and integrating it into an organisation's values, makes good business sense, because it often contributes to risk management and legal compliance. It also affects how stakeholders such as clients and investors view an organisation, and may impact on their decision to work with or support it.

It is important for construction businesses to establish links with local schools, community groups and charities, with a view to being as transparent as possible about their aims to both minimise negative impacts on the community and actively make positive impacts.

Table 11.1 provides examples of how CSR is applied in the construction industry.

▼ Table 11.1 CSR in the construction industry

Principles of CSR	Application in the construction industry
Construction design	• Ensuring buildings are well designed to suit occupants' lifestyles • Being guided by the local community to meet its needs (for example providing functional community spaces, landscaping, cycle paths, community lighting for safety) • Ensuring inclusivity (for example providing affordable homes) • Improving quality of life and wellbeing
Social, economic and environmental considerations in construction design and planning	• Supporting housing initiatives • Supporting training and development programmes, for example work experience, apprenticeships and internships • Boosting the economy by increasing local employment and paying above the minimum living wage • Encouraging flexible working • Improving working conditions for employees • Ensuring equality and diversity in the workforce by creating opportunities for minority and underrepresented groups • Supporting local community groups and charitable activities with time and resources • Reducing the environmental impact of construction work
Sustainable construction	• Using ethically and sustainably sourced building materials through the entire supply chain, for example Forest Stewardship Council certified timber • Using recycled materials • Minimising waste • Using local trades, suppliers and materials • Constructing greener buildings using low-impact materials • Working more efficiently to reduce carbon footprint • Reducing the use of fossil fuels • Reducing air, water and ground pollution • Making provision to improve the biodiversity of the local area • Providing energy-efficiency initiatives

▲ Figure 11.12 Supporting training and development programmes such as apprenticeships is an important part of CSR

Besides the moral obligations businesses have to protect the environment and people from construction activities, there are also legal duties that may align with their CSR commitments. Failure to follow legal duties can result in prosecution and damaged public relations.

Further information on CSR can be found in Chapter 3.

Test yourself

List five different ways that a construction business can implement CSR principles of sustainability in their objectives.

5 Principles of entrepreneurship and innovation

As mentioned above, an entrepreneur is an individual who starts up their own business, taking on financial risk in the hope of making a profit. To be successful, a business owner must know their market well, including its strengths and weaknesses, in order to provide a service or product for which there is a demand or that has a unique selling point (USP).

5.1 Creating a viable product or service

Market research is a valuable tool used by many businesses to analyse industry and **demographic data**, in order to gather information about consumer needs and preferences.

Typical data that a business may want to analyse in a particular location or region could include:

- age
- gender
- marital status
- education
- employment
- income.

The data gathered will help to create two different groups of people: those with no interest in the business and a possible customer base.

Once a viable business opportunity has been identified, the business' mission should be established – what does the business hope to achieve? A business' goals could be purely economic or financial, although some businesses are created for social or environmental reasons.

Research

Research the demographic data in your local area to answer the following questions:

- What is the average age?
- How many people are living in social rented accommodation?
- How many people have access to a car or van?

Key term

Demographic data: statistical data about a population in a particular location or region

Entrepreneurs who are willing to take a risk, and have the vision to solve problems with an innovative product or service, are likely to have a successful enterprise. Besides developing the initial business concept, they are also the driving force behind an organisation, and must therefore have other qualities in order to grow their business. They must be confident in their product or service and motivated by the **business model**, not just by the idea of generating a lot of money.

Developing innovative technologies, products and processes is extremely important to the future of the construction sector, in order to:

- create employment
- improve productivity
- provide value for money
- improve the way we live
- improve product service
- increase growth and profit
- make advancements in the industry.

Key term

Business model: a plan usually created by a business owner which describes the strategy or framework that an organisation will use to operate and includes the identification of products/services, revenue sources and customer base

5.2 Capital

To get a new business venture off the ground often requires an investment known as working capital. This is a sum of money remaining after all the business' debts have been covered. It does not amount to the value of the company in terms of assets or customer's unpaid bills.

The money needed to start a business could come from the owner/s, investors, shareholders or a business loan. Business loans from banks are often secured against the owner's personal assets or those of the business, depending on the type of organisation. (See section 1 of this chapter for details on different types of organisations.)

▲ Figure 11.13 Securing a business loan from an investor

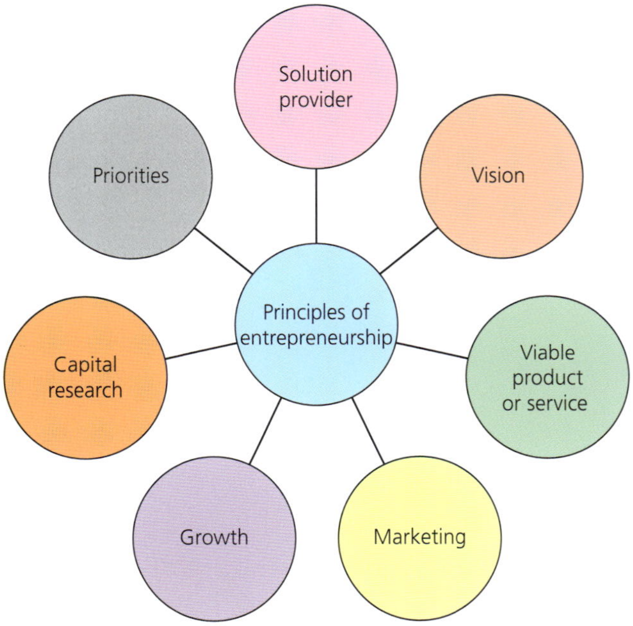

▲ Figure 11.14 Principles of entrepreneurship

5.3 Growth and marketing

For a business to thrive, potential customers need to be made aware of the product or service being sold, the people selling it, the price and the place where they can buy it. Therefore, some of the money invested in a business should be allocated to promoting the product or service through marketing. Active advertising will communicate with a bigger audience more quickly than waiting for the business to grow through word of mouth alone.

Before a customer decides to invest their money in a product or service, they may want evidence that it is as good as it sounds. As a business begins to grow, it can build customer confidence by sharing evidence of its success through showcasing satisfied customers or completed projects

Research

Research different methods of marketing and explain the impact each one may have on a construction business.

5.4 Priorities

Entrepreneurs with a concept for a new business must have a list of priorities to get their venture off the ground, become financially stable as soon as possible, and grow in the future. Even with a strong initial investment, a business will need to earn money quickly to repay any loans it may have for resources and equipment and to cover the cost of overheads and wages.

Below are some of the general priorities an entrepreneur should have when they start a business:

- Develop a concept or product.
- Research the market.
- Produce a business plan, including business aims and objectives.
- Seek investment.
- Produce business documentation, for example a contract of employment, code of conduct and health and safety policy.
- Acquire premises, such as an office, workshop or factory.
- Acquire resources, such as machinery, equipment and staff.
- Register the business.
- Arrange suitable insurance for the business.
- Market/advertise the business, for example by networking.
- Ensure the business income is more than its outgoings.
- Regularly measure performance against benchmarks (covered later in this chapter).
- Set improvement strategies to meet targets.
- Keep accurate business accounts.
- Submit annual tax returns and pay any duty due.

It is important to acknowledge that just as many entrepreneurs fail as succeed. However, there are many more who never progress beyond their initial idea because of the fear of failure. Entrepreneurs with a passion about something important to them, who have patience, confidence and tenacity, are most likely to achieve their goals.

6 Measuring success

6.1 Benchmarking

Successful organisations regularly evaluate their performance and make improvements, either to tackle areas of weakness or to demonstrate excellence.

Many new businesses fail within the first few years of trading because they cannot manage their cash flow effectively. The concepts for the business may be sound, and there may be no shortage of work, but if the business is poorly managed then the money goes out quicker than it comes in.

The term '**benchmarking**' means measuring an organisation's internal and external performance against pre-determined industry standards, competitors or completed projects. It is a powerful management tool that can increase both productivity and profits with minimum input (time and money) and maximum output (benefits). Any business targets set by an organisation must be realistic and achievable, if the business is to accomplish its aims.

Key term

Benchmarking: measuring an organisation's internal and external performance against pre-determined industry standards, competitors or completed projects

6.2 SMART objectives

SMART is a common acronym used in the built environment and construction industry for target setting. Smart objectives must be:

- **S**pecific – clear, unambiguous and understood by those who are expected to achieve them
- **M**easurable – in terms of time, productivity or cost
- **A**chievable – in terms of the size of the organisation, resources available and budget
- **R**elevant – appropriate to what is being done
- **T**ime-bound – in terms of a start and finish date (objectives with no deadline for completion are usually dismissed and never achieved).

6.3 Key performance indicators (KPIs)

Benchmarking is a recognised system that involves identifying problem areas in a business that need improvement, for example:

- project completion times
- planned budgets against actual costs
- profit margins.

Areas that can be measured in terms of a value are known as key performance indicators (KPIs). Measuring business performance against KPIs helps to establish if objectives have been met or whether new strategic targets need to be planned. If a business is not honest with the findings of KPIs, fails to learn from its mistakes and starts the cycle again, then the process of benchmarking is pointless.

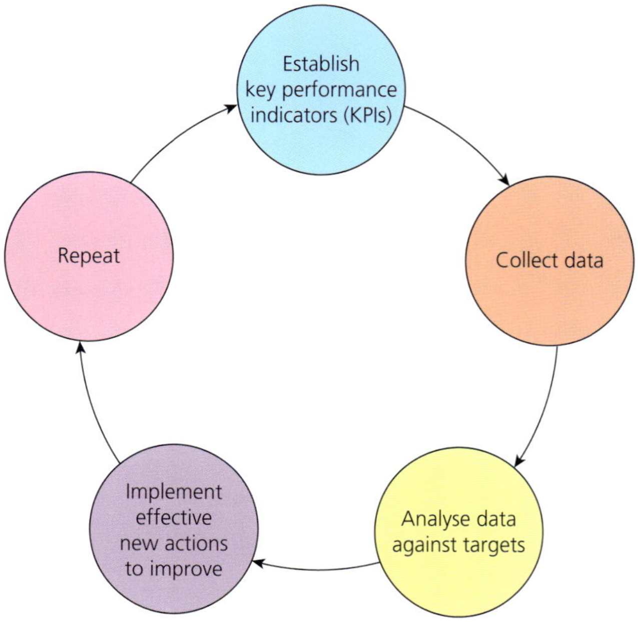

▲ Figure 11.15 The process of benchmarking

Every business selects its own KPIs based on what is important to it. For example, they could be used to measure the performance of a construction project based on data from a similar completed project. By looking at the results, a business can establish what worked well, as well as what worked less well, and find areas for improvement (for example when scheduling and sequencing work).

Health and safety

Using a business' accident and near-miss statistics over a given period is an accurate way to determine benchmarks and strategies to improve health, safety and welfare in a workplace. These benchmarks can be used to measure future performance of a business.

Gathering information for benchmarking within an organisation is much easier than trying to measure the success of a business against its competitors, because of difficulties in accessing all of the data needed. Internal benchmarking is often more valuable than external benchmarking, because it is focused and can be tailored to meet specific business objectives.

External benchmarking is used by businesses and other organisations such as the government to focus on the industry as a whole. Construction industry KPIs can be gathered from the results of national surveys of clients and construction professionals and performance data based on thousands of projects across the UK. Glenigan and Constructing Excellence publish annual sets of KPIs with the Department for Business, Energy and Industrial Strategy, to support businesses and drive improvement.

Improve your maths

There are various financial KPIs that can be used to measure the success of a business. Explain the difference between:

- gross profit margin
- net profit
- turnover
- working capital.

Test yourself

What acronym is commonly used when target setting? Explain the meaning of each letter.

Research

Look at the latest annual KPI survey published by Glenigan and Constructing Excellence.

What were the results of the contractor and client satisfaction surveys?

Contractor satisfaction survey:

- Provision of information – ? per cent
- Overall performance – ? per cent
- Payment – ? per cent.

Client satisfaction survey:

- Service – ? per cent
- Value for money – ? per cent
- Product – ? per cent.

Are these results better or worse than the previous year? Explain why you think the results of the surveys may be different to last year.

7 Project management

Throughout this book we have looked at the principles of project management, including the importance of:

- effective planning
- setting clear goals and objectives
- defining roles and areas of responsibility
- setting realistic milestones
- constraints on cost and time.

To ensure all the business objectives are measurable and achievable, the SMART technique (see section 6.2) can be applied.

Research

Research PRINCE2 (PRojects IN Controlled Environments) and explain how this can be used in project management.

8 Quality management

There are a number of quality management systems and techniques used in business to maintain the standard or quality of the work in a consistent manner, to track progress and to measure success. These include the following:

- self-assessment – a process of evaluating an organisation against a model for continuous improvement
- internal audit – completed by the organisation itself to measure the effectiveness of its quality systems. The aim of the audit is to identify any areas of weakness and develop a strategy for further improvement
- external audit – a quality management system evaluation completed by an external auditor to ensure that systems, processes and documentation are appropriate for the business. This technique is often the best and most beneficial because the results are impartial
- quality control – an internal process often used by organisations to maintain or improve standards. This system may involve the sampling, testing or reviewing of products, processes or documentation
- quality improvement – a systematic approach to measuring the success of a business against benchmarks and set new goals for continuous improvement
- ISO 9001:2015 (see section 2.5 and Chapter 7).

Assessment practice

Short answer

1 Which type of business sells its products and brand name to other businesses hoping to replicate its success?

2 What is another name for the fiscal year?

3 Which term describes collaborative working outside of an organisation?

4 Which term describes the measurement of an organisation's internal and external performance against pre-determined industry standards, competitors or completed projects?

5 Which term describes the incremental steps a business needs to take in order to achieve its overall aims, which are often closely aligned to its business plan?

Long answer

6 Describe one of the business types.

7 Explain the main disadvantage of being a sole trader.

8 Outline the main objectives of the International Organization for Standardization (ISO).

9 Explain the purpose of an organisation's corporate social responsibilities (CSR).

10 Explain what is meant by key performance indicators (KPIs) and why they are used.

Project practice

You have been working as a self-employed tradesperson for three years and have successfully networked with other sole traders on a few construction projects. One of these sole traders has suggested going into partnership. However, you have some concerns about growing the business.

Both partners are expected to invest £12,000 each, in order to purchase some innovative construction equipment to maximise efficiency and profits. You do not have the money to invest in the business, therefore you will have to take out a personal loan from your bank.

The projected turnover for the partnership for the next financial year is expected to exceed £90,000.

- Explain the risks of going into partnership, taking out the loan and increasing your turnover.
- Explain how you could reduce these risks and still expand your business.
- Present your concerns and suggestions in a digital format to one of your peers, and discuss any suggestions for improvement they may have.

12 Building services engineering (BSE) systems

This chapter looks at the layout, operation and components of a range of building services engineering (BSE) systems and explores the advantages and disadvantages of each type of system.

It then identifies the mechanical and electrical principles of components, before explaining the different types of electrical supply and distribution.

To conclude, it considers the cables, accessories and equipment used in electrical installations, as well as the pipework, ductwork and components used in plumbing, heating, gas and ventilation systems.

Learning outcomes

By the end of this chapter, you will understand:

1. building services engineering systems
2. the potential effects on building performance during installation, commissioning and decommissioning of BSE systems
3. mechanical principles of components
4. electrotechnical principles of components
5. electrotechnical supply
6. earthing arrangements
7. cables, accessories and equipment used in older electrical installations
8. pipework and ductwork, components and systems.

1 Building services engineering systems

1.1 Air-conditioning systems

Air-conditioning (AC) systems are used in both domestic and commercial properties to improve the comfort of occupants. They work by using the vapour-compression refrigeration cycle.

Refrigerants are fluorinated chemicals that are used in both liquid and gas states. They can, therefore, be classified as both liquid (when compressed) and gas (vapour). All refrigerants boil at extremely low temperatures, well below 0°C. When a refrigerant gas is compressed, it changes state to a liquid. During this process, a lot of heat and pressure are created. When the pressure is released quickly, it generates cold.

During the vapour-compression refrigeration cycle, the refrigerant vapour enters a compressor, which compresses it, generating heat. The compressed vapour then enters a condenser, where the useful heat is removed and the vapour condenses to a liquid refrigerant. From here, the liquid refrigerant passes through an expansion valve, where rapid expansion takes place. The warm liquid is converted to a super-cold vapour/liquid mix, which creates the refrigeration effect. The vapour/liquid mix passes through an evaporator, where final expansion to a vapour takes place. This vapour then enters the compressor for the cycle to begin again.

The ability of refrigerants to change their state quickly with such wide temperature changes allows them to be used in refrigeration plants, air-conditioning systems and heat pumps.

Key term

Refrigerant: a working fluid used in the refrigeration cycle of air conditioning systems and heat pumps

Health and safety

Fluorinated gases (F-gases) used in heat pumps and refrigeration systems can cause freeze burns. You must be suitably qualified to work on these systems

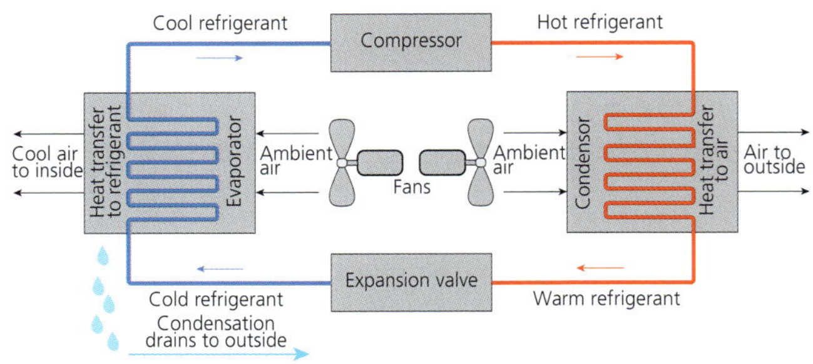

▲ Figure 12.1 An air-conditioning system

Test yourself

How does the vapour-compression refrigeration cycle work?

Types of air-conditioning system

Ducted systems

Ducted air-conditioning systems pump cooled air from a central cooling unit through a series of ducts which terminate in vents.

Components of a ducted system include:

- a circulation fan, which moves air to and from rooms
- an air-conditioning unit, which uses cooling and dehumidification processes in summer or heating and humidification processes in winter
- supply ducts, which direct conditioned air from the circulating fan to the space to be air-conditioned
- an air diffuser, which distributes the conditioned air evenly in the room
- return air grilles, which allow air to enter the return duct
- filters, which remove dust and bacteria from the air.

Split systems

Split-system air conditioners consist of inner and outer units connected by pipes, allowing refrigerant to flow to and from each unit. They can be fitted with a heat pump, ensuring they can be used all year round by providing both cool and warm air.

Single-split systems consist of an outdoor unit for every indoor unit. Multi-split systems allow numerous indoor units to be connected to a single, larger outdoor unit.

Variable refrigerant flow systems

Variable refrigerant flow (VRF) systems are air-cooled and refrigerant-based, using outdoor condenser units and indoor fan coil units in the same way as more traditional air-conditioning systems. However, instead of one large unit pumping air to the whole space, a VRF system incorporates several smaller air handlers which can be controlled individually and piped back to one system.

1.2 Electrotechnical systems

Electrotechnical systems carry, distribute and convert electrical power in a property for use in:

- socket outlets for appliances with a 3 A or 13 A three-pin plug
- fixed appliances, such as cookers, showers and immersion heaters
- lighting systems
- protection services, such as intruder alarms, surveillance systems, fire alarms and access controls
- refrigeration and ventilation systems
- telecommunication systems, such as telephones, internet, home entertainment and connections for other BSE systems
- heating systems, such as gas boilers, electrical wall heaters and fan convectors.

Health and safety

To avoid electrocution, always follow the safe isolation procedure when working on electrical systems.

Test yourself

What is the purpose of an electrotechnical system?

Research

Research the key requirements of the IET Wiring Regulations and produce a table. You could use the following headings in your table:

- Part 1 Scope
- Part 4 Protection for safety
- Part 5 Selection and erection of equipment
- Part 6 Inspection and testing
- Part 7 Special installations

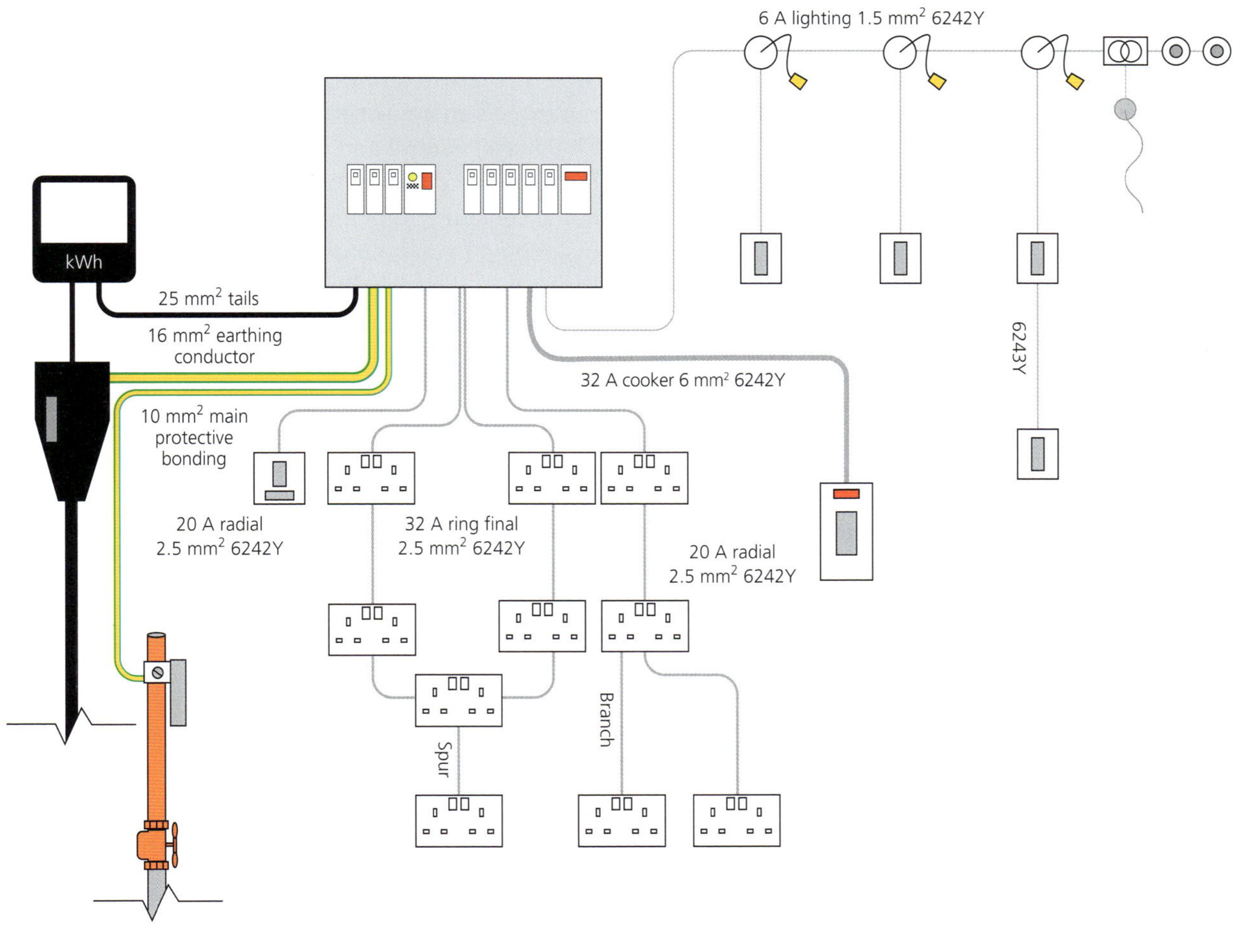

▲ Figure 12.2 Typical wiring diagram

Lighting

Lighting circuits are generally rated at 6 A but can, in some installations, be rated at 10 A or 16 A. The ratings are based on calculation of the number of lights and the type of lights included in the circuit. Consideration must be given to the type of lighting point used for higher-rated circuits.

Lighting circuits are intended to supply lighting points, but in some cases they may also supply small power electrical equipment, such as bathroom fans and shaver supply units.

Lighting circuits can be wired in different ways:

- three-plate – commonly used in domestic properties for circuits wired in **composite cable**, such as thermosetting insulated and sheathed flat-profile cable
- conduit method – used for circuits wired in single-core cables within a suitable containment system.

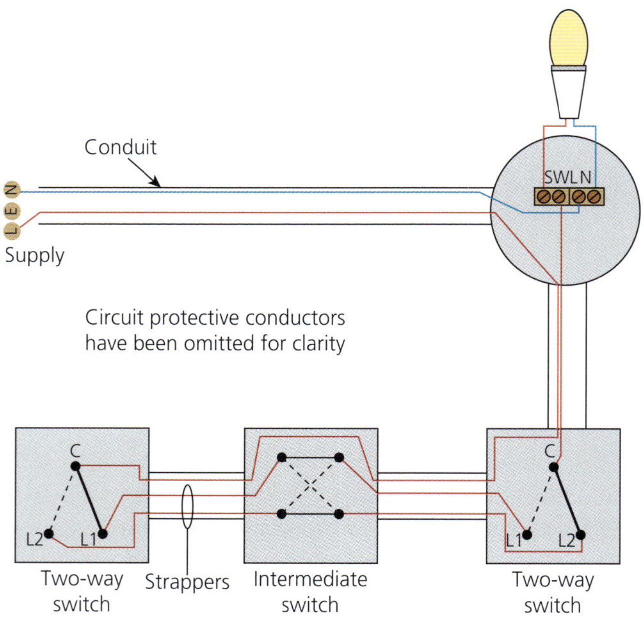

▲ Figure 12.3 Three-plate lighting circuit

Key term

Composite cable: multi-core cable, in which the cores are surrounded by a sheath that provides mechanical protection

Consumer unit: a component in an electrical system which contains the main switch, isolation for circuits and circuit protection devices

Miniature circuit breaker (MCB): a small trip switch operated by an overload, used to protect individual electrical circuits

Test yourself

Explain how lighting circuits may be wired.

Power

Power circuits generally supply socket outlets but may also supply individual appliances. They may be wired in two ways:

- ring final
- radial.

Ring final circuits provide socket outlets for appliances and are protected at the **consumer unit** with a 30 A or 32 A **miniature circuit breaker (MCB)**. They should not exceed an area of 100 m² in domestic installations. The phase, neutral and circuit protective conductors (CPCs) are connected to their dedicated terminals at the consumer unit and then form a ring circuit by looping in and out of the respective terminals at the back of the sockets, eventually returning to the point from where they began. There is no limit to the number of sockets that can be installed on a ring final circuit if proper design considerations are adhered to. Kitchens and utility rooms may require their own dedicated circuits.

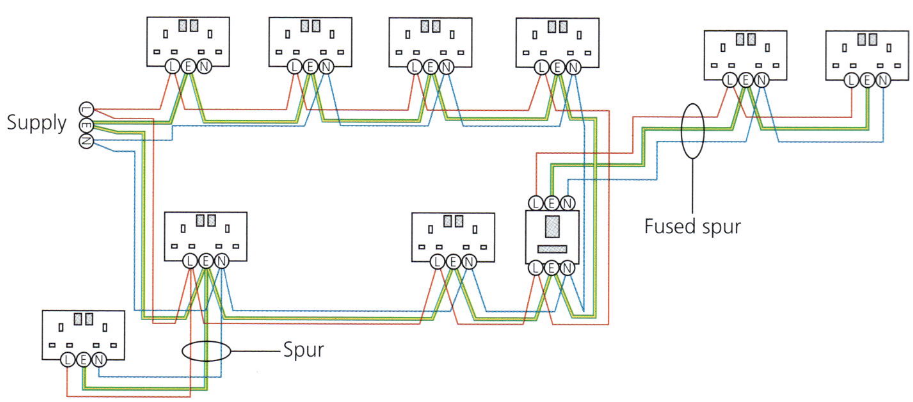

▲ Figure 12.4 Ring final circuit showing spurs

Radial circuits may be used to supply multiple socket outlets. They do not return to the consumer unit (unlike ring final circuits) and are protected at the consumer unit with a 20 A MCB.

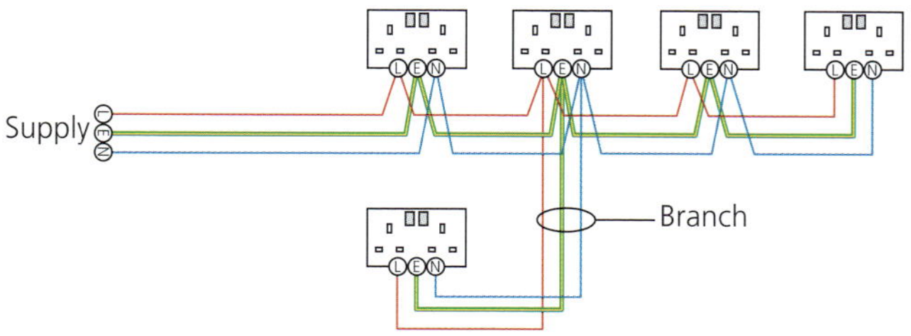

▲ Figure 12.5 Radial circuit showing branch

Heating and appliances

Radial circuits can supply individual appliances or dedicated fixed-appliance circuits, such as boilers, immersion heaters, cookers and showers. They can have any rating, depending on the cable size, but typical ratings for heating appliances include:

- 16 A for single appliances such as dishwashers
- 16 A for dedicated boiler circuits
- 16 A for immersion heaters
- 32 A for cooker installations.

Electric showers are rated between 7 and 10.5 kW. This means that a 10.5 kW shower requires a 45 A electrical circuit to supply it.

Data and control

Data cabling is cabling within a building installed to support multiple hardware systems, such as:

- telephones
- internet and television
- **CCTV**
- music servers
- printers
- **VoIP** systems.

The system contains all the wiring to transmit data around the building from routers and data cabinets to individual data points. Care must be taken when installing data and control cabling within a building to prevent **crosstalk**.

Key terms

CCTV: closed-circuit television, also known as video surveillance, is the use of video cameras to transmit a signal to a specific place, on a limited set of monitors

VoIP: voice over internet protocol, technology used to deliver voice and multimedia via an internet connection

Crosstalk: interference between telecommunication signals

▲ Figure 12.6 Data point

Case study

Phoebe and her family have just purchased a new property and are considering installing multiple hardware systems for communication and entertainment.

1. What are their options for telephones?
2. How are music servers connected to the system?
3. What smart devices can be incorporated?

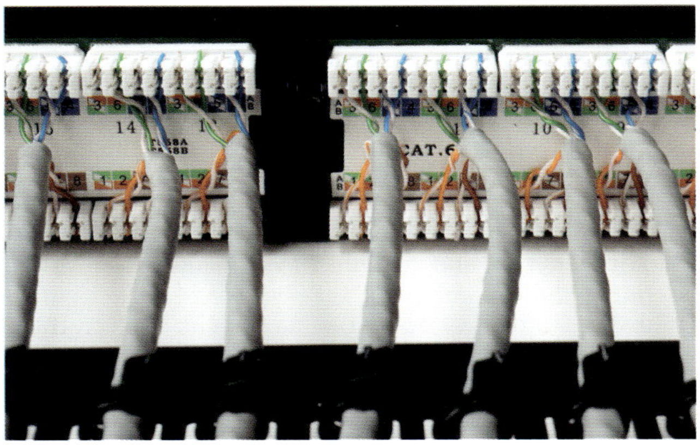

▲ Figure 12.7 Data cable panel

1.3 Gas systems

Gas systems supply gas to a property as a source of fuel for heating and cooking.

There are two different types of gas:

- Natural gas is a naturally occurring hydrocarbon gas mixture, consisting primarily of methane. It is formed when layers of decomposing plant and animal matter are exposed to intense heat and pressure under the surface of the Earth over millions of years.
- Liquefied petroleum gas (LPG) is a by-product of crude oil. It was developed with the compactness and portability of a liquid but can be readily used as a gas. It is mainly associated with leisure applications but also used in homes where no natural gas supply is available.

Health and safety

Gas is extremely flammable and can cause explosions. Care should be taken when working on gas systems and correct procedures must be followed.

In the UK, gas is extracted from the Irish and North seas from offshore drilling rigs. It is then compressed and transported across the UK in a system of pipes known as the transmission and distribution network. The gas passes through several pressure-reduction tiers in preparation for entry to properties.

Components of a gas system include:

- an emergency control valve, which isolates the gas supply
- a meter governor, which reduces/regulates the incoming gas pressure
- a meter, which measures the amount of gas used for billing purposes
- installation pipework, which supplies gas to all appliances in a property.

Industry tip

In order to install and work on gas appliances, you must be registered with Gas Safe.

Gas installation pipework must be adequately sized to meet the requirements of the appliances it serves. A typical small domestic installation has 22 mm pipework from the meter supplying appliances.

Research

Research suitable pipework and jointing methods for gas installations. List the different types and identify which jointing methods are not suitable.

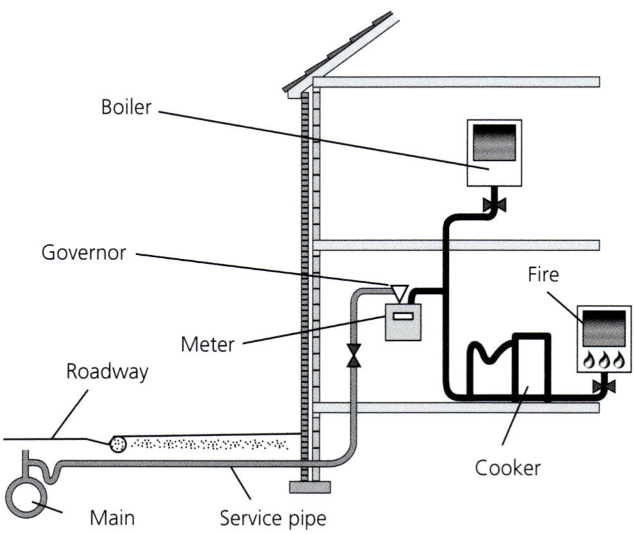

▲ Figure 12.8 Gas supply to a property

Gas boilers

Gas boilers are the most popular of all central-heating appliances. Over the years, there have been many different types, from large, multi-sectional, cast-iron domestic boilers to small, low-water-content condensing types.

Boilers that can be used with wet central-heating systems fall into distinct categories:

- Traditional boilers do not contain any form of expansion vessel or operational controls, such as a pump or filling loop. This is the simplest type of boiler, a basic heat source and heat exchanger. It requires other controls and components to form part of a functional system.

- A traditional boiler does not give instantaneous hot water. It must be installed in conjunction with a hot water storage system and heating system. It contains an expansion vessel, filling loop and pressure-relief valve, and does not require a feed and expansion cistern.
- System boilers have all the necessary safety and operational controls fitted directly to them. There is no need for a separate expansion vessel, pressure-relief valve or filling loop, and this makes the installation much simpler.
- Combination boilers provide central heating and instantaneous hot-water supply from a single appliance. Modern combination boilers are very efficient and contain all the safety controls (i.e. expansion vessel and pressure-relief valve) of a sealed system.
- Condensing boilers work by extracting heat from flue gases produced when natural gas is combusted (CO2, nitrogen and water vapour). These flue gases pass over two heat exchangers: the primary heat exchanger extracts about 80 per cent of the heat from the gases and the secondary heat exchanger extracts a further 12–14 per cent of the heat. In the secondary heat exchanger, the water vapour in the flue gases condenses to form water droplets, which are collected in a condensate trap before falling to drain via a condensate pipe. Condensing boilers produce a distinctive 'plume' of water vapour during operation. All of the boiler types listed here can be the condensing type.

▼ Table 12.1 Types of gas boiler

	Energy efficient	Cast-iron heat exchanger	Low water content	Open vented system	Sealed (pressurised) system	Open flue	Room sealed (natural draught)	Room sealed (fan assisted)	Wall mounted	Free-standing
Traditional boilers	✗	✓	✓	✓	✓	✓	✓	✓	✓	✓
Condensing boilers	✓	✗	✓	✗	✓	✗	✗	✓	✓	✓
System boilers	✓	✗	✓	✗	✓	✗	✗	✓	✓	✗
Combination boilers	✓	✗	✓	✗	✓	✗	✓	✓	✓	✓

Improve your English

Write a letter to a potential customer who is looking to upgrade their boiler, informing them of the types available.

Research

Research the operating principles of condensing boilers and produce a flowchart.

All central-heating appliances require a flue to remove the products of combustion (POC) safely to the outside. The basic concept is to produce an updraught, whether by natural means or using a fan, to eject fumes away from the building. There are two types of flue system:

- open
- room-sealed.

An open flue system is the simplest. Because heat rises, it relies on the heat of the flue gases to create an updraught. There are two different types, natural draught and forced draught.

An open flue is made up of four components:

- primary flue: creates the initial pull to clear the products of combustion
- draught diverter: draws air in to dilute the products of combustion
- secondary flue: carries the flue gases to the terminal
- terminal: allows the flue gases to evacuate to the atmosphere.

Natural-draught systems take the POC from within the room where the appliance is sited and expel them through the flue terminal using the draught created by the different densities of the flue gases and the colder air outside. There is no fan in this type of system.

Room-sealed flue systems draw air for combustion directly from the outside through the same flue assembly used to discharge the POC. They are safer than open-flue systems, since there is no direct route for the POC to spill back into the room. There are two basic types: natural draught and forced draught (fan-assisted).

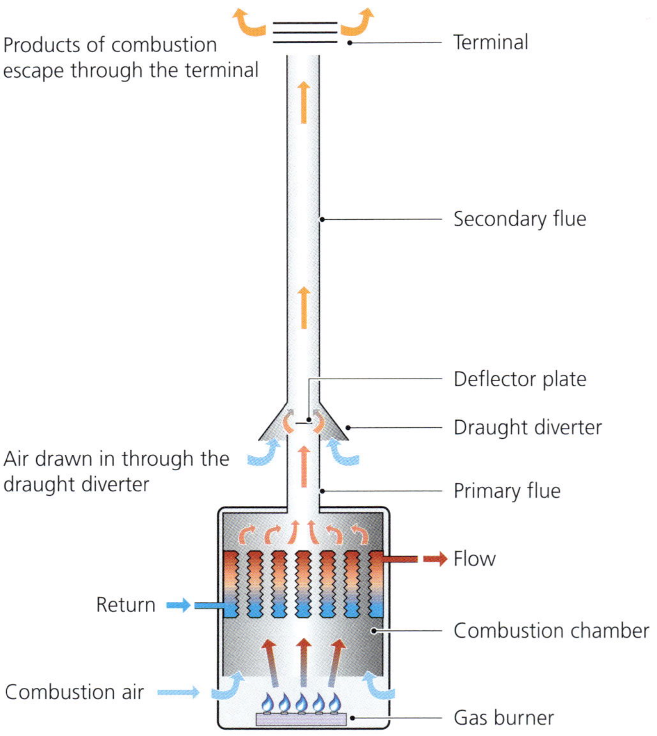

▲ Figure 12.9 Open flue

Gas fires

Gas fires are space-heating appliances used to heat the rooms in which they are installed. There is a wide range of fires available, with different flue options:

- Flueless (type A): this type of appliance is not intended for connection to a flue or any device for evacuating the POC outside the room in which the appliance is installed. The air for combustion is taken from the room, and the POC are released into the room.
- Open-flue (type B): this type of appliance is intended to be connected to a flue that evacuates the POC outside the room containing the appliance. The air for combustion is taken from the room via a series of vents.
- Room-sealed (type C): the air supply, combustion chamber, heat exchanger and evacuation of POC for this type of appliance are sealed with respect to the room in which the appliance is installed.

Types of gas fire include:

- radiant convector: a type of fire which relies on ceramics near the burner to radiate heat into the room.
- inset live fuel effect (ILFE): a type of fire which is sealed to the builder's opening within the chimney of a building. Heat is radiated from the fuel bed and produced by convection through the heat exchanger.

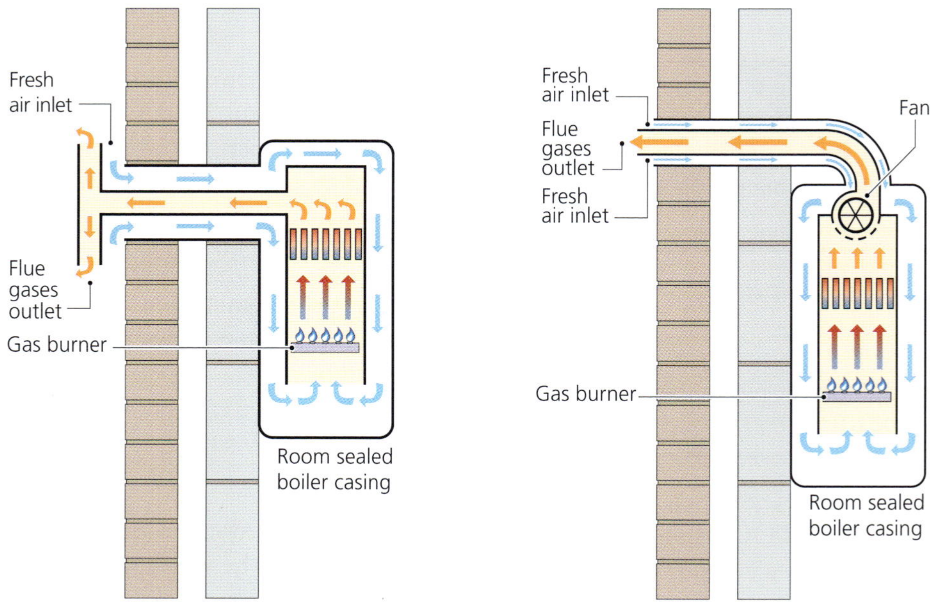

▲ Figure 12.10 Room-sealed flue systems: natural (left) and fan-assisted (right)

- decorative fuel effect (DFE): a type of fire which incorporates radiants which are made to look like coal, wood or stone placed inside or underneath an open chimney.
- flueless catalytic: a type of fire which is not connected to a chimney or flue system. Products of combustion are discharged directly into the room.

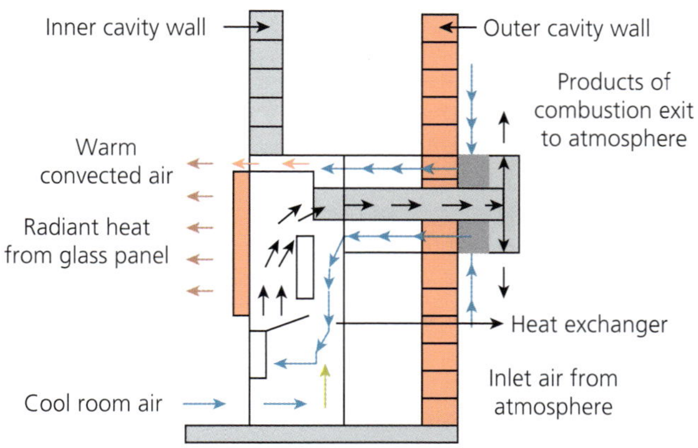

▲ Figure 12.11 Room-sealed space heater

Gas cooking appliances

Gas cooking appliances include the following:

- Freestanding cookers rest directly on the floor and comprise a hotplate with one or more burners, one or more ovens with or without thermostats, and possibly a grill and/or griddle.
- Grills cook food by means of radiant heat. They can be an integral part of a freestanding cooker or individual appliances.
- Hobs (hotplates) comprise one or more covered/uncovered burners designed to support cooking vessels.
- Ovens are closed compartments for roasting and baking food.

> **Research**
>
> Research the gas connection requirements for cooking appliances and produce a simple diagram for each type of appliance, showing the:
>
> - height of the cooker point
> - connection method.

1.4 Heating systems

Central-heating systems distribute warmth throughout the whole or part of a building from a single heat source (the boiler), for the **thermal comfort** of the occupants. Boilers can be fuelled by coal, gas, oil or electricity; they transfer their heat energy to another medium, usually water or air, which carries the heat to the areas where it is needed. Water-based systems are known as wet systems, while air- or electric-based systems are referred to as dry systems.

There is no standard heating system; all systems are tailored to suit the individual installation requirements, such as property type, property construction and fuel availability.

> **Key term**
>
> ***Thermal comfort:*** a person's satisfaction with the thermal environment (whether they feel too hot, too cold or just right)

> **Health and safety**
>
> Water in central-heating systems can be hot. Take care when draining down central-heating systems to prevent scalding.

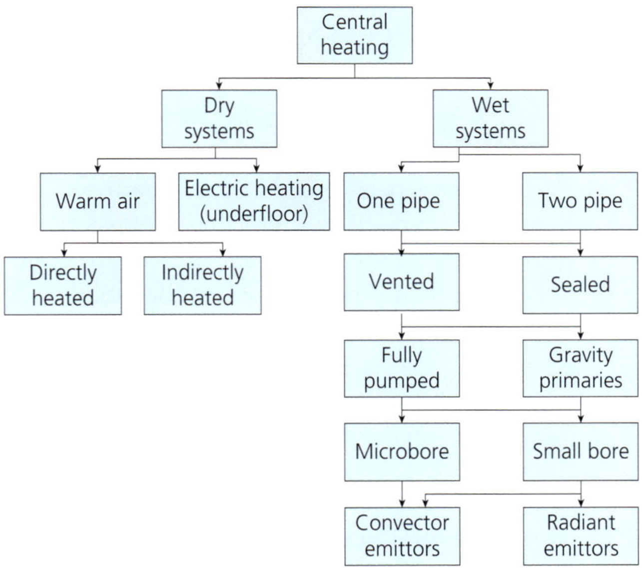

▲ Figure 12.12 Different types of central-heating system

Test yourself

Explain the following types of central-heating system:

- Full
- Background
- Selective
- Wet
- Dry

Domestic heating systems

Domestic wet central-heating systems contain several components:

- heat source (boiler/heater)
- heat emitters (radiators/underfloor heating circuits)
- thermostats for temperature control
- time control to set on/off periods.

Test yourself

List the key components of a wet central-heating system.

There are three categories of central-heating system:

- Full central heating provides heating in all rooms to a comfortable temperature.
- Background heating heats the property to a lower temperature than full central heating.
- Selective heating heats only parts of the property as required.

Heating systems fall into two different categories, based on the way the system is filled with water and the pressure at which it operates:

- Low-pressure, open-vented central-heating systems are supplied by a feed and expansion cistern located in the roof space, or in a tank room in a commercial property. They can be modern, fully-pumped systems or existing **semi-gravity systems**.
- Sealed, pressurised central-heating systems are fed directly from the mains cold-water supply and incorporate an expansion vessel to take up the expansion of water as it is heated. A filling loop is required to charge the central-heating system with water.

As mentioned above, combination boilers provide both central heating and instantaneous hot-water supply from a single appliance. They contain all the safety controls of a sealed system.

Key term

Semi-gravity system: a central-heating system that has pumped heating circulation but gravity hot-water circulation

Test yourself

Explain the difference between open-vented and sealed central-heating systems.

Research

Research the layout requirements for the following central-heating systems and note their advantages and disadvantages:

- C-plan
- S-plan
- S-plan+.

Industry tip

Properties which exceed 150 m^2 should be split into multiple heating zones with individual time and temperature controls.

Industrial/commercial heating systems

Industrial/commercial wet central-heating systems operate following the same principles as domestic systems. However, as the heat requirements are much greater, a larger single boiler or multiple boilers may be required. The boilers heat water, which is pumped around the building, giving off heat via radiators, underfloor heating, unit heaters or trench heating. As with a domestic system, time and temperature controls are required, with the addition of separate heating zones to control temperature throughout the building.

This system uses a single three-port motorised mid-position valve to control the flow of water to the central heating circuit and the hot water circuit. It is controlled by a cylinder thermostat and a room thermostat. Individual thermostatic radiator valves independently control the temperature of each room.

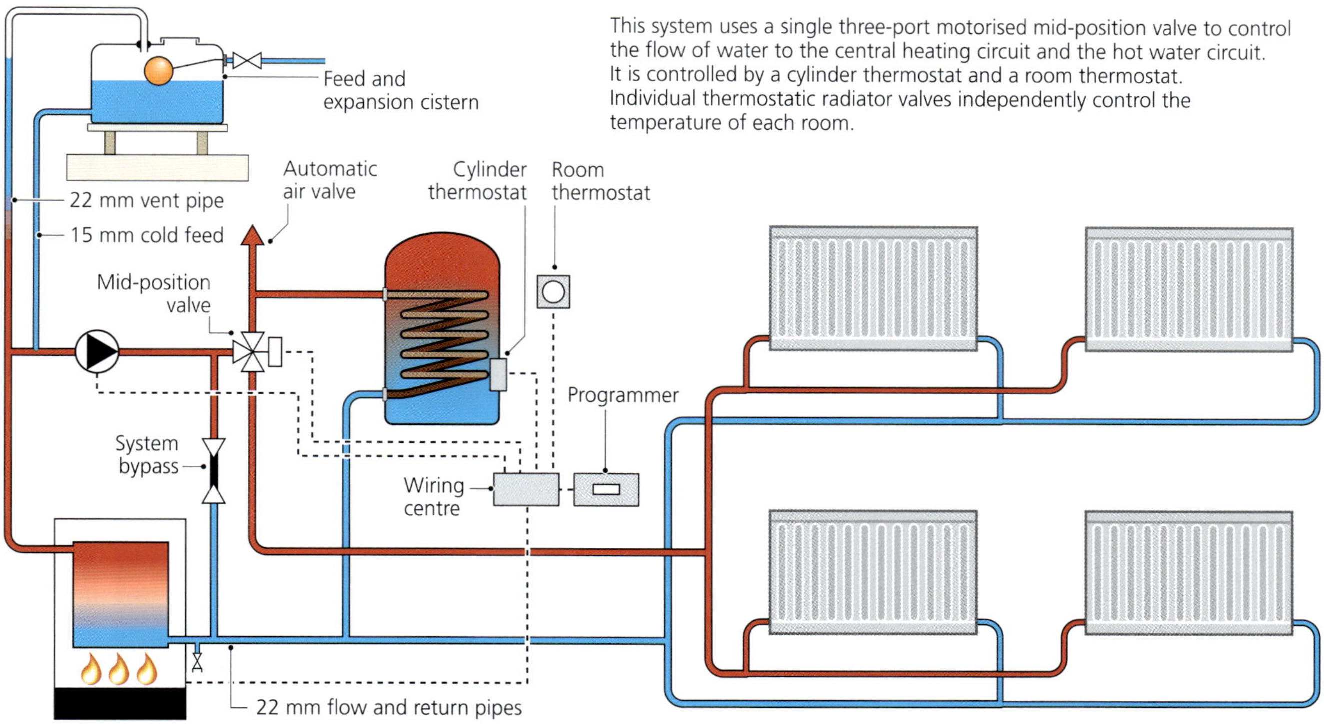

▲ Figure 12.13 Low-pressure central-heating system

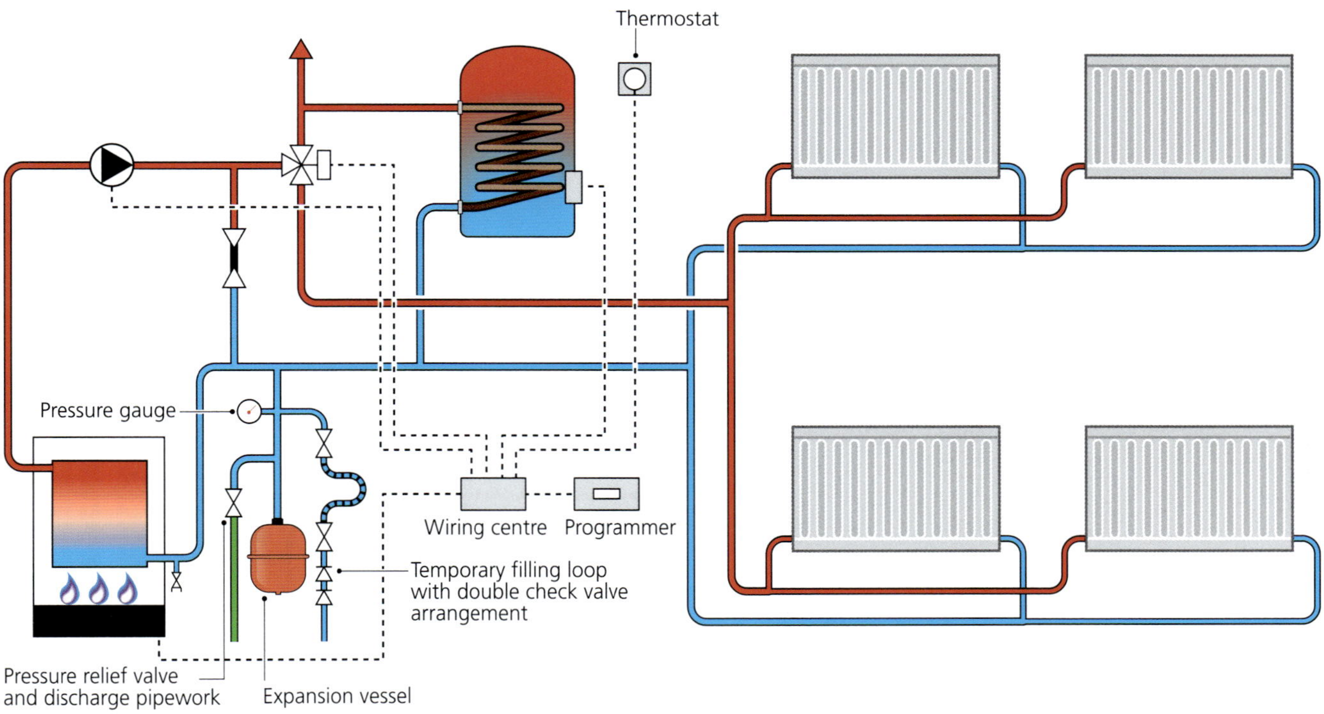

▲ Figure 12.14 Sealed central-heating system

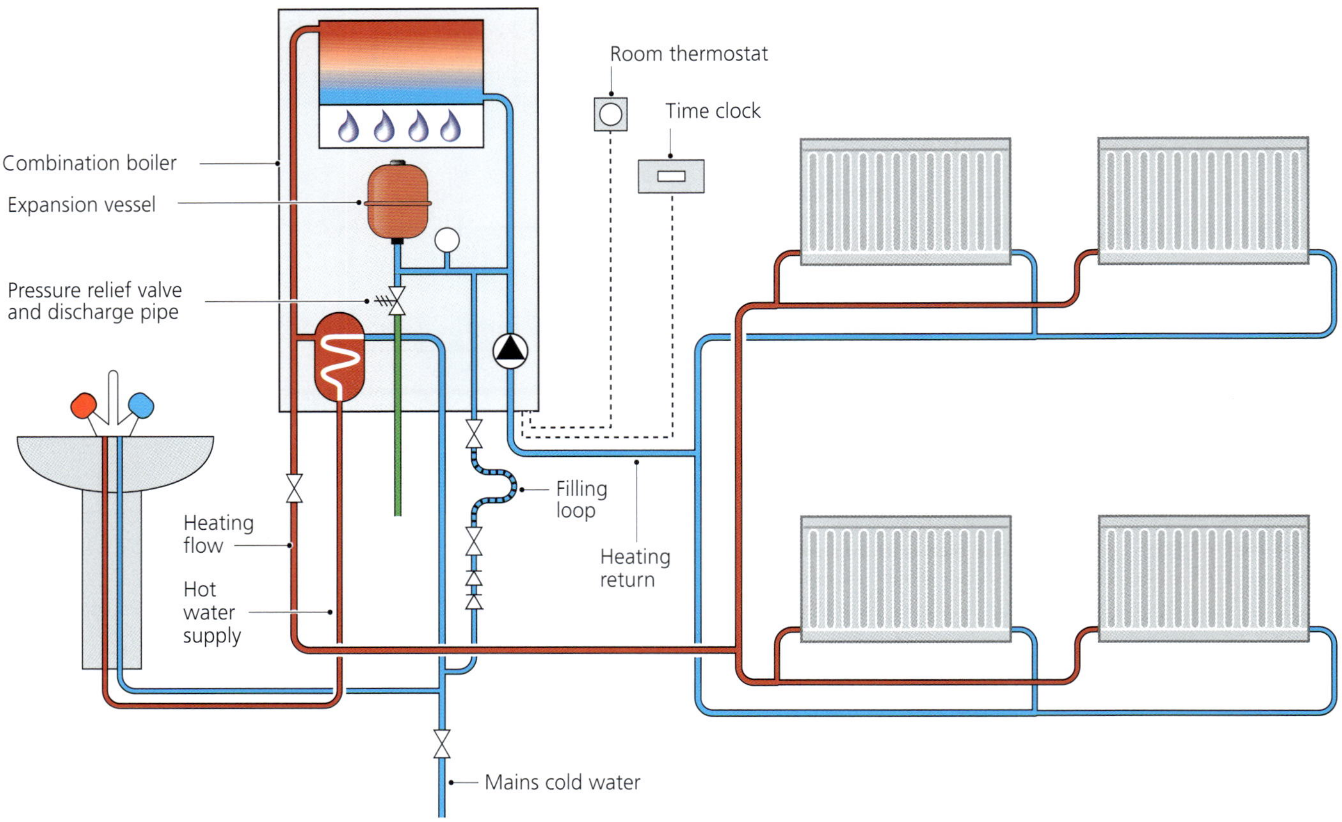

▲ Figure 12.15 Combination boiler system

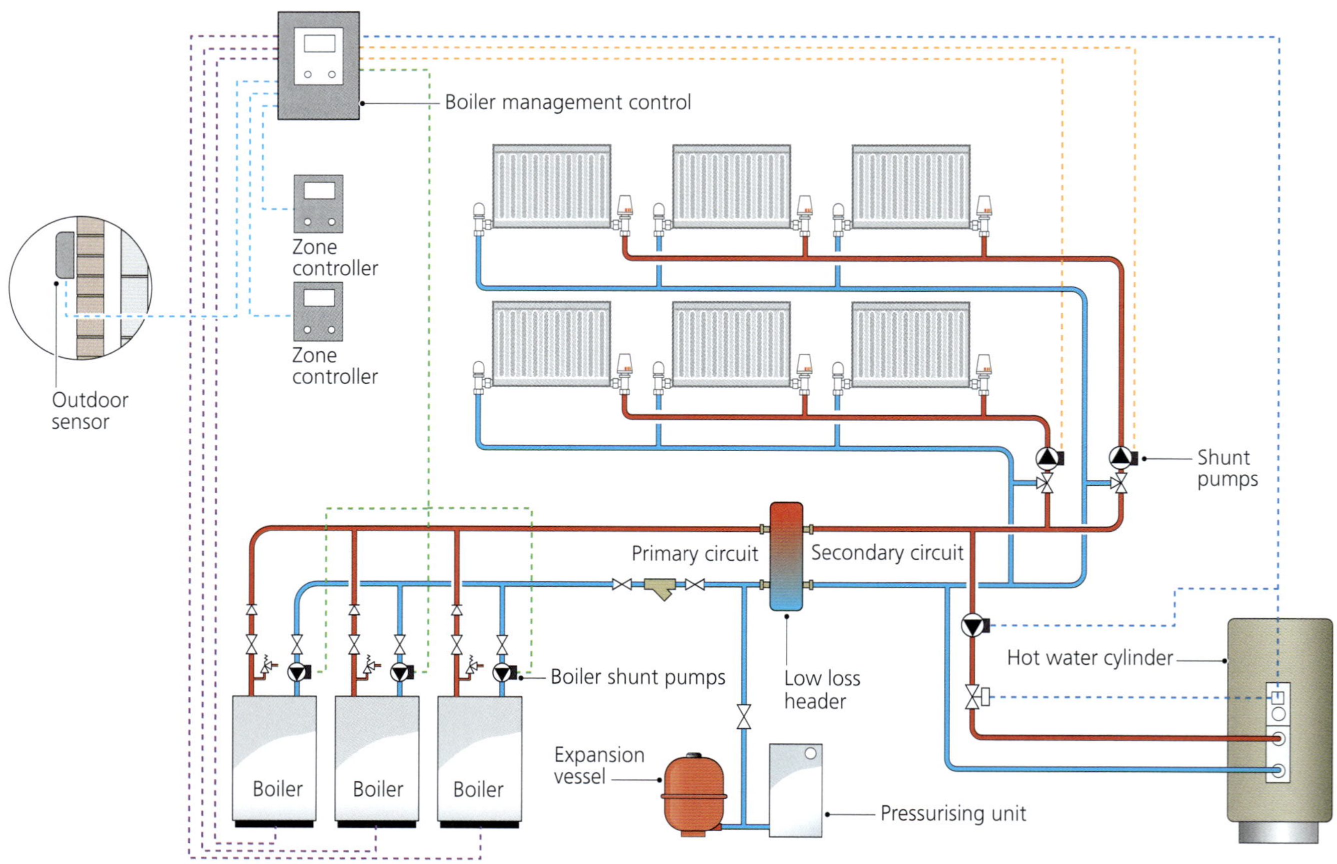

▲ Figure 12.16 Multiple boilers may be needed in industrial or commercial heating systems

Industrial/commercial electric boilers are compact, electrically powered heating devices designed for commercial and industrial buildings. They are connected to a three-phase, 400 V electricity supply. They heat water, which is pumped around the building, giving off its heat via heat emitters.

Warm-air heaters are also commonly used in industrial and commercial buildings. They are designed to be installed to heat the whole space. Alternatively, radiant tube heaters may be used to heat a specific zone of the building and are usually supplied by natural gas.

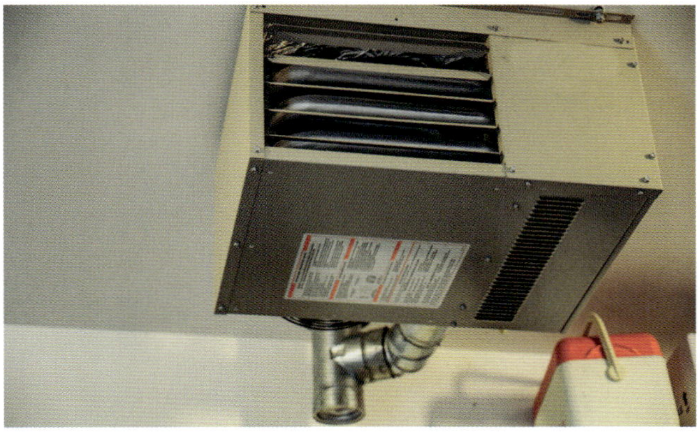

▲ Figure 12.17 Warm-air heater

1.5 Plumbing systems

Cold-water systems

Cold-water systems provide fresh, **wholesome water** to properties. This water can be used directly from the water main for cooking and drinking purposes or can be stored for use within other systems, for example to flush toilets or to supply hot-water and central-heating systems.

There are two different types of water supply to a property:

- Private water supplies are covered by the Private Water Supplies (England) Regulations 2016. The water is sourced close to the property from streams, rivers, natural springs or boreholes. It is supplied to the property via a pump and treated to ensure it is safe for drinking and hygiene purposes.
- Public/municipal water supplies come from the water undertaker's main. They are covered by the Water Supply (Water Fittings) Regulations 1999.

Key term

Wholesome water: water that is fit to use for drinking, cooking, food preparation or washing without danger to human health

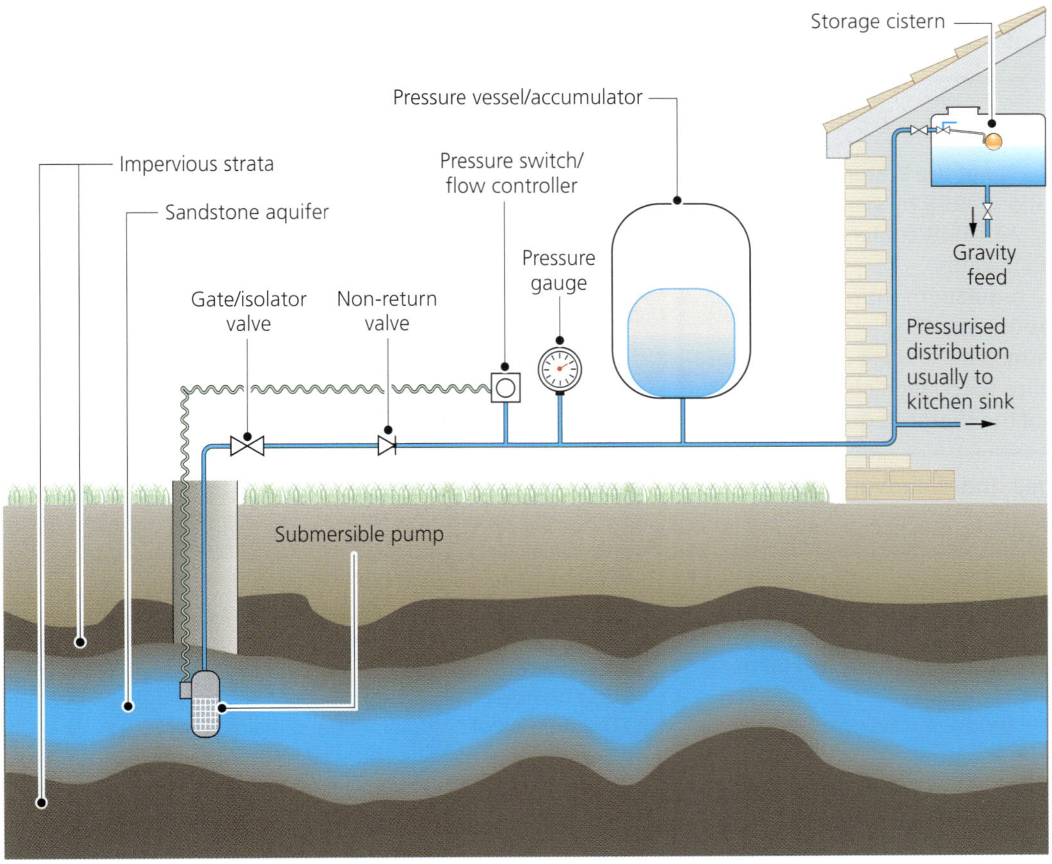

▲ Figure 12.18 Private water supply

The collection, treatment and supply of wholesome water to consumers is the responsibility of the water undertakers. The water is supplied to homes via a grid-system network of pipes which runs underneath the roads.

The connection between the mains water supply and a dwelling is made via a **ferrule**, which removes the need to isolate the supply.

The water supply from the water main into the building comprises two separate pipes:

- The communication pipe is owned and maintained by the water undertaker and leads to the boundary stop tap.
- The service pipe is owned and maintained by the owner of the building. It must be installed at a minimum depth of 750 mm and a maximum depth of 1350 mm.

Key term

Ferrule: a type of fitting for joining two pipes together; it allows a new connection to a communication pipe to be made without having to isolate the water supply

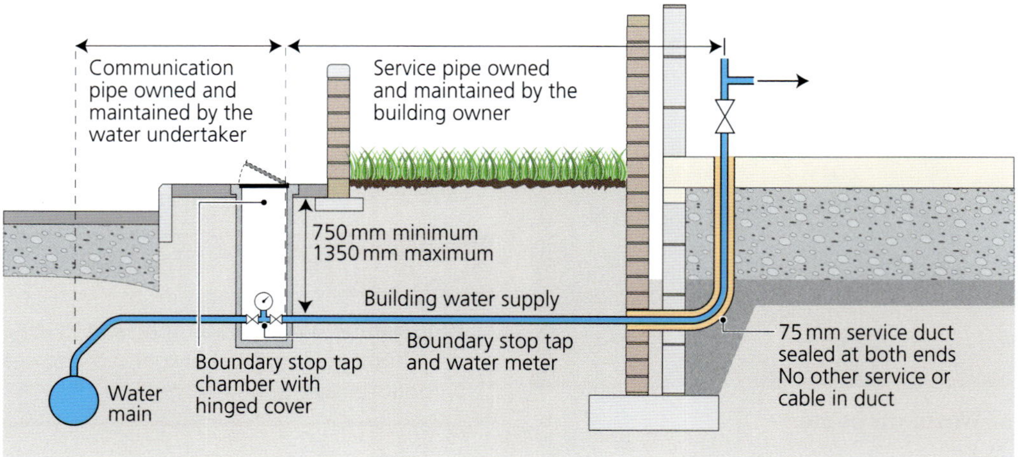

▲ Figure 12.19 Public water supply to a building

Research

Research the requirements of the Water Supply (Water Fittings) Regulations 1999 by visiting www.legislation.gov.uk/uksi/1999/1148/contents/made

Prepare notes on the following:

- What date were the regulations introduced?
- Who enforces the regulations?
- What does the term 'material change of use' mean?
- What are the penalties for convening regulations?
- What are the aims of the Water Regulations?

Direct cold-water systems

In a direct cold-water system, all cold-water taps are fed directly from the mains supply. This means that all taps are provided with wholesome water. Storage is only required for supplying cold water to the hot-water system via a cistern where an instantaneous hot-water heater or combi boiler is not fitted. This system is designed to be used in areas with high water pressure.

▼ Table 12.2 Advantages and disadvantages of direct cold-water systems

Advantages	Disadvantages
• Cheaper to install • Drinking water at all terminal fittings • Less pipework • Less structural support required in roof space for the cold-feed cistern • More suitable for instantaneous showers, hose taps and mixer fittings • Used in conjunction with a high-pressure (unvented) hot-water supply • Smaller pipe sizes may be used in most cases • Good pressure at all cold-water outlets	• Pressure may drop at times of peak demand • Property has no water if the mains are under repair • Any leak in the premises will cause a great deal of damage due to high pressure • Can be noisy • Greater risk of contamination to mains • Greater wear on taps and valves • More problems with **water hammer** • Greater risk of condensation build-up on the pipework

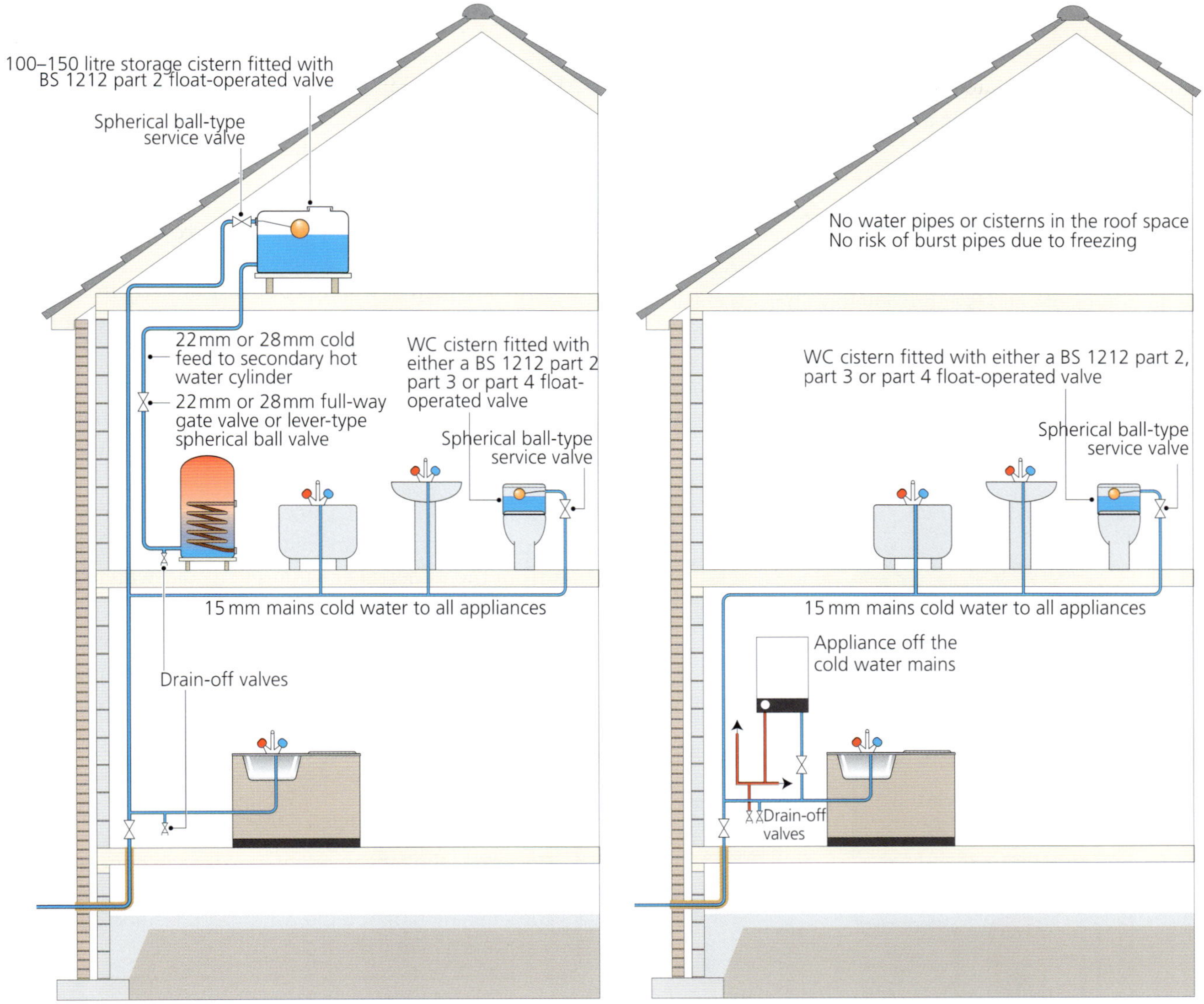

▲ Figure 12.20 Direct cold-water system (left) and direct cold-water system with combi boiler or instantaneous hot-water heater (right)

Key term

Water hammer: a knocking sound in water pipes caused by fast-flowing water being stopped abruptly when terminal fittings are closed off

Improve your maths

Produce a simple line diagram of a direct cold-water system including components. Use a manufacturer's catalogue to produce a materials list and price up the components.

Indirect cold-water systems

In an indirect cold-water system, only wholesome water outlets such as kitchen sinks are fed directly from the mains cold-water supply. The other outlets are fed indirectly via a cold-water storage cistern in the roof space. This system is designed to be used in areas with low water pressure, where the mains-supply pipework is not capable of supplying the full requirement of the system.

In larger buildings, such as office blocks, factories and hotels, it is preferable for all water except drinking water to be supplied indirectly via a protected storage cistern. This ensures there is a backup supply of water to flush toilets and provides hot water services in the event of disruption to the mains supply.

Industry tip

Cold-water storage cisterns should be installed as high as possible within a property to increase the system pressure.

Improve your maths

10 metres of head is approximately 1 bar. Calculate the approximate pressures for the following:

- a cistern installed 7 m above a terminal fitting
- a cistern installed 12 m above a terminal fitting.

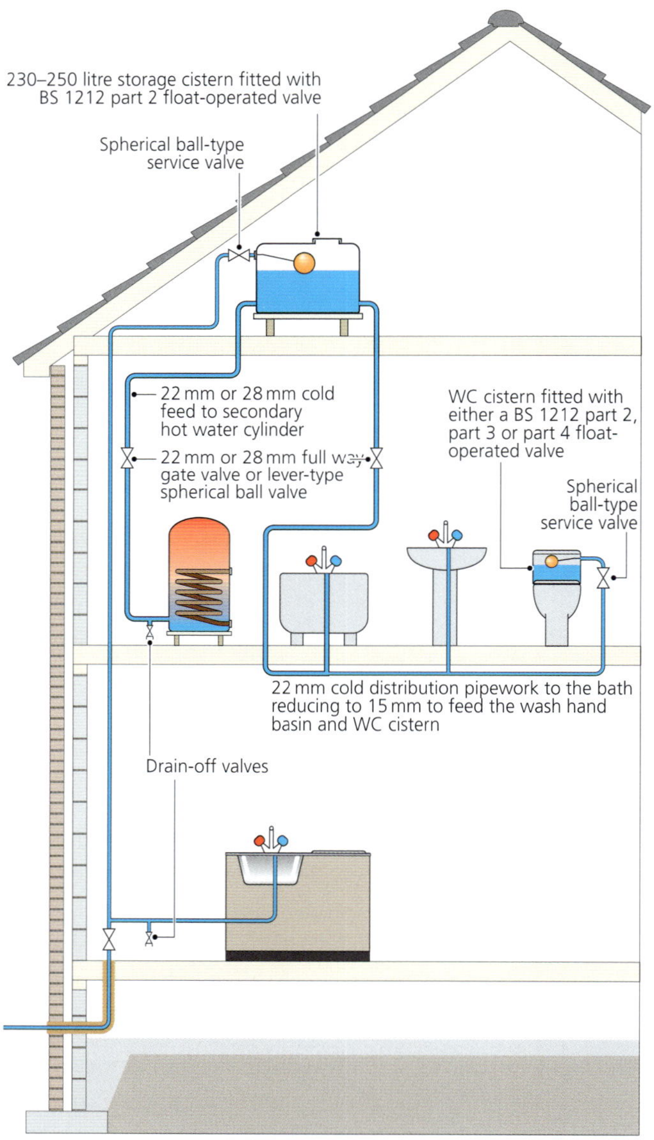

▲ Figure 12.21 Indirect cold-water system

Boosted cold-water systems

Typical mains water pressure is between 3 and 7 bar. Multi-storey buildings are too tall to be supplied using this pressure and, after taking into consideration frictional loss when pipe sizing, it may be necessary to boost their cold-water supply.

There are several types of boosted cold-water (BCW) system:

- Direct boosted: water is boosted directly from the undertaker's main to a cold-water storage cistern at the upper level.
- Direct boosted with drinking-water header: water is boosted directly from the undertaker's main to a cold-water storage cistern and drinking-water header.
- Indirect boosted to a storage cistern: water is boosted from a break cistern with water supplied from the undertaker's main to a cold-water storage cistern at the upper level.
- Indirect boosted with a pressure vessel: water is boosted from a break cistern with water supplied from the undertaker's main to individual cold-water storage cisterns on various floors.

Key term

Backflow: the movement of liquid in the opposite direction to its regular flow; this can lead to contamination of potable water supplies and create a serious health risk

Industry tip

Most shower mixers require an equal head of pressure for both hot- and cold-water supplies to ensure correct operation.

▼ Table 12.3 Advantages and disadvantages of indirect cold-water systems

Advantages	Disadvantages
• Reduced risk of water hammer and noise • Constant low-pressure supply reduces risk and rate of leakage • Suitable for supply to mixer fittings for vented hot-water supply • Reserve supply of water available in case of mains failure • Less risk of **backflow** – fewer fittings supplied directly • Showers may be supplied at equal head of pressure • Reduces demand on main at peak periods • Can be sized to give greater flow rate	• Supply pipe must be protected against backflow from cistern • Risk of frost damage in the roof space • Structural support needed for the cistern • Space taken up in the roof space • Increased cost of installation • Reduced pressure at terminal fittings

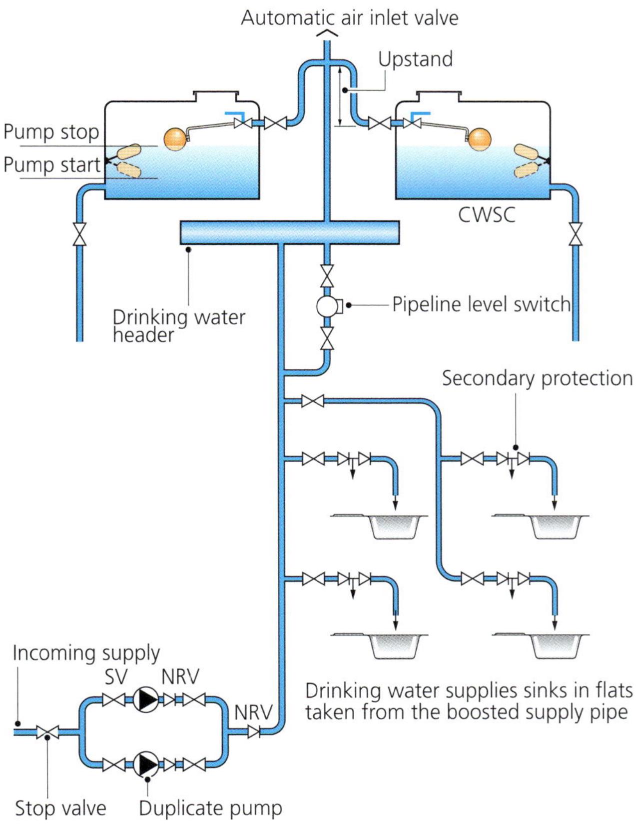

▲ Figure 12.22 Direct boosting to a drinking-water header and duplicate cisterns

Research

Research the typical layout of boosted cold-water systems used in multi-storey buildings.

Hot-water systems

Hot-water systems are used for both personal hygiene and for cleaning purposes. They should provide an adequate flow of water at the correct temperature required by the users of the building.

Water can be heated in a property by:

- burning fossil fuels (coal, oil or gas)
- using electricity
- using a solar thermal system.

Research

Visit www.planningportal.co.uk/info/200135/approved_documents and look for the following information:

- Which approved document provides guidance on hot-water supply to a property?
- What are the requirements for the provision of hot-water systems?
- When designing a hot-water system, what temperature must it not exceed?

There are two categories of hot-water system:

- centralised, where hot water is supplied from a central source within the building and supplies several outlets throughout the building
- localised, where the hot water is heated and supplied at the point of use.

These systems can be divided into two further categories:

- storage, where hot water is stored at temperature in a vessel/cylinder
- instantaneous, where water is heated on demand.

Test yourself

What is meant by the terms 'centralised' and 'localised' in relation to hot-water systems?

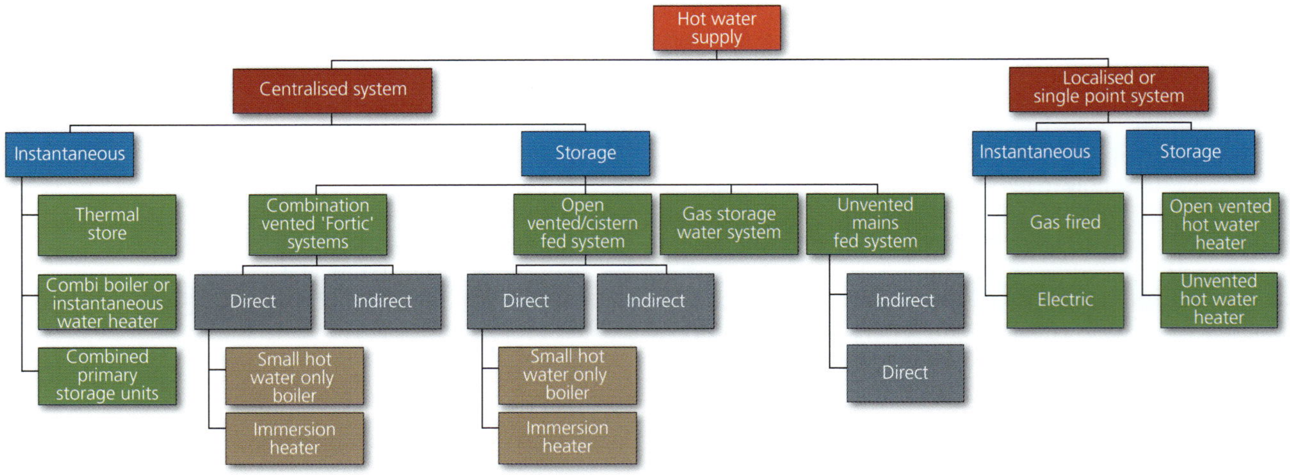

▲ Figure 12.23 Types of hot-water system

Industry tip

The type of hot-water system required for a property will depend on the:

- type and size of the property
- number of occupants
- number of hot-water outlets
- distance from the source
- type of fuel used to heat the water.

Centralised hot-water systems

Open-vented hot-water storage systems are heated by either a boiler or an electric **immersion heater**. Water is stored in a hot-water cylinder, which is fed from a cold-water storage cistern. Hot-water outlets are supplied via a system of copper or plastic pipework, sized at a minimum of 15 mm. The system contains an open vent pipe, which acts as a safety relief in the event of overheating. Water is stored at 60–65°C and should be distributed at no less than 55°C.

There are two types of open-vented hot-water system:

- direct, which contains no form of **heat exchanger**
- indirect, which contains a heat exchanger to heat up the secondary water.

Key terms

Immersion heater: an electrical element that sits in a body of water; when switched on, the electrical current causes it to heat up, which in turn heats up the surrounding water

Heat exchanger: a device or vessel that allows heat to be transferred from one water system to another, without the two water systems being allowed to come into contact with each other

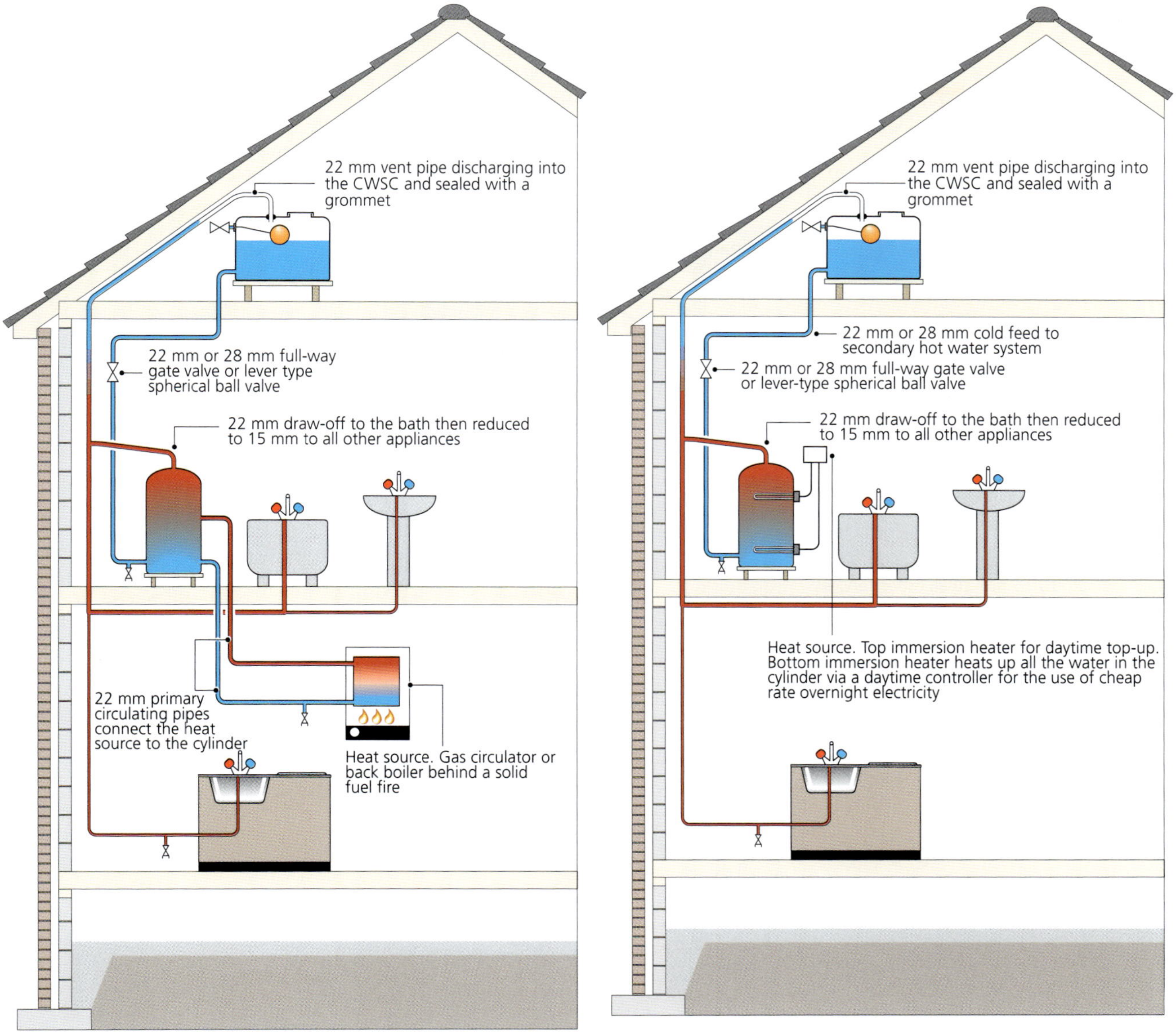

▲ Figure 12.24 Direct hot-water system (left) and direct hot-water system with immersion heaters (right)

▼ Table 12.4 Advantages and disadvantages of direct hot-water systems

Advantages	Disadvantages
• Quick heat-up of water • Cheap to install	• Risk of rusty water being drawn off at the taps if the wrong type of boiler is used • High risk of scale build-up in hard-water areas if water temperature exceeds 65°C • High risk of scalding because of the lack of thermostatic control

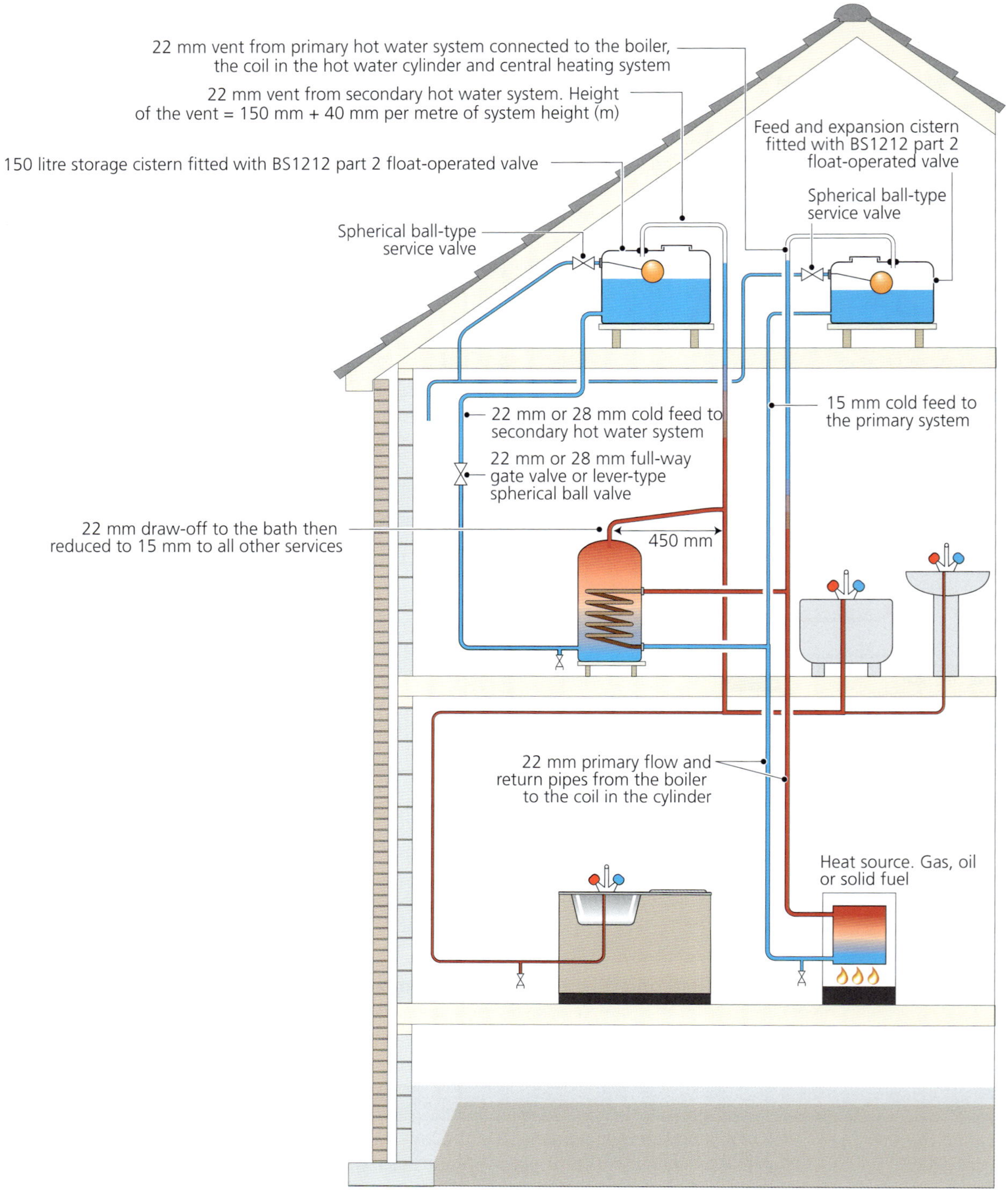

▲ Figure 12.25 Indirect open-vented (double-feed) hot-water system

▼ Table 12.5 Advantages and disadvantages of open vented hot water systems

Advantages	Disadvantages
• Storage is available to meet demand at peak times • Low noise levels • Always open to the atmosphere • Water temperature can never exceed 100°C • Reserve of water available if mains supply is interrupted • Low maintenance • Low installation costs	• Space needed for both hot-water and cold-water storage vessels • Risk of freezing • Increased risk of contamination • Low pressure and, often, poor flow rate • Outlet fittings can be limited because of the low pressure

Combination storage systems are vented hot-water systems with the cold-water storage cistern, cylinder and associated pipework combined in one unit. They come in a variety of storage capacities and, as with vented hot-water cylinders, are available in both direct and indirect models. They are ideal for properties that require stored hot water but have no loft space. Care needs to be taken when siting these units due to the low static head resulting in poor flow rates out of the taps.

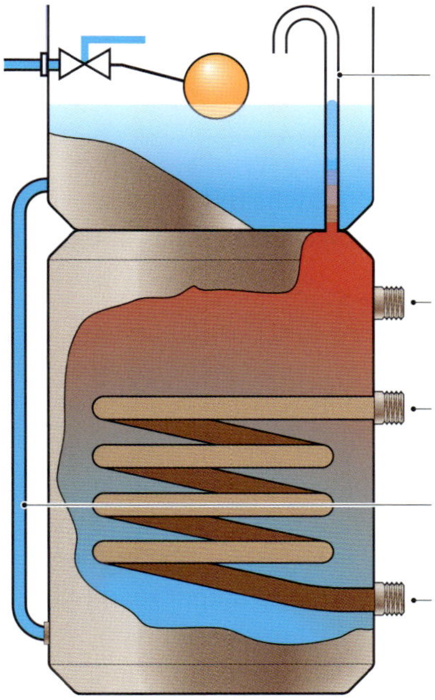

▲ Figure 12.26 Combination storage system

Research

Using the internet, research the advantages and disadvantages of combination storage systems.

Unvented hot-water storage systems are heated by either a boiler or an electric immersion heater. Water is stored in a hot-water cylinder, which is fed directly from the cold-water mains. Hot-water outlets are supplied via a system of copper or plastic pipework, sized at a minimum of 15 mm. The system contains an open vent pipe, which acts as a safety relief in the event of overheating. Water is stored at 60–65°C and should be distributed at no less than 55°C.

This system contains several components to ensure both correct operation and safety for the user:

- line strainer: filters the water supply to protect system components
- pressure-reducing valve: reduces the mains water pressure
- single-check valve: prevents backflow
- expansion-relief valve: discharges water to a safe place in the event of an increase in system pressure
- temperature-relief valve: discharges water to a safe place in the event of an increase in system temperature
- expansion vessel: takes up the expansion of water within the system during operation.

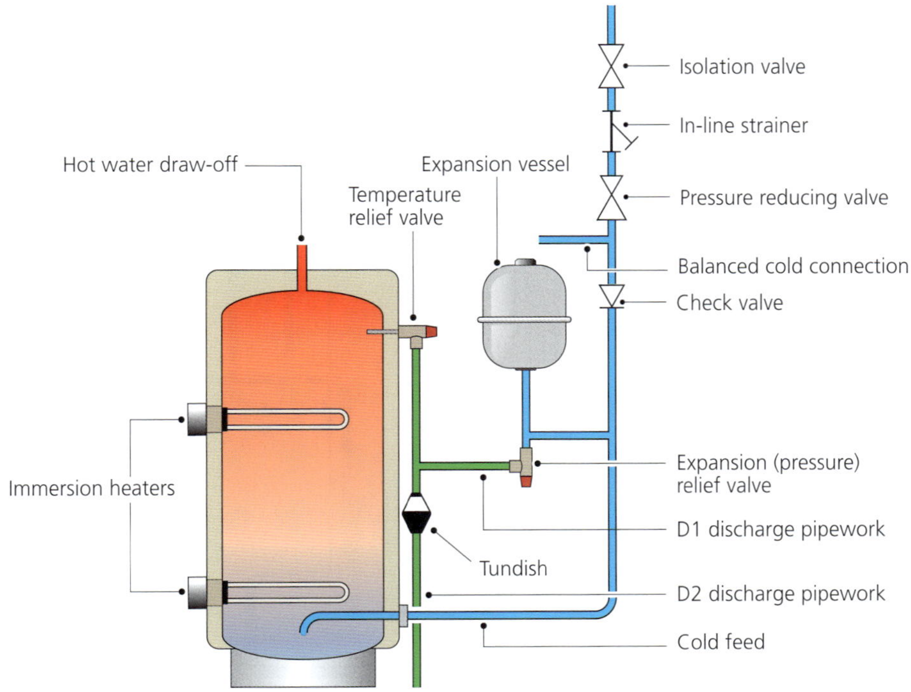

▲ Figure 12.27 Unvented hot-water storage system

▼ Table 12.6 Advantages and disadvantages of unvented hot-water storage systems

Advantages	Disadvantages
• Higher pressure and flow rates at all outlets, giving a larger choice of outlet fittings • Pressures balanced at both hot and cold taps • Low risk of contamination • Hot-water storage vessel can be sited almost anywhere in the property, making the system suitable for both houses and flats • Reduced risk from frost damage • Less space required because cold-water storage is not needed • Quicker installation, as less pipework is required • Smaller-diameter pipework used in some circumstances	• No back-up of water should the water supply be isolated • If cold-water supply suffers from low pressure or flow rate, the system will not operate satisfactorily • Discharge pipes are needed that are able to accept very hot water; there are restrictions on their length • High level of maintenance required • Higher risk of noise in system pipework • High initial cost of the unvented hot-water storage vessel

Improve your maths

Using Figure 12.27 and a manufacturer's catalogue, produce a materials list, price breakdown and total cost for all components shown.

Thermal store heaters work by heating up cold water that is passed through the vessel; the cold water can either be fed directly from the mains or from a cold-water storage cistern. A thermostatic mixing valve must be installed on the outlet, to ensure the water does not exceed 60°C.

Gas storage water heaters contain a hot-water cylinder that is heated via an integral gas burner. The unit has a flue which takes the POC safely outside. It contains a valve to control the gas supply, which is connected to a thermostat to control the temperature of the stored hot water.

Combination boilers are dual-function appliances. When a hot tap is opened, a diverter valve diverts the boiler water around a second heat exchanger, which heats water from the water undertaker's cold-water main to supply instantaneous hot water at the hot tap. In this mode, the entire heat output is used to heat the water. Temperature control is electronic: the burner is adjusted automatically to suit the output required. Typical flow rates are around 9 litres per minute (35°C temperature rise). Some combination boilers incorporate a small amount of storage, and this can double the flow rate to around 18 litres per minute.

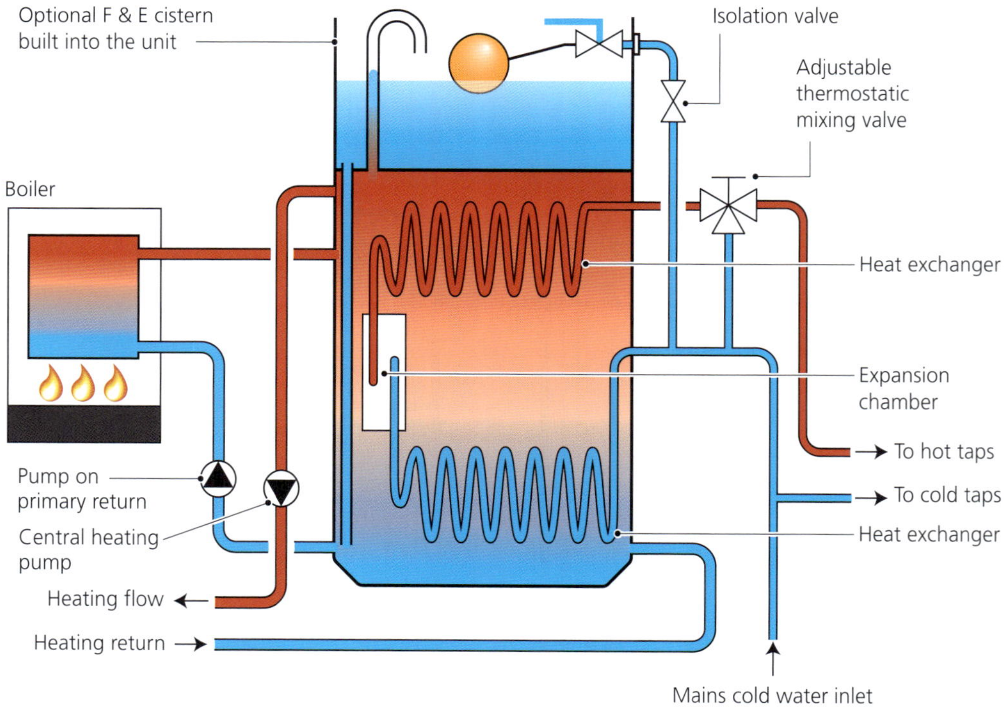

▲ Figure 12.28 Thermal store heater

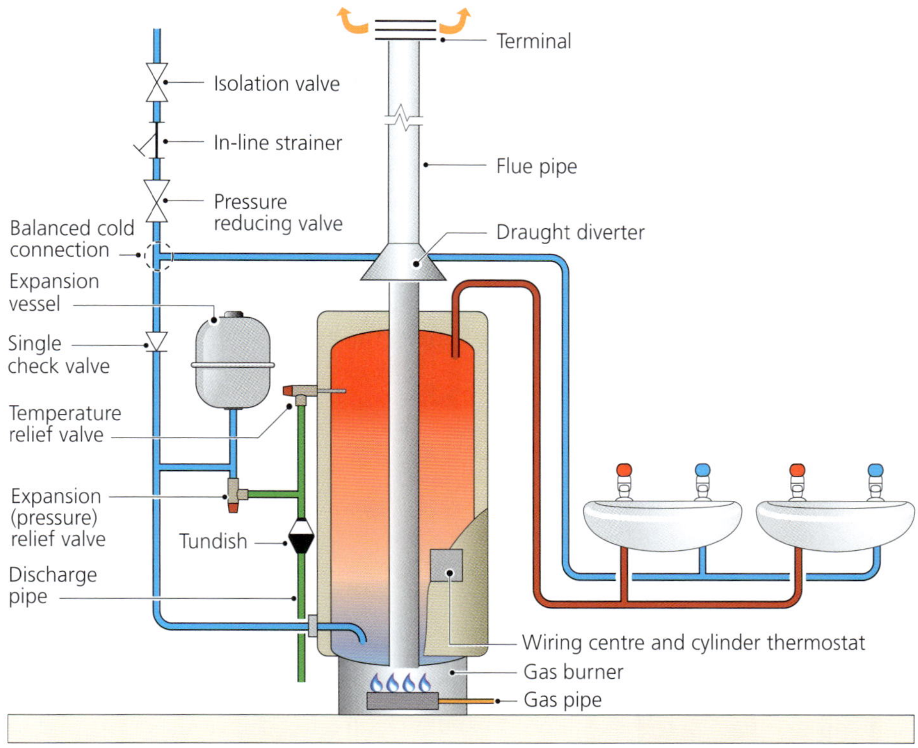

▲ Figure 12.29 Gas storage water heater

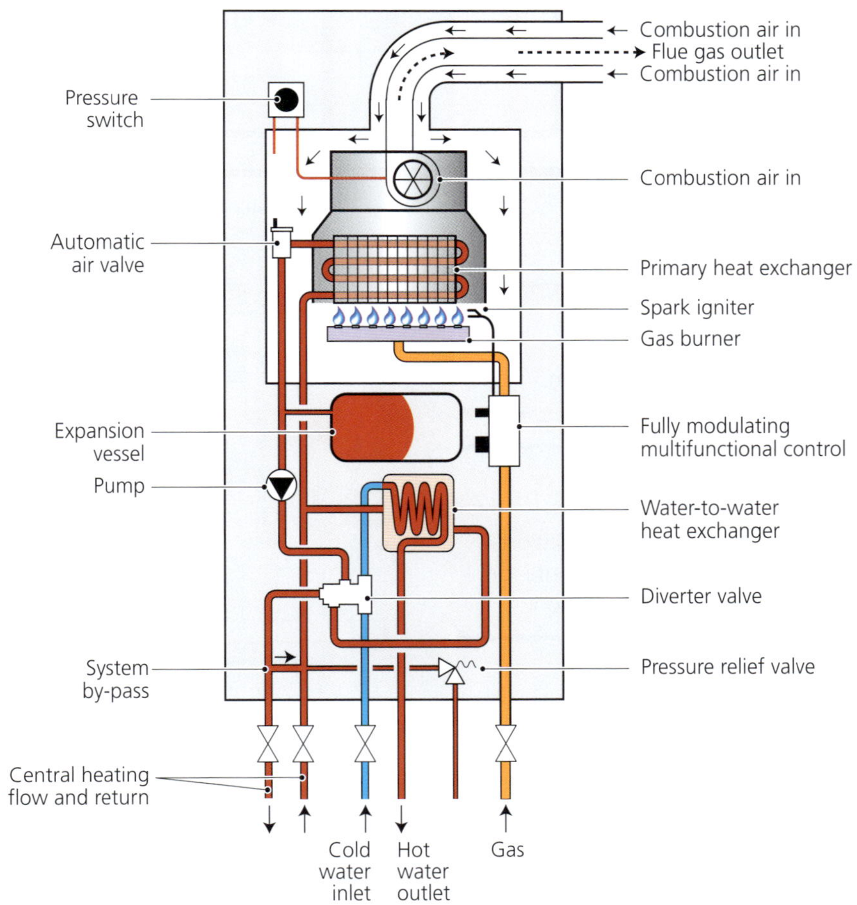

▲ Figure 12.30 Combination boiler

Secondary circulation

Industry tip

When a hot tap is opened, a certain amount of cold water is usually drawn off and allowed to run to drain before hot water arrives at the tap. This length of pipework is known as a dead leg.

Secondary circulation prevents the wastage of water due to excessive lengths of hot-water draw-off from the storage vessel to the outlet. It is a method of returning the hot-water draw-off back to the storage cylinder in a continuous loop, to eliminate cold-water 'dead legs' by reducing the distance the hot water must travel before it arrives at the taps. In all installations, secondary circulation incorporates a bronze- or stainless-steel-bodied circulating pump to circulate the water to and from the storage cylinder.

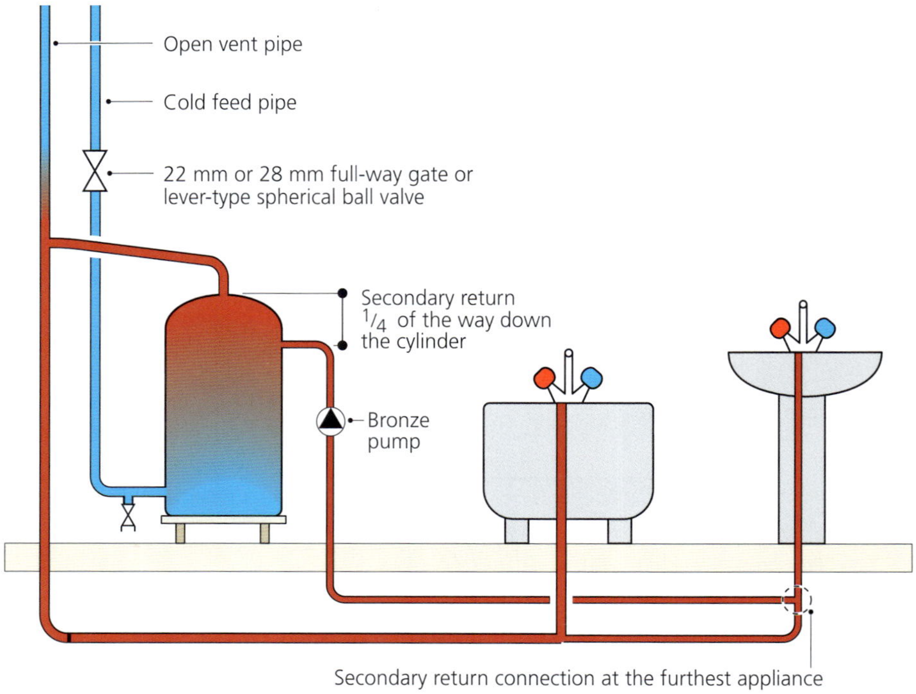

▲ Figure 12.31 Secondary circulation

Test yourself

Describe the purpose of secondary circulation.

Research

Using the Water Supply (Water Fittings) Regulations 1999 and BS EN 806, research the requirements for and methods of providing secondary circulation.

Localised hot-water systems

Cistern-type water heaters are designed to be connected directly to the rising main and allow hot water to be supplied to several outlets. The unit contains the cold-water storage cistern and an integral heater chamber, which is heated by an electric immersion heater controlled by a thermostat.

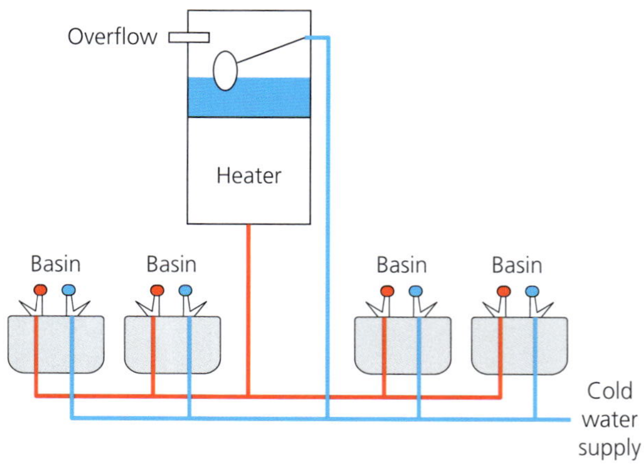

▲ Figure 12.32 Cistern-type water heater

Instantaneous water heaters can be fuelled by either gas or electricity and are generally described as inlet controlled (the water supply is controlled at the inlet to the heater). The water is heated as it flows through the heater and will continue to be heated as long as the water is flowing. When the control valve is closed, the water flow stops and the heat source shuts down.

This type of heater is generally used to supply small quantities of hot water, such as for washbasins and showers. Typical minimum water pressure is 1 bar.

There are many different types of electric shower, with varying outputs from 8.5 to 11 kW. The higher the kW output, the better the overall flow rate at a showering temperature. All electric showers feature a low-pressure heater-element cut-off, so that the temperature of the water does not cause harm if the supply pressure/flow rate is low.

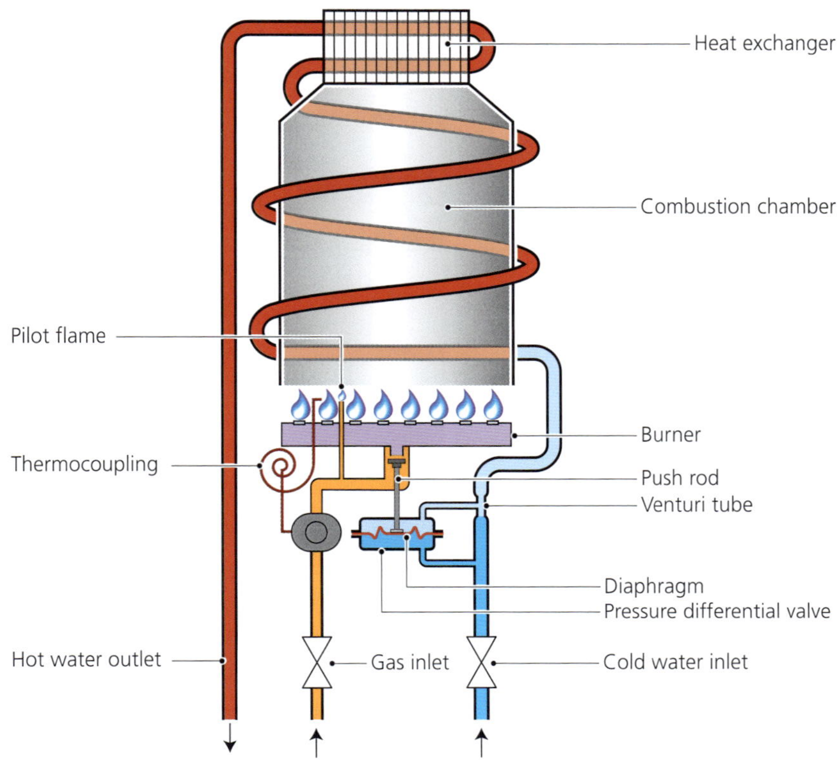

▲ Figure 12.33 Instantaneous water heater

Storage-type localised water heaters are often referred to as displacement heaters, as the hot water is displaced from the heater by cold water entering the unit. Typical storage capacities are between 7 and 10 litres.

As the name suggests, over-sink heaters are fitted over a sink. The water is heated by an electric heater element and delivered from a spout on the heater.

Under-sink heaters are fitted under the sink and work in the same way as over-sink heaters. The main difference is that they usually require a special tap or mixer tap that permits the outlet to be open to the atmosphere at all times, to allow for expansion. The inlet of water to the heater is still controlled from the tap. Typical capacities are up to 15 litres.

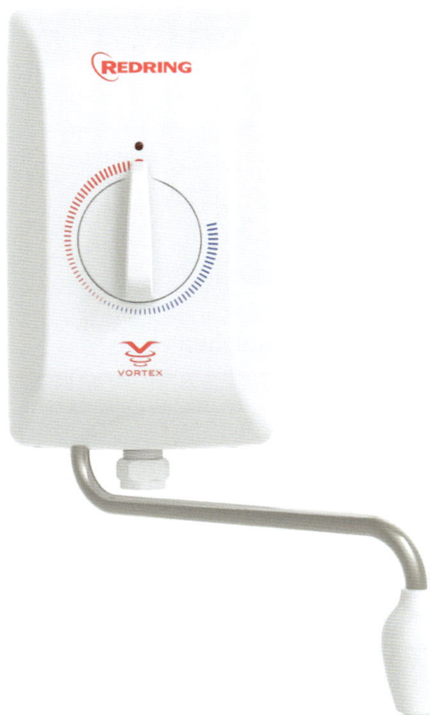

▲ Figure 12.34 Instantaneous hot-water fitting

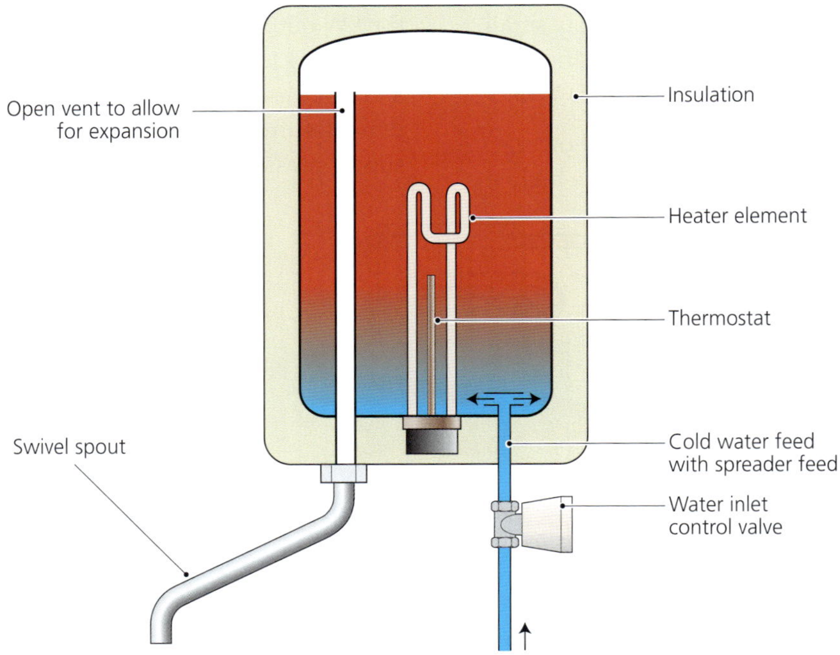

▲ Figure 12.35 Over-sink heater

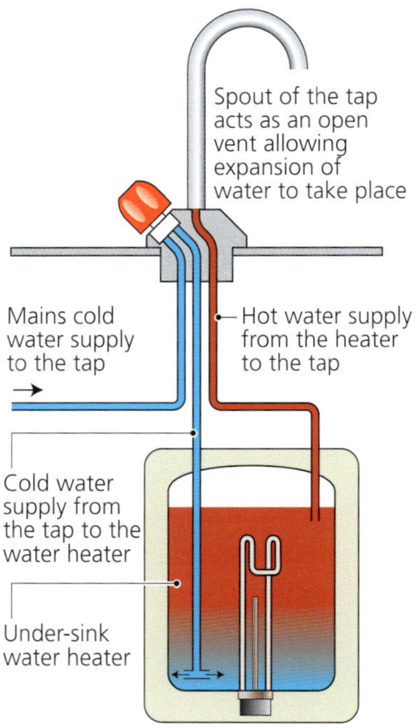

▲ Figure 12.36 Under-sink heater

Test yourself

List three types of hot-water system.

Research

Specify a hot-water system for a building you are familiar with and state the components required to install it.

Sanitation systems

Sanitation systems take waste solids and liquids away from a building, to ensure hygienic conditions are maintained within. This includes waste from toilets, baths, basins, sinks, bidets and showers.

Health and safety

Working on sanitation systems involves working with human waste. To prevent disease, the correct PPE must be worn and personal hygiene procedures followed.

Sanitation-system pipework comprises two sections:

- The **soil pipe**, also known as the soil stack, is the lower, wet part of the system that takes effluent away from a building.
- The **vent pipe**, also known as the vent stack, is the upper part of the system that introduces air to help prevent loss of **trap seal**.

Together, these pipes are known as a stack system.

Key terms

Soil pipe: the lower, wet part of sanitation-system pipework that takes effluent away from a building

Vent pipe: the upper part of sanitation-system pipework that introduces air to help prevent loss of trap seal

Trap seal: a plug of water left in the trap which prevents bad smells entering a building

Primary ventilated stack system

The primary ventilated stack is the most common system found in domestic properties. It is used where appliances are grouped closely around the stack.

Secondary ventilated stack system

The secondary ventilated stack system is installed to prevent positive and negative pressure fluctuations. It is used where appliances are grouped closely around the stack.

Ventilated branch discharge system

The ventilated branch discharge system is commonly used in commercial and industrial premises, where appliances are installed in ranges and sited a distance away from the stack. It ensures compliance with building regulations in relation to maximum permitted branch pipework lengths.

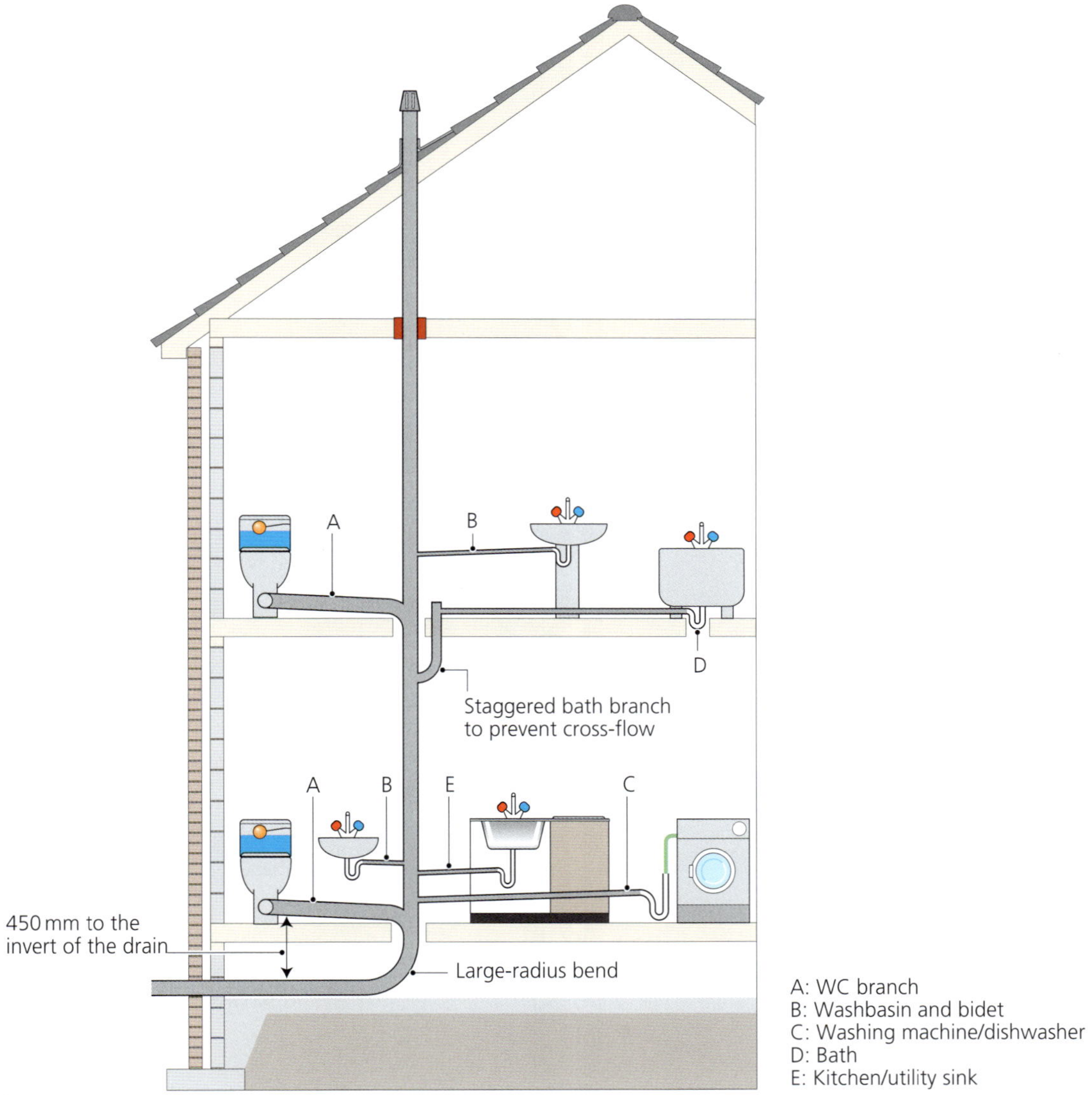

▲ Figure 12.37 Primary ventilated stack system

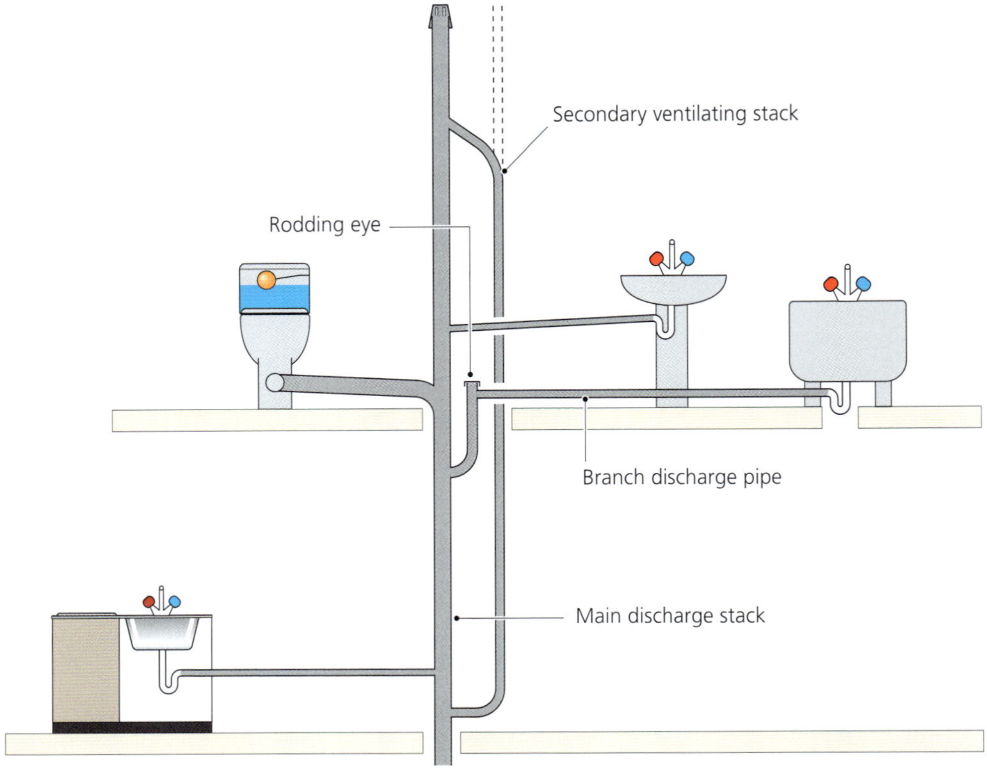

▲ Figure 12.38 Secondary ventilated stack system

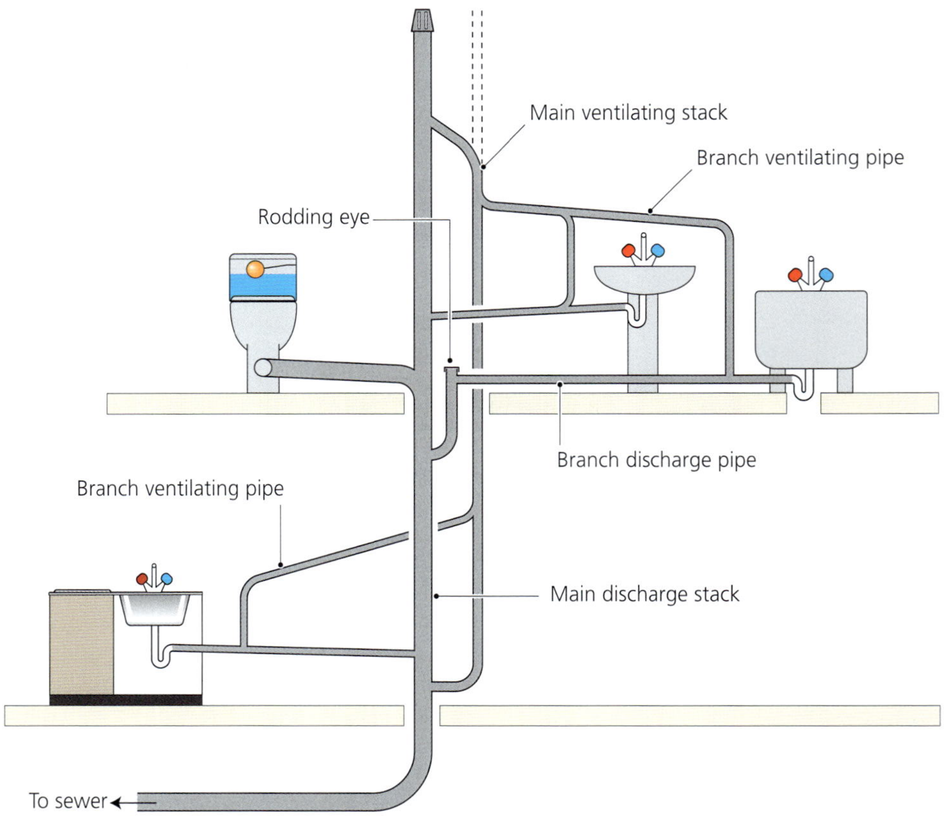

▲ Figure 12.39 Ventilated branch discharge system

Stub stack system

A stub stack system can be used where sanitary appliances are connected directly to an internal drain. This reduces the need for ventilation pipework and removes the requirement for the soil system to penetrate the building structure.

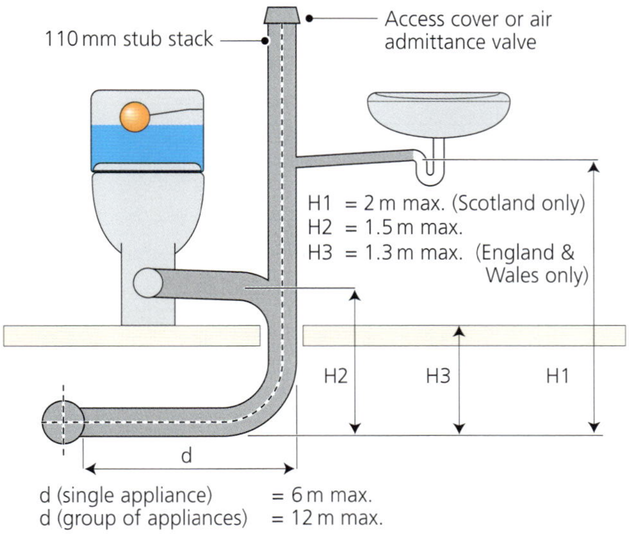

▲ Figure 12.40 Stub stack system

When installing a stub stack:

- the stack should be terminated with an air admittance valve
- the maximum distance between the invert of the drain and the base of a WC should be 1.3 m
- the maximum distance between the invert of the drain and the highest branch connection should be 2 m.

Research

Visit www.planningportal.co.uk/info/200135/approved_documents and research the requirements for above-ground drainage systems and stub stack systems.

- Which approved document provides guidance on drainage and waste disposal?
- What are the termination requirements for the vent pipe?
- What are the maximum permitted dimensions for branch discharge pipework?

Produce a set of line diagrams to illustrate your findings.

Traps

Every appliance discharging into a soil and waste system must be fitted with a trap to prevent foul air within the system entering the property.

The type of trap used will depend on the appliance it is connected to, and examples include:

- P traps and S traps: these are types of swivel trap which are used on new work and appliance replacements
- running traps: these are used for a range of appliances. Rather than individual traps on each appliance, a single running trap can be used at the end of a pipework run
- bottle traps: used on washbasins because of their neat appearance; shower traps are a combined waste and trap allowing the trap to be cleaned of potential blockages, such as hair, from the top of the waste on the shower tray
- washing machine traps: used for appliances such as washing machines and dishwashers, they are generally of P-trap configuration, with an extended neck to accommodate a washing machine/dishwasher outlet hose
- in-line traps: designed with washbasins in mind, an in-line trap is essentially an S trap where the inlet and outlet are in line
- anti-vac traps: use a small air admittance valve located after the water seal
- self-sealing traps: waterless valves that use a thin neoprene rubber membrane to create an airtight seal, preventing foul air from entering the dwelling while maintaining equal pressure within the soil and vent system.

Trap size is dependent on the size of an appliance's waste/soil pipe.

▼ Table 12.7 Minimum sizes of waste pipe

Appliance	Minimum size of waste pipe
Wash basin Bidet	32 mm
Washing machine Dishwasher Bath Kitchen sink Shower tray	40 mm
Multiple appliance branch pipework Food waste disposal units	50 mm

Research

Using the internet, research manufacturers of soil and vent pipes. Note the types of fittings and installation requirements for above-ground drainage systems.

Rainwater systems

Rainwater systems collect and carry away rain from roofs using either integrated channels or eaves-mounted gutters connected to rainwater pipes. The water is discharged into surface-water drains, combined sewers, **soakaways** or watercourses such as streams and rivers.

Health and safety

Working on rainwater systems involves working at height. The Work at Height Regulations must be followed at all times to prevent falls.

Building features determine the type of gutter bracket used to secure a gutter. For example:

- when fascia boards are fitted, fascia brackets are used
- when no fascia boards are fitted and the rafters are exposed, rafter brackets are used
- when no fascia boards are fitted and no rafters are exposed, rise and fall brackets which are directly fixed into the masonry are used.

The purpose of a rainwater system is to:

- protect the foundations of the building
- reduce ground erosion
- prevent damp and water penetration of the building.

It can also be used as part of a rainwater-harvesting system.

Key term

Soakaways: large underground holes, filled with coarse stones or purpose-made plastic crates, which allow water to filter through and soak into the ground

Test yourself

What are the key components of a rainwater system?

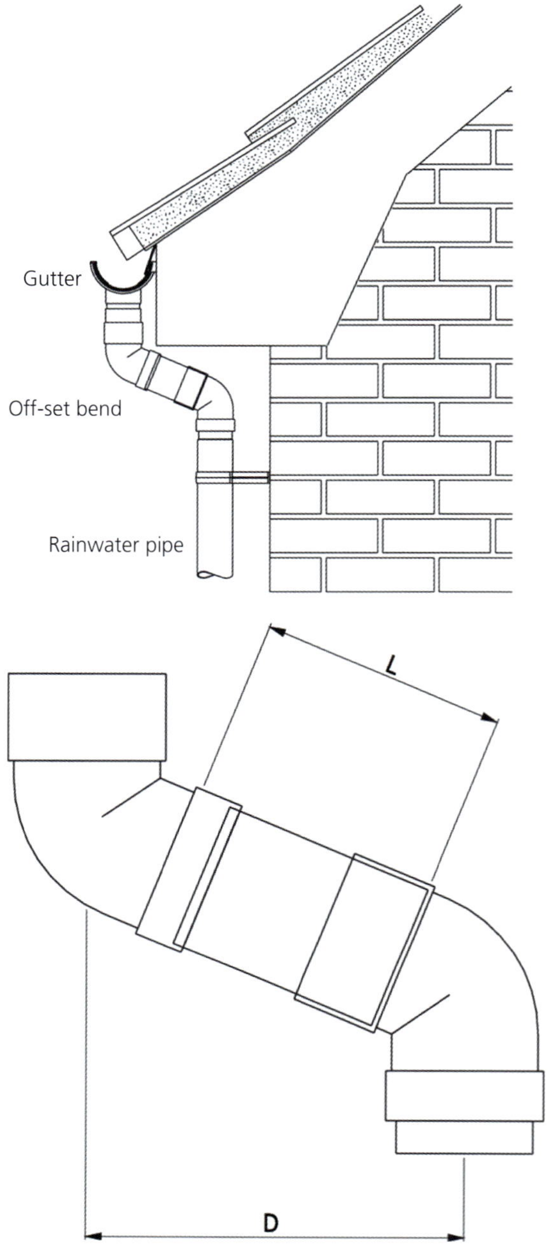

▲ Figure 12.41 Rainwater system

Improve your maths

Visit www.planningportal.co.uk/info/200135/approved_documents. With reference to Approved Document H, calculate the size of gutter required for a roof with an area of 25 m^2 and a pitch of 45 degrees.

Research

Visit www.planningportal.co.uk/info/200135/approved_documents
Research the requirements for rainwater systems by reading Approved Document H. Look for the following information:

- What are the requirements of H3?
- What is detailed in Section 1 Gutters and rainwater pipes?
- What are the requirements for materials used for gutters and rainwater pipes?

There are four main gutter profiles available, as shown in Table 12.8.

Industry tip

Gutter sizes are calculated using the effective roof area, which takes into account the plan of the roof and its pitch. Reference to manufacturer's instructions is essential when calculating gutter sizes.

▼ Table 12.8 Gutter profiles

Gutter profile	Description
Half-round gutter 	Standard gutter profile used with round rainwater pipes
Square-section gutter 	Standard gutter profile used with square-section rainwater pipes
High-capacity gutter 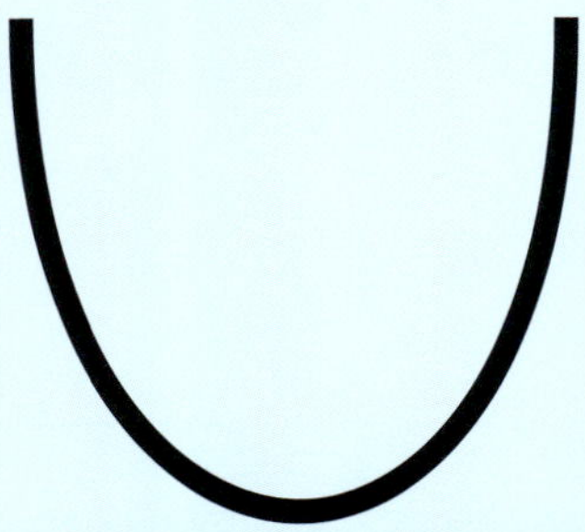	Used on larger or steeply angled roofs; also known as deep-flow or storm-flow gutter
Ornamental gutter (OG) 	A modern redesign of a Victorian gutter that provides a period look

Test yourself

What are the four main gutter profiles?

▼ Table 12.9 Gutter and rainwater fittings

Fitting	Purpose
Running outlet 	Connects the gutter and rainwater pipework
Gutter angle 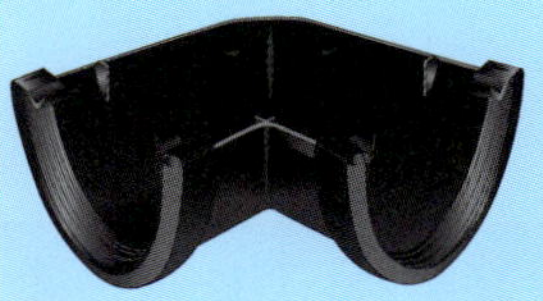	Allows a change of direction in the gutter
Gutter union 	Joins two sections of gutter
Stop end 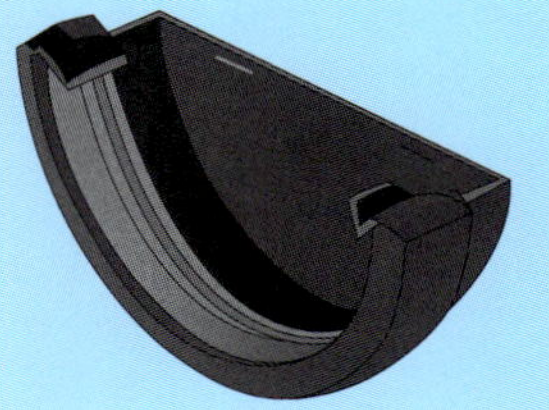	Used at the end of gutter runs
Specialist union 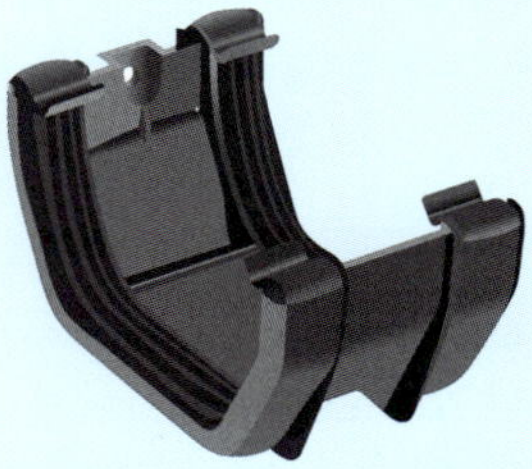	Connects different materials and gutter profiles together
Shoe 	Discharges rainwater over a gully

Gravity rainwater systems typically use PVC-U, extruded-aluminium or cast-iron gutters:

- PVC-U gutters use a snap-fit system that allows them to click together. They are sealed with integral rubber.
- Extruded-aluminium gutters are jointed in the middle and secured with screws or rivets.
- Cast-iron gutters are jointed in the middle and bolted together with gutter bolts.

Test yourself

List three materials used for gutters and rainwater pipework.

Research

Using the internet, research gutter and rainwater pipework manufacturers. Note the types of fittings available and produce a poster explaining each type.

1.6 Protection systems

Protection systems are designed to safeguard a property, its occupants and its contents against intruders and fire. They can also provide surveillance of the building and its surrounding area and provide access control.

Intruder alarm systems

There are many different types of modern intruder alarm, with a range of components and functions available. The type chosen will depend on the building being protected and the level of security required.

Before installing an intruder alarm, a survey should be carried out to ensure it is appropriate for the level of risk and meets the requirements of the client. A system design proposal should then be produced that lists the equipment to be installed and its location.

Industry tip

BS EN 50131 specifies requirements for intrusion and hold-up alarm systems.

Types of intruder alarm system

Audible intruder alarm systems

This is the most basic type of intruder alarm. When the alarm is triggered, a loud sound alerts people that an intruder has entered or is trying to enter the property.

This type of alarm system is suitable for most domestic dwellings, depending on several factors:

- location of the property: is the property remote or located in a public area?
- property use: for example a bank would require a higher level of protection
- contents of the property: for example a building with expensive ICT equipment would require a higher level of security.

Remotely monitored intruder alarm systems

When this type of alarm is triggered, it can provide notification of the intrusion to:

- a remote monitoring centre
- a building owner
- the police.

It usually also incorporates an audible alarm.

Intruder alarm system components

Intruder alarm systems consist of three main components:

- control unit
- detection device
- audible warning device.

Control unit

The control unit operates on mains-derived 230 V AC electricity, with the alarm circuits and wiring using 12 V. The system should be fitted with a standby, rechargeable battery to provide power in the event of disruption to the mains power supply.

The control panel can be programmed to perform a range of tasks, from switching the system on and off to altering the system configuration and timers. Some systems also allow selective parts to be activated.

The keypad is usually located in a convenient position to allow the user to operate the alarm system, rather than being fitted to the control panel.

For remotely monitored systems, connection to an ethernet or Wi-Fi communicator is required. The system is connected using PVC-insulated and PVC-sheathed multi-core alarm cable, with the communications connected using **Cat 5/6 data cable**.

Key term

Cat 5/6 data cable: twisted pair cable used for ethernet connection

Detection device

Detection devices detect the presence of an intruder and signal the information to the control unit:

- Door contacts comprise a magnetic reed switch fitted in a door frame with a magnet sited alongside it in the door. When the door is opened, the magnetic field is removed from the area of the reed switch, which generates an alarm signal.
- Passive infrared (PIR) sensors can be mounted in a room to detect movement in their field of view.
- Break-glass detectors comprise a microphone and an amplifier that are tuned to the frequency generated by breaking glass.
- Inertia sensors detect vibrations associated with forced entry into premises.
- Personal attack buttons are connected to an alarm system and always active. In the event of an emergency, they can be operated to cause a full alarm condition.

▲ Figure 12.42 PIR movement detector

Audible warning device

Audible warning devices emit sound when the alarm is activated, alerting people within earshot to the presence of an intruder. Typical sounds include bells, sirens and voice warnings. The audible alarm may also be accompanied by flashing lights.

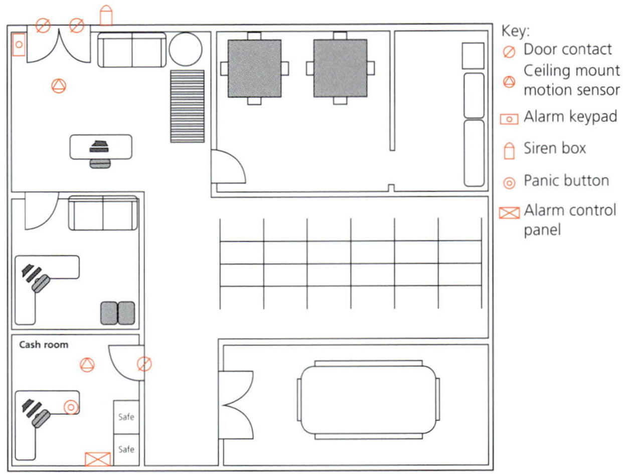

▲ Figure 12.43 Layout of an intruder alarm system

Test yourself

Name the three main components of an intruder alarm system.

Research

Using the internet, research the different intruder alarm systems available, including their connectivity. Produce an information sheet with your findings.

Surveillance systems

CCTV is a powerful deterrent to crime, working with other security systems to protect people and property. It operates on mains-derived 230 V AC electricity, with accessories using 12 V.

A typical CCTV surveillance system comprises:

- site cameras, which can be internal, external, static and/or fully functional (pan–tilt–zoom, PTZ)
- illumination, provided as either standard white or infrared (IR) light
- detectors in the form of beams or PIR sensors
- a public address (PA) system, which can be used for audio challenge
- a system controller (either digital video recorder, DVR, or network video recorder, NVR)
- a CCTV transmitter unit.

The system is connected using PVC-insulated and PVC-sheathed multi-core cable and coaxial cable, with communications connected using Cat 5/6 data cable.

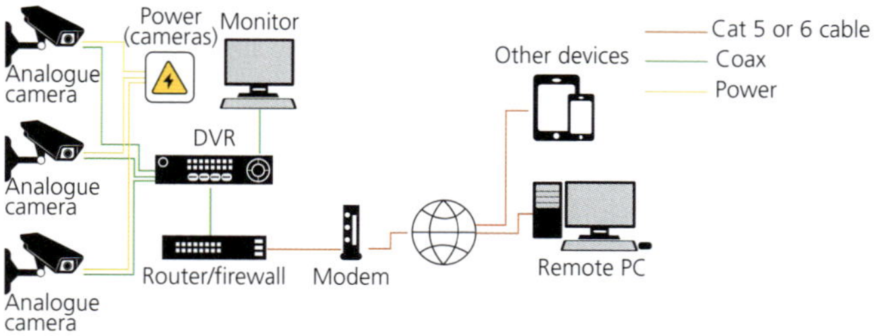

▲ Figure 12.44 CCTV system diagram

Research

Using the internet, research the levels of CCTV system available:

- entry level
- mid-range
- high end.

Produce a poster detailing the components and functions for the different levels.

Fire alarm systems

Fire alarm systems provide early detection and warning of a fire. They usually consist of a control panel linked to fire detectors and manual call points (often referred to as detection zones) and alarm circuits. The systems operate on mains-derived 230 V AC electricity and also contain a backup battery. Detectors use 24 V DC.

Prior to a fire alarm system being designed and installed, a risk assessment should be undertaken to define its main objectives.

Test yourself

What is the primary purpose of a fire alarm system?

Industry tip

BS 5839 covers fire detection and fire alarm systems for buildings.

Fire alarm systems must be fit for purpose. BS 5839 defines different categories:

- Category M systems rely on manual operation by the people using the building. The usual method of raising the alarm is to break the glass on a manual call point.
- Category L systems provide automatic fire detection (AFD) and are designed primarily to protect life. This category is subdivided, according to the areas of the building that require the installation of AFD.
- Category P systems provide AFD and are designed primarily to protect property. This category is subdivided, with P1 requiring AFD in all areas of the building and P2 requiring AFD only in specific parts of the building.

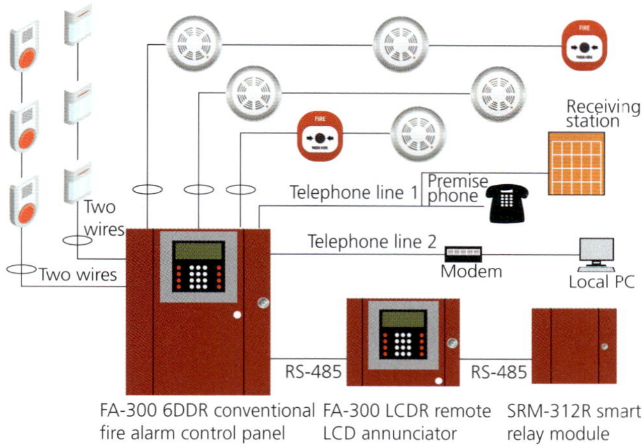

▲ Figure 12.45 Fire alarm systems

The fire alarm control panel (FACP) contains the:

- electronics that supervise and monitor the integrity of the wiring and components of the fire alarm system
- switches to allow the sounders to be activated or silenced and the detectors to be reset following an alarm.

There are usually several fire-detection zones, comprising a mixture of automatic fire detectors and manual call points. The size of a zone is limited to 2000 m^2 and should not cover more than one storey.

Fire detectors can be smoke or heat activated and are what is known as initiating devices, sending a signal to the fire alarm control panel to activate the alarm circuit.

A manual call point consists of a simple switch with a resistor in series with it. When the call point is activated, the resistor is switched across the line and a current of 50–80 mA is drawn, sending a signal to the fire alarm control panel to activate the alarm circuit.

▲ Figure 12.46 Manual call point

The system is connected using fire-performance (FP) cables, with communications connected using fire-resistant UTP Cat 5/6 data cable.

Test yourself

What are the different categories of fire alarm system under BS 5839?

Research

Using the internet, research intelligent fire alarm systems and prepare notes explaining the system components and basic operation. Present your findings to the class.

Access control systems

Access control systems ensure that only authorised persons are able to enter a location, whether that is an individual room or a whole site. They can be used anywhere within a building and its grounds, for example:

- external gates and barriers
- main entrances
- doors, turnstiles and lifts inside the building.

The primary purpose of these systems is to protect a building's occupants and/or contents from threats arising from unauthorised access. As such, they ensure that a building owner is able to meet their statutory duty of care.

A successful system will be able to maintain the security of the building while managing the access requirements of different users. It may also be possible to record the movements of those users.

Access control systems operate on mains-derived 230 V AC electricity, with accessories using 12 V. They are connected using PVC-insulated and PVC-sheathed multi-core cable, with communications connected using Cat 5/6 data cable.

Types of access control system include:

- stand alone, with a single entry point converted to an access control solution from a mechanical one
- online, where the decision to grant access is made by an electronic access control (EAC) system
- fully integrated/wireless, through incorporation into a building's existing security system.

Typical access control installations on a single door require a reader or keypad, door monitor, lock manager and press-to-exit switch.

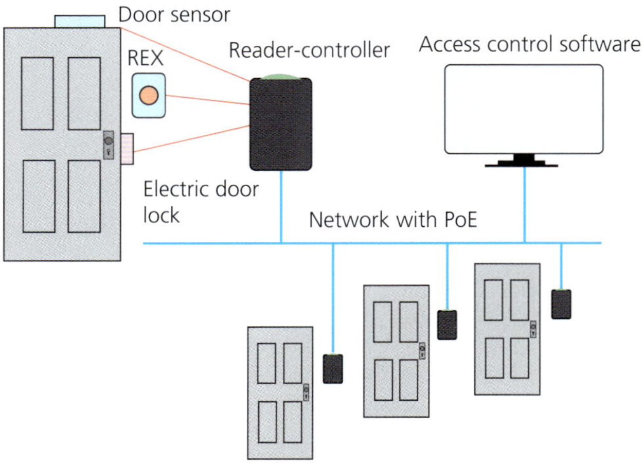

▲ Figure 12.47 Access control system

Access control systems rely on a person being recognised and validated using a **credential**, such as:

- something the person has, for example a key card, tag, token or smartphone app
- something the person knows, for example a password or PIN
- a person's biometric data, for example their fingerprint.

Key term

Credential: something used to recognise and validate a person

Test yourself

Give three examples of credentials that can be used for access control systems.

Research

Using the internet, identify different types of access control system that could be installed in an office.

1.7 Refrigeration systems

Refrigeration systems reduce the temperature of a space or substance by removing unwanted heat. They are used as part of:

- HVAC systems, to ensure the comfort of building users
- cooling systems for industrial processes
- cooling systems for keeping products fresh in the food and drink manufacturing sector (for example in cold rooms).

They comprise a combination of components and equipment connected in sequential order to produce the refrigeration effect.

1.8 Ventilation systems

Ventilation systems are used to change the air within an enclosed space in order to:

- provide fresh air
- manage the oxygen content in the air
- control levels of carbon dioxide
- prevent damp and control moisture
- remove excess heat
- remove airborne contaminants.

Ventilation can be achieved by either natural or mechanical means. The requirements for air change rates in a building are covered in the Building Regulations and vary depending on the:

- type of building
- building's occupancy levels
- activity the building has been designed for.

Natural (non-mechanical) ventilation systems

Natural ventilation provides air changes within a building using components such as air bricks, louvres, trickle vents or openable windows. It relies on wind effects and pressure and the principles of convection.

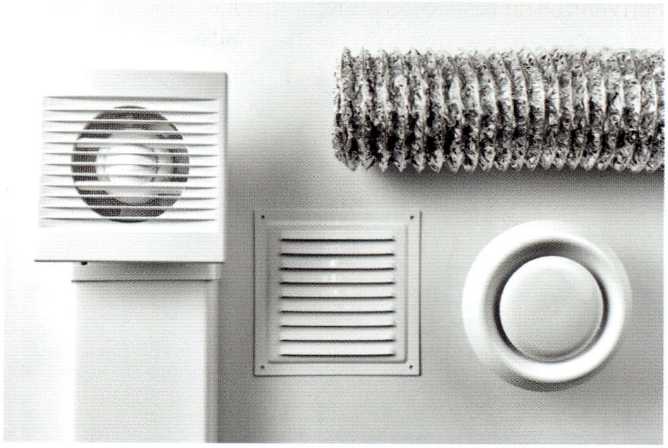

▲ Figure 12.48 Domestic ventilation components

The stack effect relies on the principles of convection. Cool air from outside enters a building at a low level where it is warmed by the occupancy, lighting, machinery and building activities. This warmed air rises within the building then discharges through vents at a high level.

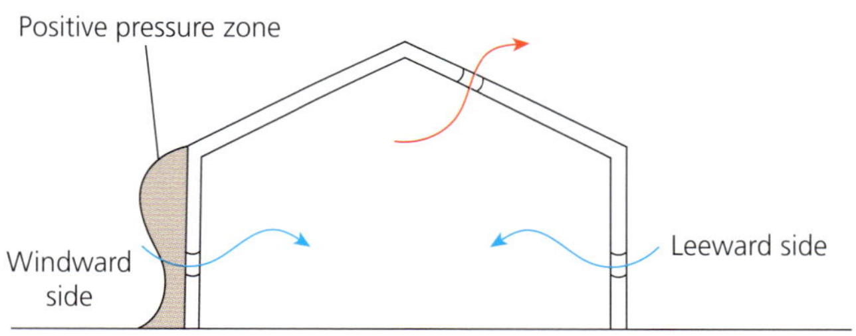

▲ Figure 12.49 Stack effect

Passive stack ventilation is an energy-efficient and environmentally friendly method of ventilation with zero running costs. The system consists of vertical ducts sized at 100 to 150 mm in diameter, with grilles installed at ceiling level which are connected to terminals above the ridge height of the roof. Air movement within the building is achieved by a mixture of the warm air and air flowing over the roof. Fresh air is drawn into the building through the trickle vents in the windows and doors.

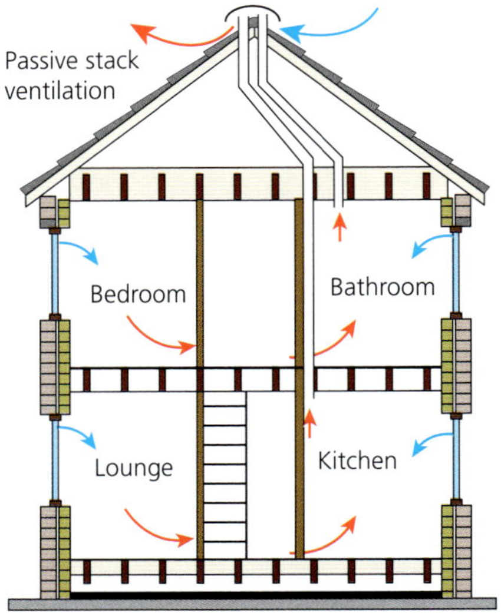

▲ Figure 12.50 Passive stack ventilation

Research

Visit www.planningportal.co.uk/info/200135/approved_documents and research the requirements for ventilation systems.

- Which approved document provides guidance on ventilation?
- What types of ventilation system are included in the approved document?
- What are the ventilation requirements for a bathroom?

Mechanical ventilation systems

There are three categories of mechanical ventilation system:

- natural inlet and mechanical extract
- mechanical inlet and natural extract
- mechanical inlet and mechanical extract.

Mechanical extract ventilation (MEV)

Mechanical-extract ventilation (MEV) systems are suitable for both domestic and commercial properties where passive stack ventilation is considered inadequate. While comprising a similar arrangement of vertical ducts and ceiling-level grilles connected to roof terminals, they also have a low-powered fan located in the roof structure that runs continuously. These systems can incorporate humidity sensors to automatically increase air flow and may be boosted by manual control when necessary.

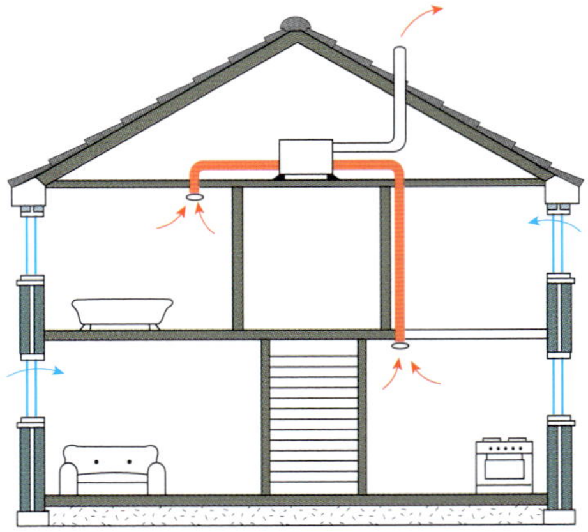

▲ Figure 12.51 Mechanical-extract ventilation (MEV) system

Mechanical ventilation with heat recovery (MVHR) systems

Mechanical ventilation with heat recovery (MVHR) systems share the same working principles as MEV systems, with the addition of a fresh air supply fan and a heat exchanger. They extract warm, stale air from inside a building and pass it through a heat exchanger, where the heat is transferred to incoming fresh air supplied by the fan. The warmed fresh air passes into the building while the cooled stale air is extracted into the atmosphere. By recovering the heat in the exhaust air, these systems improve a building's energy efficiency.

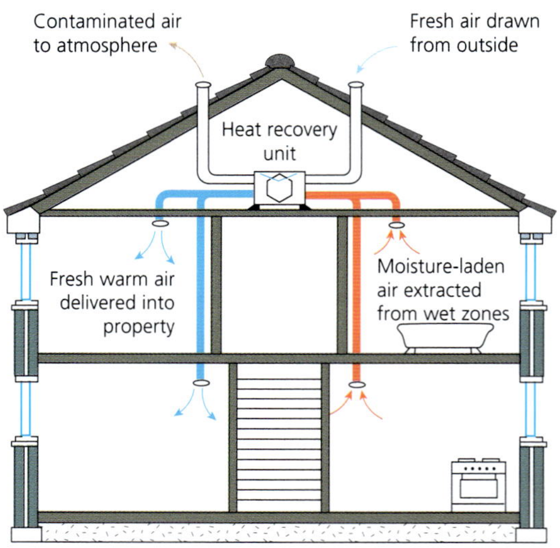

▲ Figure 12.52 Mechanical ventilation with heat recovery (MVHR) system

Research

Produce layout diagrams for the following ventilation systems and provide examples of buildings where they may be installed:

- natural inlet and mechanical extract
- mechanical inlet and natural extract
- mechanical inlet and mechanical extract.

2 The potential effects on building performance during installation, commissioning and decommissioning of BSE systems

2.1 Effects on the environment

In order to minimise negative effects on the environment, it is important to:

- specify and install energy-efficient products and systems
- measure accurately to avoid waste
- use materials correctly
- follow environmental policies.

A range of published documents support compliance, including:

- the Domestic Building Services Compliance Guide
- Approved Document L: Conservation of fuel and power.

BSE systems must be commissioned according to the manufacturer's instructions to ensure they work efficiently. This includes setting up system controls for optimum performance and instructing building users on how the system is operated.

It is particularly important to consider the environment when decommissioning systems, whether **temporarily** or permanently. This includes ensuring the:

- correct recycling of materials and waste
- correct disposal of system fluids
- correct disposal of hazardous waste, such as asbestos, refrigerants, electrical and electronic equipment (EEE) and lead.

Test yourself

List three factors that must be considered when decommissioning BSE systems.

Key term

Temporary decommissioning: the process of taking a system out of action for a short time for maintenance or upgrades

2.2 Effects on trades

Other trades may be affected by the activities involved in work on BSE systems. For example, the isolation of a cold-water system during installation work could affect other BSE trades who require a connection to the water supply and onsite construction operatives such as bricklayers and plasterers who require water for mixing their materials.

Test yourself

Explain how different trades can be affected during the installation, commissioning and decommissioning of BSE systems.

2.3 Loss of service

While undertaking work on BSE systems and components, a loss of service may occur:

- Isolating the water supply in a domestic property will result in no drinking-water outlets and a loss of sanitary conveniences.
- Isolating the electrical supply in a commercial building can result in the loss of a range of BSE systems, such as electrical outlets, heating and ventilation systems and protection systems where battery backup is not installed.

Where possible, alternative supplies should be provided to minimise disruption to end users. End users should be informed about which part of the system will be out of service, how long for and the location of/arrangements for alternative supplies.

Research

Research the effects of installation, commissioning and decommissioning of BSE systems on building performance. Produce a table for the following systems:

- plumbing
- heating
- gas
- electrotechnical
- control
- air conditioning
- ventilation.

3 Mechanical principles of components

3.1 Fans

Fans are used in a range of BSE systems and components, for example:

- in gas central-heating boilers to discharge products of combustion to the outside while at the same time drawing in fresh air
- in ducted ventilation systems either to draw air into the system or to discharge stale air.

The function of the fan will depend on its position.

The two types of fan commonly used in BSE systems are referred to as centrifugal and axial. They are defined by the direction of air flow through them. These two types can then be split into subtypes relating to flow volume/pressure characteristics, size, noise and vibration.

Centrifugal fans

In centrifugal fans, the air enters the impeller along its axis and is then discharged radially from the impeller by the centrifugal motion.

There are several different shapes of blade that can make up the impeller, including:

- backward curved
- forward curved
- radial.

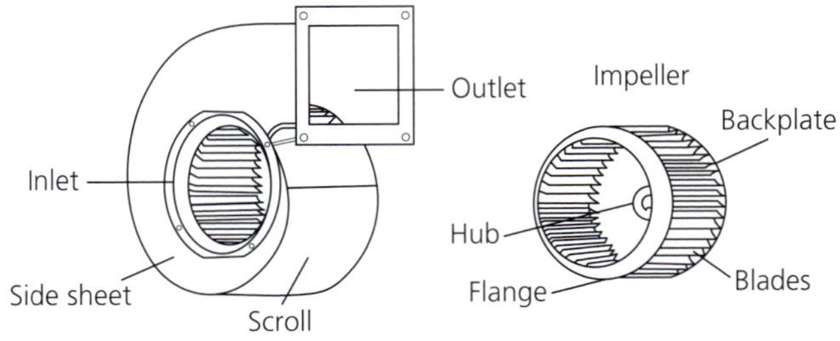

▲ Figure 12.53 Centrifugal fan

Axial fans

In axial fans, air passes through the fan in line with the axis of rotation. Types include:

- tube axial, consisting of a propeller or disk-type wheel within a cylinder
- vane axial, consisting of a disk-type wheel within a cylinder, with a set of air-guide vanes located either before or after the wheel
- mixed-flow inline fan, a development of the axial fan that combines the characteristics of both axial and centrifugal fans.

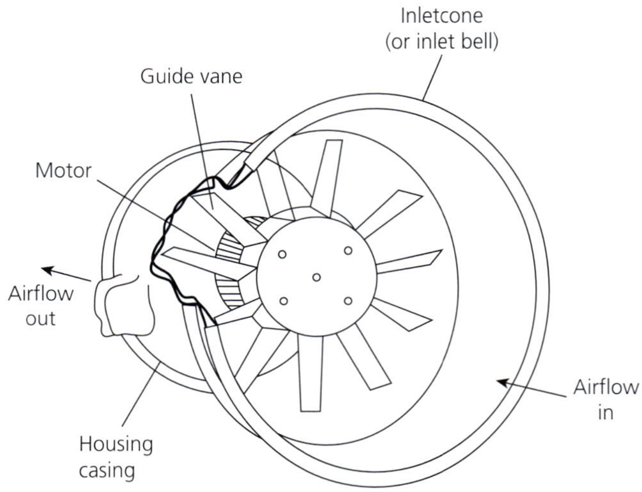

▲ Figure 12.54 Axial fan

Research

Using the internet, research centrifugal and axial fans and produce a handout detailing their operating principles and uses.

Research

Research the different types of fan installed in BSE systems and identify the implications for each system of component failure.

Pumps

Pumps are used in BSE systems to circulate fluids:

- Hot-water circulating pumps are installed on the hot-water return (pumping into the cylinder) and aid efficient circulation to and from the cylinder. They are controlled by a simple clock and manufactured from bronze to ensure they do not corrode. If a pump fails, hot water will not be pumped around the system, which will reduce efficiency and increase hot-water draw-off times.
- Shower pumps increase shower flow rate:
  - A single-impeller pump is designed to boost the mixed supply from a shower mixer valve to the shower outlet.
  - A twin-impeller pump is designed to boost the hot- and cold-water supplies to a shower mixer valve.

  Failure of a shower pump will result in insufficient flow at the shower valve/outlet.
- Central-heating circulators (or pumps) are simple electric motors with a fluted waterwheel-like impeller that circulates water around the system by centrifugal force. The circulator must be positioned with care to avoid faults that could lead to problems with corrosion by aeration.
- Cold-water booster pumps are used to pump (boost) cold-water supplies and **draw-off water** to upper floors of a building. Failure of this component will result in no water on the upper floors.

Key term

Draw-off water: water discharge from a terminal fitting such as a tap

Research

Research the different types of pump installed in the following BSE systems:

- hot-water supply
- shower
- cold-water supply
- central heating.

Identify the implications for each system of component failure.

Boilers

Boilers generate the heat required to warm the systems they are connected to. They are generally heated by one of the following:

- solid fuel
- gas
- oil
- electricity.

Connection methods vary, depending on the type of system they serve. In addition to the heat source, there is usually a series of time and temperature controls.

▼ Table 12.10 Types of boiler

Fuel type	Boiler type	Working principles
Solid fuel	High-output back boiler installed behind a real open coal fire	This type of boiler is installed directly into a chimney or open flue and connected to a central-heating/hot-water system.
	Independent boiler (freestanding)	This type of boiler is open flued and designed to provide both hot water and central heating in a range of domestic premises. There are two main types of independent boiler: • Gravity feed boilers – these incorporate a large hopper, positioned above the firebox used to feed the fuel automatically to the fire bed as required • Batch feed boilers – these are 'hand-fired' appliances requiring manual stocking of the fuel 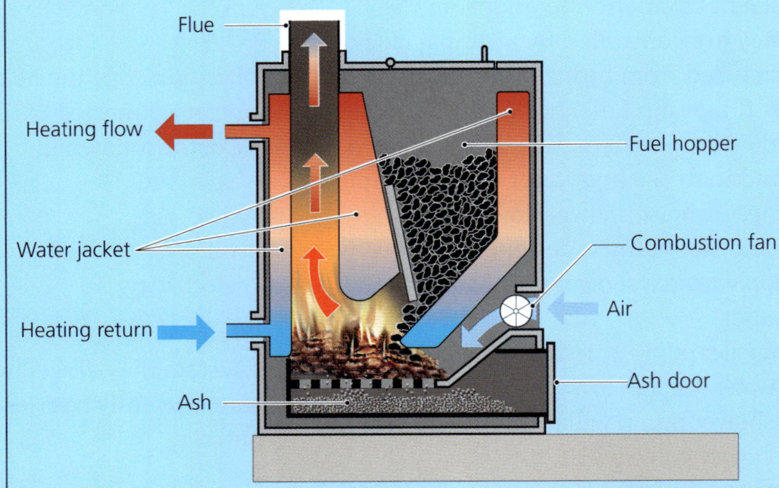
Gas	Traditional wall-mounted boiler	This type of boiler is designed for fully pumped S- and Y-plan heating systems only and uses a variety of flue types. It does not contain any form of expansion vessel or operational control, such as a pump or filling loop. It uses a high-temperature limiting thermostat (energy cut-out) to guard against overheating. 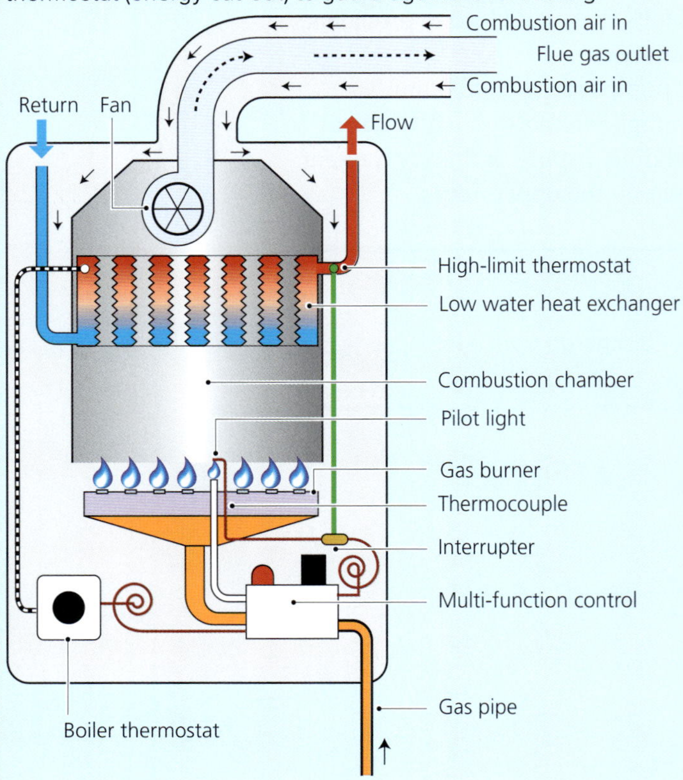

Fuel type	Boiler type	Working principles
	Combination boiler	This type of boiler provides central heating and instantaneous hot-water supply from a single appliance. They are very efficient and contain all the safety controls (i.e. expansion vessel, pressure-relief valve) of a sealed system. Most 'combis' also have an integral filling loop. 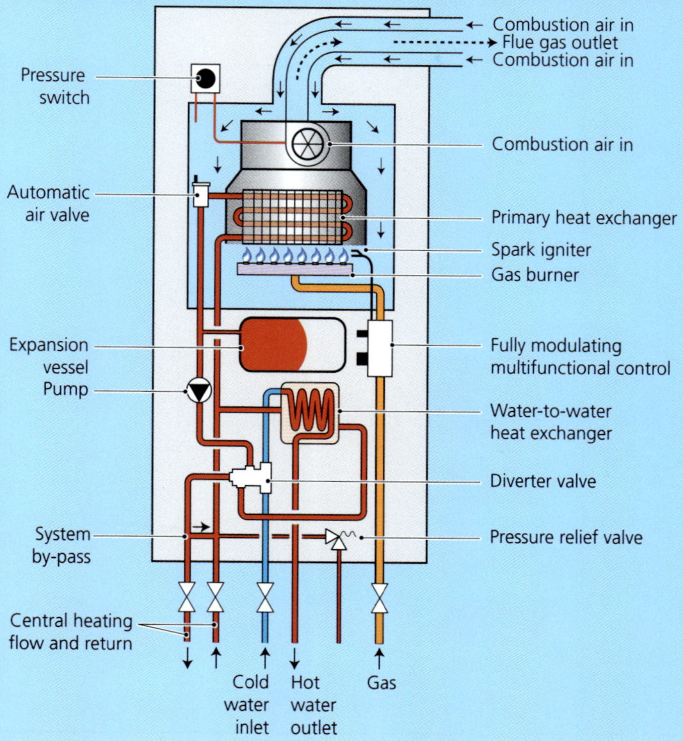
Oil	Pressure-jet boiler	This type of boiler uses an oil **burner** that mixes air and fuel: • An electric motor drives a fuel pump and an air fan. • The fuel pump forces the fuel through a fine nozzle, breaking down the oil into a mist. • This mist is mixed with air from the fan and ignited by a spark electrode. 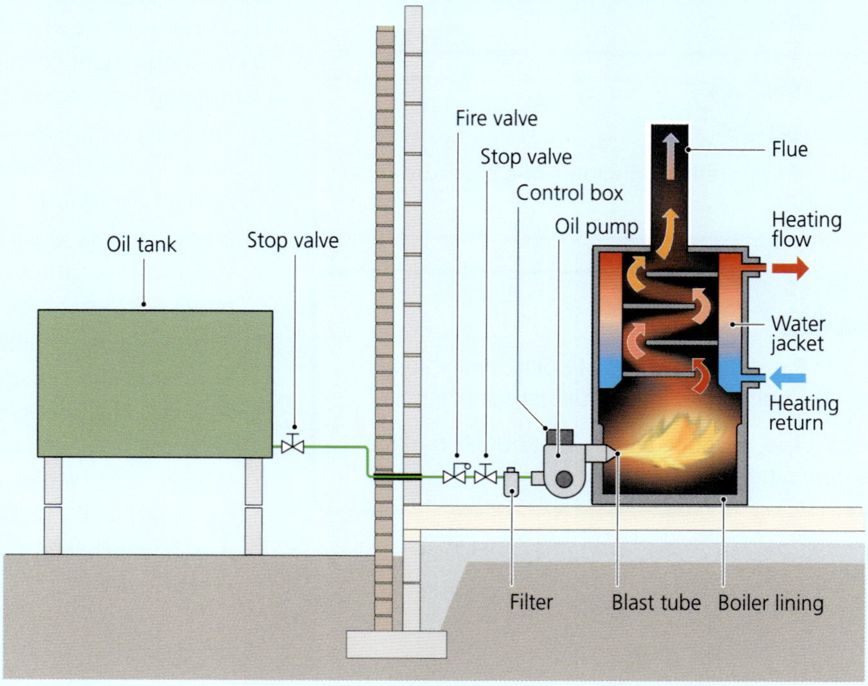

▼ Table 12.10 Types of boiler

Fuel type	Boiler type	Working principles
	Vaporising boiler	The burner in a vaporising boiler works on gravity oil feed; there is no pump. The oil flows to the burner, where a small heater warms it until vapour is given off. This vapour is then ignited by a small electrode. 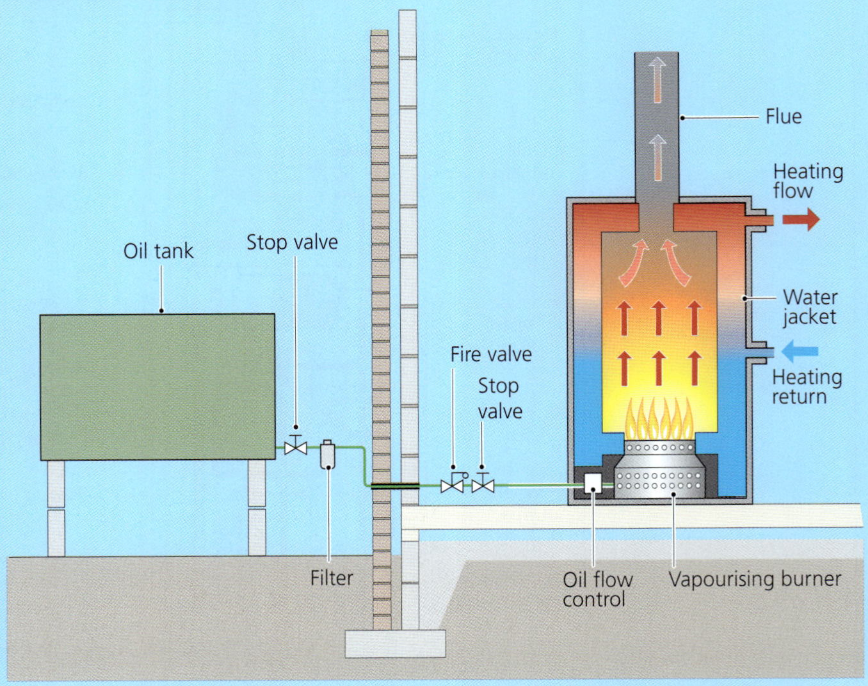

Key term

Burner: an integral part of a boiler where combustion takes place

Test yourself

Describe the operating principles of a pressure-jet oil boiler.

Research

Research the different types of boiler available for domestic, commercial and industrial applications.

For each type, produce a table that details the flue types, fuel types and kW rating.

Chillers

Chillers generate cold liquid by removing heat via a compression or an absorption refrigeration cycle. This liquid can then be distributed through a heat exchanger to chill equipment or through other HVAC processes. Components are connected via a system of pipework.

- Vapour-compression chillers are the most common type. They use a mechanical compressor to force refrigerant around the cooling system.
- Screw chillers use a **rotary-screw compressor**.
- Vapour-absorption chillers use a heat source to move refrigerant around the cooling system.
- Ammonia chillers use water as an absorbent and ammonia as a refrigerant.
- Air-cooled chillers remove heat with fans that force air across exposed condenser tubes.
- Water-cooled chillers remove heat with pumps that send water through a sealed condenser and disperse it through a cooling tower.

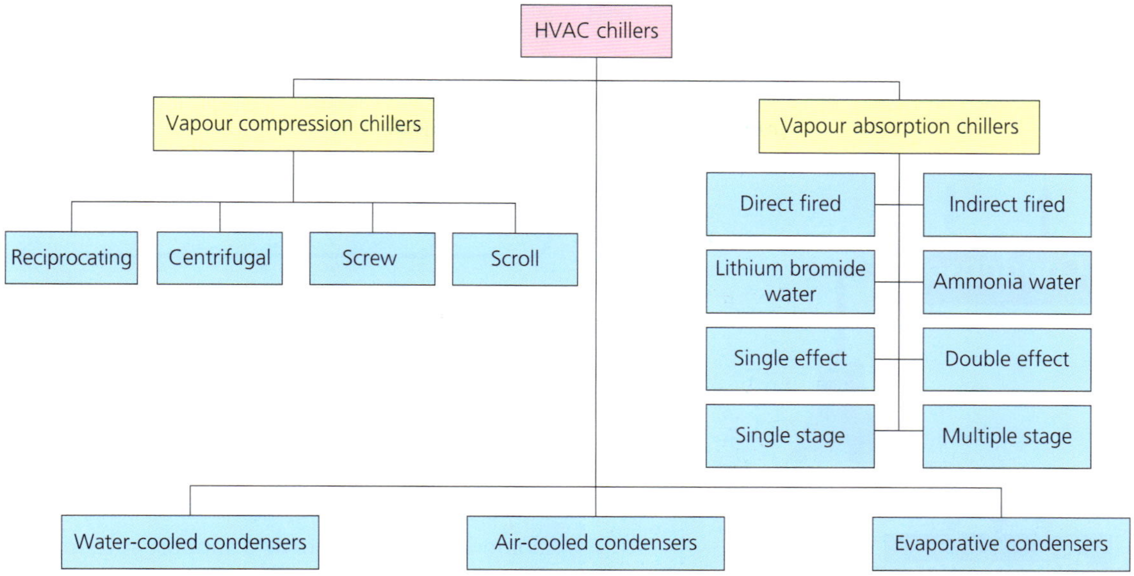

▲ Figure 12.55 Types of chiller

Key term

Rotary-screw compressor: compressor that uses two closely meshing helical screws, known as rotors, to compress a gas

Test yourself

Describe how an air-cooled chiller works.

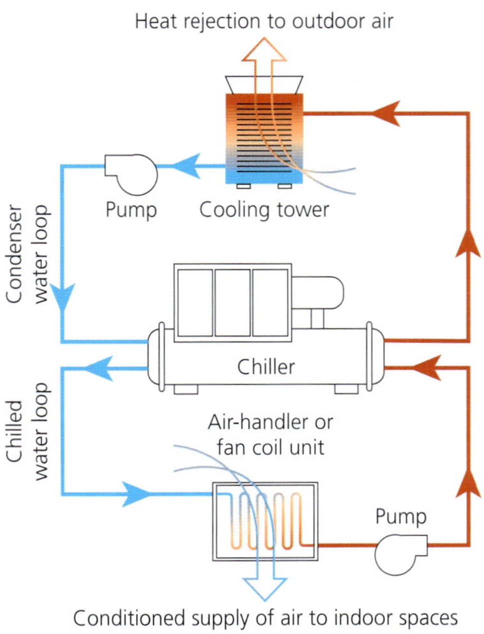

▲ Figure 12.56 Operation of chillers

Thermoelectric refrigeration is the process of pumping heat energy out of an insulated chamber in order to reduce the temperature of the chamber to below that of the surrounding air. It uses a principle called the **Peltier effect** to pump heat electronically.

Key term

Peltier effect: when the passage of a direct electric current through the junction of two dissimilar conducting materials causes the junction to absorb or reject heat

Research

Research the different types of chiller available and their basic operating principles. Produce a presentation and deliver it to your class.

Heat pumps

A heat pump warms or cools a building by moving heat from a low-temperature reservoir to another reservoir at a higher temperature. Its working principles are the same as for a refrigerator, which creates heat while making the refrigerator cold.

The process is known as the vapour-compression refrigeration cycle and involves compressing a gas (called the refrigerant) with a compressor until it becomes a liquid. This generates useful heat that can be used to warm a building. When the pressure is released through an expansion valve, very cold temperatures are generated, which can be used for cooling a building. This process is reversible.

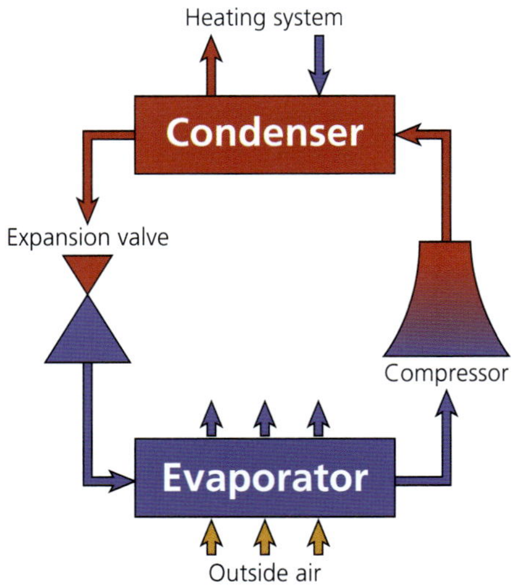

▲ Figure 12.57 Components of a heat pump

There are several different types of heat pump:

- Air-to-air heat pumps are used mostly in commercial buildings as **reverse-cycle heat pumps** that can provide both heating and cooling.
- Air-to-water heat pumps are used to heat swimming pools and to provide hot water and space heating for dwellings.
- Water-to-air heat pumps can use wells or boreholes but can also be installed with many units connected together on a common closed water loop to transfer energy from hot to cold points of a building.
- Ground-to-air heat pumps use constant ground temperatures to provide the heat source, with warm air delivered to the building.
- Ground-to-water heat pumps are the same as ground-to-air heat pumps but used with underfloor heating systems, radiators or wall heaters.

Failure of a heat pump within a system will result in no heating or cooling within a property.

Key term

Reverse-cycle heat pumps: heat pumps that can be used for both heating and cooling

Test yourself

Describe the operating principles of a heat pump.

Research

Using the internet and manufacturer information, create a presentation detailing common controls used in BSE systems. Include the:

- name of the system and its purpose
- names of the controls and what they are used for.

4 Electrotechnical principles of components

A wiring system comprises a cable or collection of cables used to deliver power or services. The cables may be supported by cable trays, baskets, clips or cleats (sometimes referred to as a support system). A cable management/containment system further supports, protects and separates cables, usually by enclosing them in conduit, ducting or trunking. There are many different types of wiring system.

4.1 Cable types

Cables have three main parts:

- The conductor carries the electrical current and is commonly made from copper, either in a single piece or multiple strands. It may also be made from other materials, such as aluminium.
- The insulation is a layer of non-conductive material that covers the conductor. It provides basic protection against electric shock, as well as being a means to identify the use of the conductor.
- The sheath is a secondary layer of non-conductive material surrounding the insulation. It holds the insulated conductors together in one cable and also provides minor mechanical protection to the inner conductor.

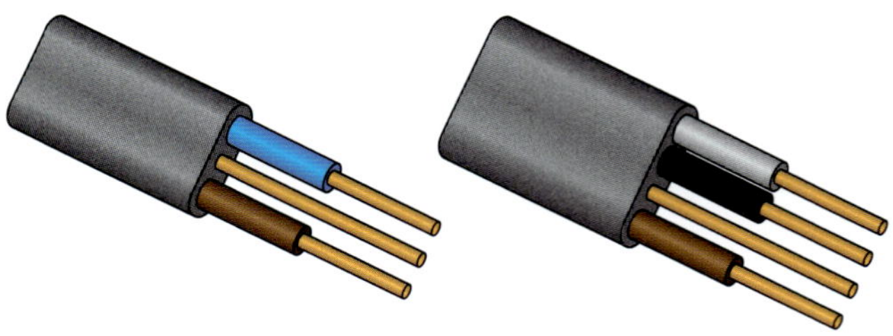

▲ Figure 12.58 Cables

Thermoplastic cables are commonly referred to as PVC (polyvinyl chloride) cables and come in various shapes, sizes and forms, including:

- single-core
- twin-core and CPC (circuit protective conductor) flat profile
- three-core and CPC flat profile
- multi-core flexible.

The type of cable used will depend on its application:

- non-fixed appliances are connected using flexible cable
- twin-core and CPC flat-profile cable are used for lighting and socket circuits.

In domestic installations, the most common cables are:

- twin-core and CPC flat profile
- three-core and CPC flat profile.

Mineral-insulated copper-clad cable (MICC) is manufactured using copper and magnesium oxide as the insulation with an outer polymer sheath. It is used in high temperatures and harsh environmental conditions, resists oxidation and has low flammability.

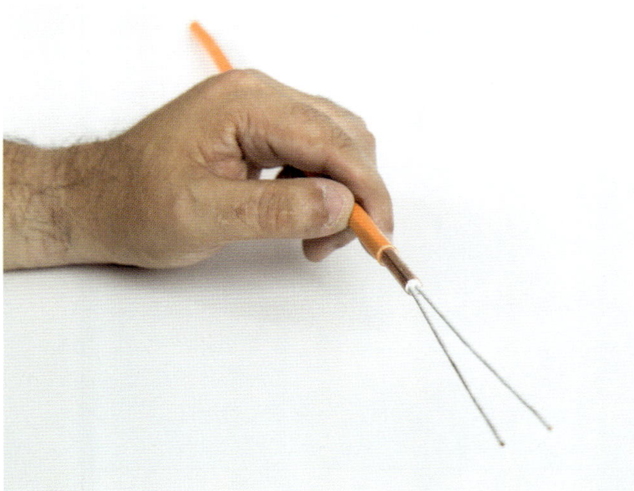

▲ Figure 12.59 Mineral-insulated copper-clad cable

Steel-wire-armoured (SWA) cable is hard-wearing and made up of black PVC sheath, cross-linked polyethylene (XLPE) insulation, copper conductors and steel-wire armouring. It is used for underground systems, cable networks and power networks, as well as indoor and outdoor applications such as power to sheds and garages. It is available in a wide range of core options.

▲ Figure 12.60 Steel-wire-armoured cable

Data cables are used in data networking and telecommunications networks to transmit electronic information from a source to a destination. Cable types include Cat 5 and Cat 6, which are twisted-pair cables used for ethernet and network applications (Cat 6 is an evolution of Cat 5 that supports higher bandwidths).

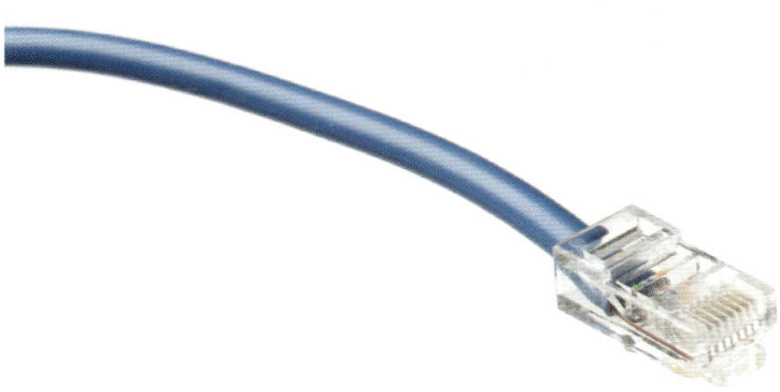

▲ Figure 12.61 Data cable

Fibre-optic cables are also used in data networking and telecommunications networks to transmit electronic information. They contain one or more transparent optical fibres that are used to carry light. The optical fibres are typically glass or plastic pipes contained in a protective tube.

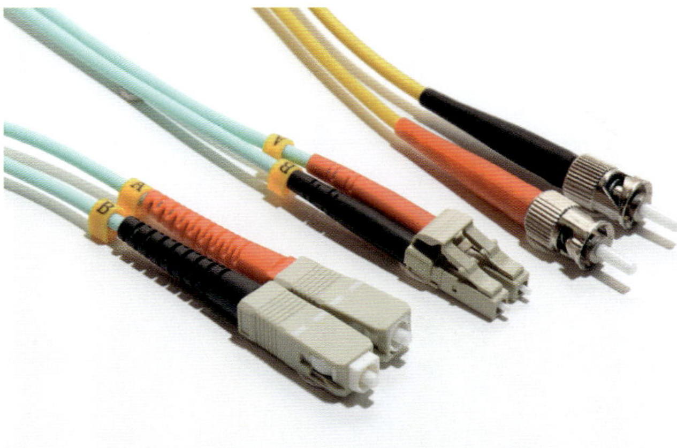

▲ Figure 12.62 Fibre-optic cable

Fire-resistant cables are used in fire detection and alarm systems, voice alarm systems, emergency lighting systems and other essential service circuits. **Insudite insulation** ensures fire/heat resistance and greater cable durability.

Key term

Insudite insulation: a type of fire- and heat-proof insulation

Test your knowledge

Explain the uses of steel-wire-armoured (SWA) cable and fire-resistant cable.

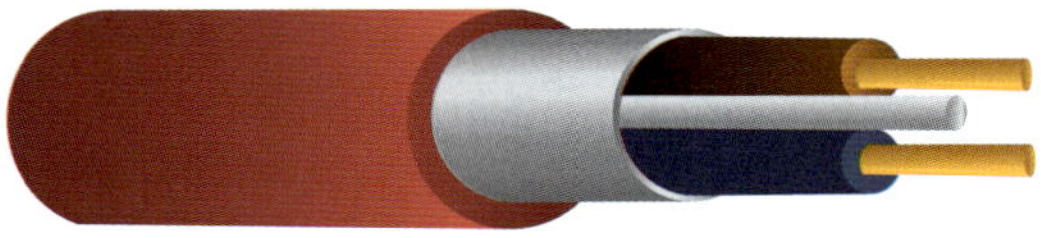

▲ Figure 12.63 Fire-resistant cable

4.2 Electrical accessories

There is a wide range of accessories that may be required when installing electrical systems. These can be split into two categories:

- first-fix accessories, such as electrical back boxes and jointing boxes
- second-fix accessories, such as sockets, switches and pull cords.

▼ Table 12.11 First-fix electrical accessories

Type	Used for
Drylining box 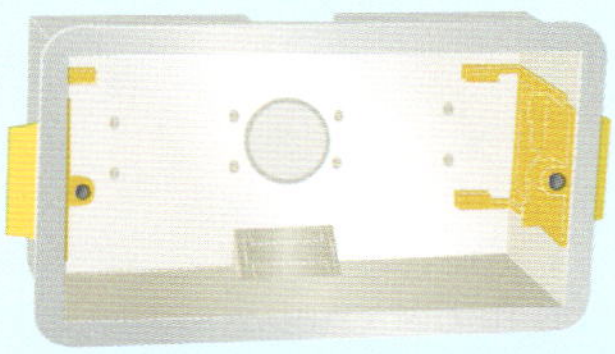	Installations in drylined (plasterboard) walls
Countersunk metal box 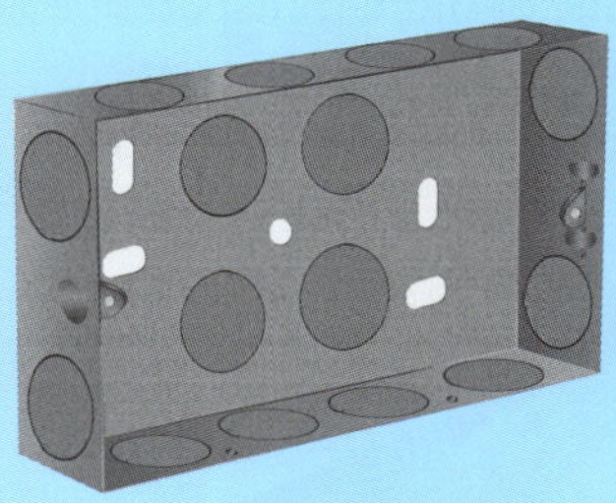	Recessed installations in brick/block/plaster
Surface-mounted box 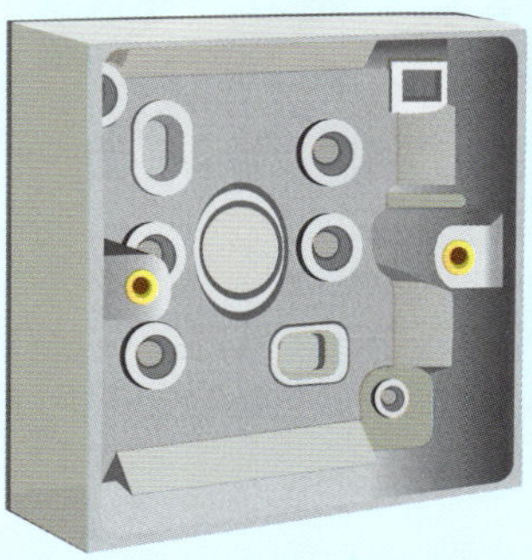	Surface-mounted installations
Junction box 	Connecting a number of cables together

Test your knowledge

List three first-fix electrical accessories.

▼ Table 12.12 Second-fix electrical accessories

Type	Used for
Fan isolator 	Isolating kitchen and bathroom fans
Two-way lighting switch 	Operating lights
Pull cord 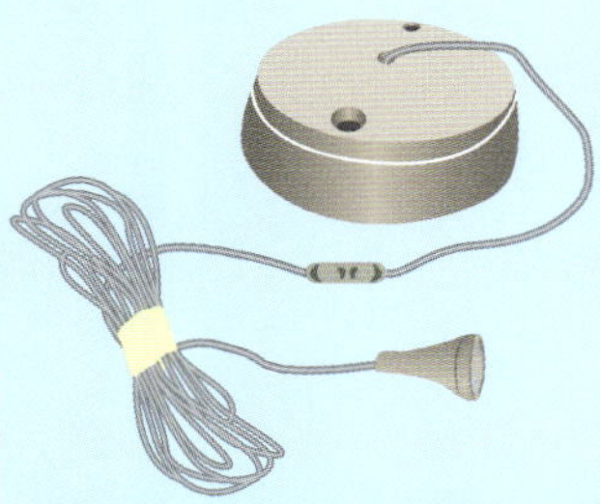	Isolating the power supply to showers and operating lights in bathrooms
Pull cord with neon indicator 	Isolating the power supply to showers

Type	Used for
Unswitched fused spur 	Connecting fixed appliances, such as WC **macerators**
Switched fused spur with neon indicator 	Connecting fixed appliances, such as boilers
Unswitched socket (one gang) 	Socket circuits with up to 13 A plugs
Switched socket (double gang) 	Socket circuits with up to 13 A plugs

Key term

Macerators: plumbing components used to convert waste from a toilet, shower or washbasin into a fine slurry that can be pumped into the sewage line; they are used where access to gravity drainage is not available

Test your knowledge

List three second-fix electrical accessories.

Research

Research the different types of electrical accessory and produce a material directory.

4.3 Containment

Conduit

Where an installation is likely to be subjected to external influences which may damage the wiring, a common method of protection is **conduit**. Conduit is available in both steel and PVC and comes in many different forms, including:

- solid-steel extruded
- solid-steel rolled
- flexible steel
- rigid PVC
- flexible PVC.

▲ Figure 12.64 Conduit

Typical conduit external diameters are 16, 20, 25 and 32 mm, and it is supplied in 3 m and 3.75 m lengths.

Key term

Conduit: a tube used for protecting electric wiring

▼ Table 12.13 Advantages and disadvantages of using conduit

Advantages	Disadvantages
• There is increased mechanical protection of the wiring. • Rewiring can be carried out relatively easily, providing all cables are replaced at the same time. • Using steel conduit can minimise the risk of fire spreading. • Conduit can be used as the circuit protective conductor (CPC).	• Installation costs are high, due to the time and materials required. • PVC conduit is directly affected by UV and can become discoloured and brittle in direct sunlight. • Both PVC and steel conduit can be affected by acids, alkalis and corrosive fumes, with steel conduit being susceptible to rust which can form on the inside.

Steel and PVC conduits use some common accessories in their installation, but they also have their own specific accessories.

▼ Table 12.14 Conduit accessories

Accessory	Description
Junction box 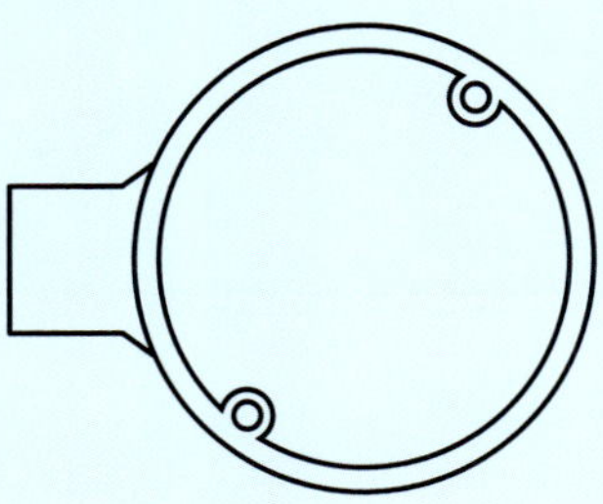	Used at the end of a conduit run

Accessory	Description
Through box 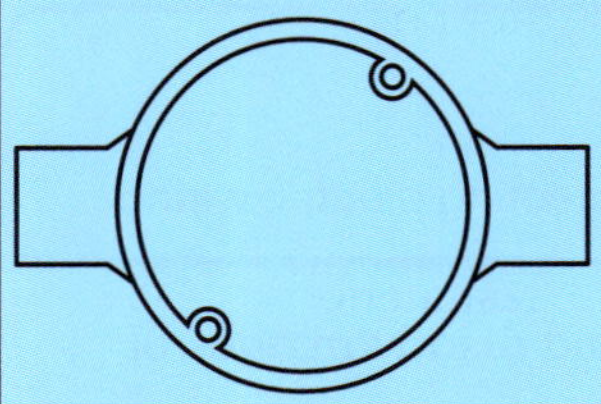	Used as a through link
Angle box 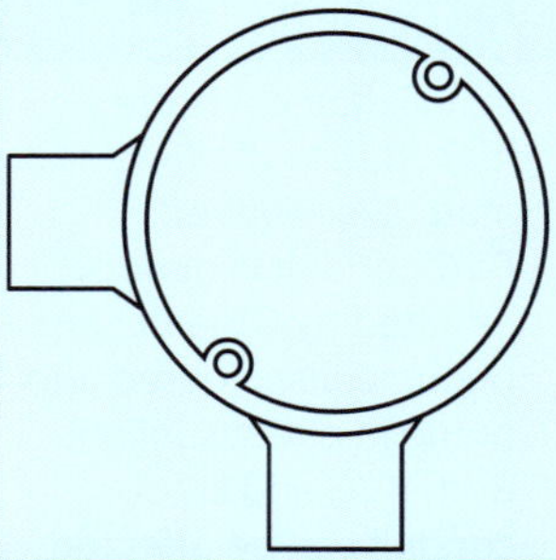	Used for going round corners or changing direction
Four-way box 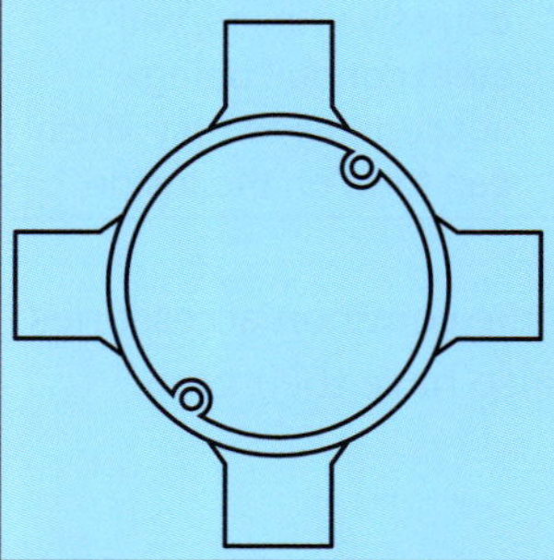	Sometimes called a cross box; has four spouts at 90° to one another
H-box 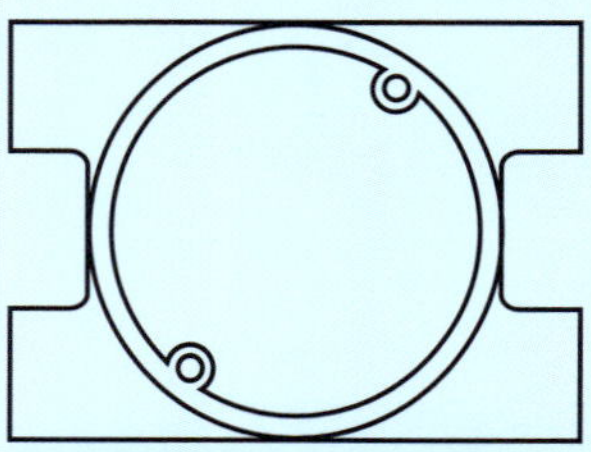	Has four spouts forming a letter H, with two on each side
Strap saddle 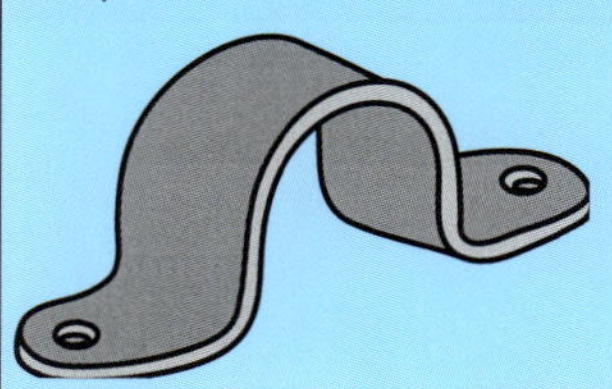	Used when the conduit is to be secured in place directly on the surface, with no gap at the back; also referred to as a stamp saddle

Trunking

Trunking is an enclosure for a wiring system with one removeable or hinged side. It enables cables to be installed easily and provides them with a level of mechanical protection.

Trunking comes in two forms:

- PVC
- steel.

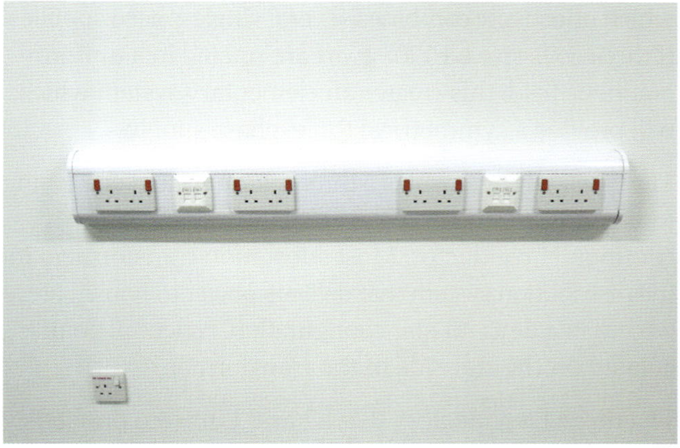

▲ Figure 12.65 Trunking

Segregated trunking

Irrespective of whether trunking is PVC or steel, it can be split into different sections or compartments. This can assist with segregation of specific circuits, for example keeping data cables separate from low-voltage power circuits, and keeping band-1 and band-2 circuits apart. This will reduce any interference between systems and ensure each system can be easily identified.

Cable tray

Sometimes the size or type of cable and the route or required bend radius mean it is not practical to use trunking or conduit. Instead, armoured cables are used in conjunction with cable tray.

Cable tray comes in several forms, including:

- heavy-gauge perforated
- light-gauge perforated
- heavy-gauge solid
- light-gauge solid.

The choice of cable tray depends on the application, but heavy gauge is used for larger cables or where there is a greater mass of smaller cables to be supported. Heavy-gauge cable tray tends to have walls with bent-over edges, whereas the walls on light-gauge cable tray are straight with no returned edge.

As with all other forms of support system, standard accessories are available, including:

- couplers
- tees
- crosses
- brackets
- coupler plates
- nut bolts
- washers
- full-threaded rods
- anchor fasteners/bullet fasteners.

Cable basket

Also referred to as basket tray, cable basket is commonly used for the structured cabling of information technology systems. The main purpose is to provide support rather than protection, as the cables are just laid in the basket. However, this means that wiring can be installed easily.

▲ Figure 12.66 Cable basket

Test your knowledge

Describe three types of containment system.

5 Electrotechnical supply

Several different electricity supply systems may be used in premises, catering for specific requirements. These may be direct current (DC) or alternating current (AC) operating at different voltages, and in the case of AC the supply may be single- or three-phase. For more on the use of DC and AC and different voltages, see Chapter 2.

DC is not used for public electricity supplies (with the exception of the links between England, the Netherlands, Ireland and France) but has some work applications, such as battery-operated works plant (for example forklift trucks). Certain parts of the UK railway system, in particular the London Underground and services in southern England, also use DC for traction supplies.

AC is the distribution system of choice for electricity suppliers all over the world, mostly due to its versatility. An AC supply is compatible with a wider range of circuit arrangements and supply voltages, and transformers enable the supply voltage to be changed up or down.

Research

Research online the voltage supplies required for a range of BSE systems and the benefits of having different voltage supplies for domestic and commercial electrical systems.

5.1 Single-phase supply systems

The simplest AC supply arrangement is a two-wire system, known as single phase. It is used in both domestic and work premises for applications such as lighting and socket outlets.

The two conductors are referred to as the line conductor (L) and the neutral conductor (N). The N conductor is connected to earth at every distribution substation on the public supply system and therefore its voltage, with reference to earth, should be no more than a few volts at any point. The voltage between the L and N conductors corresponds to the nominal supply voltage (230 V).

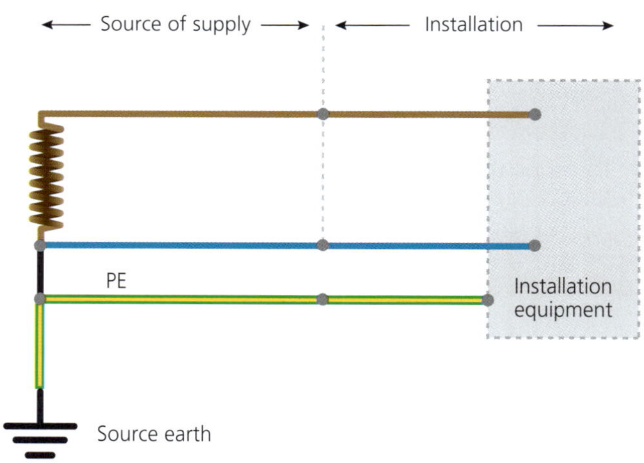

▲ Figure 12.67 Single-phase supply system

Industry tip

In rare circumstances, an installation may be supplied with a two-phase and neutral supply, for example to supply a large heating load which cannot be accommodated by a single-phase supply.

5.2 Three-phase and neutral supply systems

While single-phase AC supplies are adequate for domestic premises, the much higher loads typical of industrial and commercial premises need to use very large conductors to carry the high currents involved. The high currents also give rise to a large **voltage drop**. However, it is possible to use a multi-phase arrangement, which effectively combines several single-phase supplies. If three coils spaced 120° apart are rotated in a uniform magnetic field, this creates an elementary system which will provide a symmetrical three-phase supply.

The usual arrangement is a three-phase system employing four conductors – three separate L conductors and a common N conductor. Most substation transformers in the distribution system that delivers power to houses are wound in a **delta**-to-star configuration. A neutral point is created on the star side of the transformer.

The three L conductors (L1, L2 and L3) were previously distinguished by standard colour markings: red, yellow and blue. However, European harmonisation resulted in these conductor colours being changed to brown, black and grey.

Key terms

Voltage drop: the decrease of electrical potential along the path of a current flowing in an electrical circuit

Delta: where the windings of a transformer are arranged in a triangular formation, with the start of one winding connected to the end of another, meaning the voltage across each winding is the same as the line voltage

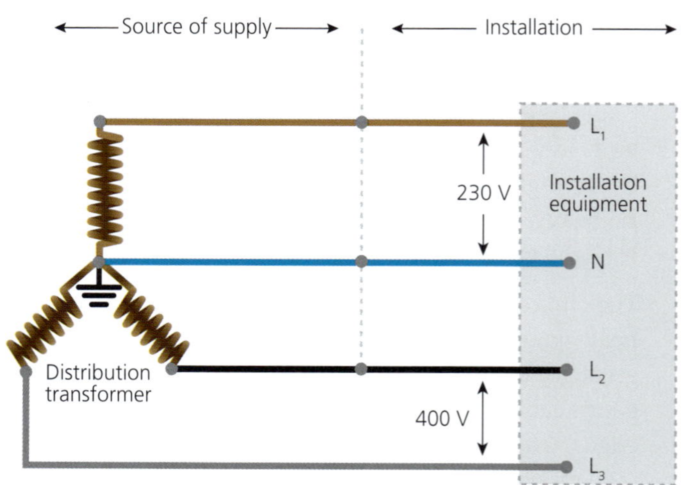

▲ Figure 12.68 Three-phase supply system

Balanced three-phase supply system

In the UK, electricity is provided from a delta-star transformer, with an earthed star point on the secondary side of the transformer. This is normally distributed as a three-phase four-wire system.

Balanced three-phase systems, as described above, will balance out, therefore giving no neutral current. Single-phase loads require a neutral connection, which will carry any out-of-balance current. This should be kept to a minimum, to reduce cable and switchgear sizes and ensure maximum transformer use. Any difference in load will be 'mopped up' by current flowing in the neutral conductor.

It must be mentioned at this point that the effects of neutral current become much more complex with different loads. Values of neutral current can be determined simply by using a scale drawing based on an equilateral triangle. If all phases are balanced, therefore equal, all three sides of the triangle will meet. If they are not balanced, there will be a gap, which represents the neutral current.

For more on supply systems, see Chapter 2.

Research

Using the internet, research balanced three-phase supply systems and produce a handout to explain how they work.

6 Earthing arrangements

6.1 Earthing

To reduce the risk of electric shock, circuits are protected using a system called automatic disconnection of supply (ADS).

The green and yellow earth cables in an electrical circuit are called circuit protective conductors (CPCs). Every exposed metallic part in an electrical circuit must be joined to a CPC to ensure it is connected to the earth path should a fault occur. This will ensure the low-resistance earth path creates a high fault current, causing quick disconnection.

Earth cables play a major part in achieving ADS in the event of a fault to earth. If a fault exists between line and earth, there is a major risk of electric shock, as someone may come into contact with metallic parts.

Test yourself

- Explain the purpose of an earthing system.
- State the colour of earth conductors found in electrical systems.

6.2 TN-C-S (PME) systems

The TN-C-S system is now common throughout the UK, as it allows the **district network operator (DNO)** to provide a low-voltage supply with a reliable earthing arrangement to many installations across the country. It is also known as protective multiple earthing (PME). It relies on the neutral being earthed close to the source of supply and at points throughout the distribution system. There is also a neutral-to-earth connection at the intake of the installation.

As the DNO uses the combined neutral and earth return path (known as a protective earthed neutral or PEN), the maximum external earth fault loop impedance declared by the DNO is 0.35 Ω. There may be a number of consumers using the supply cable. A rise in current flow will create a voltage rise in the PEN, which needs multiple connections to the general mass of earth along the supply route.

Key term

District network operator (DNO): a company licensed to distribute electricity

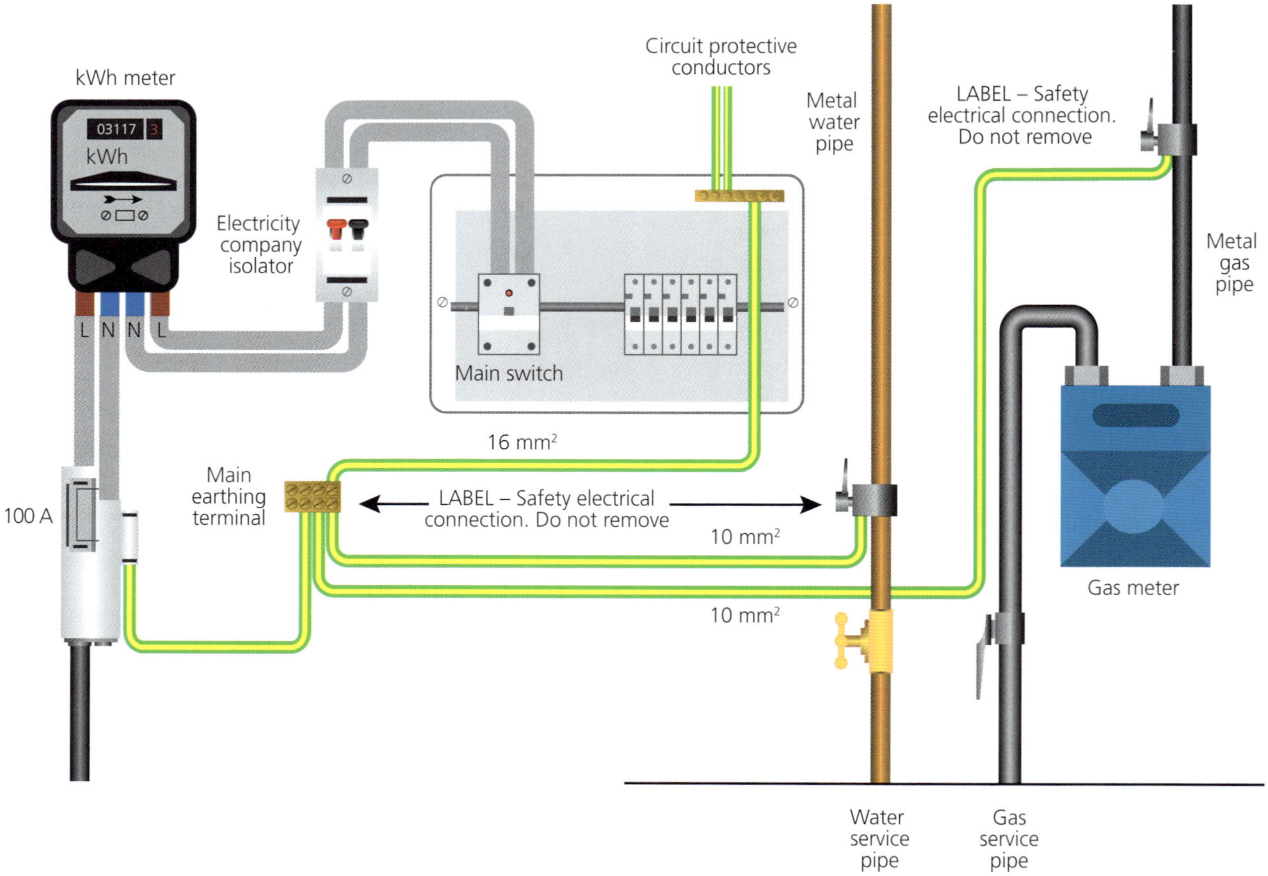

▲ Figure 12.69 TN-C-S system

There are risks. If the PEN conductor becomes an **open circuit** in the supply, current flowing through the installation will not have a path back to the substation through the supplier's cable. Instead, current could try to follow an alternative path through the earthing system of the installation, which may include service pipework or the general mass of earth. Unfortunately, it could be people who make that link between the earthed metallic equipment and the earth. As a result, certain installations such as petrol filling stations and some construction sites and caravan parks, cannot be supplied by TN-C-S arrangements. Even in domestic or commercial installations some restrictions may apply, such as certain outbuilding supplies including garages, sheds and workshops. In these cases, TT systems are preferred (see below).

Key term

Open circuit: an electrical circuit that is not complete so current does not flow

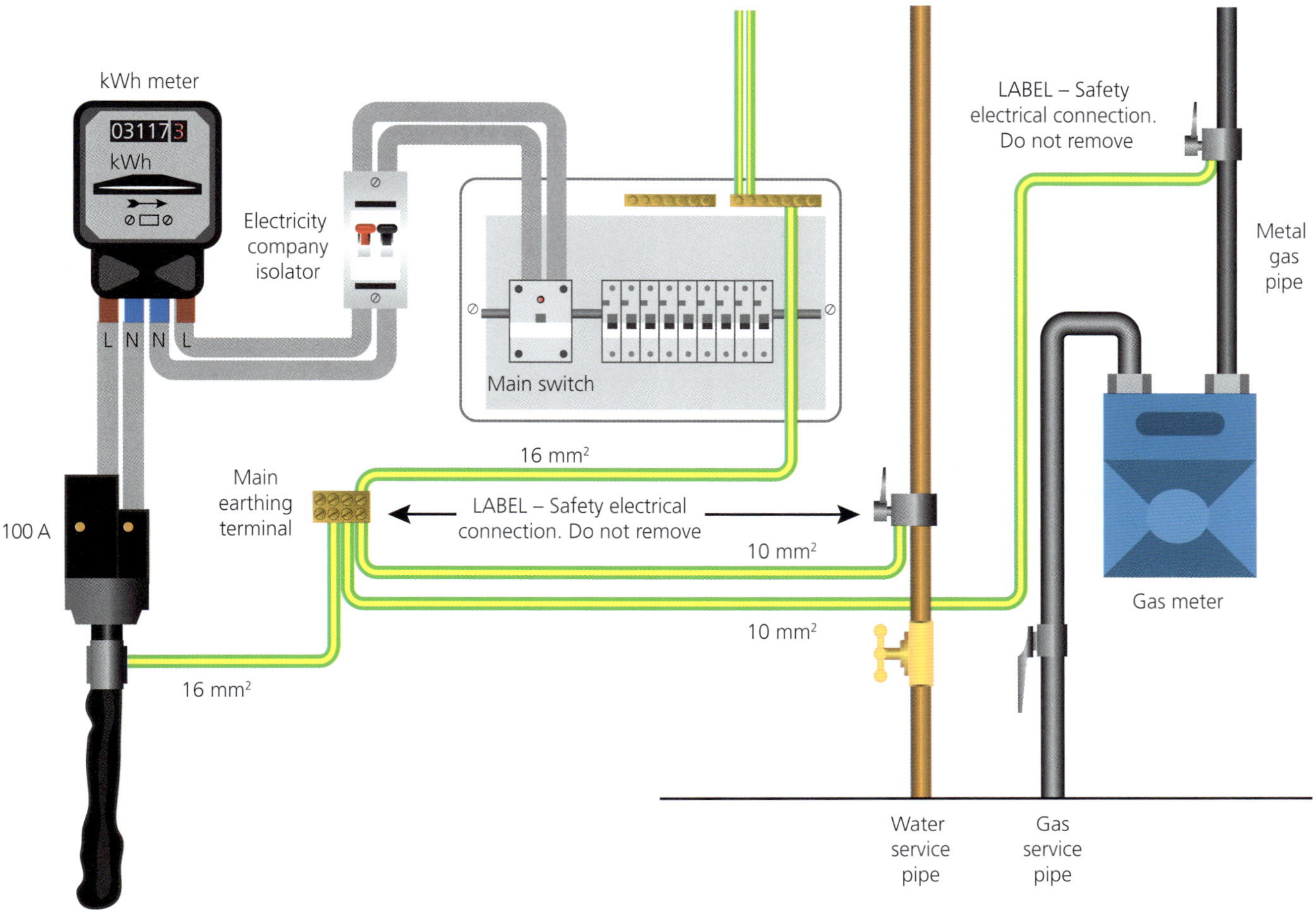

▲ Figure 12.70 TN-S system

6.3 TN-S systems

A TN-S earthing configuration has only one neutral-to-earth connection, which is as near as practicable to the source (supply transformer). In low-voltage supplies, the consumer's earth conductor is connected to the metallic sheath of the DNO's supply cable. This sheath provides a separate route back to the substation transformer. Because the return path is usually a material such as steel, the DNO will normally declare a maximum external earth fault loop impedance (Ze) of 0.8 Ω.

6.4 TT systems

This type of system is configured in much the same way as a TN-S system in terms of the earthing of the supply source. However, a TT system does not provide the consumer with an earth connection. Instead, the earth for the consumer's installation has to be supplied by the consumer, usually by driving earth rods into the ground or burying metallic plates or strips to provide a path of low-enough impedance through the ground to give protection.

TT systems are usually installed either where a TN-C-S arrangement is not permitted (for example in a petrol filling station or in rural installations where the supply is provided via overhead poles) or where there is no opportunity to provide other types of system. As the earth return path uses the general mass of earth, external earth fault loop impedance values (Ze) may be very high where different soil types exist, meaning further shock-protection measures such as **residual current devices (RCDs)** may be required to provide ADS.

Key term

Residual current device (RCD): also called ground-fault circuit interrupter (GFCI), this is an electrical safety device that quickly breaks an electrical circuit with live current leakage to earth, in order to protect equipment and to reduce serious harm from ongoing electric shock

Test yourself

Name three types of earthing system.

Research

Research the hazards associated with each type of earthing system and how they impact different building services.

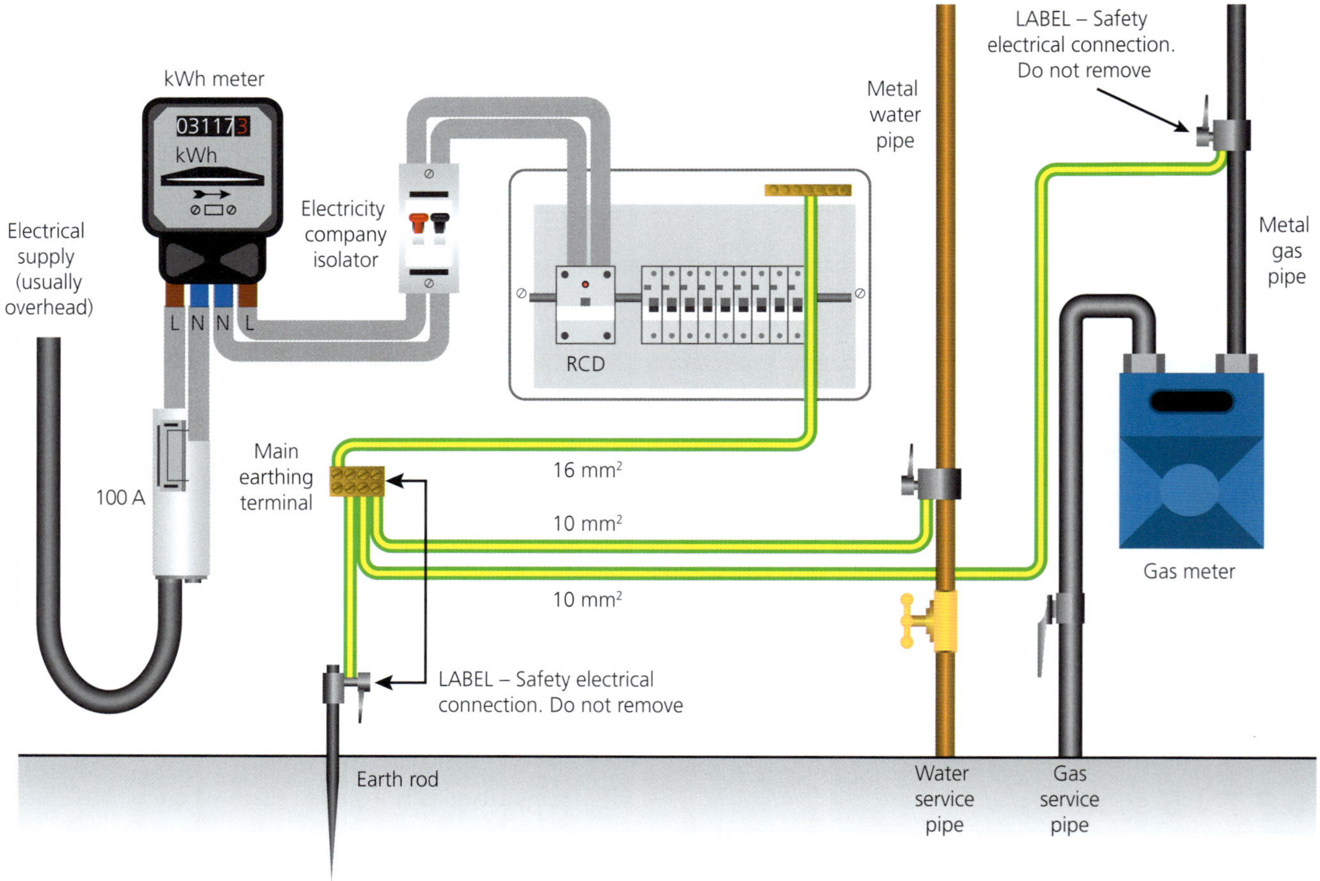

▲ Figure 12.71 TT system

7 Cabling, accessories and equipment used in older electrical installations

Test yourself

Why are installations that use lead-sheathed cable considered unsafe?

7.1 Cables, accessories and equipment

Lead-sheathed cable

In some of the earliest electrical installations in around 1880, multi-core cables were available with wax-impregnated cotton and silk insulation and lead sheaths. As this type of cable is over a hundred years old and the insulation around the inner cores will have deteriorated, any installation still using it should be considered unsafe and isolated immediately. A failure of the inner insulation could cause the outer lead coating to become live.

Vulcanised India rubber (VIR) insulated cable

Before PVC cables and up until the 1960s, electrical cables used vulcanised India rubber (VIR) for insulation and protection. Rubber is perishable so, over time, the sheathing on the cable becomes brittle and starts to crack and break away, exposing the copper in the cable. This can create the risk of electric shock, fire or immediate loss of power due to a short circuit rupturing the protective fuses.

Test yourself

What is the problem with vulcanised India rubber (VIR) insulated cable?

Cable colours

UK cable colours were harmonised with countries in mainland Europe in 2006, as shown in Table 12.15.

▼ Table 12.15 Types, colours and functions of cables

Type	Colour of cable	Function
Live	Brown	Carries electricity to the appliance
Neutral	Blue	Transfers electricity away from the appliance
Earth	Green and yellow	Connects the electrical appliance or installation with the ground (vital for safety)

During a new installation, electricians should attach the correct-coloured insulation, sheathes, or sleeving to the wires, to enable easy identification. Where there is any mixing of old and new colours, cables should be clearly marked with the relevant colour codes.

Test yourself

What are the colours of live, neutral and earth cables?

Table 12.16 shows typical cable sizes available, with corresponding current and maximum power ratings.

▼ Table 12.16 Cable sizes

Cable size	Current (A)	Maximum power (W)
1.0 mm^2	10	2400
1.25 mm^2	13	3120
1.5 mm^2	15	3600
2.5 mm^2	20	4800
4.0 mm^2	25	6000

BS 3036: rewireable fuses

A fuse is a basic protection device that is destroyed and breaks the circuit should the current exceed the fuse rating. In older equipment, it may be just a length of fuse wire fixed between two terminals. These are now becoming uncommon, as electrical installations are rewired or updated.

One of the main problems associated with rewireable fuses is the overall lack of protection, including insufficient breaking-capacity ratings caused by overcurrent or a fault in the installation.

Another major problem is unreliability, due to:

- using the wrong gauge of wire when changing the fuse
- the wire being labelled with an incorrect current rating
- the number of times and length of time that a fuse wire has been subjected to overload.

▲ Figure 12.72 Rewireable fuse

Non-fire-rated consumer units/distribution boards

Consumer units are a type of distribution board typically installed in domestic premises to provide control, distribution and protection for the various circuits within the electrical power system. They can include:

- circuit breakers (CB)
- residual current breakers with overload (RCBO)
- fuses.

They also have a main switch to isolate the entire installation.

In early installations, it was acceptable to install a consumer unit with a wooden back box, and many are still in existence. However, the IET Wiring Regulations require that consumer units in domestic premises are now manufactured from non-combustible material to contain any fire within the enclosure and to minimise the spread of fire. Non-fire-rated consumer units and distribution boards add an additional source of fuel for a fire.

Research

Answer the questions below using the information on https://electrical.theiet.org/courses-resources-career/free-resources/consumer-guidance/consumer-units

- Which British Standard states the requirements for consumer units?
- What are the requirements for protection from fire?
- What are the requirements for protection from electric shock?
- What are the common problems associated with consumer units?

Industry tip

Since January 2016, the IET Wiring Regulations have required that consumer units in domestic premises are manufactured from non-combustible material.

Test yourself

Why must consumer units be manufactured from non-combustible material?

8 Pipework and ductwork, components and systems

8.1 Pipework

Pipework allows fluids and gases to travel through a building and is used in a range of BSE systems, including plumbing, heating, gas and air-conditioning systems. It is jointed using a range of fittings.

Copper pipework is manufactured for water and gas installations to BS EN 1057 and is available in several grades, as shown in Table 12.17.

▼ Table 12.17 Grades of copper tube

Grade	Description
R220	• Softer copper tube, fully annealed and supplied in coils • Thicker walled than other grades of copper tube • Used for underground water services (sizes 15, 22, 28 mm) and microbore central-heating systems (sizes 6, 8 and 10 mm)
R250	• The most widely used grade of copper tube for plumbing and heating applications • Supplied in straight lengths of 3 or 6 m, in sizes 15, 22, 28, 35, 42 and 54 mm • Known as half-hard tempered
R290	• Hard tempered, thin walled and totally unsuitable for bending • Not normally used in the UK

Low-carbon steel pipe is available in three grades, with each identified by a different colour. The grades all have identical external diameters but varying wall thicknesses; heavy grade has the thickest pipe wall and light grade has the thinnest.

Key term

Low-carbon steel: a ferrous metal used for pipework applications; also known as mild steel

▼ Table 12.18 Grades of low-carbon steel pipe

Grade	Colour	Description
Light	Brown	• Not normally used for plumbing/heating pipework • May be used for dry sprinkler installations for fire prevention
Medium	Blue	• Used for wet central-heating systems and oil and gas pipework • Available in 6 m lengths with plain or threaded ends • Referred to in imperial sizes ½", ¾", 1", 1¼", 1½", 2" available in sizes up to 6" and 8" for use in commercial and industrial applications.
Heavy	Red	• Used for wet central-heating systems and oil and gas pipework where long system life is expected • Available in 6 m lengths with plain or threaded ends • Referred to in imperial sizes ½", ¾", 1", 1¼", 1½", 2"

Polybutylene is manufactured into pipe for pressurised plumbing systems. It can be used in hot- and cold-water installations and wet central-heating systems, and is available in sizes 10, 15, 22 and 28 mm in both straight lengths of 3 m and coils of 25, 50 and 100 m.

Advantages include:

- good flow-rate characteristics
- low noise transmission
- flexible and easy to install
- high resistance to frost damage
- 50-year guarantee
- non-corrosive.

Corrugated stainless-steel tubing (CSST) is flexible tubing used to supply natural gas in domestic, commercial and industrial buildings. It is available in sizes 15, 20, 25, 32, 40 and 50 mm in coils of 45, 75 and 90 m.

Research

Research the advantages and disadvantages of using corrugated stainless-steel tubing (CSST). Produce a table to detail your findings.

Plastic pipework is used for above-ground drainage systems and comes in a range of materials depending on the application and jointing method. Materials include polyvinyl chloride, acrylonitrile butadiene styrene (ABS) and polypropylene. Waste pipes are available in sizes 32, 40 and 50 mm. Soil pipes are available in sizes 110 and 150 mm.

Medium-density polyethylene (MDPE) is a hard-wearing plastic for water pipes, gas pipes and fittings. It is available in a variety of colours (yellow for gas, blue for water), in sizes of 20 to 63 mm, and supplied in coils of 25 to 150 m.

Industry tip

The most common pipe size used for cold-water services in domestic properties is 25 mm.

Test yourself

List three pipework materials and their use within BSE systems.

Pipework accessories

▼ Table 12.19 Pipework accessories

Type	Purpose
Isolation valves	To turn off (isolate) complete systems, parts of a system or appliances
Drain valves	To drain down systems
Stop taps 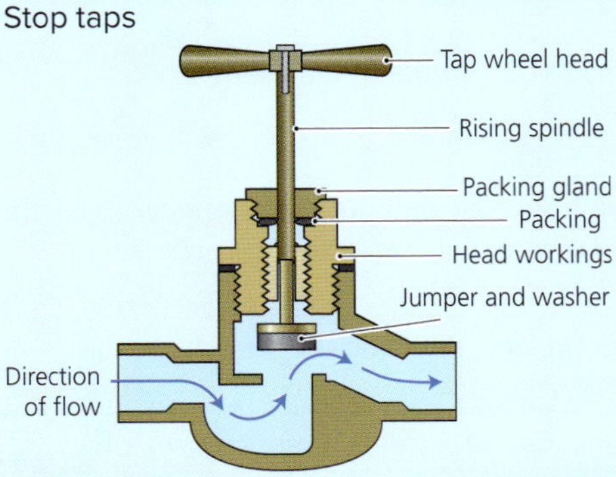	To isolate high-pressure cold-water systems
Gate valves 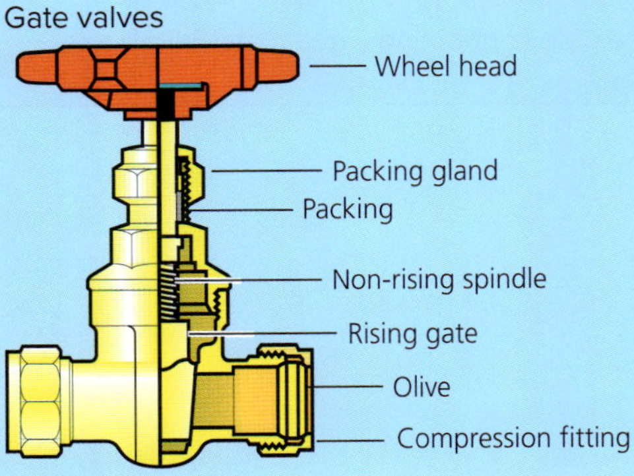	Used on low-pressure installations, such as the cold feed to vented hot-water storage cylinders and the cold-distribution pipework for indirect cold-water systems, to isolate systems and components

Type	Purpose
Spherical plug valves 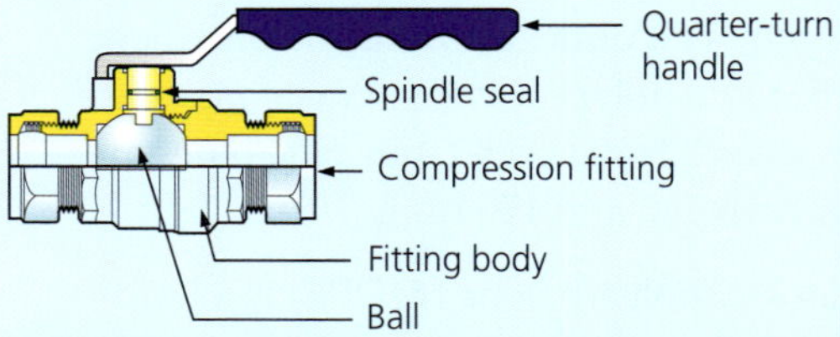 	To isolate appliances and terminal fittings such as taps and float-operated valves
Drain-off valves 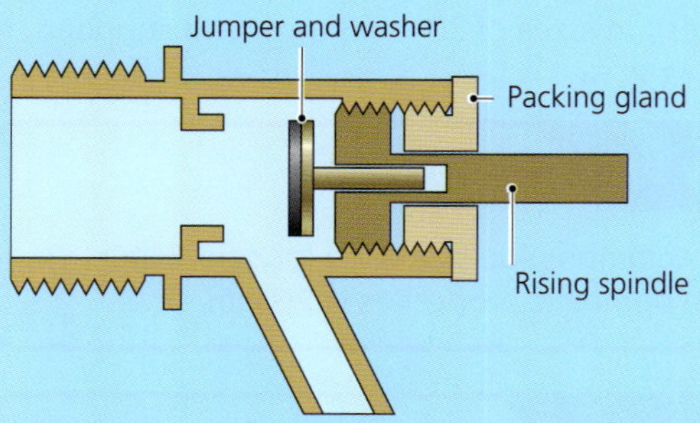 	Small valves strategically placed at low points in pipework installations to allow draining down of the system
Float-operated valves 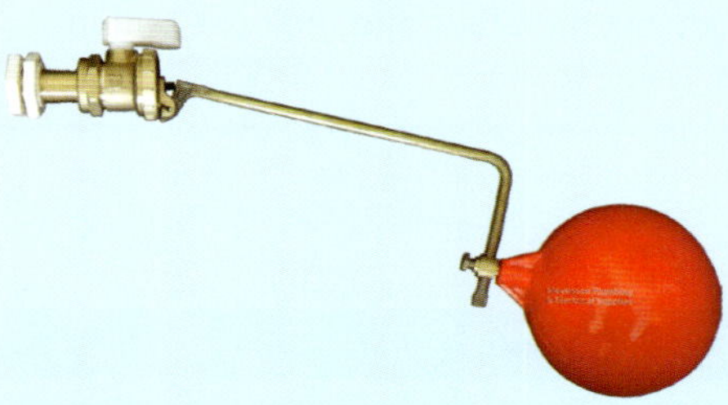	To control the flow of water into cold-water storage and feed cisterns, feed and expansion cisterns, and WC cisterns; designed to close when the water reaches a pre-set level
Radiator valves • Thermostatic radiator valves (TRVs) • Wheel-head valves • Lockshield valves	To control the temperature and flow through a radiator • To control the temperature of a room by regulating the flow of water through a radiator • To allow manual control of a radiator by being turned on or off • To regulate the flow of water through a radiator; designed to be adjusted during system balancing
Automatic air valves	To allow collected air to escape from a system but seal themselves when water arrives at the valve; fitted where air is expected to collect in a system, usually at high points
Anti-gravity valves	To prevent unwanted gravity circulation within heating and hot-water systems
Pressure-relief valves	To protect against over-pressurisation of water in a range of systems
Emergency control valves (ECVs)	To allow the gas user to shut off the supply of gas in the event of an escape; found on the service pipe connecting a gas meter to the gas mains

Test yourself

List four types of valve found in plumbing systems.

Improve your maths

Refer to Figure 12.21 showing an indirect cold-water system. Using a supplier's catalogue, produce a materials list and calculate the cost of components required.

8.2 Ductwork

Ductwork allows heated or cooled air to travel through a building and is used in heating, ventilation and air-conditioning systems.

▲ Figure 12.73 Ductwork

Types of ducting

Flexible ducting

Flexible ducting is available in a range of materials and diameters from 80 to 500 mm, depending on its application, and is also suitable for a range of temperatures.

Types of flexible ducting include:

- aluminium foil
- insulated
- acoustic
- PVC domestic
- PVC-coated fabric.

The type of ducting required will be specified by BSE designers.

Metal ducting

Metal ducting is made from different kinds of sheet metal, such as galvanised steel and aluminium, and is available in a range of sizes from 63 to 630 mm. It is jointed using a range of fittings, including:

- dampers
- T pieces
- connectors
- cap ends
- 90-degree bends
- 45-degree bends.

Fabric ducting

Fabric ducting, also known as textile air diffusers or air socks, is an alternative to cumbersome conventional metal ductwork. Each duct is designed specifically to deliver the air pattern and velocity required for the room.

Cardboard ducting

Cardboard ducting has a coating made from a water-based solution with a water-dispersal polymer, fire-retardant minerals and a final hydrophobic finish.

Research

Visit www.ductstore.co.uk
Research different types of ductwork and jointing methods. Produce a presentation for your class.

Test yourself

- Explain the different ductwork materials available.
- List three ductwork accessories and describe their purpose.

Ductwork accessories

▼ Table 12.20 Ductwork accessories

Type	Purpose
Zone dampers	To control the flow of air in an HVAC heating or cooling system to improve efficiency and comfort; also known as volume control dampers (VCDs)
Variable air volume (VAV) systems	To supply constant-temperature air while the volume of air varies
Constant air volume (CAV) systems	To supply air at a consistent and constant volume while the air temperature varies
Fire dampers	To prevent the spread of flames through ductwork systems during fire conditions
Attenuators	To reduce noise transmitted inside ventilation ductwork in an HVAC system
Heating and cooling coils	To cool or heat air in many HVAC applications
Air-extract grilles	To extract air to the outside in a mechanical ventilation system
Air-supply valves	To supply return air to habitable rooms, directing the air sideways to prevent occupant discomfort; available in a range of sizes (80, 100, 125, 150, 160, 200 and 250 mm)

Assessment practice

Short answer

1 Suggest three uses of electricity within a building.
2 Name three types of cold-water system.
3 List three components of a ducted air-conditioning system and describe their purpose.
4 List three pipework accessories and describe their purpose.
5 Explain the vapour-compression refrigeration cycle.

Long answer

6 Explain the purpose and basic operating principles of a heat pump.
7 Explain the purpose of fire alarm systems and the different categories available.
8 Explain the purpose and basic operating principles of rainwater systems.
9 Describe how secondary circulation is used in a hot-water system.
10 Explain the purpose of central-heating systems.

Project practice

Wilson plc is an established building contractor, with branches throughout the UK. Each branch works on different types of development, including new-build residential and commercial retail. It is preparing a tender submission for the building of a commercial premises for multi-purpose use. You are part of the team working to respond to the tender.

- Provide an overview of a range of BSE systems that can be included in the build.
- Explain the purpose of each system, including advantages and disadvantages where applicable.
- Explain the components and their purpose for each system.

This information should be included in a PowerPoint presentation, which will be shown as part of the tender process.

13 Maintenance principles

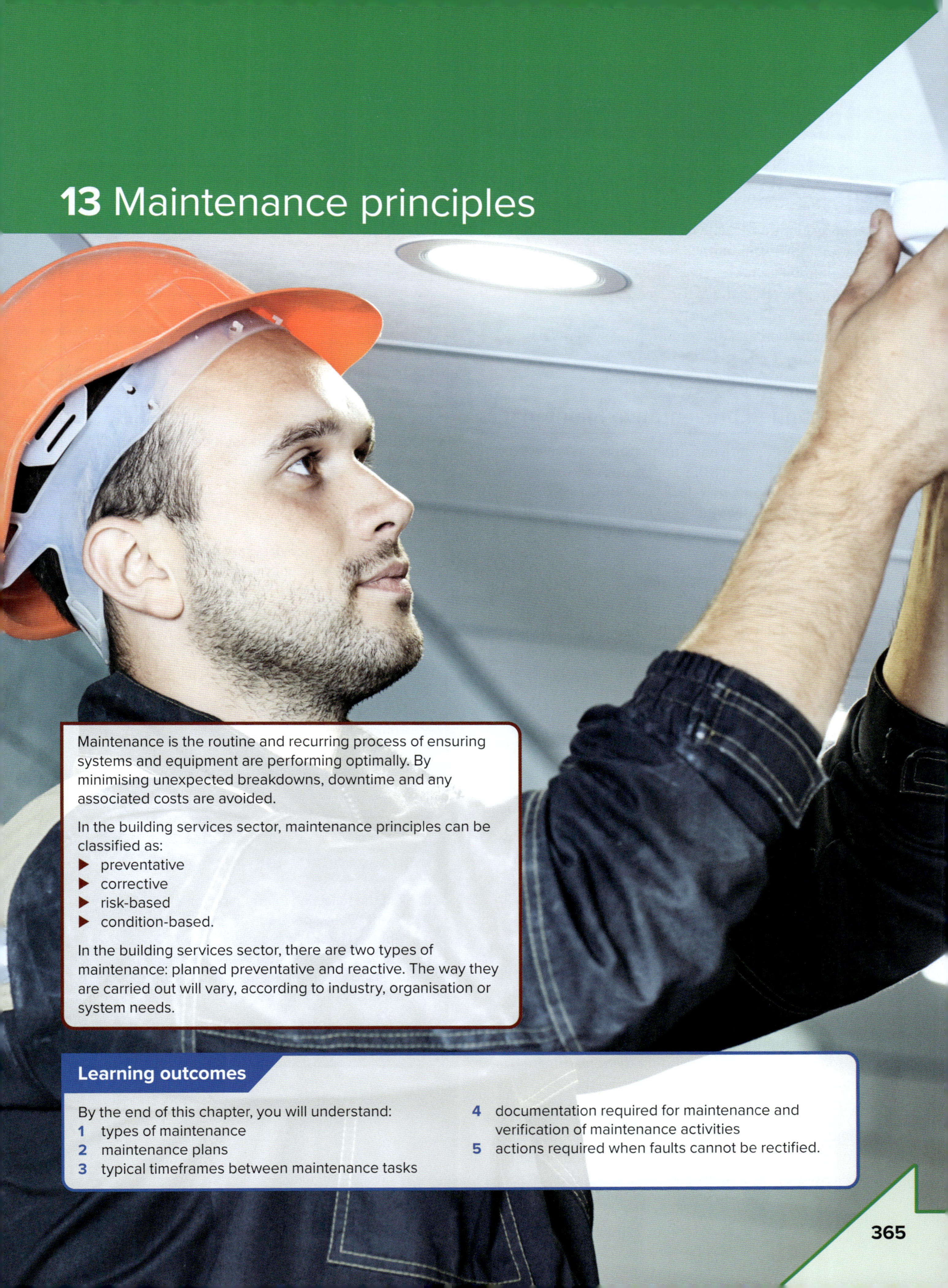

Maintenance is the routine and recurring process of ensuring systems and equipment are performing optimally. By minimising unexpected breakdowns, downtime and any associated costs are avoided.

In the building services sector, maintenance principles can be classified as:

- preventative
- corrective
- risk-based
- condition-based.

In the building services sector, there are two types of maintenance: planned preventative and reactive. The way they are carried out will vary, according to industry, organisation or system needs.

Learning outcomes

By the end of this chapter, you will understand:

1. types of maintenance
2. maintenance plans
3. typical timeframes between maintenance tasks
4. documentation required for maintenance and verification of maintenance activities
5. actions required when faults cannot be rectified.

1 Types of maintenance

There are two types of maintenance:

- planned preventative maintenance (PPM)
- reactive maintenance.

The systems and facilities found in large buildings require different maintenance schedules, for example:

- plumbing and heating systems
- fire detection and alarm systems
- wiring and lighting systems
- ventilation and air-conditioning systems
- drainage systems
- communication and data systems.

1.1 Planned preventative maintenance

Planned preventative maintenance (PPM) refers to maintenance activity that is planned, documented and scheduled. It aims to reduce downtime by having all necessary resources on hand, such as labour and parts, and a strategy to use those resources.

PPM is usually performed on larger systems and commercial/industrial installations. Certain items are serviced or replaced according to a predetermined schedule, regardless of condition, so that faults can be identified and prevented before they become a problem. This could involve out-of-hours or weekend work, depending on the installation.

When planning preventative maintenance, the following factors need to be taken into consideration:

- business type
- cost
- other business commitments and operational needs, for example downtime of a production line.

Larger organisations may choose to schedule PPM to avoid the risk of downtime with its associated costs. For example, a large warehouse storing frozen food would implement PPM for its refrigeration units, to ensure stock is constantly kept at the correct temperature and avoid loss of stock through system breakdown.

In contrast, a small wholesaler is more likely to adopt a reactive maintenance programme due to the costs associated with PPM, including the replacement of parts that are still functioning efficiently.

A similar example of PPM within the Building Services Engineering sector might be an office or education environment with an air-conditioning and heating system. PPM would ensure that the building is operating efficiently with regards to comfort heating and comfort cooling for its occupants at all times.

Planned preventative maintenance checklist

Date: ....

Prepared by: ....

Approved by: ....

No	Equipment description	Date checked	Action required	Action completed	Person responsible for checking	Signature	Due date for next check

▲ Figure 13.1 Planned preventative maintenance checklist

Industry tip

Regular maintenance of systems and equipment will ensure energy efficiency and optimum performance.

Test yourself

Consider the following types of building:

- nursing home
- large supermarket
- primary school
- domestic dwelling.

Which of these might require PPM? Explain your answers.

Improve your maths

Look at the following information for a PPM activity:

- Twenty float-operated valves (FOVs) need to be replaced every six months.
- The valves cost £12.50 each.
- Total labour time to fit the valves is three hours, at a cost of £20 per hour.

What is the total cost to the business for a calendar year?

▲ Figure 13.2 Float-operated valves (FOVs)

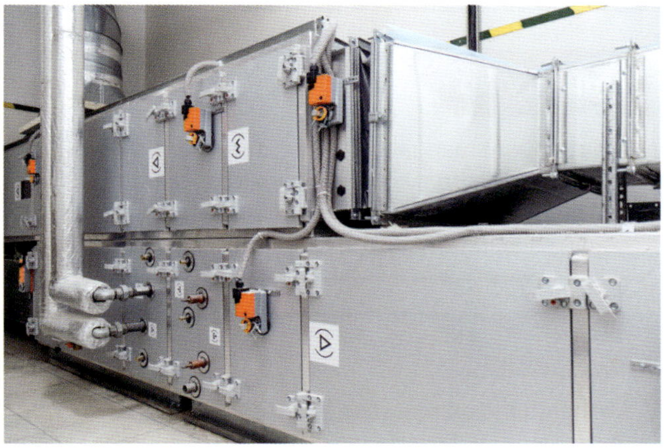

▲ Figure 13.3 Air-handling unit

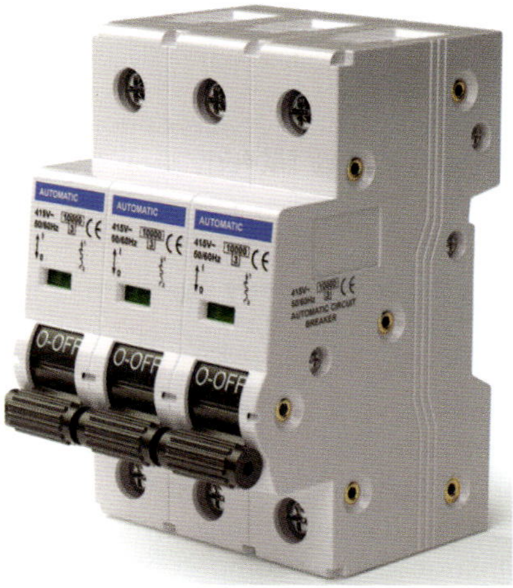

▲ Figure 13.4 Miniature circuit breaker

Manufacturers often provide maintenance recommendations. For example, it may be necessary to service equipment annually to ensure it works efficiently and performs optimally. Sometimes, maintenance is a requirement of the manufacturer's **warranty**/guarantee, to ensure continued support for equipment.

Key term

Warranty: a guarantee from a manufacturer or seller that a product will be repaired or replaced within a certain period of time if it does not function as originally described or intended

▼ Table 13.1 Examples and benefits of PPM for different BSE systems

Type of system	Examples of PPM	Benefits of PPM
Plumbing	Re-washering taps/valves Replacing FOVs	Saves water
Fire detection	Replacing smoke/heat detectors	Ensures compliance with standards and regulations (BS 5839)
Air conditioning	Recharging system	Allows the system to run efficiently, ensuring optimal conditions for end users
Heating	Servicing and replacing thermostatic radiator valves and lockshield valves	Reduces energy costs
Drainage	Checking and replacing all traps	Eliminates smells and odours from main sewers
Ventilation	Checking and replacing filters on air-handling units (AHUs)	Ensures *Legionella*-free systems
Wiring and electrical installation	Servicing and replacing miniature circuit breakers (MCBs) and fuses	Ensures the system is operating optimally, keeping users safe

Industry tip

Inspections carried out at regular intervals, coupled with prompt action to pre-empt or remedy problems, are the basis of good maintenance.

Health and safety

The Benchmark Scheme covers heating and hot-water systems, placing responsibilities on heating engineers to ensure products are 'installed, commissioned and serviced in accordance with the manufacturer's instructions by competent persons in a way that they will be safe, perform with optimum efficiency and meet the requirements of the appropriate Building Regulations'. Adherence to this scheme meets Gas Safe regulations, and will allow a Landlord Gas Safety Certificate to be granted in a rented property.

Test yourself

What are the advantages of using PPM?

1.2 Reactive maintenance

Reactive maintenance is also known as unplanned or emergency maintenance. Unlike PPM, reactive maintenance is carried out only when a system (or part of a system) fails. Examples might include:

- a WC system overflowing with water because a FOV has failed
- an intruder alarm constantly activating because of a defective sensor.

A building services engineer might receive a phone call out of hours to manage reactive maintenance jobs – this is referred to as a call-out. It is important to understand which types of problem need to be prioritised and to take action accordingly.

Unlike PPM, reactive maintenance may require additional problem-solving skills in order to maintain systems in the short term, for example due to a lack of part availability.

Industry tip

In some systems, there may be recurring breakdowns or maintenance issues with common components. Experienced engineers often carry an assortment of stock parts to maximise efficiency and limit system downtime.

Research

Find manufacturer operation and maintenance (O&M) manuals for the following assets and note the maintenance requirements:

- boiler
- fire alarm
- emergency lighting
- data server.

Improve your English

Good verbal communication involves listening carefully and speaking clearly. What three questions might you ask a caller who is reporting a problem with a faulty air-conditioning system?

Industry tip

A good O&M manual contains information on how to operate all the systems in a building, a parts list, detailed maintenance requirements, testing/commissioning certificates and record drawings.

Research

Research the advantages and disadvantages of running a PPM schedule versus a reactive maintenance programme.

2 Maintenance plans

A maintenance plan includes information on:

- specific maintenance tasks that need to be carried out
- the correct way to carry out those tasks
- the premises where the tasks will be carried out
- priority levels for the tasks
- deadlines, scheduled dates or frequency for the tasks
- duration of the contract
- budget allowances and restrictions
- equipment or parts required to complete the tasks
- manufacturer manuals or instructions.

An effective plan covers every aspect of an organisation's assets and systems, to ensure the building functions efficiently. It may also prioritise systems or areas and provide an order (or sequence) in which maintenance tasks should be carried out.

Risk management might also be applied when producing a maintenance plan for certain types of system, for example fire detection and alarm systems, data systems and electrical systems.

Key term

Risk management: the process of identifying, assessing and controlling threats to an organisation; this might involve making a strategic plan or putting a system in place to minimise impact and disruption

Test yourself

What is meant by the term 'risk management' within BSE systems?

Maintenance tasks will vary depending on the type of system. Some systems may require a visual inspection to check for any visible faults, while others may require components to be tested physically.

A building services engineer may be responsible for maintaining various systems across multiple sites and locations. This may involve travelling both locally and nationally, sometimes at short notice for reactive maintenance, and this should be factored into a maintenance plan.

Most maintenance plans have deadlines for when vital system components require maintenance. These should be scheduled to allow for ordering of parts or changing of components if required. They are often planned ahead of the manufacturer's requirements, to allow for other unplanned maintenance taking priority.

Prioritising tasks on a maintenance plan is important, as some components/systems require a higher level of maintenance than others. This may be in response to a health and safety/regulatory requirement or simply to maintain business efficiency.

The frequency of maintenance depends on the system and its components, for example some maintenance plans might be annual while others are monthly.

When maintenance plans are devised, it is important to factor in what equipment or parts may be required and allow plenty of time for them to be ordered if necessary. For example, a plumbing and heating engineer may carry FOVs or isolation valves in stock so that they are readily available. However, a refrigeration and air-conditioning engineer may not carry a specific condenser for an air-conditioning unit in their van and so will need to ensure the correct part is ordered in a timely manner.

In order to meet the manufacturer's standards, it is important for an engineer to refer to installation and servicing manuals when carrying out maintenance work. Failure to do so may affect the warranty and ultimately the efficiency of the system.

Health and safety

All businesses must adhere to government health and safety guidelines: www.hse.gov.uk. When providing maintenance plans, care must be taken to ensure operators are complying with all current regulations. For example, a company employing more than five employees would need to have a written health and safety policy. This statement might include risk assessments for all planned maintenance tasks, all PPE that will be provided and the number of first aiders required on site.

Industry tip

Installation and servicing manuals can usually be downloaded from a manufacturer's website or emailed on request.

▼ Table 13.2 Typical planned maintenance tasks across BSE systems

BSE system	Typical planned maintenance tasks
Heating	• Visually inspect pipework for damage. • Check thermostatic radiator valves (TRVs) and lockshield valves are controlling and balancing temperatures across the system. • Check the heating circulator is circulating water around all radiators, boiler and cylinder. • Check inhibitor levels to prevent limescale and corrosion. • Bleed the system to remove trapped air. • Check drain-off valves (DOVs) are working correctly in order to fully drain system if required.
Boiler	• Check for correct combustion as per manufacturer's instructions. • Check gas rates as per manufacturer's instructions. • Check flue gases as per manufacturer's instructions. • Issue Landlord Gas Safety Certificate for rented property (Gas Safe engineer required). • Complete Benchmark checklist to maintain manufacturer's warranty.

BSE system	Typical planned maintenance tasks
Water	• Visually inspect for leaks. • Check all float-operated valves (FOVs)/isolation valves are at the correct levels and working. • Re-washer taps and valves to prevent leaks and water wastage. • Check correct flow rates and pressures. • Check for *Legionella* (this is also a regulatory requirement for landlords).
Firefighting	• Visually inspect fire extinguishers for damage (at least annually). • Check hose reels for damage. • Check fire blankets are in the correct location, visible and undamaged. • Wet riser (sprinkler system): check that glycol (anti-freeze) is at correct ratio to prevent pipe leakage due to extreme low temperatures. • Dry riser: visually inspect pipework for damage. • Check fire curtains are working and unobstructed. • Check the operation of self-closing fire doors.
Fire detection and alarm	• Check the operation of heat, smoke and carbon monoxide detectors. • Check fire alarm buttons trigger the alarm. • Check the control panel for any error messages and ensure all detection equipment is functioning. • Check the control panel is linked to the local fire service. • Check audio alarms/bells/sirens function correctly and are audible.
Intruder alarm	• Check motion detectors are working. • Check closed-circuit television (CCTV) and security cameras (internal and external). • Check that pressure pads are working. • Check the alarm control panel for any error messages and ensure all detection equipment is functioning. • Check audio alarms/bells/sirens sound when the alarm is triggered. • Check links to the keyholder/security company/police.
Wiring and electrical installation	• Visually inspect wiring and components for damage. • Test the consumer unit residual current device (RCD) is functioning correctly. • Test the miniature circuit breaker (MCB) to ensure it is functioning correctly. • Carry out **portable appliance testing (PAT)**. • Carry out a polarity test to ensure all single-pole devices (fuses/switches/circuit breakers) are connected to the correct pole: live to live, neutral to neutral. • Carry out an insulation resistance test. • Carry out an earth loop impedance test.
Air conditioning	• Visually inspect the system for damage. • Ensure F-gas compliance. • (For older systems) maintain R22 refrigerant, or replace with compliant new F-Gas to meet current regulations. As an ozone-depleting substance, R22 is being phased out and it is now illegal to replenish it in a system when carrying out repairs and maintenance. • Check the CO_2 equivalent charge rate to determine how often the system is serviced. • Carry out a TM44 energy-efficiency assessment for systems with a rated output of 12 kW. • Inspect the condenser and evaporator for damage.
Ventilation	• Visually inspect the system for damage. • Clean vents and ductwork. • Replace damaged vents. • Check air-handling units (AHUs) for the correct static pressure used in the coils; clean and replace filters as required.
Drainage	• Visually inspect pipes for damage. • Check all traps for correct seal levels. • Carry out an air test of the stack system (as per Part H of the Building Regulations). • Carry out a performance test of the system (as per Part H of the Building Regulations).

BSE system	Typical planned maintenance tasks
Lighting	• Visually inspect all fittings for any damage and correct operation. • Check correct MCBs/fuses are being used. • Check the starters in fluorescent tubes. • Ensure emergency lighting is functioning correctly in case of mains failure. • Replace older bulbs with energy-saving bulbs/filaments. • Check for correct lumens (as per correct location). • Check lighting diffusers. • Clean lighting lenses and covers.
Communications and data	• Visually inspect. • Check sockets and data ports for any damage and correct operation. • Check server rooms and any control panels within for any faults indicated; service and rectify as per manufacturer's instructions. • Check ventilation and air conditioning of server rooms is working correctly. • Check bandwidth is within tolerance as per requirements and manufacturer's specifications. • Check data is being stored in compliance with GDPR.

▲ Figure 13.5 Thermostatic radiator valve (TRV)

▲ Figure 13.6 Sprinkler system

▲ Figure 13.7 Server rooms need to be ventilated and air conditioned to keep them at the correct temperature

Key terms

Portable appliance testing (PAT): a process by which electrical appliances are routinely checked for safety. The format term is 'in-service testing for electrical equipment'

Research

What does GDPR stand for and why is it important to be compliant?

Test yourself

What appliances or components would be maintained within an electrical wiring system?

3 Typical timeframes between maintenance tasks

The frequency of maintenance tasks will vary according to the system. Some systems require priority maintenance to meet regulatory requirements. For example:

- in a rented property, the boiler requires regular servicing to meet Benchmark and warranty requirements, and an annual service to meet the requirements of the Landlord Gas Safety Certificate. This must be carried out by a Gas Safe engineer. Electrical systems should be inspected every five years by a qualified and competent electrician.
- Maintenance requirements for drinking water systems will need to be carried out to check for *Legionella* and will vary depending on the system and on the outcome of the risk assessment. For open systems such as cooling towers, evaporative condensers and spa pools, routine testing should be carried out every three months.
- With good maintenance, air-handling units can have a lifespan of up to 25 years. To keep them functioning efficiently throughout their life, it is recommended that filters are replaced and coil chambers cleaned every three months. Without any maintenance, the estimated lifespan of an AHU can drop to as low as 1–5 years.
- There should be a minimum of one smoke alarm per floor, with a heat detector in any kitchen. A registered engineer is required to fit enough areas where a fire could start. Fire detection and alarm systems require regular, ideally weekly, testing. The alarm system's wiring should also be tested every five years.

With any systems, manufacturers' instructions should always be followed for each component or appliance to ensure a fault-free maximum lifespan. Sometimes components will fail despite regular maintenance due to manufacturer's defects.

4 Documentation required for maintenance and verification of maintenance activities

When carrying out PPM and reactive maintenance, BSE engineers use a variety of documentation both to perform the tasks and to record the work completed.

Paper-based documentation, such as manuals and logbooks, should remain on site and be easily accessible.

Electronic documentation, on tablets and mobile devices, can be uploaded and stored centrally for later retrieval.

Types of documentation required for maintenance and verification of maintenance activities include:

- manufacturer's instructions
- maintenance checklists
- servicing logbooks
- maintenance schedules
- job sheets
- condition reports.

4.1 Manufacturer's instructions

Manufacturer's instructions come in three formats:

- Manufacturer's literature acts like a catalogue for a manufacturer's components or systems. It provides details such as specifications, performance data, running costs and dimensions.
- Installation instructions describe the installation process for an appliance or system and give specific details of site and input service requirements (for plumbing this might be flow rates or pressure required; for electrical systems it might be the amps and voltage required). They also detail commissioning procedures.
- Servicing/maintenance instructions detail the procedures and tasks involved in performance testing and replacing of components (for example changing air filters during maintenance of an air-conditioning system). They also specify correct performance data, which will indicate to the building services engineer whether components or systems are working to the correct standards. This document is important as part of both PPM and reactive maintenance.

4.2 Maintenance checklists

A maintenance checklist identifies tasks to be carried out on specific components of an appliance or an entire system. It might form part of the manufacturer's servicing/maintenance instructions.

Maintenance checklists can be sub-divided to cover individual sections of larger systems, for example:

- a checklist for a fire detection system might cover sprinklers, fire extinguishers, fire blankets, hose reels, heat/smoke detectors and alarms/buzzers
- a checklist for a heating system might cover pipework, radiators, TRVs, lockshield valves, DOVs, the expansion vessel, pressure gauge and filling loop.

4.3 Servicing logbooks

Servicing logbooks record the maintenance, servicing or commissioning of a component, appliance or system. They inform a servicing engineer about which prior tasks, repairs or replacements have been carried out.

A gas boiler service would be documented in the Benchmark Scheme logbook, to ensure the manufacturer's warranty remains intact. This contains details such as model number, serial number, installation date, service dates and performance data such as rating, gas pressure and flue temperatures.

Heating and hot water system complies with the appropriate Building Regulations?			Yes
Optional: Building Regulations Notification Number (if applicable):			
Time, temperature control and boiler interlock provided for central heating and hot water			Yes
Boiler Plus requirements (tick the appropriate box(s))			
Boiler Plus option chosen for combination boiler in ENGLAND		Weather compensation	Smart thermostat with automisation and optimisation
		Load compensation	Flue Gas Heat Recovery
Time and temperature control to hot water		Cylinder thermostat and programmer/timer	Combination boiler
Zone valves	pre-existing	Fitted	Not required
Thermostatic radiator valves	pre-existing	Fitted	Not required
Automatic bypass to system	pre-existing	Fitted	Not required
Underfloor heating	pre-existing	Fitted	Not required
Water quality			
The system has been flushed, cleaned and a suitable inhibitor applied upon final fill, in accordance with BS7593 and boiler manufacturers' instructions			Yes
What system cleaner was used?		Brand:	Product:
What inhibitor was used?		Brand:	Product:
Primary water system filter	pre-existing	Fitted	Not required
CENTRAL HEATING MODE measure and record (as appropriate)			
Gas rate (for combination boilers complete DHW mode gas rate)		m³/hr	or ft³/hr
Central heating output left at factory settings?		Yes	No
If no, what is the maximum central heating output selected?			kW
Dynamic gas inlet pressure			mbar
Central heating flow temperature			°C
Central heating return temperature			°C
System correctly balanced/rebalanced?			Yes
COMBINATION BOILERS ONLY			
Is the installation in a hard water area (above 200ppm)?		Yes	No
Water scale reducer/softener	pre-existing	Fitted	Not required
What type of scale reducer/softener has been fitted?		Brand:	Product:
Water meter fitted?		Yes	No
If yes- DHW expansion vessel	pre-existing	Fitted	Not required
Pressure reducing valve	pre-existing	Fitted	Not required
DOMESTIC HOT WATER MODE Measure and record			
Gas rate		m³/hr	or ft³/hr
Dynamic gas inlet pressure at maximum rate			mbar
Cold water inlet temperature			°C
Hot water has been checked at all outlets		Yes	Temperature °C
CONDENSATE DISPOSAL			
The condensate drain has been installed in accordance with the manufacturers' instructions and/or BS5546/BS6798			Yes
Point of termination		Internal	External (only where internal termination impractical)
Method of disposal		Gravity	Pumped
ALL INSTALLATIONS			
Record the following	At max rate: CO ppm	CO_2 %	CO/CO_2 Ratio
	At min rate (where possible) CO ppm	CO_2 %	CO/CO_2 Ratio
Where possible, has a flue integrity check been undertaken in accordance with manufacturers' instructions, and readings are correct?			Yes
The operation of the boiler and system controls have been demonstrated to and understood by the customer			Yes
The manufacturers' literature, including Benchmark Checklist and Service Record, has been explained and left with the customer			Yes
Commissioning Engineer's signature			
Customer's signature (To confirm satisfactory demonstration and receipt of manufacturers' literature)			

* All installations in England and Wales must be notified to Local Authority Building Control (LABC) either directly or through a Competent Persons Scheme. A Building Regulations Compliance Certificate will then be issued to the customer.

▲ Figure 13.8 Benchmark Scheme logbook

4.4 Maintenance schedules

There are two types of maintenance schedule:

- manufacturers' maintenance schedules, for example a drainage system might need a schedule for annual inspection and maintenance
- individualised maintenance schedules which are bespoke to a specific system and risk-rated depending on the type of system, for example a hot-water system in a gym might require monthly maintenance to check for *Legionella* and ensure temperatures are correct at the point of use.

4.5 Job sheets

A building services engineer uses a job sheet to inform them of the client's name, site location, date and time, priority of the task and work to be carried out.

A job sheet could require a:

- visual inspection to check for wear and tear or external damage
- more detailed check of components not functioning efficiently
- full-system check of individual components, comparing performance data in the service instructions against how the system is working.

4.6 Condition reports

A condition report could form part of the job sheet and highlights any defects with a component, appliance or system. This information could come from visual inspections or performance tests.

A condition report also covers maintenance carried out previously and what parts or components have been changed.

The information on a condition report feeds into a maintenance plan or schedule and, if used as part of PPM, parts can be changed before they fail.

4.7 SFG20 maintenance specification

SFG20 is the UK's industry standard for building maintenance specification, developed by the Building Engineering Services Association (BESA). It provides a comprehensive, continuously updated library of over 1,500 maintenance schedules tailored to specific assets, building types, and regulatory requirements.

Each SFG20 schedule outlines:

- what needs maintaining
- how often it should be done
- who should perform the task
- the required skill level (mechanical, electrical, gas-safe)
- priority level using a colour-coded system (statutory, mandatory, discretionary)

The system translates complex legislation and codes of practice into clear, actionable tasks, helping facilities managers, contractors, and building owners stay compliant, reduce risk, and avoid over- or under-maintenance. It integrates with CAFM and BIM systems.

Test yourself

What are the advantages of using correct documentation for maintenance activities?

5 Actions required when faults cannot be rectified

On occasion, a building services engineer may not be able to rectify, replace or recommission a component, appliance or system due to a range of factors, for example:

- The spare part is not carried in the van.
- The spare part is **special order** only.
- The spare part is no longer available.
- There is a lack of knowledge of the system due to it being outdated or new technology.
- Incorrect information has been provided by the customer/client.
- There is a lack of time due to a higher priority breakdown.

When parts are not readily available, an engineer will need to place an order with a merchant or specialist supplier. If the part is in stock, it may be available immediately. However, if it is not in stock, it might be a special order or **back order** item, and the engineer has no control over how long this might take to deliver.

Key terms

Special order: an order for items or components that are custom made or configured to a client's specifications

Back order: an order for items or components that are not available due to a lack of supply; they might still be in production or the manufacturer may need to make more of them

Test yourself

When ordering parts for maintenance tasks, what is the difference between a special order and a back order?

When an engineer cannot rectify a fault immediately, this can have negative implications for the customer, client or business, for example:

- additional costs
- downtime of systems
- loss of income
- increased hazards
- loss of services.

The costs of PPM are usually known in advance, because the maintenance is scheduled and expected. However, the costs for reactive maintenance can vary greatly depending on the:

- time of day (out-of-hours call-outs are more expensive)
- time of year (call-outs on bank holidays are more expensive)
- type of system that requires maintenance (some require specialist engineers).

It is difficult to put a price on reactive maintenance, as the engineer might not be able to diagnose the reason for failure or determine the parts required until they are on site.

Some breakdowns have an impact on the downtime of systems:

- A leaking flow pipe on a central-heating system will require a drain down of the whole system in order to carry out a repair. While the drain down is taking place, the heating system will be out of action.
- An electrical system needs to be isolated while an electrician changes components in a consumer unit. This will have an impact on electrical items being used within the building.

Downtime of certain systems can have a costly impact. For example, downtime of an electrical supply in a food manufacturing plant will stop production. This could have a knock-on effect, making it difficult for the business to fulfil orders and meet deadlines.

Downtime of systems should always be a last resort but might be unavoidable if there is a lack of spare parts or it is unsafe to run the system while faulty.

Some businesses may have insurance to cover loss of income due to downtime, but this is an additional cost to the business. Having PPM for BSE systems can help with avoiding downtime, because the maintenance has been planned in advance. Some businesses have a higher risk and require increased PPM to ensure systems have minimal or no downtime.

Case study

You have been called out to a small nursing home to repair a faulty gas boiler. On inspection, you realise you do not have the part required. Upon further investigation, you realise the part is special order and the system will have to be shut down until the repair is carried out.

Compose an email to the nursing home manager stating why the system cannot be reinstated immediately. You need to give an indication as to when the part might be available and when the system will be back up and running.

How else could you help/advise the nursing home while the heating is off?

When certain systems break down, an engineer might need to make the system safe before carrying out any repairs. For example, if a gas appliance is leaking, a Gas Safe engineer might have to isolate the gas supply at the meter point to make the system safe, due to the increased risk of explosion. Similarly, an electrician would need to isolate the system at the consumer unit where there is a faulty RCD.

Failure to identify hazards or risks when leaving a system operating under faulty conditions can have serious implications, including risk to life in some cases.

If an engineer is forced to take a system out of action for long periods of time, it is good practice to have a back-up or secondary service available. This might include portable storage heaters to cover boiler failure, mobile air-conditioning units to cover a centralised RAC system, and portable generators to provide an electrical supply.

An engineer could also use the following to help bring systems back into service as soon as practicable, thus helping to reduce inconvenience to a customer or business:

- manufacturer's technical support
- online research
- virtual assistants.

Even an experienced building services engineer may require help or support in diagnosing a breakdown in a system. The first point of contact should always be the manufacturer's technical support helpline. This allows the engineer to describe the fault and get expert advice from the manufacturer, who will know how to rectify problems that are not in the servicing manual.

Some engineers use online search engines to research problems and find alternative solutions. These solutions might come from manufacturer or industry body forums, where similar problems are logged and resolutions suggested by other engineers. When researching online and visiting manufacturers' websites, engineers may be able to access virtual assistants or live chat, where they are able to speak with experts online in real time.

While carrying out maintenance and fixing breakdowns, particularly where work cannot be completed immediately, it is essential to keep the customer or business fully updated so they understand the costs and timeframes involved in bringing the system back into operation and to manage their expectations.

Test yourself

What sources of information might a building services engineer refer to when diagnosing a fault on a system?

Research

Listed below are some professional bodies operating within the BSE sector.

- Air Conditioning and Refrigeration Industry Board (ACRIB)
- Joint Industry Board for Plumbing Mechanical Engineering Services (JIB-PMES)
- Association of Plumbing and Heating Contractors (APHC)
- Electrical Contractors' Association (ECA)
- Joint Industry Board for the Electrical Contracting Industry (JIB)
- Building Engineering Services Association (BESA) (formerly the Heating and Ventilating Contractors' Association)
- Chartered Institute of Plumbing and Heating Engineering (CIPHE)
- Security Systems and Alarms Inspection Board (SSAIB)
- British Approvals for Fire Equipment (BAFE)
- Institute of Electrical and Electronics Engineers (IEEE)
- Gas Safe Register.

Choose three and find out:

- what purpose they serve
- why they are important
- how they support their sector.

Assessment practice

Short answer

1 What is the difference between planned preventative maintenance and reactive maintenance?

2 Which system would have a maintenance plan that includes inspection of a condenser and evaporator?

3 For how long is a Landlord Gas Safety Certificate valid?

4 Which documentation might a building services engineer complete to advise on future repairs or replacements?

5 What does the term 'downtime' mean with regards to a BSE system?

Long answer

6 Describe what maintenance tasks might be on the PPM schedule for the following systems:

- air-handling units
- drainage
- domestic wiring.

7 Write a detailed maintenance plan for an intruder alarm and CCTV system in an office.

8 Describe typical timeframes for a building services engineer carrying out work on fire detection and alarm systems.

9 You are required to design a job sheet for engineers to use when carrying out work on site. What information should be included?

10 Evaluate the potential implications for a telesales business that has been told its electrical systems are being taken offline for 48 hours while essential maintenance is carried out.

Project practice

You have been asked to produce a planned preventative maintenance (PPM) schedule for firefighting equipment and fire detection and alarm systems for a small office block.

- Decide what format the schedule would take and explain why.
- Describe all the components that make up the two systems.
- Outline what is required to meet industry and manufacturing standards.
- State which health and safety regulations you must consider when carrying out maintenance on firefighting equipment and fire detection and alarm systems.

14 Tools, equipment and materials

A wide range of tools, equipment and materials are used on construction sites. Depending on the application, they can be relatively simple or technologically advanced and complex. In all cases, the same principles must be applied to ensure their suitability and the safety of the operatives using them.

This chapter discusses methods for checking that tools, equipment and materials are fit for purpose for construction activities, as well as identifying proper care and maintenance procedures, to support reliability, productivity and safety on site.

Learning outcomes

By the end of this chapter, you will understand:

1. methods used to ensure tools, equipment and materials are fit for purpose
2. maintenance of tools, equipment and materials.

1 Methods used to ensure tools, equipment and materials are fit for purpose

To ensure tools and equipment are fit for purpose, every operative using them must inspect them before use, clean them after use and follow appropriate maintenance routines.

Materials must be selected in accordance with the work specification, inspected for good condition and used in accordance with manufacturer's instructions and established best practice.

It is important to be observant in order to identify potential problems with tools, equipment and materials. This will:

- improve site safety
- extend the life and usefulness of valuable resources
- help ensure the completed project fully matches the design brief.

1.1 Tools and equipment

In order to support reliability, productivity and safety on site, the operational efficiency of all tools and equipment must be maintained. The principles for ensuring items are fit for purpose apply equally to simple items such as screwdrivers and complex items such as self-levelling laser levels.

▲ Figure 14.1 Regular safety inspections of equipment are essential

Let us look at some examples of methods for ensuring items are fit for purpose.

Portable appliance testing (PAT)

Portable appliance testing (PAT) is an ongoing procedure that ensures electrical appliances are safe to use. Tools and equipment powered by electricity are inspected visually and tested electrically to identify possible faults. Typically, a checklist is followed to ensure all potential safety issues are addressed.

Portable tools and items of equipment that are frequently moved from place to place are at greater risk of damage to exposed parts, such as attached power leads, projecting plug attachment points and surface switches.

Even a new tool or piece of equipment should be visually checked before use, to ensure there is no damage that could make it unsafe.

Health and safety

Portable tools and equipment are often supplied with specifically designed cases to transport and store them. These cases provide valuable protection and help to keep the items safe to use.

Testing frequency varies depending on factors including location and type of equipment. For example, portable equipment such as extension leads in offices, shops or hotels should be tested every two years; all 110 V construction tools and equipment on site should be tested every three months.

Items are usually labelled with the date of the last test. This simple practice provides a record for supervisors to monitor ongoing safety and a reassurance to operatives that the item is safe to use. Separate written records of tests and inspections should also be kept.

Industry tip

Portable appliance testing does not only apply to tools and equipment. For example, the kettle in a site canteen or the printer in a site manager's office should also be tested.

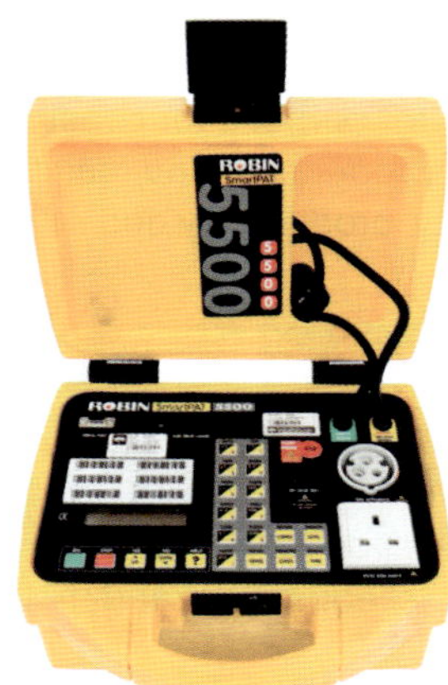

▲ Figure 14.2 A portable appliance test (PAT) instrument

Test yourself

Which features of a portable appliance are more easily damaged compared to a fixed appliance?

Calibration of instruments

Tools and equipment used for measurement or testing can be relatively simple, such as a tape measure or thermometer. However, in construction engineering applications they are often sophisticated and complex in design.

Where measurements must be accurately checked and maintained, the instrument of measurement must be calibrated periodically to maintain its reliability and effectiveness. In this context, **calibration** refers to the process of comparing, and if necessary adjusting, readings on an instrument to match an established and accepted external reference.

Measuring and testing are an important part of the installation, commissioning and maintenance of many critical systems within the built environment. For example, a high-pressure gas-distribution system has pipework junctions, valves and other components that are typically connected using **high-tensile** steel bolts. These must be tightened to precise levels of **torque**. If the specified torque is not met and the bolts are tightened to a lower or higher torque value, there could be either leaks of gas or catastrophic failure of components, which in either case can be dangerous.

The tool used to measure torque is known as a torque wrench, and the readings it displays must be calibrated to ensure its reliable and accurate function.

▲ Figure 14.3 Gas-distribution pipework with high-tensile steel bolts at junctions

Key terms

Calibration: comparison of a measurement device against a traceable reference

High-tensile: very strong under tension

Torque: a force that causes rotation or twisting

Research

Research different types of torque wrench. Produce a simple sketch to show the design features of at least two types.

Test yourself

Why must high-tensile bolts be tightened to the correct torque when used in a high-pressure gas-distribution system?

Table 14.1 lists some other instruments that require calibration.

▼ Table 14.1 Instruments that require calibration

Instrument	Application
Laser distance meters	These are used for accurate measurement between stations. They easily provide straight-line distance measurements over undulating terrain.
Coating thickness testers	These measure the thickness of protective paint coatings and membrane applications, to ensure consistency in coverage.
Holiday detectors	These are used to detect flaws and porous areas in protective coatings applied to metal surfaces. The flaws may not be visible to the naked eye but must be addressed to avoid rust or corrosion.
Electrical testers	A range of specific testers are used for electrical applications, such as PAT testers, socket testers, insulation testers and cable testers.
Gas detectors	These are used to detect leaks in gas-delivery systems. Specific types can analyse flue gases to confirm burner efficiency in boilers for heating and hot-water systems.

Research

Research holiday detectors and write a short report on how they work.

Instrument calibration is usually carried out by specialist companies or the instrument manufacturer. This ensures measurement standards are validated and documented as part of a system of quality verification. The calibration procedure will differ depending on the type of instrument, the environment in which it is used, and allowable tolerances in measuring and testing values.

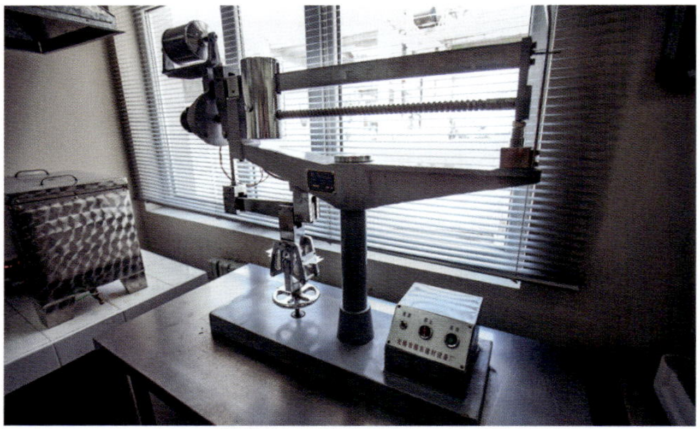

▲ Figure 14.4 Instruments used to test the hardness of concrete must be calibrated to maintain their accuracy

Instruments are often identified by a unique serial number. This allows records to be generated, providing details about the most recent calibration date, the calibration interval, the next calibration date and any other relevant information specific to individual instruments and tools.

Visual inspections

Health and safety

The Health and Safety Executive (HSE) website states: 'Inspection is necessary for any equipment where significant risks to health and safety may arise from incorrect installation, reinstallation, deterioration or any other circumstances. The need for inspection and inspection frequencies should be determined through risk assessment.'

Visually inspecting tools and equipment before use is an important habit to develop. A brief but careful examination of an item can quickly establish whether it needs repair or replacement. Not only will this prevent potential injuries, but it can also reduce the risk of failure or incorrect function of tools and equipment during a work task, which could lead to errors and production delays.

For example, inspection of an optical or laser level could find signs of impact damage to the casing. The accuracy of this type of instrument could be affected by an impact, and if an operative carried on using the potentially defective item without further action, it could lead to incorrect levels and **datum** positions being established. This could have serious consequences. Constructing a building to incorrect reference levels may lead to costly alterations or even the need to completely rebuild the structure in order to fulfil the project specification.

▲ Figure 14.5 An optical level should be inspected before use

Key term

Datum: a fixed point or height from which reference levels can be taken

Industry tip

If you need to use an instrument that appears to have suffered an impact and you are not sure whether it is still usable, bring it to the attention of your supervisor or line manager to get advice.

On a more basic level, simple hand tools should be inspected prior to use to confirm their suitability. For example, a screwdriver should not be used for piercing holes in materials. This could damage the tip or blade, causing it to slip when exerting pressure on a screw head. A brief visual inspection to confirm that the screwdriver is fit for purpose could reduce the risk of injury to the operative.

▲ Figure 14.6 A brief visual inspection of this screwdriver shows it is not fit for purpose

Before using hand tools, conduct a visual inspection to check for the following:

- The surface of the tool is free from grease or oil.
- Handles are not loose or cracked.
- Blades or cutting parts are sharp and not cracked or chipped.
- Chisels and punches do not have mushroomed heads.
- The general condition is good.

Industry tip

Some steel chisels are referred to as cold chisels and can be used for cutting openings in masonry for pipes or cables to go through. The top of the chisel can form a mushroom shape through being struck repeatedly. This should be ground off from time to time to avoid hand injury to the user.

Test yourself

Why should even a simple tool like a hammer be inspected before use?

Before using portable power tools, conduct a visual inspection to check for the following:

- The outside of the tool is free from oil, grease and accumulated dirt.
- The power supply is in good condition (cable, air-line, battery).
- Insulated casing of the tool is not cracked or damaged.
- Shields or guards are in place and not damaged.
- There are no leaks of fuel, such as petrol or diesel.
- Blades or bits are not damaged or worn.

Improve your English

Select two different points from each of the bullet-point inspection lists above. For each point, write an explanation of the benefits to an operative using the inspected tool or equipment.

Remember that whatever tools or equipment are required for a work task, the correct PPE must always be used. Always visually inspect PPE before use to make sure it is fit for purpose; after use, clean PPE if necessary and store it carefully.

▲ Figure 14.7 Check PPE is fit for purpose before use

Industry tip

Do not forget to check the labels on safety helmets that give a date when the item is due for replacement. The plastic used in helmets can deteriorate over time through exposure to sunlight.

Daily checks

Many checks on equipment must be carried out on a daily basis, to ensure the safe and smooth running of a construction site.

Daily checks are especially important for site vehicles and moving machinery, in order to maintain safe operational standards and protect all workers on site. Checking vehicle fuel levels, lubricant levels and tyre pressures on a daily basis contributes to reliability and efficiency, as well as extending operational life by reducing wear and tear.

▲ Figure 14.8 Site machinery being checked

Lifting and access equipment must be checked before use to confirm it is fit for purpose and in a safe condition. Never assume that because an item seemed safe to use yesterday, it will still be safe to use today.

Health and safety

Because access equipment like scaffolding has no moving parts, it is easy to think that no changes in condition occur from day to day. However, continuous heavy rain and movement caused by high winds can cause fixing clips to loosen. Check access equipment for safe condition carefully, especially after bad weather.

To ensure a consistent approach to daily checks, inspection checklists may be produced. Ticking off each item verifies that the corresponding part of the tool or equipment will contribute to its overall correct and safe operation. Checklists should be comprehensive but simple to follow and arranged in a logical order.

Research

Search online for an inspection sheet template for air-operated (pneumatic) tools. Write down the inspection points that relate to:

- safety
- operational efficiency.

The operative completing the checklist has accountability to the user of the tool or equipment, by confirming that the item is safe to use and should operate as expected. As such, daily checks, although a matter of routine, should never be undertaken casually.

▲ Figure 14.9 Do not undertake routine inspections casually

Thorough inspections contribute to increased workplace safety, lower maintenance costs and less downtime.

Test yourself

Name three benefits of carrying out thorough inspections.

Cleaning

Work on construction sites often generates high levels of dust and airborne debris. Carefully inspecting tools and equipment for cleanliness is a vital part of avoiding deterioration in function and maintaining safety during use.

Items should be cleaned after use and before storing them, to reduce the likelihood of rust and corrosion. This also allows subsequent users to identify any damage more easily before beginning a work task.

▲ Figure 14.10 Clean tools are easier to inspect for damage

Industry tip

Skilled operatives typically carry a range of essential tools to the work location in a tool bag or case. It is good practice to clean the interior of the bag or case from time to time, to prevent a build-up of dirt that could affect the tools being transported.

Cleaning also reduces the rate of wear in moving parts. Dust and dirt that are allowed to accumulate can become abrasive when mixed with lubricants such as oil and grease, and this can cause a significant increase in the rate of wear of bearings, **bushes** and other moving parts in powered tools and equipment.

Key term

Bushes: (in machinery) plain bearings or sleeves that reduce friction

Operational checks

Even after a tool or piece of equipment has been inspected and confirmed as fit for purpose and safe, the operator must look out for possible faults or defects that can occur during use. For example:

- an electrical power tool could become hot when in use, indicating it is not operating within design performance parameters
- a tool with many rotating parts may show signs of a developing fault by creating increased noise during operation.

Operatives should be familiar with manufacturers' instructions, so that tools and equipment are used correctly and developing problems are quickly recognised and dealt with.

In order to understand safe operational procedures and necessary operational checks on tools and equipment, it is important to check risk assessments, method statements (RAMS) and manufacturers' instructions before starting a work task. For example, cutting or grinding equipment may require periodic emptying of a dust-collection vessel.

When tools and equipment fail safety checks

If defects are discovered in a tool or piece of equipment during inspection or use, the item should be removed from use immediately and the relevant supervisor should be informed. A tag or label should be attached to the item, stating the nature of the problem, and it should be kept in a secure location to avoid inadvertent use. The item should be repaired or replaced as soon as possible.

Industry tip

It is important to only use drill bits and cutting blades that are in good condition, in order to ensure the user's safety and comfort, as well as to improve work efficiency and maintain quality of work. If safety checks show that bits or blades are excessively worn or damaged, they should be replaced without delay.

▲ **Figure 14.11 Worn or rusty drill bits should be replaced without delay**

All power tools must be treated with respect. When replacing drill bits or cutting blades, the power source must be isolated:

- For mains-powered tools, unplug the tool from the source of electricity.
- For petrol- or diesel-powered tools, never attempt adjustments or replacement of parts when the engine is running.
- For battery-powered tools, remove the battery. (Do not assume that battery-powered tools create less force when operating than mains-powered tools – they can cause serious injuries.)

Health and safety

Only trained and competent personnel should change the abrasive wheels on a disc cutter, angle grinder or bench grinder.

Test yourself

Why is it important to check that drill bits and cutting blades are not worn or damaged?

Keeping records

Written records on the condition of tools and equipment can be useful in allowing operatives to determine if they are in a suitable condition to permit ongoing use. These are often known as condition reports, and they can be particularly important when monitoring the safety of hired tools and equipment.

When hired items are delivered to site, they usually have details attached confirming the inspection and testing that has been undertaken to verify their safe condition. On return to the hire company, the condition of the item will be assessed by visual inspection and, if necessary, specified testing for faults and defects. A condition report can then be used to record:

- the results of the inspection and testing to confirm safety and serviceability
- any damage and ongoing wear and tear
- adjustment or replacement of parts.

Recording inspection and testing results is especially important in the case of electrical equipment, since faults that could lead to injury or death may not be visible to the user and only become apparent when appropriate testing is undertaken by qualified competent personnel.

Complex equipment, such as mobile elevated work platforms or site lifts, may periodically require more thorough inspections to maintain safe condition and serviceability. Programmed reporting provides a history of routine and in-depth inspections that can indicate when a tool or piece of equipment is due for replacement.

▲ Figure 14.12 More thorough inspections must be carried out periodically for complex equipment

A large contractor may own many tools and items of equipment. As valuable company assets, these must be catalogued. An asset register usually includes the following information for each asset:

- a description
- an identification number
- the date of acquisition
- ownership records
- its current value
- its condition and defects
- maintenance requirements and inspection intervals
- information about spares.

A comprehensive asset register is a useful source of information when a contractor is assessing the capability of a company to take on new work. It allows decisions to be made on whether new equipment should be purchased or hired in order to fulfil a contract. The register must therefore be accurate and kept up to date.

Industry tip

The terms 'condition report' and 'asset register' are more commonly used with reference to the condition of a building and the assets or equipment necessary for its operation.

Research

Visit **www.legislation.gov.uk/uksi/1998/2306/contents/made** and research Regulation 6 of the Provision and Use of Work Equipment Regulations (PUWER) 1998.

Write down the circumstances that demand inspection of equipment to ensure health and safety conditions are maintained.

Safety with electricity

Since the use of electricity is an everyday part of our lives, it can be easy to take it for granted as an energy source. However, you should always remember that electricity can be dangerous when not treated with respect, with the potential to cause severe injury or even death. Make sure you are properly trained before using electrical equipment in the workplace. Dismantling or adjusting an electrically powered item should only be undertaken by someone who is authorised and qualified.

The use of electrical equipment on site is regulated by the Electricity at Work Regulations 1989 (for more on these, see Chapter 1). These regulations require precautions to be taken against the risk of death or personal injury from electricity in work activities.

Improve your English

Research the Electricity at Work Regulations 1989 and write an account in your own words of the main employer responsibilities regarding the use of electrical equipment in the workplace.

When working with or near equipment powered by electricity, be aware of the types of danger that can arise:

- burns and electric shock, which depending on the voltage can kill you
- faults in equipment or wiring, which can cause a fire
- electrical sparks from faulty equipment, which can cause flammable gas to explode.

Industry tip

Building sites in the UK can be wet places to work. Avoid allowing electrical cables to trail in water. Wherever possible, arrange for them to be slung overhead. This will also reduce trip hazards.

Test yourself

What dangers can arise because of faulty electrical equipment?

Voltages

A range of voltages are used, depending on the equipment and circumstances.

In our homes, the usual voltage supplied for domestic appliances is 230 V, commonly referred to as 240 V. The difference in these two voltage figures is because voltages are referred to as 'nominal', which means they can vary slightly.

230 V is often used in workshop environments to power hand tools and fixed machines. Protection for users is provided by a residual current device (RCD). This will disconnect the supply quickly if a fault or unsafe condition occurs.

Research

Find out how a residual current device (RCD) works. Explain your findings to someone in your learner group.

On site, a lower voltage of 110 V is recommended, since lower voltages are safer. A piece of equipment called a transformer is used to reduce a 230 V supply to the safer voltage of 110 V.

Each voltage level is colour-coded, to make identification easier:

- 110 V – yellow
- 230 V – blue.

Industry tip

For fixed heavy machinery in a workshop setting, a higher voltage of 410 V can be used. This is often referred to as 'three-phase' supply and is colour-coded red.

Battery-powered tools

Battery-powered tools are safer than mains-powered tools, since they operate at lower voltages. They are available in a wide variety of voltages, from 3.6 V for a small powered screwdriver all the way up to 48 V for large masonry drills.

▲ Figure 14.13 Battery-powered drill

Although powered by a battery, these are still powerful pieces of equipment that can cause injury if not used with care. For example, a battery-powered drill produces a lot of torque, which can injure an operative's wrist and forearm.

1.2 Materials

Fit for purpose

Materials and components must meet defined quality standards during manufacture. A range of tests are carried out to ensure items leave the manufacturer in a condition that is fit for purpose.

▲ Figure 14.14 Manufacturers conduct tests to maintain quality standards

These tests might establish **compressive strength**, water absorption, hardness, size, shape, colour and much more. In more technically complex components, testing could include electrical resistance, flow rates, pressure containment and other performance- and safety-related factors.

> **Key term**
>
> ***Compressive strength:*** the ability of a material or structure to withstand loads that tend to reduce size

Between the manufacturing process and the point of use on site, many activities take place, including transport and storage. If not undertaken with care, these activities can cause damage or create conditions that lead to the development of flaws and defects in materials and components. Operatives on site must be aware of these possibilities and conduct appropriate inspections of materials and components before use.

> **Industry tip**
>
> Materials and components should be checked for good condition when they are delivered to site. Damaged or unsuitable items should not be accepted and can be logged and returned to the supplier.

The materials and components for particular work tasks are usually detailed in a work specification, which is a contract document. This means that there are legal requirements to ensure the right materials and components are used for a job. Ignoring the specification could have consequences related to the safe condition or operation of the completed work, its durability or its appearance.

Therefore, materials and components must be checked before use to confirm they match the specification. They should then be inspected for possible faults or defects to confirm that they are fit for purpose. Remember, damage due to poor transport, storage or handling can make items unsuitable for use.

In addition, care must be taken to avoid damage to materials and components during installation, which could render them unfit for purpose. The removal and replacement of damaged items leads to increased costs and lost time, which can have a significant detrimental impact on the overall work programme.

▲ Figure 14.15 Careful installation avoids damage to materials and components

Table 14.2 gives examples of damage to or defects in materials and components, to demonstrate the benefit of checking items are fit for purpose (the list is not exhaustive).

▼ Table 14.2 Examples of damage to or defects in materials and components

Area	Materials/components	Potential damage or defects that could render items unfit for purpose
Construction	Sheet materials (for example plywood, plasterboard and insulation)	• Chipped edges, scored surfaces, cracks or splits • Distortion • Damage by moisture or chemicals
	Timber	• Warped, twisted, bent, crushed or cracked • Damage by moisture or chemicals
	Steel	• Twisted, bent, distorted or cracked • Rust or corrosion
	Masonry	• Chips and cracks • Bent, misshapen, dimensionally inaccurate, discoloured or moisture saturated
BSE	Electrical cables	• Split or punctured insulation • Excessively bent or stretched
	Electrical components (for example switches, sockets, meters, distribution boards and circuit breakers)	• Cracked casings and damaged insulation/switches/fixings • Rating mismatch or indicator failure
	Plumbing components (pipework, junctions, water-storage tanks and cylinders)	• Crushed, kinked, punctured, corroded, rusted or dimensionally inaccurate • Size mismatch, rating mismatch (for example flowrates, thermal output), incorrect valve function
	Heating and ventilation components	• Damage to boiler components or refrigeration unit • Leaks or sensor faults • Incorrect flue gas rating, incorrect filters

Industry tip

It is important to develop the habit of checking items are fit for purpose by visually inspecting them for damage and defects. This will improve safety, productivity and efficiency on site.

Test yourself

State one benefit of checking materials are fit for purpose.

Hazards associated with materials and components

All operatives must be aware of site hazards and the methods used to identify and manage them. This includes hazards associated with the handling and use of materials and components.

The Control of Substances Hazardous to Health (COSHH) Regulations 2002 require the identification of:

- hazardous substances used on site
- processes on site that may produce hazardous substances.

The risks to site workers and members of the public must then be assessed and controlled.

▲ Figure 14.16 COSHH symbols

Hazardous substances take different forms, such as powders, liquids or gases:

- Some substances give off harmful vapours/fumes and dust, which can be inhaled (breathed in).
- Other substances are described as irritant or corrosive and can cause damage to the skin when it is exposed to them.

- Certain substances that can be injected, enter cuts in the skin or be absorbed through the skin are toxic (poisonous).
- Hazardous substances that are handled can leave deposits on the skin that can be transferred to the mouth when eating and then swallowed (ingested).

Read the manufacturer's or supplier's safety data sheets carefully. These provide information on how to safely use and store hazardous substances so that risks are minimised.

Research

Lead is a soft, durable metal that can easily be shaped (malleable), making it a good material for applications such as weatherproofing the junction between a chimney stack and a pitched or sloping roof.

Research the effects of lead on human health and write a report detailing your findings. What precautions should be followed to use lead safely?

Industry tip

Gloves form a basic part of PPE. However, specialist gloves might be needed to protect against some hazards. Safety data sheets for specific materials and substances provide information on appropriate PPE.

Test yourself

In what ways can dangerous substances enter the body?

Some common hazards associated with materials and components are listed in Table 14.3.

▼ Table 14.3 Common hazards associated with materials and components

Associated hazard	How it can be caused and what to do
Back injuries	Incorrect manual handling of heavy and awkward items such as heating units or large heavy sheets of material. • Use kinetic lifting technique and seek assistance when necessary (see Chapter 1 for more on lifting techniques). • Plan movement of heavy materials and components to avoid obstruction in the path of travel. • Where possible, use mechanical lifting equipment to avoid manual handling.
Crushing injuries to hands and feet	Trapping fingers between heavy items. Dropping heavy items onto feet. • Handle heavy materials and components in manageable quantities. • Always use appropriate PPE (gloves and safety boots). • Where possible, use mechanical lifting equipment to avoid manual handling.
Inhalation of dust (damage to the respiratory system)	Handling and cutting sheet materials and insulation can create dust and fibres which can be hazardous when inhaled (breathed in). • Always wear correct PPE or RPE (respiratory protective equipment). • Avoid cutting in enclosed spaces. • Consider dust control measures such as water misting or dust-extraction systems.
Inhalation of fumes (damage to the respiratory system)	Harmful fumes and vapours can be given off by paints, adhesives, chemical preservative treatments, soldering, brazing, welding and fuels. • Always use correct PPE or RPE. • Ensure adequate ventilation and avoid working in enclosed spaces.
Skin irritation	Irritant or corrosive materials and components can cause damage to the skin. Toxic substances can enter the body through cuts and abrasions or be absorbed through the skin. • Always use the correct PPE. • Use barrier creams. • Consult safety data sheets and COSHH sheets. • Wash hands thoroughly after working with irritants or corrosives.

▲ Figure 14.17 Safe kinetic lifting technique

Under the Health and Safety at Work etc. Act (HASAWA) 1974, it is the employer's responsibility to ensure the safe use, handling, storage and transport of components, materials and substances. (See Chapter 1 for more on HASAWA.)

However, all operatives must support their employer and others in the workplace by being aware of potential hazards, being observant on site, and handling and using materials and components correctly.

Industry tip

If you notice someone on site working in an unsafe way, do not be afraid to remind them of everyone's responsibility to work safely. If an unsafe practice continues, bring it to the attention of your supervisor.

Quantity

Calculating accurate quantities of materials and components for a project is important for efficiency and productivity. It:

- ensures the right amounts are delivered to site
- avoids shortages, which can cause production delays
- avoids surpluses, which can cause waste and a negative impact on the environment
- helps keep a project within budget.

As well as calculating how many items are needed, it may be necessary to establish:

- linear measurements
- area
- volume
- weight
- percentages.

For more on measuring and calculating quantities for different work tasks and trades, see Chapter 6.

Improve your maths

List as many construction trades and skills as you can think of. Produce a table to show which types of measurement you think each trade or skill would use in their work activities.

Specialist requirements

Some materials and components used in construction are very common, such as cement or timber. However, these common materials can be produced with variations or adaptations for specialist requirements.

For example, cement can be produced as:

- **sulphate**-resisting cement – a type of modified Portland cement that can be used in conditions where concrete or mortar is exposed to sulphate attack
- rapid-hardening cement – the one-day strength of this cement is equal to the three-day strength of ordinary Portland cement.

Key term

Sulphate: a salt of sulphuric acid

Timber can be produced and supplied as:

- stress-graded timber – classified for strength and stiffness for use as structural support in buildings
- glulam (glued laminated) timber – used to produce structural columns and beams as a sustainable alternative to reinforced concrete and steel.

Operatives may need additional training in order to use specialist materials effectively and efficiently. For example liquid roofing materials allow the installation of roof coverings where complex geometry and roof shapes make conventional methods difficult to apply. When installed by trained operatives, these roof coverings are quick to install, attractive and durable.

▲ Figure 14.18 A finished liquid roof

With the demand for environmentally friendly construction methods, more specialist materials and components are being developed for various purposes:

- insulation purposes (for example, Icynene spray foam, for application in traditional stone buildings)
- heating and ventilation systems (for example, refined computational fluid dynamics to model movement and temperature of air within spaces, for efficient system design)
- smart electricity generation and management systems (for more on this, see Chapter 8).

Research

Research ground source and air source heat pumps and how they operate. Suggest the specialist requirements for each type of heating system.

2 Maintenance of tools, equipment and materials

Tools and equipment require appropriate storage and regular maintenance to make them last longer and ensure they are efficient and safe. Materials need to be stored and handled so that they remain functional until they are ready to be brought into use.

2.1 Storage

Tools and equipment

Tools and equipment should be stored carefully, to protect them from adverse weather conditions and accidental damage. It is vital that those with metal parts and electrical components are protected from moisture.

Storage areas should be secure to guard against theft, as tools and equipment are often high-value items. Appropriate storage includes lockable steel containers (such as shipping containers) in a secure outdoor storage compound, or fixed lockable cabinets inside temporary site buildings.

It may be appropriate to keep an inventory of tools and equipment in a central store, with the requirement for an operative's signature when removing and returning items to the secure storage facility.

▲ Figure 14.19 Steel containers can be used for secure storage on site

Whatever storage facility is used, it should be managed so that tools and equipment are not randomly deposited or casually stacked. The provision of racks, stands and shadow boards is good practice to keep order and monitor the stock of valuable items. This also reduces the risk of accidental damage.

Improve your English

Search online for images of site storage. Copy and paste two different examples into a Word document and write an account of the tools and equipment that could be stored in each.

Industry tip

Shadow boards have outlines of tools marked on them to show where the tools should be stored. This makes it easy to see when tools are taken from stock or missing.

Test yourself

Suggest two examples of secure storage.

Materials

Storage facilities for materials should be:

- secure, to safeguard against theft
- protected against the weather
- located to allow easy unloading of deliveries
- organised to allow operatives to find required materials quickly and efficiently.

Materials should be stored carefully, in order to avoid injuries to operatives and accidental damage, which can add to costs and negatively affect productivity.

The following are examples of appropriate storage methods:

- Lengths of pipe for plumbing or heating purposes can be stored horizontally in racks according to diameter.
- Electrical items should be stored in a dry, dust-free environment, preferably with stable temperature conditions.
- Small, high-value items, such as door furniture or plumbing fittings, can be kept in lockable cabinets.
- Sheet materials should either be stacked flat, raised off the floor on spaced levelled bearers, or vertically, leaning at a slight incline towards a solid wall or frame.
- Bulk materials, such as sand, and heavy materials, such as bricks and blocks, can be stored in fenced outdoor compounds with protection from the weather provided by tarpaulins or plastic sheeting. These must be arranged to allow easy access by delivery vehicles and mechanical handling machinery.

Industry tip

Some materials such as cement and plaster have a limited shelf life. Storage should be arranged so that the oldest stock is used first to avoid it becoming unusable over time. This is known as the 'rotational stock control' or 'first in, first out' system (FIFO).

▲ Figure 14.20 Plastic pipes of various sizes stored neatly in a rack on site

Case study

An extensive range of materials are to be correctly stored on the construction site where you are working. Your supervisor requires you to produce directions for operatives to correctly store the items listed below.

Give instructions on how each type of component or material should be transported and stored on site:

- hot-water cylinders
- steel lintels
- sawn-timber joists
- ceramic drainage pipes
- box-profile roofing sheets.

When writing your instructions, take account of site factors such as uneven ground, trip hazards, weather conditions and bright sun reducing visibility as items are moved.

2.2 Maintenance

Whether simple or complex, all tools and equipment will deteriorate more quickly if they are not looked after, cleaned and maintained. Keeping them clean and in good working order can improve safety, make work tasks easier and make items last longer.

As mentioned previously, the accuracy of tools and equipment used for measuring, checking and testing can be reduced if they are not calibrated, checked and maintained regularly. Sensitive equipment can be affected by the harsh conditions experienced on construction sites, and they should be protected from exposure to dust and moisture where possible. Store them in their protective cases when not in use (after drying them if wet) and protect them from situations where they could suffer impact damage.

▲ Figure 14.21 Keep tools and equipment in protective cases when not in use

Test yourself

Why is it important to clean and maintain tools and equipment?

Maintenance for simple tools like hammers and screwdrivers is relatively straightforward, consisting of keeping them clean and storing them in dry conditions.

Hand cutting tools, such as chisels and saws, should be kept sharp as part of a maintenance routine, so that they function efficiently. Blunt tools can increase the likelihood of them slipping or moving unpredictably, which could lead to damage to the work or injury to the operative.

As mentioned frequently in this chapter, regular inspection should be a habitual practice. A quick visual inspection to make sure a hammer head is securely attached to its handle, or the tip of a screwdriver is not worn or damaged, is an important part of maintenance. Metal parts of tools should be treated with a rust inhibitor periodically.

More complex tools with moving parts, such as powered cutting equipment or drills, may require lubricating with oil or grease in accordance with the manufacturer's instructions. Some more specialist items of equipment require the application of specific lubrication products. Check operation manuals and equipment labels to ensure the correct lubrication is used, since using the wrong products can shorten the useful life of expensive items.

Equipment that has filters to capture dust or keep lubricating oil clean may have a specified maintenance programme, and this should be implemented carefully. Clogged filters cannot perform their function and may damage the equipment beyond repair.

Industry tip

It would be impossible to complete work on construction sites without appropriate tools and equipment. Taking good care of them will improve safety, efficiency and the quality of work.

Assessment practice

Short answer

1 What does 'PAT' stand for in relation to electrical items?

2 Who is responsible for calibrating instruments?

3 What is a holiday detector used for?

4 What action should you take if a measuring or testing instrument shows signs of impact damage?

5 State two dangerous occurrences that can be caused by faulty electrical equipment.

Long answer

6 List the steps that must be taken if a tool or item of equipment fails an inspection and explain why they are important.

7 Explain why routine daily checks and inspections should never be undertaken casually.

8 Detail why operatives should make sure they are familiar with operating instructions before using equipment.

9 Explain why a specification containing details of materials to be used for a job is called a contract document.

10 Describe the possible consequences of inaccuracy when calculating quantities of materials and components.

Project practice

You have been assigned to supervise a team of four operatives to remove a refrigeration unit from an air-conditioning system in an office block. The unit is leaking refrigerant fluid, which you have been informed is Difluoromethane (R32).

- Research R32 and download a safety data sheet for it.
- List the hazards associated with handling R32.
- In language that is easy to understand, write a guide to describe what your team should do if anyone is contaminated by the leaking fluid. Detail the action that must be taken for each of the hazards you have identified from the safety data sheet.

About the exams and employer-set project

The T Level in Building Services Engineering for Construction is made up of two components:

- A core component (350), which all students will complete.
- An occupational specialist component (or components), which is a trade-specific pathway that you will select with your lecturer.

This textbook covers the core component only, although the content you have covered and the skills you have learned will provide the foundation for your development in the occupational specialist component.

You can find out more about the qualification and how you will be assessed in the specification, which is available on the WJEC/Eduqas website.

Assessment

The core component is assessed in two ways:

- Papers 1 and 2, which are traditional written exam papers.
- An employer-set project, which is an extended real-world project that allows you to demonstrate the skills you have learned across the core component.

Core component exams

The two exams will test your knowledge and understanding of the content covered in the core component.

The 'Assessment practice' learning features, which appear at the end of each chapter, have been designed to help you apply your skills and knowledge in a similar style to what you can expect in the two exams.

Paper 1

This exam will include a range of short-answer questions, structured questions and extended-response questions.

The exam is 2.5 hours long and worth 35 per cent of the final grade for the core component.

The paper will cover:

- Health and safety in construction (Chapter 1)
- Construction design principles (Chapter 3)
- Construction and the built environment industry (Chapter 4)
- Sustainability principles (Chapter 5)
- Building technology principles (Chapter 7)
- Tools, equipment and materials (Chapter 14)

Paper 2

This exam will also include a range of short-answer questions, structured questions and extended-response questions.

The exam is 2.5 hours long and worth 35 per cent of the final grade for the core component.

The paper will cover:

- Construction science principles (Chapter 2)
- Measurement principles (Chapter 6)
- Information and data principles (Chapter 8)
- Relationship management in construction (Chapter 9)
- Digital technology in construction (Chapter 10)
- Construction commercial/business principles (Chapter 11)
- Building services engineering (BSE) systems (Chapter 12)
- Maintenance principles (Chapter 13)

Exam hints and tips

- Always read the instructions carefully. Think about what the command/keyword is asking for. These could include: identify, select, state, describe, explain, discuss, analyse, evaluate, justify.
- Concentrate on one question at a time and ask yourself the following:
  - Do I understand what the question is about?
  - How many marks is the question worth? You should try to work to one minute per mark.
  - How many parts are there to the question?
  - Can I provide a well-constructed answer?
  - How am I going to answer the question?
  - Do I need to include examples?
  - Do I need to relate my answer to a particular context?
  - Do I need to use technical terminology?

- It is important that the person marking your paper can not only read your handwriting but can also understand what it is you are trying to tell them. If they cannot read it or understand it, they cannot award you marks!
- If you make a mistake, cross it out neatly and then start again. There may be extra pages at the back of the exam paper, or you can ask for extra paper. If you use the extra pages or paper, then you must make it clear where your answer can be found.
- When you have finished answering the questions, and if you have time, go back over your answers.
- Read carefully what you have written and ask yourself:
  - Have I answered the question?
  - Have I answered all parts of the question?
  - Have I met the demands of the command/keyword, for example, have I explained?
  - Have I used the correct technical terminology?

Employer-set project

This is an externally-set and externally-marked project. The scenario, which will be written by WJEC/Eduqas in collaboration with employers, will allow you to use the skills, techniques, concepts and knowledge that you have learned across the core component. You will receive the scenario from your lecturer, along with detailed guidance.

The employer-set project is worth 30 per cent of the final grade for the core component.

The 'Project practice' learning features, which appear at the end of each chapter, have been designed to help you apply your skills and knowledge in a similar style to what you can expect in the employer-set project.

In addition to the core component content, the project also links to the core skills:

- Core skill A: Applying a logical approach to problem solving
- Core skill B: Primary research
- Core skill C: Communication
- Core skill D: Working collaboratively with other team members and stakeholders

The learning features in this book allow you to develop these skills.

Answers

Assessment practice

Chapter 1

Short answer

1 The Health and Safety Executive (HSE)

2 Mandatory, safe condition, prohibition, warning, fire fighting

3 Reporting of Injuries, Diseases and Dangerous Occurrences Regulations (RIDDOR) 2013

4 Tablets and medicines

5 During the site induction

Long answer

6 Under this legislation, employers must calculate the amount of vibration that employees may be exposed to at work. At a specific level (referred to as the 'exposure action value'), employers must introduce technical and organisational measures to reduce the risk of personal injury to an acceptable level.

7 A safe system of work is a formal set of procedures that must be followed when hazards cannot be eliminated completely.

8 Toolbox talks are short training sessions arranged at regular intervals at a place of work to discuss health and safety issues; they give safety reminders and inform personnel about new hazards that may have recently arisen. These are usually delivered to small groups of workers, in an area of the workplace where they should not be disturbed. These talks usually cover a single aspect of health and safety, such as good housekeeping.

9 A confined space is a workplace which may be substantially but not always entirely enclosed, where there is a foreseeable serious risk of injury because of the conditions or from hazardous substances. Excavations, loft spaces, sewers or wells could be described as confined spaces, because they are enclosed with restricted access and egress.

10 Working in a confined space should be avoided wherever possible, and work should be completed in another way without entering the space. If this cannot be done and there is still a significant risk of injury, then the work must be properly risk assessed, planned and organised, with appropriate control measures in place before it starts. Everyone involved in working in a confined space must be competent and specifically trained to undertake their tasks.

Chapter 2

Short answer

1 – kg/m^3
  – Radius is half the diameter and is 0.42 m so $\pi \times 0.42^2 = 0.55\ m^2$

2 Answers may include
  – Aluminium
  – Copper
  – Silver

  Brass or bronze would not be acceptable as these are alloys and not pure metals.

3 Force of load is:

  $180 \times 9.81 = 1765.8\ N$

  Force needed is:

  $f \times d = f \times d$

  So:

  $$\frac{1765.8 \times 0.5}{3} = 294.3\ N$$

4 As the load is a weight, gravity is not required. The time must be in seconds so 1.5 minutes = 90 seconds, so:

  $$\frac{5820 \times 28}{90} = 1810.67\ W$$

5 Firstly, the current needs to be determined:

  $$I = \frac{20}{58} = 0.34\ A$$

  So:

  $P = 20 \times 0.34 = 6.8\ W$

6 Stored hydro is where stored water, in a lake or reservoir, is released when needed to turn turbines. This can produce electricity almost immediately. Water is then pumped back into the lake or reservoir when the system is not required.

7 The RCD monitors the current flow in the circuit Line and Neutral. If the two values are balanced, the RCD will not trip. If a fault occurs and some current leaks to earth, less returns in the Neutral and the values become unbalanced. If the value of unbalanced current exceeds the residual current rating of the device, it will trip disconnecting the circuit.

8 If somebody, for example at work, is in a position where they are subjected to glare from a lamp, where they can see the full harsh light from the lamp, they will likely squint due to the harsh bright light. Exposure can cause discomfort and possibly

headaches. Glare can be avoided by deflecting the bright light by diffusers or louvres that soften and scatter the light.

9 Attenuation is the reduction of sound reverberation. This can be achieved in several ways including:
   - Using soft furnishings/fabrics
   - Using baffles such as hanging special panels made of soft material/fabrics from the ceiling
   - Using soft materials or fabrics on the walls or carpeting the floor.

10 Wall area is:

   $3 \times 3 \times 2 = 18\ m^2$

   $2 \times 3 \times 2 = 12\ m^2$

   Less door:

   $2 \times 0.9 = 1.8\ m^2$

   So:

   $18 + 12 - 1.8 = 28.2\ m^2$

   So heat loss through wall is:

   $28.2 \times 0.4 \times (28-(-5)) = 372.24\ W$

Chapter 3

Short answer

1 Conservation of fuel and power: Approved Document L

2 Permitted development rights

3 Any three of the following:
   - excavation of the foundation
   - laying of foundation concrete
   - installation of damp-proof course (DPC) and damp-proof membranes (DPM)
   - laying of drains
   - completion of the roof structure
   - completion of first-fix installations (before plastering or drylining)
   - testing of drains
   - completion of the project.

4 The RIBA Plan of Work

5 Any one of the following:
   - panelised – timber frame
   - steel-frame construction
   - structural insulated panels (SIPs)
   - volumetric (pod/modular).

Long answer

6 A greenfield site is an undisturbed piece of land that has never been developed (for example with roads, buildings or other structures). However, a brownfield site may have an existing structure (above or below ground), or may have been previously developed and the structure or building removed. Developing a brownfield site can be more expensive compared with a greenfield site, because of the costs involved in preparing the site and the removal of any contaminated soil. Brownfield sites are often preferred for use by local planning departments as they have less impact on the natural environment.

7 Vernacular construction is where the design of houses is sympathetic or particular to a region, relying on locally sourced materials and traditional skills that have developed over generations. Given the distinct regional characteristics of such buildings, it is problematic to design and specify this form of construction anywhere other than where it is usually found. It would be difficult to achieve the same standards of workmanship without employing labour and sourcing materials from further away.

8 A BIM designer works with and advises clients on the implementation of Building Information Modelling (BIM).

9 CDM Regulations, budget, site analysis, planning, site of special scientific interest (SSSI), animals or infestations, protected site

10 Step 1: Pre-planning application

   Step 2: Full planning application

   Step 3: Consultation process

   Step 4: Decision-making process/outcome

   Step 5: Appeal (if necessary)

Chapter 4

Short answer

1 When annual turnover exceeds £85,000

2 Any one of the following:
   - government
   - public limited company (PLC)
   - commercial
   - private.

3 Any one of the following:
   - the building owner's manual and user guide
   - guidance documents on defects reporting and aftercare
   - operational and maintenance manuals
   - a building regulations completion certificate
   - the health and safety file (including construction drawings/BIM)
   - the building log book
   - testing and commissioning certificates, for example a Building Regulations Compliance Certificate for gas installations
   - the building warranty/insurance certificate and policy booklet.

4 A snagging list

5 Any one of the following:
- political
- economic
- social
- technological
- legal
- environmental.

Long answer

6 Self-employed subcontractors are responsible for every part of running a business, for example estimating, invoicing, ordering and accounts. They do not receive some of the benefits that employed people enjoy, for example holiday pay (including bank holidays), sick pay, maternity or paternity leave or pay.

7 Building Information Modelling (BIM) uses digital technology to share construction documentation and provide a platform for effective and efficient collaboration between designers and the construction team at every stage of a building project. It is adaptable to suit the size and complexity of each project and allows technical information to be shared throughout the management and construction teams.

8 Clients sometimes withhold a percentage of money due to the contractor at each stage of the building work; this is known as a retention. The exact percentage of money to be withheld has to be agreed between the client and contractor before work starts and is usually 3–5 per cent. The retention acts as financial security for the client, to make sure the contractor finishes the building work and resolves any snagging within a reasonable amount of time after completion of the work; this is known as the defects liability period. If the contractor does not return to complete the work within the period agreed in the contract, the client has reasonable grounds to use the money to instruct other contractors to undertake the outstanding work. On the other hand, if the contractor does complete the project and resolve all of the defects identified within the defects liability period, the client must release the outstanding retention payment without delay.

9 Answer to include:
- protecting clients, customers and the public
- keeping up to date with the latest regulation changes, product developments and technological advancements
- developing product knowledge
- working more efficiently
- improving knowledge and skills
- enhancing the company image
- career progression.

10 A tender package may include the following documents:
- letter inviting the contractor to submit a bid for the work
- outline of the proposal
- form of tender and timeline to return the completed bid
- form of contract and conditions (including the process for payments and interim valuations)
- programme of work
- design drawings
- specifications
- site-specific information or issues
- preliminaries
- special planning-permission requirements
- bill of quantities (cost framework)
- tender return document.

Chapter 5

Short answer

1 Any two of the following:
- Building Research Establishment Environmental Assessment Method (BREEAM)
- Leadership in Energy and Environmental Design (LEED)
- Timber Research and Development Association (TRADA)
- WELL Building Standard
- PAS 2035
- PAS 2038.

2
- building management systems
- automated controls
- smart controls
- smart meters.

3 Any three of the following (or any other appropriate answer):
- self-healing concrete
- green roofs
- smart glass
- grey-water recycling
- reed beds
- soakaways
- smart cement.

4 The Domestic Building Services Compliance Guide

5 Grade I: buildings of exceptional interest; Grade II*: particularly important buildings of more than special interest; Grade II: buildings of special interest, warranting every effort to preserve them

Long answer

6 When planning and delivering a construction project, sustainability is achieved by:
   - using renewable and recyclable resources
   - sourcing materials locally
   - protecting resources
   - reusing and refurbishing materials
   - reducing energy consumption and waste
   - creating a healthy and eco-friendly environment
   - protecting the natural environment.

7 Five retrofit roles:
   - Retrofit assessors undertake retrofit assessments for dwellings in accordance with PAS 2035.
   - Retrofit coordinators provide a project management role.
   - Retrofit advisors provide advice to clients and homeowners on the retrofit process.
   - Retrofit designers prepare a safe and effective retrofit design.
   - Retrofit evaluators monitor the impact of installed EEMs to ensure the intended outcomes have been met.

8 Social sustainability focuses on wellbeing and quality of life. It involves recognising the needs of everyone impacted by construction projects, from design to demolition. This includes construction workers, local communities, project supply chains and users of the building.

9 Passivhaus, or 'Passive house' in English, is an energy performance standard intended primarily for new buildings. It ensures buildings are so well constructed, insulated and ventilated that they require little energy for heating or cooling.

10 Grey water is waste water that has been used for washing and is generated from hand basins, washing machines, showers and baths. Rather than sending it down the drain, it can be reused for watering plants and flushing toilets, thereby reducing mains water usage.

   Rainwater harvesting involves collecting rainwater from roofs, filtering it and storing it for reuse, thereby saving energy, lowering carbon emissions and reducing mains water consumption.

Chapter 6

Short answer

1 The volume of topsoil is calculated in cubic metres (m^3).

2 Answer to include:
   - The work should be completed within an agreed budget and timescale.
   - The design brief should be wholly fulfilled.

3 A number system based on the number ten, tenth parts and powers of ten

4 Any two of the following:
   - setting out the outline of a building
   - positioning internal walls, doors and windows
   - establishing floor and roof heights.

5 New Rules of Measurement

Long answer

6 Accurate calculation of quantities of materials and components allows for assessment of the time needed to use and install those materials, contributing to the creation of a realistic work programme. Completion targets can then be set to support efficiency and productivity and satisfy the client's requirements for timely handover of the building.

7 Tolerances allow for acceptable variations in the strength of materials, the performance of a heating or ventilation system, temperature ranges in which materials can be used and many other situations. Exceeding allowable tolerances for a given construction task will mean that the work does not meet the specification. The function of the structure could be compromised and other elements of the building may not be able to be installed properly.

8 If a building that is not dimensionally accurate or not built to square is handed over to the client, significant problems can emerge when the installation of items such as floor tiles is attempted, or fittings and modular components such as kitchen units are positioned. Necessary remedial work would be potentially very costly to the contractor and disruptive to the client.

9 If the height of floors is not measured accurately during installation, a staircase will be placed at the wrong angle of slope, and it may not sit properly where the top of the stairs meets the upper floor.

10 U-value expresses the rate of transfer of heat through a structure (or more correctly through one square metre of a structure), taking account of the difference between internal and external temperatures. It is expressed in watts per metre squared kelvin (W/m^2K). The lower the U-value, the lower the rate of heat transfer.

   Measured resistance to heat transfer is expressed as an R-value. The higher the R-value, the more resistance a material provides to heat transfer.

Chapter 7

Short answer

1 Approved Document B

2 Gas Safe

3 Pad foundations

4 50 mm

5 Any one of the following:
   - masonry screws/bolts
   - screws and plugs
   - chemical fixings
   - masonry nails
   - cartridge fixings.

Long answer

6 Off-site construction involves manufacturing modules in factories and then transporting them to site where they are assembled to form major elements of a building. It offers many benefits:
   - Automated systems improve accuracy and quality control.
   - Large numbers of components can be manufactured quickly and efficiently with reduced waste.
   - By carrying out a large proportion of the construction work in a controlled factory environment, the delays associated with adverse weather conditions on site are avoided.
   - Up to 95 per cent of a building can be constructed in a factory, therefore reducing the number of trades needed on site during the assembly and final finishing of the project.
   - There are cost savings from mass production in factories.

7 First fix is the phase of work completed after the structure has been erected and before plastering commences. Once the various trades have completed their first-fix work and the plastering/drylining is finished, second-fix work can commence. Second fixing often involves the installation of items and equipment that could have easily been damaged or affected by earlier stages of construction work.

8 Benefits of autonomous vehicles include:
   - reduced costs by improving productivity and site progress, because they are able to operate 24 hours a day without a driver
   - fewer carbon emissions than vehicles driven by humans, due to less erratic movements and more efficient use
   - better safety and accuracy.

9 Components of a solid floor:
   - compacted hardcore – provides a firm level base to bear the weight of the floor materials above
   - sand blinding – protects the polythene damp-proof membrane (DPM)
   - DPM – prevents rising damp and weed growth
   - concrete slab – provides most of the structural integrity
   - insulation – prevents heat loss through the floor
   - screed – provides a flat, level surface (also used to cover underfloor heating pipes).

10 Answer should include:
   - increased risk of harm to workers and others
   - product failure
   - premature wear
   - invalidation of warranty or product guarantee
   - minimum building standards not met therefore unable to be signed off by building control
   - breaches of health and safety legislation.

Chapter 8

Short answer

1 Numbers, measurements, words and descriptions

2 The ability of data systems to exchange and use information

3 Quantity surveyors preparing costings; contractors and subcontractors preparing tenders; systems designers planning building services requirements

4 Enterprise asset management

5 Some systems are dependent on different weather conditions and must be checked over time.

Long answer

6 When a set of data is used to describe or analyse other data, this is referred to as metadata. It speeds up the analysis of information required for technical design processes and could inform an architect regarding multiple intricate design elements of a building.

7 Generalisation is essentially a process of 'pulling back' from the mass of raw data to gain a broader, more general view and reduce the extent of analysis required. It avoids the need to process large volumes of complex data, which could be time consuming and confusing.

8 BIM allows authorised users to access a range of important information at all stages of construction and beyond. It uses digitally processed information to analyse design elements of a building, including 3D modelling.

Using BIM, complex design ideas can be transformed into a medium that is easier for all personnel to work with. It allows collaboration between every designer, engineer and contractor working on a project, providing comprehensive information about different workflows.

The Common Data Environment (CDE) is a single central source of information used within a BIM system. Relevant documents and data are brought together in a shared digital environment that can be accessed by all authorised personnel collaborating on the project.

9 When a building is handed over from the contractor to the client for occupation and brought into use, data can be harvested to answer important questions, such as:

- How successful was the delivery of the project by the contractor?
- Was the project delivered and handed over on time – if not, why not?
- Does the building meet its design brief fully now it is in use?
- What changes could be made to improve performance?

This data can be useful to contractors who build repeat structures or specialise in specific project types.

10 When calculating outcomes, data can be used to analyse past or current activities to model what could happen in the future. An example might be calculating the outcome of using timber of specific dimensions to produce a load-bearing element of a building, such as a suspended floor. Consistent data has been recorded over many years which can be applied to producing reliable structural calculations, ensuring that the proposed design is safe and capable of bearing the required loads. Other examples of data might be performance charts for heating systems.

Cost calculations draw on data that is collated, analysed, modelled and interpreted by the Building Cost Information Service (BCIS) of the Royal Institution of Chartered Surveyors (RICS). All sizes of construction company depend on accurate data to produce costings that will result in profitability for the contractor and a fair price for the client.

Chapter 9

Short answer

1 Stakeholder

2 To ensure a good company reputation and increase the chances of repeat business

3 Equality Act 2010

4 Company handbook/policy

5 Workplace pension scheme

Long answer

6 Answer to include the following:

- Distributive negotiation is used to haggle over a common single interest at stake, known as a fixed sum. A fixed sum is best described as a pie that parties are battling over for a bigger slice, with exchange offers back and forth.
- The win–lose approach is probably the most common negotiation method used to settle disputes between two parties. An agreement is more difficult to reach, because one side has to compromise in order for the other to experience a positive outcome.
- The lose–lose approach – following negotiations, all concerned parties end up worse off and not achieving their desired result. All participants should try to minimise their losses and to make sure they are fair.
- The compromise approach is used to settle disputes quickly by one party settling for less than it may have hoped for, in order to reduce strained negotiations and maintain or fix a relationship.
- Integrative negotiation is used by parties with common interests, in order to collaborate in finding a mutually beneficial solution.
- The win–win approach – negotiators with shared interests work together to find resolutions they are both satisfied with, rather than seeking to fulfil self-interests. This avoids disagreements and helps to maintain strong relationships between parties, while achieving a fair outcome for both sides.

7 An example of unethical behaviour in a business: during the development of a new construction site, a contractor wilfully damages protected trees (TPO – Tree Preservation Order) with a digger and contaminates a local stream with harmful construction waste. The polluted watercourse damages the natural environment of fish and other animals.

8 Networking is a low-cost process used by many organisations to make initial introductions with likeminded people, share information and form long-lasting business relationships.

9 Arbitration may be used as a conflict-management technique in the workplace, if an employee feels they have been unfairly dismissed from their job. This process would involve the appointment of an unbiased conciliator to meet with both parties separately and together, to weigh up both sides of an argument. They will then make a proposal based on the relative merits of each side, to find an amicable solution to the matter and bring it to a close as quickly as possible.

10 Strengths of non-verbal communication:
  - It can be referred back to.
  - There is a permanent record of the communication.
  - The same information can be distributed easily without diluting it.
  - The sender does not have to meet the recipient to pass on the information.
  - It can be used to communicate if the recipient has a hearing impairment.

Chapter 10

Short answer

1 Machine to machine

2 By monitoring energy use and matching it to demand

3 Any two of the following:
  - improvements in manufacturing efficiency
  - improvements in safety
  - streamlining of materials delivery and supply-chain activity.

4 Light detection and ranging (LIDAR)

5 For performing repetitive tasks, such as construction of building modules, and tasks that might be hazardous to operatives

Long answer

6 AI uses methods of analysis that identify patterns and correlations, in order to draw appropriate conclusions more quickly than humans are able to.

7 Future AI systems may be able to sense in real time whether a building's occupants are tired, too hot or too cold, and automatically make adjustments to heating and ventilation to enhance comfort and wellbeing.

8 Digital CAD systems can speed up the design process considerably, leading to greater productivity.

Building design ideas created using specialist software can be presented as a three-dimensional (3D) model, allowing stakeholders to examine and refine the design concept.

Digital tools are able to generate accurate materials and components lists from 3D models or 2D drawings. These can be linked digitally to materials costs databases, in order to produce an up-to-date costing for a project.

Speeding up the process of designing and costing a project, coupled with the ability to update or amend project details quickly, streamlines development and improves productivity.

Since CAD drawings are produced digitally, they can easily be stored, retrieved, shared and copied.

9 Equipment and machinery can be fitted with digital sensor technology to operate as part of the IoT. The range of movement and operational area of excavators and other machines can be controlled automatically, and programming can be updated to match new instructions from remote control locations.

10 Simulation is used to create a two- or three-dimensional model, which can provide technical data on structural and systems performance. Animation provides an engaging and lifelike view of a building, allowing a viewer to travel through a structure to assess the various elements visually.

Chapter 11

Short answer

1 Franchise

2 A financial year

3 Networking

4 Benchmarking

5 Objectives

Long answer

6 Business types include:
  - sole traders – must be registered with HMRC as a sole trader. As a proprietor, they have full control of their entire business or enterprise, although they could employ other people to work for them. A sole trader is responsible for the day-to-day management of their company, including filing self-assessment tax returns with HMRC and paying National Insurance contributions. Sole traders are personally liable for any losses or debts that the company may incur if things go wrong.

- partnerships – are owned by two or more individuals. Partners share the costs, duties and risks of managing a business together, although they may delegate certain responsibilities. Each partner is personally responsible for paying tax and National Insurance contributions, based on their share of the profits. There are three different types of business partnership: ordinary partnership, limited partnership and limited liability partnership (LLP).
- limited companies – private (Ltd) or public (PLC) – must be registered with HMRC and Companies House and file tax returns. The owners and shareholders have limited liability, so their personal assets are protected against any business debts up to the value of the money that they have invested. In a PLC, money can be raised for the business through investors (for example the public) buying shares on the stock exchange.
- small and medium-sized enterprises (SMEs) – in the UK, small and medium-sized enterprises (SMEs) are defined as follows:
  - Small – employs on average no more than 50 people and has an annual turnover of £10.2 million or less
  - Medium-sized – employs on average no more than 250 people and has an annual turnover of £36 million or less
- not-for-profit organisations– are charitable businesses that do not make a financial profit. They often seek to provide a public service or social benefits for individuals or communities in need.
- community interest companies (CICs) – aim to provide a benefit for the community or trade with a social purpose. Returns to the company owners and investors are allowed, as long as they are balanced and reasonable, and a dividend cap must be in place.
- franchises – allow a business to expand quickly with lower capital outlay by selling the rights to the business model to self-employed entrepreneurs, referred to as 'franchisees'.

7 A sole trader is personally liable for any losses or debts that the company may incur if things go wrong. Creditors can legally recover any money owed by the debtor from their business or their personal assets such as their house. If the debt is considerable and the individual is unable to pay, they could be made personally bankrupt. This could result in further difficulties, for example being unable to obtain any type of credit such as a mortgage, credit card or finance on a vehicle because they are deemed to be a high risk.

8 The main goal of the ISO is to facilitate trade. It offers solutions to global challenges and supports innovation by providing guidelines to streamline processes and improve quality and safety across a range of businesses and products.

9 The purpose of an organisation's corporate social responsibilities (CSR) is to actively make a positive contribution to the community and wider society, and to minimise any negative impacts caused by their business. Having a CSR strategy integrated into its values also affects how stakeholders (such as clients and investors) view an organisation and whether or not they decide to work with or support it.

10 Areas that can be measured in terms of a value are known as key performance indicators (KPIs). Measuring business performance against KPIs helps to establish if objectives have been met or whether new strategic targets need to be planned. If a business is not honest with the findings of KPIs, fails to learn from its mistakes and starts the cycle again, then the process of benchmarking is pointless.

Chapter 12

Short answer

1 Any three of the following:
   - socket outlets for appliances with a 3 A or 13 A three-pin plug
   - fixed appliances, such as cookers, showers and immersion heaters
   - lighting systems
   - protection services, such as intruder alarms, surveillance systems, fire alarms and access controls
   - refrigeration and ventilation systems
   - telecommunication systems, such as telephones, internet, home entertainment and connections for other BSE systems
   - heating systems, such as gas boilers, electrical wall heaters and fan convectors.

2 Direct, indirect and boosted

3 Any three of the following:
   - a circulation fan (fan coil unit) – moves air to and from rooms
   - an air-conditioning unit – uses cooling and dehumidification processes in summer or heating and humidification processes in winter

- supply ducts – direct conditioned air from the circulating fan to the space to be air-conditioned
- an air diffuser – distributes the conditioned air evenly in the room
- return air grilles – allow air to enter the return duct
- filters – remove dust and bacteria from the air.

4 Any three of the following:
- isolation valves – to turn off (isolate) complete systems, parts of a system or appliances
- drain valves – to drain down systems
- stop taps – to isolate high-pressure cold-water systems
- gate valves – used on low-pressure installations, such as the cold feed to vented hot-water storage cylinders and the cold-distribution pipework for indirect cold-water systems
- spherical plug valves – to isolate appliances and terminal fittings such as taps and float-operated valves
- drain-off valves – small valves strategically placed at low points in pipework installations to allow draining down of the system
- float-operated valves – to control the flow of water into cold-water storage and feed cisterns, feed and expansion cisterns, and WC cisterns; designed to close when the water reaches a pre-set level
- radiator valves – to control the temperature and flow through a radiator
- thermostatic radiator valves (TRVs) – to control the temperature of a room by regulating the flow of water through a radiator
- wheel-head valves – to allow manual control of a radiator by being turned on or off
- lockshield valves – to regulate the flow of water through a radiator; designed to be adjusted during system balancing
- automatic air valves – to allow collected air to escape from a system but seal themselves when water arrives at the valve; fitted where air is expected to collect in a system, usually at high points
- anti-gravity valves – to prevent unwanted gravity circulation within heating and hot-water systems
- pressure-relief valves – to protect against over-pressurisation of water in a range of systems
- emergency control valves (ECVs) – to allow the gas user to shut off the supply of gas in the event of an escape; found on the service pipe connecting a gas meter to the gas mains.

5 During the vapour-compression refrigeration cycle, the refrigerant vapour enters a compressor, which compresses it, generating heat. The compressed vapour then enters a condenser, where the useful heat is removed and the vapour condenses to a liquid refrigerant. The liquid refrigerant passes from the condenser into an expansion valve, where rapid expansion takes place, converting the warm liquid into a super-cold vapour/liquid mix, which creates the refrigeration effect. The vapour/liquid mix passes through an evaporator, where final expansion to a vapour takes place. This vapour then enters the compressor for the cycle to begin again.

Long answer

6 A heat pump warms or cools a building by moving heat from a low-temperature reservoir to another reservoir at a higher temperature. Its working principles are the same as for a refrigerator, which creates heat while making the refrigerator cold.

The process is known as the vapour-compression refrigeration cycle and involves compressing a gas (called the refrigerant) with a compressor until it becomes a liquid. This generates useful heat that can be used to warm a building. When the pressure is released through an expansion valve, very cold temperatures are generated, which can be used for cooling a building.

7 Fire alarm systems provide early detection and warning of a fire. They usually consist of a control panel linked to fire detectors and manual call points (often referred to as detection zones) and alarm circuits.
- Category M systems rely on manual operation by the people using the building. The usual method of raising the alarm is to break the glass on a manual call point.
- Category L systems provide automatic fire detection (AFD) and are designed primarily to protect life. This category is subdivided, according to the areas of the building that require the installation of AFD.
- Category P systems provide AFD and are designed primarily to protect property. This category is subdivided, with P1 requiring AFD

in all areas of the building and P2 requiring AFD only in specific parts of the building.

8 Rainwater systems collect and carry away rain from roofs using either integrated channels or eaves-mounted gutters connected to rainwater pipes. The water is discharged into surface-water drains, combined sewers, soakaways or watercourses such as streams and rivers.

The purpose of a rainwater system is to:
- protect the foundations of the building
- reduce ground erosion
- prevent damp and water penetration of the building.

These systems can also be used as part of a rainwater-harvesting system.

9 Secondary circulation prevents the wastage of water due to excessive lengths of hot-water draw-off from the storage vessel to the outlet. It is a method of returning the hot-water draw-off back to the storage cylinder in a continuous loop, to eliminate cold-water 'dead legs' by reducing the distance the hot water must travel before it arrives at the taps. In all installations, secondary circulation incorporates a bronze- or stainless-steel-bodied circulating pump to circulate the water to and from the storage cylinder.

10 Central-heating systems distribute warmth throughout the whole or part of a building from a single heat source (the boiler), for the thermal comfort of the occupants. Boilers can be fuelled by coal, gas, oil or electricity; they transfer their heat energy to another medium, usually water or air, which carries the heat to the areas where it is needed.

Chapter 13

Short answer

1 Planned preventative maintenance (PPM) is scheduled to take place at a particular time, for example monthly or annually. It is designed to prevent breakdowns and improve efficiency. Reactive maintenance is unplanned and carried out when a system (or part of a system) fails.

2 A refrigeration or air-conditioning system

3 One year

4 Servicing logbook

5 When a system is out of action while maintenance/servicing is completed

Long answer

6 Unvented hot-water system: inspection of line strainer, pressure-reducing valve, pressure-relief valve, temperature/pressure relief, high limit stat and expansion vessel

Drainage system: visually inspect pipes for damage, check all traps for correct seal levels, perform an air test of the stack system (as per Part H of the Building Regulations), carry out a performance test of the system (as per Part H of the Building Regulations)

Domestic wiring system: visual inspection of wiring and components for damage, testing of consumer unit residual current device (RCD), testing of the miniature circuit breaker (MCB), carrying out portable appliance testing (PAT), carrying out a polarity test, insulation resistance test and earth loop impedance test

7 The maintenance plan should include:
- Visually inspect all components.
- Clean passive infrared (PIR) sensor.
- Check all cameras.
- Clean lenses.
- Check correct focus.
- Check alarm bell or siren is triggering and audible.
- Check hard drives for recording capability.
- Check functionality of key pads/touch pads.
- Check emergency lighting.
- Check outside alarm box.
- Check for autodial to keyholder/security company/police.

8 This will vary according to the manufacturer's instructions.

Commercial fire alarms need to be inspected and maintained by a professional every six months, as per BS 5839.

9 A job sheet should include the:
- client's name
- site location
- date and time
- priority of the task
- work to be carried out.

10 The potential implications will vary, according to whether the offline electrical systems are isolated to one building or area, or affect the whole business. If the maintenance is completed out of hours, for example at night or over the weekend, there will be less inconvenience.

Total closure of a building, particularly for a business that relies heavily on communications and data systems, could lead to a loss of both existing income and potential new business.

Some businesses might have contingency plans that allow their staff to work from home, however again this might have huge financial implications due to equipment compatibility and GDPR compliance.

It is essential that all data is backed up before systems are taken offline, as there is the potential to lose data when systems have downtime for maintenance.

Chapter 14

Short answer

1 Portable appliance testing

2 Specialist companies or the manufacturer

3 To detect flaws and porous areas in protective coatings applied to metal surfaces

4 Bring it to the attention of your supervisor or line manager to get advice

5 Any two of the following:
- burns and electric shock, which depending on the voltage can kill you
- faults in equipment or wiring, which can cause a fire
- electrical sparks from faulty equipment, which can cause flammable gas to explode.

Long answer

6 If defects are discovered in tools or equipment during inspection or use:
- The item should be removed from use immediately and the relevant supervisor should be informed. This prevents injury to operatives.
- A tag or label should be attached to the item, stating the nature of the problem. This ensures any other operatives who might use the tool are aware of the problem.
- The item should be kept in a secure location. This will prevent inadvertent use.
- The item should be repaired or replaced as soon as possible.

7 The operative completing the check/inspection has accountability to the user of the tool or equipment, by confirming that the item is safe to use and should operate as expected.

8 Operatives should be familiar with manufacturers' instructions, so that tools and equipment are used correctly and developing problems are quickly recognised and dealt with.

9 This means that there are legal requirements to ensure the right materials and components are used for a job. Ignoring the specification could have consequences related to the safe condition or operation of the completed work, its durability or its appearance.

10 Calculating accurate quantities of materials and components for a project is important for efficiency and productivity. It:
- ensures the right amounts are delivered to site
- avoids shortages, which can cause production delays
- avoids surpluses, which can cause waste and a negative impact on the environment
- helps keep a project within budget.

Glossary

Accident book: a formal document used to record details of accidents that occur in the workplace, whether to an employee or visitor

Active monitoring: monitoring people, procedures, premises and plant in the workplace in order to identify potential hazards before an accident or incident occurs, so that control measures can be taken to prevent harm

Alternator: an electrical generator that converts mechanical energy into electrical energy

Ambiguous: unclear and difficult to understand

Appraisals: scheduled routine meetings between an employee and their employer to review their work performance against their job description

Arson: the criminal act of deliberately setting fire to property

Autonomous vehicles: vehicles able to operate without human intervention

Backflow: the movement of liquid in the opposite direction to its regular flow; this can lead to contamination of potable water supplies and create a serious health risk

Basements: habitable rooms or spaces constructed below ground level

Benchmarking: measuring an organisation's internal and external performance against pre-determined industry standards, competitors or completed projects

Bilingual: fluent in two languages

Bimetallic strip: a temperature-sensitive component comprising two different metals bound together; when heated, each metal expands at a different rate, causing the strip to bend and activate a switch

Biometric: involving the detection and analysis of individuals' unique physical characteristics as a means of verifying identity

Bonded: the arrangement of staggered joints

Building Information Modelling (BIM): the use of digital technology to share construction documentation and provide a platform for collaboration

Built to square: where all the corners of a square or rectangular building are accurately set at 90°

Bump caps: a type of PPE designed to protect the user's head when there is a low risk of bumping it; it is not designed to take an impact from an object falling from height

Burner: an integral part of a boiler where combustion takes place

Bushes: (in machinery) plain bearings or sleeves that reduce friction

Business model: a plan usually created by a business owner which describes the strategy or framework that an organisation will use to operate and includes the identification of products/services, revenue sources and customer base

Business plan: a written document that defines a business' goals and the strategies and timeframes to achieve them

Calibration: comparison of a measurement device against a traceable reference

Capital: the amount of funds or liquid assets owned by a business

Carbon: a chemical element that can be released into the atmosphere when fossil fuels are burned

Cat 5/6 data cable: twisted pair cable used for ethernet connection

Catenary wires: strong wires which are tied at each end and used to support other objects, such as cables which may stretch or break under their own weight when hung between two buildings

Cavity walls: external, load-bearing, structural walls consisting of two individual leaves (skins) of masonry with a gap (cavity) between them

CCTV: closed-circuit television, also known as video surveillance, is the use of video cameras to transmit a signal to a specific place, on a limited set of monitors

CE marking: a mark on a product that identifies it has been designed and manufactured to meet EU safety, health or environmental requirements; CE is an abbreviation of a French term *'conformité européenne'*, meaning 'European conformity'

Chases: cuts in a masonry or plaster wall to conceal pipes or cables; once fitted, they are covered with plaster or similar

Chemically inert: will not react with chemicals

Chronic: continuing for a long time

Civil engineering: a profession involving the design, construction and maintenance of infrastructure that supports human activities, for example roads, bridges, airports and railways

Climate change: a large-scale, long-term change in the Earth's weather patterns and average temperatures

Cob: a blend of subsoil (clay or earth), sand and straw mixed with water to make an organic material historically used to construct walls for homes and agricultural buildings in Devon, Cornwall and Wales

Collaborative: involving people or groups of people working together with a single common interest or aim

Commercial: relating to buying and selling

Commissioning: the process of ensuring that a building system is performing or working as it has been designed

Common law: legislation made in the civil courts rather than statute law that is made in Parliament

Common rafters: rafters that run from a ridge board to the wall plate at 90 degrees on plan

Companies House: a government body that registers and stores information on all the limited companies in the UK and makes it available to the public

Composite cable: multi-core cable, in which the cores are surrounded by a sheath that provides mechanical protection

Computer-aided design (CAD): using computer software to develop designs for buildings and structures

Concessions: something granted in response to a demand

Conductors: materials that have atoms less densely packed together and allow electron flow

Conduit: a tube used for protecting electric wiring

Consumer unit: a component in an electrical system which contains the main switch, isolation for circuits and circuit protection devices

Contingency: provision for an unforeseen circumstance; financial contingencies are often considered when planning for a construction project

Conventions: agreed, consistent standards and rules

Corporate social responsibility: the commitment of an organisation to carry out its business activities in a socially and environmentally responsible way

Corporation: a business owned by its shareholders, who often appoint a board of directors to manage the day-to-day running of its activities; the business is a legal entity and the shareholders have no personal liability for its actions and finances

Credential: something used to recognise and validate a person

Credit agreement: a legal contract made between a person or party borrowing money and a lender; it states the terms and conditions of the acceptance of credit, including how the debt will be repaid

Creditors: individuals or organisations that are owed money because they have provided goods, services or a monetary loan; HMRC would also be described as a creditor where a tax duty is owed

Crosstalk: interference between telecommunication signals

Data interoperability: the ability of data systems to exchange and use information

Datum: a fixed point or height from which reference levels can be taken

Defamation: the act of damaging someone's good reputation through a false written or verbal statement, also known as libel (written) or slander (spoken)

Demographic data: statistical data about a population in a particular location or region

Department for Business, Energy and Industrial Strategy (BEIS): the department of the UK government responsible for the UK's business, energy and industrial strategy

Design brief: a working document which specifies what a client wants; it makes clear all the design requirements of a project

Digital: in electronic form

Dissipated: energy consumed by converting to heat energy

District network operator (DNO): a company licensed to distribute electricity

Draw-off water: water discharge from a terminal fitting such as a tap

Duty holders: people with a legal responsibility under health and safety law

Dynamic: characterised by frequent change or motion

Eaves: the part of a roof that overhangs the internal skin of the external walls

Ecological balance: where living organisms, such as plants, animals and humans, co-exist in a sustainable environment

Economic downturn: when the economy has stopped growing and is on the decline, resulting in reduced financial turnover

Efficiency: the ratio of output power compared to input power expressed as a percentage

Effluent: liquid waste or wastewater

Egress: an exit or way out

Elevation: a view of the front, back or sides of a building

Encasing services: creating a framework of materials to cover pipes, cables and other services in a building; also referred to as boxing-in

Encryption: the process of converting data or information into a code to prevent unauthorised access

Environmental activists: people who campaign for the protection of the natural environment

Equality: a state where all members of a society or group have the same status, rights and opportunities

Ethics: moral values that govern a person's behaviour towards others

Feasibility: how easy or difficult it is to do something

Ferrous metal: a metal that contains iron and is magnetic

Firewall: a protective software program or hardware device that monitors, filters and may block data entering and leaving a network

First fix: a phase of construction work completed before plastering

Fulcrum: the pivot point of a lever

Gas Safe: a professional organisation that controls the health and safety of work completed on gas systems in the UK

Generalisation: (in data processing) creating layers of summarised information from mass data

Geofencing: using technology to pinpoint the location of equipment and create a virtual boundary

Geometric: consisting of defined angles, patterns and shapes

Geotechnical engineering: a field of civil engineering that deals with the behaviour of earth materials such as soil and rock

Graining: a method used by decorators to create woodgrain effects on different surfaces

Grey water: water that has not been purified for the purpose of drinking, for example recycled water from a sink

Grievance: a feeling of having been treated unfairly

Gypsum: a natural mineral, often used in building products such as plaster and plasterboard

Haggle: to negotiate for the best terms of an agreement or financial arrangement

Heat: does not, in BSE terms, mean hot, it simply means heat energy, which can be hot or cold

Human resources department: department of an organisation that deals with recruiting, administrating and training staff

Immersion heater: an electrical element that sits in a body of water; when switched on, the electrical current causes it to heat up, which in turn heats up the surrounding water

Immersive: creating a 3D image which appears to surround the viewer

Incorporated: legally registered as a limited company

Infrastructure: the basic systems and services required for the proper functioning of society

Insudite insulation: a type of fire- and heat-proof insulation

Joules: the unit of measurement for energy. Where energy is expressed as mechanical energy, it is known as work

Kinetic energy: energy derived from motion

Kinetic lifting: the act of manual handling

Liable: legally responsible

Life cycle assessment: assessing the total environmental impact of a building, considering all stages of the life of the products and processes used during its construction

Limited partner: a part-owner of a business whose financial liabilities cannot exceed their investment

Linear measurement: the distance between two given points along a line

Local Authority Building Control: local authority department responsible for inspecting building work against building regulations and signing-off completed projects

Lone workers: those who work by themselves without close or direct supervision of their employer or work colleagues; they may work in a remote location or in a workplace with members of the public

Low-carbon steel: a ferrous metal used for pipework applications; also known as mild steel

Low-emissivity (low-E) glass: a type of thermally efficient glass that is covered with a microscopic coating on its surface to minimise the amount of infrared and ultraviolet light that can pass through it to keep a building warm in the winter and cool in the summer

Low-rise buildings: buildings with up to four storeys

Luminaires: complete electric lighting units, including the casing, lamp and any internal controlling devices or electronic equipment (known as control gear or drivers)

Macerators: plumbing components used to convert waste from a toilet, shower or washbasin into a fine slurry that can be pumped into the sewage line; they are used where access to gravity drainage is not available

Malware: software that is specifically designed to disrupt, damage or gain unauthorised access to a computer system

Market share: the percentage of total sales or output that a business has in a specified market; sometimes referred to as market leadership

Metric units: decimal units of measurement based on the metre and the kilogram

Minerals: a solid, naturally occurring, inorganic substance

Miniature circuit breaker (MCB): a small trip switch operated by an overload, used to protect individual electrical circuits

Mobile bowsers: a wheeled trailer fitted with a tank for carrying oil

Networking: an activity where businesses and people with a common interest meet to share information and develop contacts

Ohm's law: a law that states the relationship between current, voltage and resistance in an electrical circuit

Open circuit: an electrical circuit that is not complete so current does not flow

Overheads: regular repeated costs associated with the day-to-day running of a business, for example rent and insurances

Parameters: limits which define the scope of a system, process or activity

Partnership agreement: a legally binding contract that sets out terms and conditions for each partner in the business

Passive infrared (PIR) sensor: an electronic sensor that measures infrared (IR) light radiating from objects in its field of view

Passivhaus: ('Passive house' in English) an energy performance standard intended primarily for new buildings, which ensures that buildings are so well constructed, insulated and ventilated that they require little energy for heating or cooling

Peltier effect: when the passage of a direct electric current through the junction of two dissimilar conducting materials causes the junction to absorb or reject heat

Philosophy: values and beliefs that act as guiding principles for behaviour

Plant operators: people in control of heavy construction machinery and equipment

Porosity: the measure of a substance's ability to hold water or allow water to pass through it

Portal frame: a large structural frame made from load-bearing timber and steel beams and columns

Potential difference: the difference in voltage from one terminal to another

Pre-cast: formed into a shape in a factory before being delivered for use on site

Prefabs: buildings manufactured using factory-made components or units that are transported and assembled on site

Preliminaries: pre-construction information that outlines items that are necessary for a contractor to complete the works but are not actually part of the works, for example general plant, welfare facilities and site security

Pre-stressed concrete: a type of concrete that has been compressed during production to improve its strength; this is normally achieved by 'tensioning' (stretching) high-tensile steel wires in the concrete

Prime cost: the actual value of goods and services without any additional costs added, for example profits margins

Principal designer: a designer appointed by the client to take the lead in planning, managing, monitoring and co-ordinating health and safety during the pre-construction phase of a project involving more than one designer

Procurement: the process of agreeing business terms and acquiring goods, products or services from suppliers

Proofread: to check a piece of written communication for errors such as spelling, grammar, punctuation and accuracy before it is shared or published

Proprietor: an individual who owns a business

Public-sector projects: projects funded by the government

Ratio: the amount or proportion of one thing compared to another

Reasonably practicable: a term used in health and safety law to describe realistic steps that should be taken to comply with the law in terms of time, effort and money

Rebar: reinforced steel bar commonly used in concrete to act as a frame to stop it moving and cracking

Rebate: a profile often used in timber products like doors and windows

Refrigerant: a working fluid used in the refrigeration cycle of air conditioning systems and heat pumps

Render: a pre-mixed layer of sand and cement, similar to mortar, used to make masonry walls flat and prepare them for top coats of finishing plaster

Renewable energy: energy that comes from natural sources or processes that are replenished or replaced, such as water, sun and wind

Renewable resources: resources that can be replaced over time by natural processes, for example wind energy or solar energy

Repository: a central location where something can be stored

Resistance: the measure of how well a material conducts electricity in ohms (Ω); the lower the value of ohms, the better it conducts

Respiratory protective equipment (RPE): personal protective equipment that protects the user's respiratory system

Responsible person: (in law) usually the employer (assuming control of the workplace), a person with control of the premises or an owner; that person can then nominate other competent people to perform legal responsibilities, for example to manage fire safety

Retrofit: the process of adding new components to older structures

Retrofitting: the process of adding new features and technologies to existing buildings

Reverse-cycle heat pumps: heat pumps that can be used for both heating and cooling

Risk assessment: a formal process of identifying significant workplace hazards, whom they affect and control measures that could be used to eliminate or reduce risk to an acceptable level

Risk management: the process of identifying, assessing and controlling threats to an organisation; this might involve making a strategic plan or putting a system in place to minimise impact and disruption

Rotary-screw compressor: compressor that uses two closely meshing helical screws, known as rotors, to compress a gas

Sabbatical: an extended period of unpaid leave from work, taken in agreement with an employer, often used for holidays, travelling or pursuing interests

Safety data sheets: written documents produced by manufactures and suppliers of hazardous substances that contain important information about how products should be transported, used, stored and safely disposed of after use, any special conditions you should be aware of and how to deal with the substance in an emergency

Schematics: a diagram representing the elements of a system using graphic symbols

Self-employed: the state of working for oneself rather than an employer; a self-employed person is responsible for paying their own tax and National Insurance contributions on any earnings

Semi-gravity system: a central-heating system that has pumped heating circulation but gravity hot-water circulation

Sentry: a person who supervises workers from the access/egress points of a confined space

Shadow flicker: when the rotating blades of a wind turbine create moving shadows

Shareholders: investors in a company who have purchased at least one share of a company's stock

Site induction: an occasion when all new visitors and workers on a construction site are informed about potential hazards and the control measures used to protect people from harm

Smart technology: computers, smartphones and tablets with software that connects to the internet in order to control, report, monitor and analyse devices and appliances remotely

Snagging list: a document used to record faults and defects discovered in building work or materials

Snagging: corrective work undertaken by a contractor or their subcontractors that has been identified by the client or their representative

Soakaways: large underground holes, filled with coarse stones or purpose-made plastic crates, which allow water to filter through and soak into the ground

Soil pipe: the lower, wet part of sanitation-system pipework that takes effluent away from a building

Special order: an order for items or components that are custom made or configured to a client's specifications

Specification: a detailed description of the materials and working methods that must be used for a project

Spring tides: exceptionally high tides that occur twice monthly at the time of the new moon and the full moon, when the moon's orbit aligns with the sun to create a greater gravitational pull; they are known as spring tides because they act like a spring

State pension: a regular sum of money paid by the state to people of or above official retirement age

Statutory sick pay: payment made by employers to employees when they are too ill to work

Storage heater: an electric heater that stores thermal energy; it heats up internal ceramic bricks when electricity is cheaper at night and then releases heat gradually during the day, acting in the same way as a convection heater

Stress-graded: timber that has been visually or machine-assessed for its strength and stiffness, and certified according to its structural classification to British Standard rules

Subsidence: the sinking of a structure into the ground

Substation: equipment that transforms voltage to a suitable level for consumers

Sulphate: a salt of sulphuric acid

Surface water: water that collects on the ground or above surface structures and buildings, normally in the form of rain

Temporary decommissioning: the process of taking a system out of action temporarily for maintenance or upgrades

Tendering: the process of inviting bids from contractors to carry out specific projects

Terminal: a connection point to an external circuit

Testimonials: statements of recommendation produced by satisfied customers or clients that confirm the quality of a product or service

Thermal: related to heat or temperature

Thermal comfort: a person's satisfaction with the thermal environment (whether they feel too hot, too cold or just right)

Third-party verification: the process of getting an independent party to confirm that the project meets standards

Tolerances: allowable variations between specified measurements and actual measurements

Toolbox talks: short training sessions arranged at regular intervals at a place of work to discuss health and safety issues; they give safety reminders and inform personnel about new hazards that may have recently arisen

Top-down: proceeding from the most senior to the least senior

Transformers: devices that convert voltages and current, proportionately, to different values

Turbine: a machine that uses a moving stream of air, water, steam or hot gas to turn a wheel and generate power

Visualisations: digital or virtual representations of a structure

VoIP: voice over internet protocol, technology used to deliver voice and multimedia via an internet connection

Voltage drop: the decrease of electrical potential along the path of a current flowing in an electrical circuit

Warranty: a guarantee from a manufacturer or seller that a product will be repaired or replaced within a certain period of time if it does not function as originally described or intended

Water hammer: a knocking sound in water pipes caused by fast-flowing water being stopped abruptly when terminal fittings are closed off

Water table: the level below which the ground is saturated with water; this level can rise with rainfall and fall with periods of dry weather

Water and sewerage undertakers: suppliers of wholesome water and sewerage services to properties

Watt: SI unit of power

Weirs: low dams across a river, which increase the force of the water as it flows over the top; sections of a weir can be raised or lowered to regulate the force of the water

Whistleblowing: the act of reporting information about wrongdoing

Wholesome water: water that is fit to use for drinking, cooking, food preparation or washing without danger to human health

Wind turbines: a vaned wheel that is rotated by wind to generate electricity

Workers: people who do not have a permanent contract of employment with an employer but are contracted for work or services

Workflow: a sequence of activities needed to complete a work task

Index

A

acceleration 49–50, 55–6
access control systems 336–7
access equipment 36–40, 382
accessibility 205, 279
accident book 23–4
accidents 2, 16, 22–4, 219
acoustics principles 80–2
active monitoring 16–17
activity costing 185
acute toxicity signage 35
Addison Act 1919 197
affordable housing 91, 113
aggregate 53, 155, 208, 221
air
  characteristics of 75–6
  quality 148
  see also ventilation
air-conditioning systems 75, 302–3, 370
air source heat pumps 167
alloys 52
aluminium 51, 54, 66
ammeter 70
animation 276
apex 213
appraisals 261
apprenticeships 295
Approved Codes of Practice (ACOPs) 6–8
Approved Documents 96, 152, 202–5, 227
arbitration 256
Archimedes' principle of displacement 60
architects 115, 120–1
area, measuring 49
argon gas 224
arson 25
Art Deco style 197
artificial intelligence (AI) 157, 270–3
asbestos 2–3, 4, 160
Association of Plumbing and Heating Contractors (APHC) 131
atmospheric pressure 50
augmented reality (AR) 134, 278
autonomous decision making 271
autonomous vehicles 200–1
axial fans 340

B

backflow 316
basements 207
batch costing 185
battered walls 207
beams 71–2
bedrock 82–3
benchmarking 298–9
bending (force) 71–2
bidding process 118–20
bill of quantities 120, 142–3
bimetallic strip 54
biofuel 62
biomass energy 161–2
biomass heating systems 78, 168–9
biometrics 270, 336
block plan 191
blockwork 76, 98, 219, 243
boilers
  types of 341–4
  working principles 340–4
  see also heating systems; hot water systems
bonds 226
boosted cold water systems 316–17
boreholes 95, 102, 313, 346
boundaries 224
bricklayers 128
bricks
  bonds 226
  parts of 218
  recycling 161
brickwork 97, 176, 210, 218, 225–6
  failures in 92
  see also walls
bridging 217
British Standards (BS) 205, 243
British Standards Institute (BSI) 205
brownfield sites 87–8, 94, 95
BTU (British Thermal Unit) 58
buildability 99
building control officers (BCOs) 104, 115, 123,
Building Cost Information Service (BCIS) 239
Building Information Modelling (BIM) 105, 106, 115, 120, 124–5, 133–5, 190, 232, 236–7, 249, 277
building line 103
building management system (BMS) 177
Building Regulations 2010 71, 89–90, 136, 152–3, 291
Building Research Establishment (BRE) 147
Building Research Establishment Environmental Assessment Method (BREEAM) 147, 155
Building Research Establishment's Home Quality Mark (HQM) 291
bump caps 30
business objectives 248, 288–91
business plan 288
business strategies 136

C

cable basket 353
cables 346–7, 358–9
calibration 379–80
calorie 58
capital 287–9, 296
carbon dioxide 44, 61, 78, 149, 161, 167, 189
carbon emissions 4, 89–90, 114, 136, 154, 162, 165
carbonic acid 84
carbon monoxide 25, 44
carpenters 127–8
Carriage of Dangerous Goods and Use of Transportable Pressure Equipment Regulations 2009 36
catenary wires 51, 52
cavity wall insulation 81, 176
cavity walls 128, 176, 196–8, 217–9
CCTV (closed-circuit television) 305, 334
CDM Regulations *see* Construction (Design and Management) (CDM) Regulations 2015
CE marking 31
cement 53, 128, 151, 156, 208, 220, 225, 389
central heating systems *see* heating systems
centralised hot water systems 318–23
centre of gravity 59

centrifugal fans 340
ceramics 53
certification 237–8
changing rooms 5
chasing 73
chillers 344–5
chimneys 84, 151
circuit breakers 70–1, 304–5, 359
circuit diagrams 243–4
circuit protective devices 70–1
circuits *see* electrical circuit principles
civil engineering 110
civil engineers 122–3
cladding 136, 225
Classification, Labelling and Packaging (CLP) Regulation 2010 33
Clean Air Act 1993 151
cleaning of equipment 382–3
clerk of works 115, 126
client representatives 126
climate change 61, 85, 106, 114, 148–9
Climate Change Act 2008 4, 114, 148–9, 151
coal 61
  see also fossil fuels
cob 196, 198
Code for Sustainable Homes 89
codes of conduct 294
cold water systems 313–17
  isolation 18–20
collaborative working 249, 294
columns 72
combination boilers 19, 307, 322, 343
combined heat and power (CHP) systems 171–2
combustion products 204
commissioning systems 200, 237–9
Common Data Environment (CDE) 134, 237
common law 6–7
common rafters 213
communication 257–61
community interest companies (CICs) 287
compact fluorescent lamps (CFLs) 174
Companies House 283–5
compliance 291
compression 71–2
compressive strength 386
computer-aided design (CAD) 115, 121–2, 124, 134–5, 246, 273, 280
computer-aided manufacturing (CAM) 280
computer numerical control (CNC) 280
conciliation 256
concrete 53, 55, 150, 217–18, 225, 226
  failures in 92
condensation 74, 203, 216
condensing boilers 307
condition reports 373
conduction 75
conductors 63–6
conduit 304, 351–2
confidentiality 241–2
confined spaces 42–5
  see also excavations
Confined Spaces Regulations 1997 4, 45
conflict 251, 256
conservation 156
  see also energy conservation
Construction (Design and Management) (CDM) Regulations 2015 5, 7–8, 15, 115, 144
construction drawings *see* drawings
Construction Industry Training Board (CITB) 14
construction methods 94, 96–9, 196–9
construction phase plan 7–8
construction site
  inductions 14–15
  visitors 15
  welfare facilities 5
Construction Skills Certification Scheme (CSCS) 14, 127
consumer unit 21, 304–5
contaminated soil 94
contingency 140
continuing professional development (CPD) 132–3, 289
contract managers 126–7
contractors 101, 116–19, 131, 139–42
contracts 137–40
Control of Asbestos Regulations 2012 3
Control of Noise at Work Regulations 2005 3
Control of Pollution (Oil Storage) (England) Regulations 2001 153
Control of Pollution Act 1974 4
Control of Substances Hazardous to Health (COSHH) Regulations 2002 2, 12–13, 153, 235–6, 387
Control of Vibration at Work Regulations 2005 3
control systems *see* smart controls
convection 74
convection heaters 74, 77
copper 51, 66, 161
corporate social responsibility (CSR) 94–5, , 136, 294–5
corporation tax 284
corrosive signage 34, 387
corrugated stainless-steel tubing (CSST) 361
costing techniques 184–5
credit agreements 142
creditors 284
critical path analysis 105–6
crosstalk 305
customer 115
customer service 250, 292–3

D

daily checks 381–2
damp-proof course (DPC) 210
damp-proof membrane (DPM) 221–2
data
  analysis 271
  Common Data Environment (CDE) 237
  confidentiality 240–1
  demographic 267, 296
  interoperability 230
  interpreting 235
  legislation 242
  management 230, 240–2, 246, 271–3
  metadata 230
  sources of 231–4, 246
  usage 234–9
data cabling 305–6
Data Protection Act 2018 242, 265
data warehouse 135
debtor 284
decimal system 186
decommissioning 18, 339
  see also isolation
decorators 130
delivery notes 141–2
demographic data 267, 296
demolition 147, 155, 156, 159
design
  brief 180, 236
  Digital Plan of Work (DPoW) 134
  impact 90–1
  principles 95–100
  process 101–6
  specification 87
design engineers 124
desktop survey 101–2
detail drawing 191
digital controllers 245
digital engineering 275–8
Digital Plan of Work (DPoW) 134
digital technologies 177–8, 244–6, 271–81
direct cold water systems 19, 314–15

discrimination 253
displacement 60
dissimilar metals 54
distribution board 21
distribution network operators (DNOs) 63
distributive negotiation 255
Distributor Takeback Scheme (DTS) 159
diversity 252
documentation 137–43
doors 224
dormer windows 215
downtime 374
drainage 73–4, 83–4
drawings 143, 190–3, 242–4
  paper sizes 192
  symbols 243
drinking water 5
drones 202, 277
drylining 199
ductwork 175–6, 363
duty holders 2, 6–7

E

ear protection 29
earth forces 84–5
Ecodesign for Energy-Related Products Regulations 2010 124
ecological balance 291
Edwardian houses 196
efficiency 58–9
efflorescence 92
effluent 150
electrical and electronic equipment (EEE), waste management 158–9
electrical circuit principles 65–70
Electrical Installation Certificate (EIC) 129
electrical plans 244
electrical resistance 50
electric current *see* current
electricians 129
electricity
  distribution systems 63–4
  electrical accessories 348–50
  energy consumption 164
  generation 60–2, 162
  isolation 21–2
  power management 272
  smart meters 178
  supply 354–5
  transmission systems 62–3
  voltages 53–4, 62–4, 70, 383
  see also renewable energy sources
Electricity at Work Regulations 1989 4
electric vehicles 114, 174–5, 205
electrolytic corrosion 53–4
electromotive force 50, 65
electrons 53–4
electrotechnical systems 303–5
elevation drawing 191
emergency control valve (ECV) 20
emergency procedures 24–6
employment
  employee duties and responsibilities 16, 265
  employer duties and responsibilities 7, 14–15, 16–18, 23–4, 45, 265
  employer liability insurance 7
  legislation 253–4, 262–5
  rights 253–4, 262
Employment Act 2008 253–4
Employment Rights Act 1996 253
encasing services 199–200
energy 50, 58
Energy Act 2020 4
energy conservation 172–6, 219
energy efficiency measures (EEMs) 148
energy efficiency ratings 175–6
Energy Performance (England and Wales) Regulations 2012 154
energy performance certificate (EPC) 88–9
Energy Performance of Buildings Directive 154
Energy-related Products (ErP) Directive 175
English bond 226
enterprise asset management (EAM) 233
entrepreneurship 287, 288, 296–7
environment
  climate change 61
  ecological balance 291–2
  effects on materials 53–4
  environmental hazards signage 35
  legislation 4
  performance measures 154–7
  pollution 4, 61, 155
environmental impact 87–90, 154–7
Environmental Protection Act 1990 151, 159
environmental risk assessment 101
equality 252–4, 264
Equality Act 2020 253
equipment
  access 36–40
  calibration 379–80
  cleaning 382–3
  daily checks 381–2
  health and safety 378
  inspections 16–18, 40
  legislation 3, 17–18, 40
  measuring 188
  portable appliance testing (PAT) 17, 378
  storage 390
  testing and maintenance 366–73, 378
  waste management 158–9
estimates 137, 140
ethics 266–7, 293
excavations 43, 210, 274
explosive gas signage 34
eye protection 29

F

fans 338, 340
feasibility study 93
fenestration 226
ferrous metal 51, 52, 92
F-gases *see* fluorinated greenhouse gases
fibreglass batts 217
fire alarms 334–5
Fire and Rescue Authority 115
fire extinguishers 27–8
fire fighting signage 33
fire safety 24–8, 219, 370
first aid 31–2
first fix 127–130, 199–200, 348
fixings 225–7
flammable signage 35
flash floods 83
flat roofs 215–17
Flemish bond 226
flexible working 265
flooding 83–4
floor insulation 176
floor plan 191
floors
  solid 221–2
  suspended 222–3
flow 49
flue systems 308–9
fluids, displacement 60
fluorinated greenhouse gases 159, 302
Fluorinated Greenhouse Gases Regulations (2018) 159
footings 72
foot protection 29
force 50, 56–7, 59–60, 187
  effects on buildings 71–3
fossil fuels 61, 95, 162
foundations 72, 73–4, 83, 209–14

see also substructure
frames 224
franchises 287–8
frog bricks 218
frontage line 103
fulcrum 56
fuses 70, 359

G

galvanic corrosion 53–4
Gantt charts 105
gas 61
gas fitters 129–30
Gas Safe 20, 78, 130, 227
Gas Safety (Installation and Use) Regulations 1998 20
gas supply 20–1, 130, 306
 isolation 20–1
gas systems 306
 boilers 306–7
 cooking appliances 309
 fires 308–9
General Data Protection Regulation (GDPR) 2021 242, 265
geofencing 274
Georgian houses 196
geotechnical engineering 123
girders 72
glass 66, 76–7, 79, 81, 155, 226
 misting 93
 smart 150
global warming 84, 162
gold 66
granite 76, 225
gravity 55, 57
 centre of 59
green buildings 106, 147
greenfield sites 87, 94
greenhouse gases 61, 167
green roofs 150
grey water 150, 173–4
grievance policy 253, 264
ground source heat pumps 166
ground structure 82–3
gutter profiles 331–2
gypsum 53, 128–9

H

halogen lamps 79, 174
handover package 143–4
hand protection 29
hard hats 30
hazardous substances 2–3, 12–13, 34–5, 153, 387–8
Hazardous Waste (England and Wales) Regulations 2005 153
Hazardous Waste Regulations 2005 160
hazards 10–11, 14–15, 44, 385
Header bond 226
head-mounted display (HMD) 277–8
health and safety
 accidents and near misses 2, 16, 23–4
 emergency procedures 24–6
 equipment 378
 file 144
 hazards 10–11, 14–15, 44, 385
 inspections 5–6, 16–17
 legislation 2–4, 23–4, 32–6
 method statements 11
 monitoring 16–17
 risk assessments 2, 8, 10–11, 25
 risk management 2–4, 9, 368–9
 safe system of work 4, 9
 temperature 74
 welfare facilities 5
 see also Approved Documents
Health and Safety (First Aid) Regulations 1981 31–2
Health and Safety (Safety Signs and Signals) Regulations 1996 32–5
Health and Safety at Work etc. Act (HASAWA) 1974 2, 6, 14, 206, 264
Health and Safety Executive (HSE) 2, 5–9, 23, 115
health hazard signage 35
health problems 12
heat exchanger 318
heating and ventilation fitters 129
heating methods 77
heating systems 309–13
heat loss 76–7, 189–90
heat principles 74
heat pumps 344–5
heat recovery ventilation (HRV) systems 174
heat transfer 74–5, 189–90
height, working at 3, 36
Heritage Protection Bill 156
hierarchical structure 115
horizontal forces 72
hot water systems 317
hot works 25
HSE Legal (L) Series 8
human resources (HR) 264
Human Rights Act 1998 254
hydroelectric energy 61, 163, 170
hydrology 102

I

illuminance 50
immersion heater 318
imperial units of measurement 186
inclusion 252
indices 48
indirect cold water systems 19, 315–16
industrial heating systems 311–13
information technology (IT) 244–6
infrastructure 90–1, 110, 224, 232
innovation 290
inspections 5–6, 16–17
Institution of Engineering and Technology (IET) 8
insulated concrete formwork (ICF) 98
insulation
 buildings 176, 189–90
 floors 222
 pipework and ductwork 175–6
 roofs 216–17
 see also cavity walls
insulators 65, 66
insurance 6–7, 144, 372–3
internal pressure 50
International Electrotechnical Commission (IEC) 245
International Organization for Standardization (ISO) 205, 291
International System of Units (SI) 48–50, 58
internet of things (IoT) 177, 270–2
intruder alarms 333–4, 369
iron 51
ironmongery 204, 205
isolation
 cold water systems 18–20
 combination boiler 19
 electricity supply 21–3
 gas supply 20–1
 and loss of service 338–9
 unvented hot water systems 19
 vented hot water systems 19–20

J

job roles 110, 121–31
job sheets 371
joiners 127–8
joists 72–3
joules 58
just-in-time deliveries 157

K

kerbs 224
key performance indicators (KPIs) 234, 298–9
kilowatt hour (kWh) 58
kinetic energy 58, 162–3, 170
kinetic lifting 41, 389
K-values 76

L

ladders 36–7
lamp choice 79–80
landfill 83, 160-1
lead 386
Leadership in Energy and Environmental Design (LEED) 147, 155
lead time 99
lean construction 157
Legionella bacteria 8
Legionnaires' disease 8
legislation
  common law 6–7
  data management 242
  employment 253, 262–5
  environmental 4
  equality 253
  failure to comply 5–6
  health and safety 2–4, 23–4, 32–6
  non-statutory guidance 6
  sustainability 151
  waste management 4, 158–61
  water supply 313
  welfare facilities 5
  see also planning permission
leg protection 29
length 49
levers 56–7
liability insurance 6–7
life cycle assessment 107, 147
life cycle cost analysis 185–6
lifting, manual *see* manual handling
lifting aids 41–2
Lifting Operations and Lifting Equipment Regulations (LOLER) 1998 17–18, 40
light-emitting diodes (LEDs) 78, 80, 174
lighting
  circuits 304
  energy-efficient 174
  principles 78–80
lime 53, 128
limestone 84
limited companies *see* private limited companies
limited liability partnership (LLP) 284–5
linear measurement 183–4
linear temperature expansion coefficients 55
lintels 73, 218–19
liquefied petroleum gas (LPG) 306
liquid roof 387–8
listed buildings 96, 156–7
living roofs 150
loadings 187, 203, 205, 231
Local Authority Building Control (LABC) 104, 115, 123, 145
local community 90–1
localised hot water systems 323–5
location drawing 191
locked fuse 21
lockers 5
loft insulation 176
lofts 215
log book 143
lone workers 4, 31
low-carbon steel pipe 359
low-emissivity (low-E) glass 224
low-pressure central-heating system 310–11
low-rise buildings 96
luminaires 78, 100
luminous intensity 49

M

macerators 21, 350
maintenance 365–73, 389–90
Management of Health and Safety at Work Regulations 1999 4, 9–10, 45
mandatory signage 32
manual handling 3, 40–2, 59, 387
Manual Handling Operations Regulations (MHOR) 1992 3, 41
manufacturers 117, 235
manufacturers' instructions 206–7, 339, 368–9, 371
marble 225
marketing 297
market share 289
masonry 208–9
  see also blockwork; brickwork
mass 49, 55, 58
materials
  defects or failures in 91, 385
  environmental effects 53–5
  properties 51–3, 65–6, 383–4
  quantity 387
  recycling 157, 160–1
  source and usage 154–7
  storage 389
  thermal conductivity 76
  see also waste management
maternity leave 263–4
mathematics 48–50
measurement
  conventions 190–3
  equipment 188
  principles 180–2
  scales 190–3
  techniques 188–90
  tolerances 193
  units of 48–50, 186–7
mechanical building services (design) engineers 124
mechanical design engineers 124
mechanical-extract ventilation (MEV) 337
mechanical loads 55–60
mechanical ventilation with heat recovery (MVHR) 337
mechanisation 96
mediation 256
medium-density polyethylene (MDPE) 360
mental health 148, 258
metadata 230
metals
  dissimilar metals 54
  electrolytic and galvanic corrosion 53–4
  failures in 92
  properties 51–2
  recycling 160–1
meters 178
method statements 11
metric units 186
micro combined heat and power (micro CHP) systems 171
micro-hydro systems 170
micro-wind turbines 169–70
mildew 75
minerals 53
miniature circuit breaker (MCB) 21, 71, 304
Minor Electrical Installation Works Certificate (MEIWC) 129
mixed reality (MR) 278
mobile bowser 153
mobile elevating work platforms (MEWPs) 39
mobile scaffold towers 37

modern building methods 96–9, 196
modular construction 98, 198–9
mullioned windows 196
musculoskeletal disorders (MSDs) 41

N

National Grid 62–3
National House Building Council (NHBC) 145
National Insurance 263
National Living Wage 262
National Minimum Wage 262
National Planning Policy Framework 91
National Vocational Qualifications (NVQs) 127
near misses 16, 23
negotiation 254–6
networking 267
net zero 114
newton-metres 56, 58, 59–60
Newton's second law 55
noise 3, 29, 82
 see also acoustics principles
non-skilled operatives 131
non-verbal communication 258–9
notching 73
not-for-profit organisations 287
nuclear fission 161–2

O

ohmmeter 70
Ohm's law 65–9
oil 61, 153
onsite construction 196–8
organisational culture 289–90
oriented strand board (OSB) 199
overheads 112, 137, 185
overheating 204, 318
oxidising chemicals signage 34

P

paid leave 263–4
paint, failures in 92
panel heaters 77
panelised frame construction 97
paper sizes 192
partition walls 199–200, 220–1, 226–7
partnerships 284–5
Party Wall etc. Act 1996 224
PAS 2035:2019 148–9
PAS 2038:2021 149
passive-aggressive behaviour 257–8
passive infrared (PIR) sensor 177
Passivhaus 155
payment schedule 138
payslips 263
pebbledash 196
penalty clauses 139
pensions 263
permits to work 15
permitted development 88, 102
Personal Protective Equipment (Enforcement) Regulations 2018 31
personal protective equipment (PPE) 3, 4, 29–30
Personal Protective Equipment (PPE) at Work Regulations 1992 3, 29
Personal Protective Equipment Regulations 2002 18
PESTLE factors 135–6
photovoltaic (PV) cells 62
pipework 359–61
insulation 175
pitched roof 213–16
Planning (Listed Buildings and Conservation Areas) Act 1990 4, 156
Planning and Energy Act 2008 4
planning permission 88, 102–3, 117, 156–7, 224, 254–5
plant operators 42, 117
plaster 53
plasterboard 199–200, 217, 220, 226–7
plasterers 128
plastic
 failures in 93
 pipework 360
 recycling 161
plumbers 129
plumbing systems 313–32, 359–61
pneumonia 8
podiums 38
pollution 4, 61, 155
polybutylene 360
polyvinyl chloride (PVC) 54–5, 63, 66
porosity 208, 209
portable appliance testing (PAT) 17, 378
post-occupancy evaluation (POE) 232–3
potential difference 70
potential energy 58
power (mechanical) 50, 58, 65, 68–9, 70
power circuits 304–5
powers (indices) 48
PPE *see* personal protective equipment (PPE)
prefabricated construction 149–50
preliminaries 120
pressure-jet boilers 342
Pressure Systems Safety Regulations (PSSR) 2000 18
pre-stressed concrete 223
prime cost 137
principle contractor 7–8, 15
principle designer 7–8
private limited companies 285–6
Private Water Supplies (England) Regulations 2016 313
procurement 117–18, 234
product data 235–6
products of combustion (POC) 308
profitability 93–5
programme of work 105, 126
programming 244–5
prohibition signage 6, 32
project management 249–50
project planning 104–6, 237, 249
protection systems 333–6
Provision and Use of Work Equipment Regulations (PUWER) 1998 3, 18, 40
public liability insurance 7
public limited companies (PLCs) 286
Publicly Available Specifications (PAS) 205
public sector projects 205
pulleys 57
pulverised fuel ash 208
pumps 340
purchase orders 140, 141–2
pylons 63

Q

qualifications and training 14–15
quality 290–1
quality control 198, 300
quotations 141

R

radial circuits 304
radiant heaters 75
radiation, infrared 75
rain 84
rainwater harvesting 172–3
rainwater systems 330–32
ratio 190
reactive monitoring 16–17
rebar 51, 52
rebate 227
record keeping 382
recycling 155, 158, 160–1
redundancy 265

reed beds 150
refrigerants 302
refrigeration systems 336, 343–4
regulatory bodies 131–2
Regulatory Reform (Fire Safety) Order 2005 24
render 176
renewable energy sources 61, 124–5, 131, 154, 162–5, 272
Reporting of Injuries, Diseases and Dangerous Occurrences Regulations (RIDDOR) 2013 2, 16, 23–4, 41
reservation point 255
residual circuit breaker with overload (RCBO) 71, 358
residual current devices (RCDs) 71
resistance 62, 65–6, 70
resistivity 65–6
respiratory protective equipment (RPE) 28, 30, 338
responsible person 25, 141
rest facilities 5
restoration work 275–6
retainer walls 207
retention 139
retrofit assessors/installers 125, 149
retrofitting 114, 148–9
reverberation 80
RIBA (Royal Institute of British Architects) Plan of Work 100, 236
risk assessments 2, 4, 8, 10–11, 25
risk management 2–4, 9, 368–9
robotics 198, 279–80
roof lantern 99–100
roofs 150, 196–7, 213–17
roof trusses 207
rot 75
Royal Institution of Chartered Surveyors (RICS) 239
rubber 52, 66
R-values 76

S

sabbatical 265
safe condition signage 32
safe system of work 4, 9
safety data sheets 12
safety signage 14, 32–5, 257, 259
safe working load 72, 189
sanitation systems 326–9
scaffolders 117
scaffolding 37
scale rule 192
scales 190–3
schedule of rates 139–40
schematics 243–4
  see also drawings
screed 184
sealed central-heating system 310–11
second fix 128, 129–30, 200, 348–9
sectional drawing 191
security, of dwellings 204, 333–6
self-driving vehicles 200–1
self-employment 111–12
self-healing concrete 150
semi-detached houses 197
sentry 45
services
  connecting 95
  encasing 199–200
  locating 183
sewerage 73, 152
shareholders 111
shearing 72
showers 316
signage 32–6, 257, 259
simulation 275
site *see* construction site
site analysis 102
site inductions 8, 14–15, 127, 202
Site Management Safety Training Scheme (SMSTS) 14
site managers 119, 126–7
site of special scientific interest (SSSI) 101–2
site plan 191
Site Supervision Safety Training Scheme (SSSTS) 14
site survey 183
site waste management plan (SWMP) 89
SI units *see* International System of Units (SI)
small and medium-sized enterprises (SMEs) 286–7
smart cement 151
smart controls 177, 245, 271–2
smart glass 150
smart meters 178
SMART objectives 298
smartphones 245
smart technology 133
snagging 139, 143
soakaways 83, 151
social media 267
Social Security (Claims and Payments) Regulations 1979 23
sockets 348–9
soil 74, 82–3, 94
solar energy 62, 163
solar heat gain 154
solar photovoltaic (PV) panels 62, 95, 151, 158, 163, 168–9
solar thermal systems 165–6
sole traders 111, 283–4
sound 80
  see also acoustics principles
specification 94, 119
specific heat capacity 50
speed (velocity) 49
spring tides 84
square, built to 181–2
stability of structures 71–4
stack effect 75, 336
staging boards 39
stakeholders 248
state pension 263
statutory sick pay (SSP) 262–3
steel 51, 55, 66
steel-frame construction 98, 208
stepladders 38
stone 225
storage
  equipment 388
  materials 389
storage heaters 75, 77
storms 84
straw-bale construction 97
Stretcher bond 226
structural insulated panels (SIPs) 98, 199
structural members 72–3
subcontractors 112, 116
subsidence 74, 209
substations 63
substructure 72, 209–13
sulphate attack 92, 93
superstructure 87, 213
supervisors 126, 130–1
suppliers 91, 117, 141–2
supply chain 114–17
surface water 83–4, 151, 224, 330
surveillance systems 334
surveyors 115
sustainability 147, 291, 295
  legislation 151
sustainable homes 89, 106, 198
switched fuse 21, 350
symbols
  drawings 243
  units of measurement 49–50

T

take-off sheets 136–8
tamping 53, 211

team work 251–2
technical drawings 143
tectonic plates 84
temperature
  and bricklaying 96
  health and safety 74
  unit of measurement 49
  see also heat loss
tender documentation 119–20
tension 71–2
terraced houses 197
testimonials 250
theodolite 188
thermal conductivity 76
thermal expansion 54
thermoplastics 52
thermosetting plastics 52
thin-joint blockwork 98
three-dimensional (3D) modelling 273, 277
three-dimensional (3D) printing 201, 279
tidal energy 164
tilers 128
tiles 225
timber
  cladding and fabric 225
  classification 207–8
  failures in 92, 387
  recycling 161
  stress graded 207
  sustainability 148
timber frame 97, 207–8
Timber Research and Development Association (TRADA) 148
time 49
time off 263–4
toilets, on site 5
tolerances 193
tongue and groove (T&G) 225
toolbox talks 15
tools *see* equipment
topography 93–4
toroidal transformer 71
torque 56, 58, 59–60
torsion 71–2
Town and Country Planning (Environmental Impact Assessment) Regulations 2017 88
Town and Country Planning Act 1990 156–7
traditional building methods 96–9, 196
training and qualifications 14–15
transformation process 58–9
transformers 63
transport 155
trees 73–4
trenches *see* excavations
trestles 39
trunking 352
trusses 73
trust 250
Tudor houses 196
turbines 61
turnover 112

U

ultra-violet (UV) radiation 55
units of measurement 48–50, 58, 186–7
unplasticised polyvinyl chloride (uPVC) 52, 55, 224
unvented hot water systems, isolation 19
uranium 162
U-values 76, 99, 154, 189–90

V

vaporising boilers 343
VAT (value added tax) 111
velocity (speed) 49, 56
vented hot water systems 19–20, 320–1
ventilation 75, 99, 153, 216, 337–8
verbal communication 257–8
verbal estimates 140
vernacular construction 94
vertical forces 72
vibrating concrete poker 211
vibration, exposure to 3
vibration white finger 211
Victorian houses 196
virtual reality (VR) 277–8
visualisations 134, 257, 259–60
VoIP (voice over internet protocol) 305
voltages 53–4, 62–4, 70, 383
voltmeter 70
volume 49
volumetric modular construction 98

W

wages 262–3
walkie-talkies 260
walkover survey 102
walls
  cavity 129, 217–19
  external 213, 217, 227
  internal 220–1
  partition 220–1, 227
  party 224
  solid 217
warm-air heaters 313
warning signage 33–6
warranties 144, 366
washing facilities 5
Waste Electrical and Electronic Equipment (WEEE) Regulations 2013 4, 158–9
waste management 4, 89, 153, 155, 158–61
water
  drinking 5
  grey water 150, 172–4
  hydrology 101–2
  legislation 313
  quality 148
  rainwater harvesting 172–3
  reed beds 150
  source and usage 155
  supply 313
  surface 83–4, 151, 218
  vapour 75
  wholesome 203, 313–14
Water Act 2014 152
watercourses 83–4, 170
water heaters 323–4
  see also hot water systems
water source heat pumps 167
Water Supply (Water Fittings) Regulations 1999 313
water table 83, 102
wattle and daub 196, 198
wattmeter 70
watts 48, 50
wave energy 61, 164
weather 84
weep holes 219
weight 55
weirs 61
welfare facilities 5
WELL Building Standard 148
wet heating systems 78
whistleblowing 252–3
whole building concept 106
wholesome water 203, 313–14
Wildlife and Countryside Act 1981 4
wind 84
windows 99, 196, 215, 224
wind turbines 61, 154, 163, 169–70
wiring diagrams 142, 244
woodworm 92
work (mechanical loads) 57–8
Work at Height Regulations 2005 3, 10, 18
written communication 258–9

Photo credits

The Publishers would like to thank the following for permission to reproduce copyright material.

p.1 © Eleventh Hour Photography/Alamy Stock Photo; **p.6** © Health and Safety Executive; **p.15** © Bogdan Vasilescu/Fotolia; **p.17** © Matjoe/Shutterstock.com; **p.21** *r* Maksym Yemelyanov/stock.adobe.com, *br* Reece Safety Products Ltd; **p.22** *tr* Martindale Electric; **p.23** © Health and Safety Executive; **p.24** *l* © Bertrand Cadilhac/ Shutterstock.com, *r* © Grand Warszawski/stock.adobe.com; **p.25** © Kumar Sriskandan/Alamy Stock Photo; **p.26** © wellphoto/Fotolia; **p.27** *t* © Jenny Thompson/stock.adobe.com, *c* © Alan Stockdale/Fotolia.com, *b* © Hartphotography/stock.adobe.com; **p.28** *t* © Mr.Zach/Shutterstock.com, *c* © Jijomathai/stock.adobe.com, *br* © Mr.nutnuchit phutsawagung/EyeEm/stock.adobe.com; **p.30** © Kamolrat/stock.adobe.com; **p.31** © Ross Swift/ Alamy Stock Photo; **p.32** *tl* © Poco_bw/iStock/Getty Images, *bl* © St John Ambulance, *tr* © YongyuthChanthabutr/ stock.adobe.com, *r* © Daseaford/stock.adobe.com, *br* © Viktorijareut/stock.adobe.com; **p.33** *t* © Viktorijareut/ stock.adobe.com, *c* © Ricochet64/stock.adobe.com, *b* © Ricochet64/stock.adobe.com; **p.34** all © Ricochet64/stock. adobe.com except last row © olando/stock.adobe.com; **p.35** all © Ricochet64/stock.adobe.com; **p.38** © Euro Towers Ltd; **p.39** *tl* © Southern Plant & Tool Hire Ltd, *tr* © Andersphoto/stock.adobe.com, *r* © Sergejus Bertasius/123RF. com; **p.40** © Totz_photo/Shutterstock.com; **p.42** *tl* © Africa Studio/stock.adobe.com, *l* © Sveta/stock.adobe.com, *r* © shime/Fotolia.com; **p.43** © Linus/stock.adobe.com; **p.44** © EdNurg/stock.adobe.com; **p.45** © Chitsanupong/ stock.adobe.com; **p.47** © Tong2530/stock.adobe.com; **p.59** © SPP/stock.adobe.com; **p.75** © Burdun/stock.adobe. com; **p.81** © evgris/stock.adobe.com; **p.83** © Kadmy/stock.adobe.com; **p.84** © VectorMine/stock.adobe.com; **p.86** © Chlorophylle/stock.adobe.com; **p.87** *l* © Sveta/stock.adobe.com, *r* © Candy1812/stock.adobe.com; **p.88** *tl* © Silent Corners/stock.adobe.com, *br* © Alekss/stock.adobe.com; **p.89** *l* © UrbanImages/Alamy StockPhoto, *r* © Phil Wills/Alamy Stock Photo; **p.90** © Alicia G. Monedero/Shutterstock.com; **p.91** © Duncan/stock.adobe.com; **p.92** *tl(1)* © Baharlou/stock.adobe.com, *tl(2)* © Ian Shipley IND/Alamy Stock Photo, *c* © Aboutmomentsimages/stock. adobe.com, *bl(1)* © Evening_tao/stock.adobe.com, *bl(2)* © Petert2/stock.adobe.com; **p.93** *tl* © Fasphotographic/ stock.adobe.com, *l* © Kim sayer/Alamy Stock Photo; **p.94** *l* © Pelvidge/stock.adobe.com, *r* © Beataaldridge/stock. adobe.com; **p.95** © Rdonar/stock.adobe.com; **p.96** © FotoVoyager/E+/Getty Images; **p.97** *c* © Christian Delbert/ stock.adobe.com, *bl(1)* © Oleg/stock.adobe.com, *bl(2)* © Kingspan Insulation Ltd; **p.98** *tl(1)* © Bildlove/stock.adobe. com, *tl(2)* © Timmins Engineering & Construction Ltd, *l* © Simon Turner/Alamy Stock Photo, *bl(1)* © Kingspan Insulation Ltd, *bl(2)* © Kara/stock.adobe.com; **p.99** © Brizmaker/stock.adobe.com; **p.100** © Laurence Berger/ Shutterstock.com; **p.102** *tl* © Pellinni/stock.adobe.com, *l* © John Copland/Shutterstock.com, *r* © Frank/stock. adobe.com; **p.103** © Avpics/Alamy Stock Photo; **p.104** © Torbay Council, Torbay, Devon, UK; **p.105** © Sculpies/ stock.adobe.com; **p.109** © HENADZ/stock.adobe.com; **p.110** *l* © ceebeestock/stock.adobe.com, *bl* © Dan Kosmayer/stock.adobe.com, *r* © Simon Turner/Alamy Stock Photo; **p.113** © Avpics/Alamy Stock Photo; **p.114** *l* © Sonyachny/stock.adobe.com, *tr* © Roy Fenton Wylam/stock.adobe.com, *br* © Anton Gvozdikov/Shutterstock.com; **p.117** *tl* © LightfieldStudios/stock.adobe.com, *bl* © Monkey Business/stock.adobe.com, *r* © Cineberg/stock.adobe. com; **p.122** *tl* © Sergey Nivens/stock.adobe.com, *l* © Cyberstock/Alamy Stock Photo, *br* © Travelview/stock.adobe. com; **p.123** *l* © Microgen/stock.adobe.com, *br* © KomootP/stock.adobe.com; **p.125** © Ronstik/stock.adobe.com; **p.126** © Kaspars Grinvalds/stock.adobe.com; **p.127** *l* © Rido/stock.adobe.com, *bl* © Kings Access/stock.adobe.com, *br* © Monkey Business/stock.adobe.com; **p.128** *tl* © Kzenon/stock.adobe.com, *tr* © Alexander/stock.adobe.com; **p.129** © Ladanifer/stock.adobe.com; **p.130** © Gas Safe Register; **p.131** © Collins Photography/stock.adobe.com; **p.133** © HENADZ/stock.adobe.com; **p.134** © Zapp2photo/stock.adobe.com; **p.135** © konstruktor1980/stock.adobe. com; **p.136** © HM Government published by the RIBA; **p.138** © The Joint Contracts Tribunal Ltd; **p.139** © KMPZZZ/stock.adobe.com; **p.144** *l* © Used with the permission of The Chartered Institution of Building Services Engineers, *r* © National House Building Council; **p.146** © Polifoto/stock.adobe.com; **p.148** © Jandrie Lombard/ stock.adobe.com; **p.150** © Elke Hotzel/stock.adobe.com; **p.151** © Inna/stock.adobe.com; **p.152** © maximilian_100/ stock.adobe.com; **p.153** © JohnBorda/stock.adobe.com; **p.154** © Khamkhor/stock.adobe.com; **p.159** © Ungvar/ stock.adobe.com; **p.163** © Blueringmedia/stock.adobe.com; **p.167** © Desintegrator/Shutterstock.com; **p.174** ©

Pioneer111/stock.adobe.com; **p.175** *t* © Steve Lagreca/Shutterstock.com, *b* © Oleksandr Delyk/stock.adobe.com; **p.177** © Stanisic Vladimir/stock.adobe.com; **p.178** © Robert/stock.adobe.com; **p.179** © Kadmy/stock.adobe.com; **p.180** *l* © Monkey Business/stock.adobe.com, *r* © Robert Kneschke/stock.adobe.com; **p.181** © Hybrid Images/Connect Images/Alamy Stock Photo; **p.182** *l* © DragonImages/stock.adobe.com, *r* © Michaeljung/stock.adobe.com; **p.183** *tl(1)* © Zero Creatives/Connect Images/Alamy Stock Photo, *tl(2)* © Hansenn/stock.adobe.com, *bl(1)* © Fotolia RAW/stock.adobe.com, *bl(2)* © Andy Dean/stock.adobe.com; **p.184** *t* © Philippe Devanne/stock.adobe.com, *b* © Avalon/Construction Photography/Alamy Stock Photo; **p.185** © AnnaStills/stock.adobe.com; **p.186** *l* © SteF/stock.adobe.com, *r* © StockPhotosArt/stock.adobe.com; **p.188** *l* © Mint Images Limited/Alamy StockPhoto, *r* © Carsten Rehder/dpa picture alliance/Alamy Stock Photo; **p.189** © Scharfsinn86/stock.adobe.com; **p.190** *l* © Braozova/stock.adobe.com, *r* © Ionia/stock.adobe.com; **p.192** © Bokic Bojan/Shutterstock.com; **p.193** © NorGal/stock.adobe.com; **p.195** © ceebeestock/stock.adobe.com; **p.197** *l* © esdras700/stock.adobe.com; *r* © Ron Ellis/Shutterstock.com; **p.198** *tl* © Peter Turner Photography/Shutterstock.com, *bl* © Miriam Heppell/Alamy StockPhoto, *tr* © oatautta/stock.adobe.com; **p.199** © Peter Jordan_NE/Alamy Stock Photo; **p.200** © Andrey_Popov/Shutterstock.com; **p.201** *l* © Built Robotics, *tr* © Construction Robotics, *br* © PERI AG; p.202 Gorodenkoff/stock.adobe.com; **p.207** © Sveta/stock.adobe.com; **p.208** © IrinaMoskalev/123RF.com; **p.209** © Jason/stock.adobe.com; **p.211** *l* © Readin Wilson/Shutterstock.com, *r* © bandung/stock.adobe.com; **p.212** © Ali/stock.adobe.com; **p.215** *tl* © Moving Moment/stock.adobe.com, *bl* © Wirestock/stock.adobe.com, *tr* © U. J. Alexander/stock.adobe.com, *br* © Marko Beric/Alamy Stock Photo; **p.219** *tl* © Keystone Lintels Ltd, *br* © ARC Building Solutions Ltd; **p.221** ©jpldesigns/stock.adobe.com; **p.222** © helgidinson/stock.adobe.com; **p.223** *tl* © David Darkin/Shutterstock.com, *l* © Simple Shots/Ron Rovtar Photography/Alamy Stock Photo, *r* © len44ik/stock.adobe.com; **p.225** © sergeevspb/stock.adobe.com; **p.229** © ImageFlow/stock.adobe.com; **p.230** © Gorodenkoff/stock.adobe.com; **p.231** *t* © Goodluz/stock.adobe.com, *b* © Chris Pearsall/Alamy Stock Photo; **p.232** © Karel Miragaya/123RF.com; **p.233** *l* © JoeFoxBelfast/Radharc Images/Alamy Stock Photo, *r* © AddictiveStock Creatives/Alamy Stock Photo; **p.235** © Cookiecutter/stock.adobe.com; **p.236** © WoGi/stock.adobe.com; **p.237** Avalon/Construction Photography/Alamy Stock Photo; **p.238** © Andrey Popov/stock.adobe.com; **p.239** © Atstock Productions/stock.adobe.com; **p.240** © Blue Planet Studio/stock.adobe.com; **p.241** *b* © Asadykov/stock.adobe.com, *t* © Gorodenkoff/stock.adobe.com; **p.242** © Funtap/stock.adobe.com; **p.245** *l* © Michal Sanca/Shutterstock.com, *r* © Gen_A/stock.adobe.com; **p.247** © Rido/stock.adobe.com; **p.248** *t* © Halfpoint/stock.adobe.com, *b* © Kzenon/stock.adobe.com; **p.250** *l* © Syda Productions/stock.adobe.com, *tr* © Wordley Calvo Stock/stock.adobe.com, *br* © JackF/stock.adobe.com; **p.251** *l* © Justin Kase zsixz/Alamy Stock Photo, *r* © BigPixel Photo/Shutterstock.com; **p.252** *l* © Gorodenkoff/stock.adobe.com, *tr* © Frank/stock.adobe.com, *br* © cineberg/stock.adobe.com; **p.253** © TommyStockProject/stock.adobe.com; **p.254** © pressmaster/stock.adobe.com; **p.257** *l* © trgowanlock/stock.adobe.com, *r* © Bangkok Click Studio/Shutterstock.com; **p.258** © eccolo/stock.adobe.com; **p.259** © insta_photos/stock.adobe.com; **p.261** © pressmaster/stock.adobe.com; **p.262** © Monkey Business/stock.adobe.com; **p.263** © Kevin Wheal/Alamy Stock Photo; **p.264** © Anoushkatoronto/stock.adobe.com; **p.265** © iQoncept/Shutterstock.com; **p.266** © Alex Kosev/Shutterstock.com; **p.269** © Zephyr_p/stock.adobe.com; **p.270** *l* © ChompopsonG/stock.adobe.com, *r* © Arie Storm/Alamy Stock Photo; **p.271** *l* © Simon Turner/Alamy Stock Photo, *r* © Andrey Popov/stock.adobe.com; **p.272** *l* © Simon Turner/Alamy Stock Photo, *r* © Hans-Peter Merten/Mauritiusimages GmbH/Alamy Stock Photo; **p.273** © Cyberstock/Alamy Stock Photo; **p.274** *l* © Hans Engbers/Shutterstock.com, *r* © Justin Kase zsixz/Alamy Stock Photo; **p.275** © Bits and Splits/stock.adobe.com; **p.276** *l* © Yon Marsh/Alamy Stock Photo, *r* © Murattellioglu/stock.adobe.com; **p.277** *l* © Stanley Cabigas/Alamy Stock Photo, *r* © Escapejaja/stock.adobe.com; **p.278** © Gorodenkoff/stock.adobe.com; **p.279** © Yoshio Tsunoda/NipponNews/Aflo Co. Ltd./Alamy Stock Photo; p.280 *tl* © Steven May/Alamy Stock Photo, *bl* © Solkafa/Shutterstock.com, br © Alotofpeople/stock.adobe.com; **p.282** © Hipproductions/stock.adobe.com; **p.283** © highwaystarz/stock.adobe.com; **p.284** © MK photograp55/Shutterstock.com; **p.285** © Isabelle Plasschaert/Alamy Stock Photo; **p.286** © Chris Leachman/Alamy Stock Photo; **p.288** © bernardbodo/stock.adobe.com; **p.289** © Rido/stock.adobe.com; **p.290** © Seventyfour/stock.adobe.com; **p.292** © Andrew Paterson/Alamy Stock Photo; **p.293** *l* © MilanMarkovic78/Shutterstock.com, *r* © Isuzek/E+/Getty Images; **p.294** © Albert Pego/Shutterstock.com; **p.295** © goodluz/stock.adobe.com; **p.297** © fizkes/stock.adobe.com; **p.301** © zzMidnightzz/Shutterstock.com; **p.305** © Proxima Studio/stock.adobe.com; **p.313** © Lost_in_the_Midwest/stock.adobe.com; **p.325** © Dimplex; **p.334** © Coprid/stock.adobe.com; **p.336** © Matt/stock.adobe.com; **p.337** © Amixstudio/stock.adobe.com; **p.351** © Tom Leahy; **p.353** *l* © Alan Oliver/Alamy Stock Photo, *r* © Snake Tray; **p.362** https://www.

stevensonplumbing.co.uk/part-2-ballcock-with-float.html; **p.363** © Vchalup/stock.adobe.com; **p.365** © Kadmy/stock.adobe.com; **p.367** *tl* © Wathanyu sowong/123RF.com, *bl* © Vladdeep/stock.adobe.com, *r* © MaksymYemelyanov/stock.adobe.com; **p.371** *tl* © Perytskyy/stock.adobe.com, *bl* © Kungfu01/Shutterstock.com, *r* © Dotshock/123RF.com; **p.373** © Heating and Hotwater Industry Council; **p.377** © NDABCREATIVITY/stock.adobe.com; **p.378** *l* © Michaeljung/stock.adobe.com, *r* © Marc Tielemans/Alamy Stock Photo; **p.379** © Ionutanisca/stock.adobe.com; **p.380** *l* © Alexey Rezvykh/Alamy Stock Photo, *r* © Mr Twister/stock.adobe.com; **p.381** *l* © Freshly/stock.adobe.com, *r* © Nipastock/stock.adobe.com; **p.382** *l* © Kzenon/stock.adobe.com, *tr* © Kamolrat/stock.adobe.com, *br* © Vladimir Zhupanenko/stock.adobe.com; **p.383** © Photostocklight/stock.adobe.com; **p.384** © RZUK_Images/Alamy Stock Photo; **p.385** © Sbarabu/Shutterstock.com; **p.386** *l* © Sigrid Gombert/CavanImages/Alamy Stock Photo, *r* © Kampan/stock.adobe.com; **p.387** © Ricochet64/stock.adobe.com; **p.390** *l* © Contour Roofing (Essex) Ltd, *r* © Evannovostro/stock.adobe.com; **p.391** © Worrawong Kittithummawong/Shutterstock.com; **p.392** © Tomasz Zajda/stock.adobe.com.

The Publishers would like to thank City & Guilds for permission to reproduce photos and artworks from the following textbooks:

Carpentry & Joinery for the Level 1 Diploma (6706)

Site Carpentry and Architectural Joinery for the Level 2 Apprenticeship (6571), Level 2 Technical Certificate (7906) & Level 2 Diploma (6706)

Site Carpentry & Architectural Joinery for the Level 3 Apprenticeship (6571), Level 3 Advanced Technical Diploma (7906) & Level 3 Diploma (6706)

Bricklaying for the Level 1 Diploma (6705)

Bricklaying for the Level 2 Technical Certificate & Level 3 Advanced Technical Diploma (7905), Level 2 & 3 Diploma (6705) and Level 2 Apprenticeship (9077)

Plumbing Book 1 for the Level 3 Apprenticeship (9189), Level 2 Technical Certificate (8202) & Level 2 Diploma (6035)

Plumbing Book 2 for the Level 3 Apprenticeship (9189), Level 3 Advanced Technical Diploma (8202) and Level 3 Diploma (6035)

Electrical Installations Book 1 for the Level 3 Apprenticeship (5357), Level 2 Technical Certificate (8202) & Level 2 Diploma (2365)

Electrical Installations Book 2 for the Level 3 Apprenticeship (5357), Level 3 Advanced Technical Diploma (8202) & Level 3 Diploma (2365)

Level 2 NVQ Diploma in Plumbing and Heating

Level 3 NVQ Diploma in Plumbing and Heating 6189